HANDBOOK OF POULTRY SCIENCE AND TECHNOLOGY

HANDBOOK OF POULTRY SCIENCE AND TECHNOLOGY

Volume 2: Secondary Processing

Editor
Isabel Guerrero-Legarreta, Ph.D.

Consulting Editor
Y.H. Hui, Ph.D.

Associate Editors
Alma Delia Alarcón-Rojo, Ph.D., Christine Alvarado, Ph.D., Amarinder S. Bawa, Ph.D., Francisco Guerrero-Avendaño, Ph.D., DVM, Janne Lundén, Ph.D., DVM, Lisa McKee, Ph.D., Yoshinori Mine, Ph.D., Casey M. Owens, Ph.D., José Angel Pérez-Álvarez, Ph.D., Joe M. Regenstein, Ph.D., Marcelo R. Rosmini, Ph.D., Jorge Soriano-Santos, Ph.D., J. Eddie Wu, Ph.D.

A JOHN WILEY & SONS, INC., PUBLICATION

Published by John Wiley & Sons, Inc., Hoboken, New Jersey.
Published simultaneously in Canada.

***Library of Congress Cataloging-in-Publication Data*:**

Handbook of poultry science and technology/editor, Isabel Guerrero-Legarreta; consulting editor, Y.H. Hui; associate editors, Alma Delia Alarcón-Rojo ... [et al.].
p. cm.
Includes index.
ISBN 978-0-470-18537-7 (2-vol. set)
ISBN 978-0-470-18552-0 (v. 1)
ISBN 978-0-470-18553-7 (v. 2)
1. Poultry–Processing. 2. Poultry plants. I. Guerrero-Legarreta, Isabel.
TS1968.H36 2009
664′.93–dc22

2009014025

Printed in the United States of America
10 9 8 7 6 5 4 3 2 1

CONTENTS

CONTRIBUTORS

Editor-in-Chief

Isabel Guerrero-Legarreta, Departamento de Biotecnología, Universidad Autónoma Metropolitana, México D.F., México

Administrative Editor

Y.H. Hui, Science Technology System, West Sacramento, California

Associate Editors

Alma Delia Alarcón-Rojo, Facultad de Zootecnia, Universidad Autónoma de Chihuahua, Chihuahua, México

Christine Alvarado, Department of Animal and Food Sciences, Texas Tech University, Lubbock, Texas

Amarinder S. Bawa, Defence Food Research Laboratory, Siddartha Nagar, Mysore, India

Francisco Guerrero-Avendaño, Archer Daniels Midland Co., Mexico

Janne Lundén, Department of Food and Environmental Hygiene, Faculty of Veterinary Medicine, University of Helsinki, Helsinki, Finland

Lisa H. McKee, Department of Family and Consumer Sciences, New Mexico State University, Las Cruces, New Mexico

Yoshinori Mine, Department of Food Science, University of Guelph, Guelph, Ontario, Canada

Casey M. Owens, Department of Poultry Science, University of Arkansas, Fayetteville, Arkansas

José Angel Pérez-Alvarez, Grupo Industrialización de Productos de Origen Animal (IPOA Research Group), Departamento de Tecnología Agroalimentaria,

Escuela Politécnica Superior de Orihuela, Universidad Miguel Hernández, Orihuela, Alicante, Spain

Joe M. Regenstein, Cornell Kosher and Halal Food Initiative, Department of Food Science, Cornell University, Ithaca, New York

Marcelo R. Rosmini, Faculty of Agricultural Sciences, Universidad Católica de Córdoba, Córdoba, Argentina

Jorge Soriano-Santos, Departamento de Biotecnología, Universidad Autónoma Metropolitana–Unidad Iztapalapa, México D.F., México

Jong-Yi Eddie Wu, Foster Farms, Turlock, California

Contributors in Addition to the Editors and Associate Editors

Liliana Alamilla-Beltrán, Departamento de Graduados en Alimentos, Escuela Nacional de Ciencias Biológicas, Instituto Politécnico Nacional, México D.F., México

Alicia de Anda-Salazar, Universidad Autónoma de San Luis Potosí, Facultad de Ciencias Quimicas, Centro de Investigación de Estudios de Posgrado, San Luis Potosí, México

Walter Onofre Benejam, Faculty of Agricultural Sciences, Universidad Católica de Córdoba, Córdoba, Argentina

Johanna Björkroth, Department of Food and Environmental Hygiene, Faculty of Veterinary Medicine, University of Helsinki, Helsinki, Finland

L. Castigliego, Department of Animal Pathology, Prophylaxis and Food Hygiene, University of Pisa, Pisa, Italy

José Jorge Chanona-Pérez, Departamento de Graduados en Alimentos, Escuela Nacional de Ciencias Biológicas, Instituto Politécnico Nacional, México D.F., México

Octavio Dublán-García, Facultad de Química, Universidad Autónoma del Estado de México, Toluca, México

Juana Fernández López, Grupo Industrialización de Productos de Origen Animal (IPOA Research Group), Departamento de Tecnología Agroalimentaria, Escuela Politécnica Superior de Orihuela, Universidad Miguel Hernández, Orihuela, Alicante, Spain

Susana Fiszman, Instituto de Agroquímica y Tecnología de Alimentos, Burjassot, Valencia, Spain

José L. Flores-Luna, Food Safety Management Systems (Consultant), México D.F., México

Alicia Grajales-Lagunes, Universidad Autónoma de San Luis Potosí, Facultad de Ciencias Quimicas, Centro de Investigación de Estudios de Posgrado, San Luis Potosí, México

Maria Dolors Guàrdia, IRTA, Finca Camps i Armet s/n, Monells, Girona, Spain

Luis Guerrero Asorey, IRTA, Finca Camps i Armet s/n, Monells, Girona, Spain

Alessandra Guidi, Department of Animal Pathology, Prophylaxis and Food Hygiene, University of Pisa, Pisa, Italy

Gustavo F. Gutiérrez-López, Departamento de Graduados en Alimentos, Escuela Nacional de Ciencias Biológicas, Instituto Politécnico Nacional, México D.F., Mexico

Marja-Liisa Hänninen, Department of Food and Environmental Hygiene, Faculty of Veterinary Medicine, University of Helsinki, Helsinki, Finland

Juan Francisco Hernández Chávez, Tecnológico de Estudios Superiores de Ecatepec, Laboratorio de Alimentos, Ecatepec de Morelos Estado de México, México

Elvia Hernández-Hernández, Departamento de Biotecnología, Universidad Autónoma Metropolitana–Unidad Iztapalapa, México D.F., México

Syeda K. Hussain, Department of Poultry Science, Auburn University, Auburn, Alabama

Elza Iouko Ida, Food Science and Technology Department, Londrina State University, Londrina, Paraná, Brazil

Lekh R. Juneja, Taiyo Kagaku Co. Ltd., Yokkaichi, Japan

Mahendra P. Kapoor, Taiyo Kagaku Co. Ltd., Yokkaichi, Japan

Adriana Lourenço Soares, Food Science and Technology Department, Londrina State University, Londrina, Paraná, Brazil

Ernesto Mendoza-Madrid, Departamento de Graduados en Alimentos, Escuela Nacional de Ciencias Biológicas, Instituto Politécnico Nacional, México D.F., México

Casilda Navarro-Rodríguez de Vera, Grupo Industrialización de Productos de Origen Animal (IPOA Research Group), Departamento de Tecnología Agroalimentaria, Escuela Politécnica Superior de Orihuela, Universidad Miguel Hernández, Orihuela, Alicante, Spain

Francisco Alfredo Nuñez-González, Facultad de Zootecnia, Universidad Autónoma de Chihuahua, Chihuahua, México

Omar A. Oyarzabal, Department of Biological Sciences, Alabama State University, Montgomery, Alabama

M.C. Pandey, Defence Food Research Laboratory, Siddartha Nagar, Mysore, India

María de Lourdes Pérez-Chabela, Departamento de Biotecnología, Universidad Autónoma Metropolitana–Unidad Iztapalapa, México D.F., México

Edith Ponce-Alquicira, Departamento de Biotecnología, Universidad Autónoma Metropolitana–Unidad Iztapalapa, México D.F., México

Baciliza Quintero Salazar, Centro de Investigación y Estudios Turísticos, Universidad Autónoma del Estado de México, Toluca, México

Molay K. Roy, Taiyo Kagaku Co. Ltd., Yokkaichi, Japan

S.N. Sabapathi, Defence Food Research Laboratory, Siddartha Nagar, Mysore, India

Ana Salvador, Instituto de Agroquímica y Tecnología de Alimentos, Burjassot, Valencia, Spain

Elena José Sánchez-Zapata, Grupo Industrialización de Productos de Origen Animal (IPOA Research Group), Departamento de Tecnología Agroalimentaria, Escuela Politécnica Superior de Orihuela, Universidad Miguel Hernández, Orihuela, Alicante, Spain

Teresa Sanz, Instituto de Agroquímica y Tecnología de Alimentos, Burjassot, Valencia, Spain

Carmen Sárraga, IRTA, Finca Camps i Armet s/n, Monells, Girona, Spain

Estrella Sayas-Barberá, Grupo Industrialización de Productos de Origen Animal (IPOA Research Group), Departamento de Tecnología Agroalimentaria, Escuela Politécnica Superior de Orihuela, Universidad Miguel Hernández, Orihuela, Alicante, Spain

Esther Sendra-Nadal, Grupo Industrialización de Productos de Origen Animal (IPOA Research Group), Departamento de Tecnología Agroalimentaria, Escuela Politécnica Superior de Orihuela, Universidad Miguel Hernández, Orihuela, Alicante, Spain

Massami Shimokomaki, Food Science and Technology Department, Londrina State University, Londrina, Paraná, Brazil

Marcelo L. Signorini, Consejo Nacional de Investigaciones Científicas y Técnicas, Instituto Nacional de Tecnología Agropecuaria, Estación Experimental Rafaela, Departamento de Epidemiología y Enfermedades Infecciosas, Provincia de Santa Fe, Argentina

Vandana Sohlia, Defence Food Research Laboratory, Siddartha Nagar, Mysore, India

Alfonso Totosaus-Sánchez, Tecnológico de Estudios Superiores de Ecatepec, Laboratorio de Alimentos, Ecatepec de Morelos Estado de México, México

Carol W. Turner, Department of Family and Consumer Sciences, New Mexico State University, Las Cruces, New Mexico

Elina J. Vihavainen, Department of Food and Environmental Hygiene, Faculty of Veterinary Medicine, University of Helsinki, Helsinki, Finland

Manuel Viuda-Martos, Grupo Industrialización de Productos de Origen Animal (IPOA Research Group), Departamento de Tecnología Agroalimentaria, Escuela Politécnica Superior de Orihuela, Universidad Miguel Hernández, Orihuela, Alicante, Spain

Jorge Welti-Chanes, Departamento de Graduados en Alimentos, Escuela Nacional de Ciencias Biológicas, Instituto Politécnico Nacional, México D.F., México

Ana Paola Zogbi, Faculty of Agricultural Sciences, Universidad Católica de Córdoba, Córdoba, Argentina

PREFACE

Poultry has been and still is a major animal product in our diets. With the advances in preservation techniques for fresh poultry and processed products, consumer preferences for poultry and poultry products are higher than ever. Information on the science and technology of processing this important food commodity is essential to the work of government, academia, and industry.

Many good professional reference books are available. The preference for any particular one depends on the needs of the users. Most are single-volume books, with some covering general and others specific topics. Excluding encyclopedias, multivolume reference books in the discipline are uncommon for many reasons, such as cost, wide coverage, and standard technical challenges, including but not limited to the involvement of a large number of professionals and pressure of a timely publication. On the other hand, most big technical libraries in the world (government, academia, and industry) prefer comprehensive multiple-volume books because they reduce the needs for several books. From this perspective, our two-volume set is designed especially for libraries, although books of this nature will always serve as useful reference sources for students, researchers, instructors, and R&D personnel. The first volume covers the primary processing of fresh poultry and preservation of raw poultry meats. The second volume covers the secondary processing of raw poultry meats to processed retail products.

Volume 1 emphasizes primary processing and covers poultry and their slaughter practices, with an emphasis on classification, biology, production, transportation, slaughtering, pre- and postmortem handling, and carcass evaluation and cutting. The preservation methods for raw poultry meat are also described, such as heat, cold, chemical compounds, irradiation, and high pressure. Emphasis is placed on refrigeration and freezing since these preservation techniques are of major importance. The remaining topics include the engineering principles of packaging, quality attributes of poultry meat (taste, texture, tenderness, juiciness), safety of products and workers, sanitation, and government requirements for hazard control and risk analyses. Details are also provided for Jewish and

Muslim practices for slaughtering and processing poultry and poultry products. Eggs are always an integral part of a discussion related to poultry and poultry products. Coverage related to eggs includes health, nutrition, and the science and technology of processing eggs. Accordingly, the coverage in Volume I is divided into five sections. The table of contents provides the topics for the 38 chapters.

Volume 2 deals with secondary processing of poultry and poultry products covering the transformation from basic raw poultry meat into safe and wholesome products tailored for consumers. These products are available in many forms, including but not limited to such popular poultry items as sausage and deli meats. Some of these items are raw, some cooked but not ready to eat, and some cooked and ready to eat. Thus, the major goal of this volume is to present the technical knowhow needed for manufacturing such products. To do so, this volume presents a sequence of topics divided into seven sections.

Volume 2 begins with the basic principles in formulating and processing poultry products, including mechanical deboning, marination, emulsion basics, formulation, and breading. Many processed poultry products for consumers contain nonmeat ingredients, and this topic is discussed in detail. This is followed by the practical applications and techniques in manufacturing patties, sausages, bacon, ham, luncheon meats, nuggets, pâté, and other products. To produce a high-quality poultry product, one must be familiar with the color, flavor, and texture of raw and cooked poultry meats, and these quality attributes are described in detail. Obviously, the wholesomeness and safety of the product is a primary concern for all government agencies around the world. Because of the many outbreaks of foodborne diseases from contaminated poultry products, 9 of 39 chapters in this volume are devoted to sanitation and food safety system in the United States, covering topics such as contaminants, microbiology, pathogens, analytical techniques, and the requirements for sanitation, hazards identifications, and risks factors involved.

Although many topics are included in these two volumes, we do not claim the coverage to be totally comprehensive. The work is the result of the combined expertise of more than 150 people from industry, government, and academia: professionals from Argentina, Brazil, Canada, Finland, India, Italy, Japan, Malaysia, Mexico, Spain, and the United States. An international editorial team of 15 members from six countries led these experts. Each contributor or editor was responsible for researching and reviewing subjects of immense depth, breadth, and complexity. Care and attention were paramount to ensure technical accuracy for each topic. In sum, these two volumes are unique in many respects. It is our sincere hope and belief that they will serve as essential references on poultry and poultry processing.

We wish to thank all the contributors for sharing their expertise throughout our journey. We also thank the reviewers for giving their valuable comments, leading to improvements in the contents of each chapter. In addition, we thank members of the production team at John Wiley & Sons, Inc., for their time, effort, advice, and expertise. All these professionals made this two-volume treatise possible. You are the best judge of the quality of their work and we trust that you will benefit from the fruits of their labor.

I. Guerrero-Legarreta
Y.H. Hui
A.D. Alarcón-Rojo
C. Alvarado
A.S. Bawa
F. Guerrero-Avendaño
J. Lunden
L. McKee
Y. Mine
C.M. Owens
J.A. Pérez-Alvarez
J.M. Regenstein
M.R. Rosmini
J. Soriano-Santos
J. Eddie Wu

PART I

SECONDARY PROCESSING OF POULTRY PRODUCTS

1

PROCESSED POULTRY PRODUCTS: A PRIMER

ISABEL GUERRERO-LEGARRETA
Departamento de Biotecnología, Universidad Autónoma Metropolitana, México D.F., México

Y.H. HUI
Science Technology System, West Sacramento, California

INTRODUCTION

For centuries, poultry and poultry products have been a popular food for human beings. Similar to other foods, we have been able to preserve them from spoilage by salting, drying, smoking, and fermenting. Eventually, sugar or another sweet raw ingredient was used as a preservation tool. Then canning, refrigeration, and frozen storage became available. For the last 30 years, advances in science, technology, and engineering, accompanied by intensive research and development, have resulted in an array of preservation techniques that rival our culinary

Handbook of Poultry Science and Technology, Volume 2: Secondary Processing, Edited by Isabel Guerrero-Legarreta and Y.H. Hui

interests. Such techniques include, but are not limited to, the use of packaging, chemical and biological agents, heat, cold, fermentation, irradiation, pressures, and reduced atmospheres. Processing and formulation of the raw materials have also improved tremendously in terms of sizing, emulsification, marination, curing, smoking, drying, and so on. Other chapters in the volume address such scientific developments and advances. In this chapter we present an overview of some economics aspects of poultry products, such as commercialization, pricing, consumer preferences, and certain special dietary needs.

TRENDS IN POULTRY PRODUCT COMMERCIALIZATION

Processed poultry meat and their fabricated products are consumed worldwide. However, sociopolitical and economic circumstances modify the scope of commercialization. Let us look at these trends in various population groups outside the United States. They will serve as a general frame of references to facilitate programs in the United States for the exportation of poultry and poultry products.

European Union Broiler imports, mainly from Brazil, are expected to increase in 2008. The European Union (EU) may become a net importer of chicken meat for the first time in 2008. Chicken meat consumption is expected to increase in 2008 as an inexpensive source of protein. France is the principal exporter of chicken meat in the EU region, but France reduced its market share in 2007, due to Brazil's successful competition in the Middle East and sub-Saharan Africa. Brazil is also the main poultry product supplier to the EU (salted, marinated, dried, smoked, cooked, etc). Thailand is the second largest exporter to the EU. Brazilian poultry meat exporters have recently purchased several European poultry processors, including some in the UK. With respect to consumption, and despite rising prices, chicken meat remains a low-cost protein source.

Canada Canadian broiler chicken imports in 2006 were 82% from the United States and 17% from Brazil. After Mexico and the Russian Federation, Canada is the third most important export market for U.S. poultry meat. The U.S. Department of Agriculture does not allow imports of Brazilian chicken; the Canadian Food Inspection Agency has strict import control procedures to ensure that Brazilian chicken in Canada does not enter the United States.

Latin America In Venezuela, chicken remains the most available and cheapest source of animal protein. About 80 to 90% of the poultry produced in Venezuela is purchased fresh by households. The rest goes to the processing sector (hams, sausages, frozen nuggets, etc.). Poultry products offered through the government's distribution network are cheaper than current controlled prices; however, the government imports poultry from Brazil.

Poultry meat exports from Argentina in 2007 were 110,000 metric tons (plus 50,000 metric tons of chicken paws, which are expected to be exported to China and Hong Kong). Chile is the largest market, since it demands products very similar to those produced for the local market (mainly breasts and leg quarters). South Africa is also one of the country's primary buyers of frozen whole broilers, leg quarters, and mechanically deboned meat. Saudi Arabia mainly imports from Argentina individually quick-frozen breasts; the Russian Federation imports broilers, mechanically deboned meat, and wings. Exports of processed poultry products are primarily cooked poultry meat, frozen cooked or fried pieces, and hamburgers to Germany, the Netherlands, France, the UK, and Chile.

The outlook in 2007 for Brazil's export of broiler and turkey meat was expected to be an increase of 4 and 10% over the preceding year, mainly to the United Arab Emirates, the EU, Hong Kong, and Saudi Arabia. This consists primarily of broiler cuts and processed meat. As for domestic consumption, competition of other meats does not affect broiler consumption, since it is more affordable than beef and pork. In addition, it is expected that the demand of the food-service industry, institutional, and fast-food sectors for products such as frozen chicken meals, precooked meals, and chicken burgers will increase; an increase in consuming more highly processed broiler products has also been observed. Large Brazilian poultry processors are responding to these changes by shifting their sales strategies toward broiler parts (mostly leg quarters and breast meat) and further-processed value-added products such as precooked meals, chicken nuggets, and chicken burgers.

The Chilean poultry production has expanded significantly during the last decade. Poultry meat availability for domestic consumption has grown over 200% during the last two decades; it has become the most important source of animal protein in Chile since 2006, mainly as a result of the higher prices of beef. Chile exports poultry products to 26 different countries. In 2007 two new markets have been opened for the Chilean poultry industry: the Russian Federation and the United States.

Guatemala has the capacity to cover all domestic demand; the U.S. exports provide almost 30% of local consumption. Seventy-eight percent of Guatemala's broiler production plants are technically advanced, including the biggest producer, which also owns the largest chicken fast-food franchise covering the Central America market; 93% of chicken imports are U.S. products, and 75% of the turkey imports are also American. Processed products such as ready-to-fry fillets and patties are starting to build a market. Poultry cuts have a very efficient distribution chain, being sold directly to markets, hotels, restaurants, and supermarkets.

Mexican consumers' concerns about cholesterol and other health issues are creating more marketing opportunities for chicken meat. Poultry meat is cheaper than other protein sources. Consumers prefer fresh whole chickens rather than chicken cuts; however, purchases of chicken cuts are increasing, primarily in supermarkets. Chicken and turkey meat are the primary poultry products imported by Mexico. The processing industry imports most mechanically separated poultry (chicken and turkey) and poultry cuts as inputs for the domestic sausage and cold-cut industries. Imports of chicken cuts (mainly leg quarters) and mechanically separated chicken is decreasing due to high international chicken prices, which have diverted much of the U.S. exportable supply to other markets, such as China and Russia. The United States is the main supplier of chicken meat to Mexico, although Chilean imports are increasing. In recent years, the Mexican export trend toward value-added products has been increasing, mainly chicken and turkey sausages.

India India broiler meat production grew at over 15% per annum in recent years. Due to wide acceptance and affordable prices, poultry meat is the major meat consumed in India. High mutton prices, religious restrictions on beef and pork, and the availability of fish in coastal regions made poultry the preferred meat. Indians typically prefer fresh chicken meat. The processed poultry meat market (chilled/frozen dressed/cut chicken) is limited, confined primarily to institutional sales. Poultry meat exports are negligible; in recent years, some southern Indian industries have been exploring the possibility of exporting poultry meat to the Middle East and Southeast Asian markets.

Thailand Thailand imports cooked chicken from the EU and Japan; China is likely to reduce its exports to Thailand as a result of growing domestic consumption. The volumes of Thai uncooked items, such as boneless leg and skinless boneless breast, exported to Japan will increase. Cooked chicken products are normally made-to-order meat products that are processed or prepared by heat, such as grilling, steaming, and boiling. Some of these cooked meat products are seasoned with salt, Japanese sauce, and other flavorings.

Japan Broiler meat makes up about 90% of the Japanese poultry meat market, including both domestic production and imports. Spent laying hens account for about 10% of the poultry meat market; consumption of ducks, turkeys, and other poultry is limited. In general, leg meat (boneless) is preferred over breast meat. The Japan food-service sector utilizes large quantities of imported generic cuts, mainly from Brazil, including cuts to be processed into prepared products after entry into Japan. Some U.S. bone-in leg cuts are utilized as well in this industry. Japan also imports cooked products from Thailand and China, mainly to prepare yakitori products (skewed grilled chicken).

Republic of Korea Due to the continued oversupply of domestic broiler meat, imports were reduced in 2007. The domestic industry prefers Brazilian deboned leg meat to bone-in leg meat from the United States, due primarily to the larger size of American imports. About 80% of total imports were frozen wings, and the remainder was heat-treated meat used to fabricate a traditional product known as *samgyetang*. The bulk of imported heat-treated meat was from China, with only a small fraction originating in Thailand. Korean broiler meat exports ranged between 1000 and 3000 metric tons over the last decade. Imported chicken cuts are used primarily in the food and processed service sector, for fabricating seasoned chicken dishes, chicken nuggets, seasoned wings, patties, and so on. Local chicken is sold primarily as chilled whole birds, with a smaller amount as cuts to both the food-service sector and retail markets. Meat derived from domestic spent hens is used mostly as raw ingredients in further-processed products such as sausages and hams.

People's Republic of China Broiler production in China in 2007 was 12.5 million metric tons due to the high local demand; this year, China also imported turkey meat, a nontraditional product. Poultry meat is traditional in the Chinese culture: kungpao chicken, Peking duck, stir-fries, Cantonese-style air-dried duck, deep-fried chicken nuggets. It is expected that in 2008 China will export poultry to Japan, Hong Kong, and South Korea.

Malaysia Malaysia has one of the highest per capita consumption rates in the world for chicken. About 30% of broilers are channeled through modern processing plants; 60% are sold as dressed birds in markets. Chicken meat is the most popular and cheapest source of meat protein among Malaysians, due to the fact that there are no dietary prohibitions or religious restrictions against chicken consumption; for this reason, quick-service restaurants based on processed chicken products (e.g., battered products, hamburgers, nuggets) are popular in Malaysia; further-processed products are also distributed to wholesalers, supermarkets, hypermarkets, catering institutions, restaurants, and hotels. Consumers are very sensitive toward health and halal matters. It is important to note that the majority of Muslim consumers will not accept poultry products not certified halal by the Malaysian religious authority. The Malaysian poultry industry is well positioned to supply halal-processed poultry to other Islamic countries and Muslim consumers worldwide. Chicken parts, mainly wings and mechanically deboned chicken, are frequently imported; the major suppliers are the Netherlands and Denmark. Poultry processing is also of importance in Malaysia, where products such as chicken frankfurters, cocktail sausages, burgers, and nuggets are the most popular items. All turkey meat is imported; the United States is the dominant supplier.

Taiwan The United States is the predominant supplier of poultry to Taiwan; however, it exports only limited quantities of poultry products, mostly fresh and frozen meat and prepared chicken. During 2005, domestic production, including nonbroiler chickens, ducks, geese, and turkeys, showed a considerable decline compared to 2004, although the meat production is now recovering.

Saudi Arabia The Saudi Ministry of Agriculture classifies poultry production farms as either specialized (commercial) or traditional; 97% of poultry meat produced in Saudi Arabia consists of broiler chicken. The item consumed most is broiler meat. Brazil has been the leading frozen broiler meat supplier to Saudi Arabia, followed by France, Argentina, and South Africa. Domestically produced poultry is generally marketed fresh or chilled. During the summer months, when sales drop, major operations will freeze some production. One major operation is producing chicken frankfurters and burgers for local market. Boneless chicken imports are dominated by Brazil, which took over the market from China and Thailand. Consumption of whole turkey is seasonal, while duck is consumed primarily in Chinese restaurants and in some Arabic restaurants.

VARIETY AND PRICES

Due to a number of possible processing methods, poultry products are available to the consumer in a wide range of products. Poultry products can be classified in various ways, depending on the specific objective of the food item: degree of comminuting (e.g., whole birds), processing characteristics (e.g., cured), shelf life expected (e.g., fresh, requiring refrigeration), and particular specifications (e.g., low fat). More details are as follows:

1. According to degree of comminuting:
   - Whole birds
   - Mechanically deboned meat
   - Products retaining their integrity, such as bone-in or deboned legs, breast, and thighs.
   - Restructured or formed products, such as cooked hams and hamburgers
   - Coarsely ground products, such as poultry summer sausages
   - Finely ground, or emulsified, products, such as frankfurters
2. According to processing characteristics:
   - Cured
   - Emulsified (sausages)
   - Battered (nuggets)
   - Marinated
   - Smoked
   - Roasted
   - Dried and semidried
3. According to expected shelf life:
   - Fresh; need refrigeration
   - Intermediate-moisture food (reduced water activity)
   - Dried and semidried
   - Fermented
4. According to particular specifications:
   - Low fat
   - Low sodium
   - Special diets
   - Religious specifications (halal, kosher)
   - Ethnic dishes (e.g., yakitori, kungpao chicken, Peking duck, Cantonese-style air-dried duck, deep-fried chicken nuggets, samgyetang, chorizo)

TABLE 1 Main Characteristics of Processed Poultry Products

Whole or Parts	Bone Removal	Preparation	Flavor	Post-Processed Products
Breasts (whole and split)	Bone-in	Canned	Asian	Bacon
Breast chops	Boneless	Cooked	BBQ	Bologna
Breast cutlets		Breading	Buffalo style	Bratwurst
Breast scaloppini		Deep fried	Cajun	Breakfast sausages
Breast strips		Dry roasted	Citrus	Burgers
Breast tenderloins		Frozen	Dijon mustard	Dinner sausages
Drumsticks		Grilled	Hickory	Ham
Ground meat (with various lean/fat ratios)		Marinated	Honey	Kebab
Necks		Ready-to-cook	Honey smoked	Luncheon meat
Thighs		Roasted	Honey-pepper	Meatballs
Whole birds		Rotisserie-like	Lemongrass	Nuggets
Wings		Smoked	Maple	Pâté
		Sun dried with gravy	Mesquite	Patties
			Smoke	Salami
			Teriyaki	Sausages
			Zesty Italian	Sausage rolls
				Summer sausages

Therefore, applying one or several of the above-mentioned characteristics, a wide variety of poultry products (chicken and turkey) is obtained. The main characteristics of poultry products that have the highest market share in the United States are shown in Table 1. Combining these processing characteristics enables an enormous variety of poultry products to be produced: fresh bone-in turkey breast, hickory smoked bone-in turkey breast and honey mesquite smoked chicken breast, hickory smoked chicken ham, turkey bologna, and so on.

The possibility of diversifying products gives the poultry industry the potential to market its products to a wide variety of population sectors. In general, poultry meat is cheaper than pork or beef, making poultry, especially chicken, cheaper and more accessible to large sections of the population. However, for products that require specific processing, such as hickory smoking and honey roasts, the product price is increased considerably.

CULTURAL AND RELIGIOUS ACCEPTANCE

Because poultry meat and products are cheaper than other meats, such as beef and pork, they are widely accepted and consumed in all parts of the world. Asian countries base most of their ethnic dishes on poultry: China's Peking duck, kung-pao chicken, and steamed and grilled lemon chicken; Thailand's curry chicken, chicken Satay, and chicken macadami; India's chicken Tikka and chicken Biryani, and murg do pyaaza; Africa's roasted and barbecued chicken; Mexico's mole and pibil chicken; Brazil's Xim-xim; Colombia's ajiaco. As for new developments in poultry processing, the wide variety of products and ready-to-cook dishes developed by transnational companies have found wide acceptance in modern societies.

Religious acceptance of food products is based on fulfilling certain laws of a particular religion, especially Jewish and Islamic requirements. In the Jewish religion, the dietary laws that govern kosher foods are found in Deuteronomy and Leviticus in the Torah; these laws have been followed for over 3000 years. Meat must come from a kosher animal, as outlined by Deuteronomy 14. Kosher birds are chickens, turkey, duck, and geese.

Halal poultry is slaughtered in accordance with requirements of the Shariah (Islamic law). Chickens are also halal birds. The animal must be slaughtered and processed with the highest standards of cleanliness, purity, and wholesomeness.

PRODUCTS FOR SPECIAL DIETS

Worldwide concern about heart disease, hypertension, and diabetes, among others, has resulted in many "healthy" new products developed by poultry meat processors. The poultry industry has developed light turkey sausages, light chicken and turkey ham, and other processed meat that fulfill consumer expectations. These products relate to a suitable dietary supply without including chemical compounds that have been reported as possibly harmful, such as saturated fats and high sodium. Among these products are low- or reduced-fat products, reduced-sodium products, and products fortified with omega-3 long-chain fatty acids.

Poultry is relatively high in polyunsaturated fatty acids, due to the fact that their diet is generally rich in these fatty acids. However, further fat reduction is highly appreciated by the concerned consumer. Low- and reduced-fat meat products are made by substituting for the animal fat a variety of carbohydrates that mimic the palatability of a fat-added product and provide a texture similar to that of fat globules in batters. Among the most common fat replacements are tapioca starch, wheat flour, and carrageenan.

It has been reported that the intake of enough omega-3 fatty acid [e.g., eicosapentaenoic acid (EPA), docosahexaenoic acid (DHA)] prevents conditions such as atherosclerosis, coronary heart disease, hypertension, cancer, and diabetes. The omega-3 fatty acids are present in marine fishes, mainly of temperate waters. Poultry products fortified with these acids are expected to help prevent such diseases. Several turkey and chicken products, such as nuggets, frankfurters, and hamburgers, have been fortified with omega-3 acids in Canada and the European Union.

Reduction in sodium intake prevents blood hypertension, an extremely common condition in most industrialized countries that is related to cardiovascular disease. Although poultry meat itself is relatively low in sodium content, sodium is increased during further processing. At the same time, salt is an imp ortant flavoring in meat products; therefore, there is a compromise between reducing sodium content in a processed product and its palatability. About 30% of the sodium chloride is now replaced by magnesium or potassium without notably affecting product palatability.

It is important to emphasize that the use of food additives in processed poultry products is tightly regulated in many countries, especially the United States. The information provided here explains the scientific and technical feasibility. Other resources should be used to obtain more details on government regulations.

REFERENCE

USDA Foreign Agricultural Service. 2006, 2007, 2008. http://www.thepoultrysite.com/articles.

PART II

METHODS FOR PROCESSING POULTRY PRODUCTS

2

GELATION AND EMULSION: PRINCIPLES

Ana Paola Zogbi and Walter Onofre Benejam
Faculty of Agricultural Sciences, Universidad Católica de Córdoba, Córdoba, Argentina

GELATION

Gelation is an orderly aggregation of denatured molecules that gives rise to a three-dimensional solid network that traps an aqueous solvent consisting of immobilized water within a matrix (Totosaus and Guerrero-Legarreta, 2006). This state is an intermediary between solid and liquid states (Girard, 1991). Network formation from protein is seen as a result of a balance between protein–protein and protein–water interactions and attractive and repulsive forces between close polypeptide chains (Fennema, 1996). Gelation is used to improve absorption of

Handbook of Poultry Science and Technology, Volume 2: Secondary Processing, Edited by Isabel Guerrero-Legarreta and Y.H. Hui

water, thickening, and the fixing effect of particles and the stabilization of emulsions and foams (Smith, 1994). It is influenced by heat, pH, ionic strength, and protein concentration (Smith, 1994; Wong, 1995). In most cases, gel formation requires heat treatment followed by cooling and is possibly favored by a slight acidification or calcium ion addition (Ponce et al., 2000). Most gels found in food are heat induced (Foegeding, 1988), and in many meat products they determine textural attributes, performance, quality, and other sensory characteristics, as in emulsified sausages (Wirth, 1992).

Gelation Mechanisms

In most cases, heat treatment is essential to achieve gelation, yet several proteins could also be gelled only by moderate enzymatic hydrolysis, calcium ion addition, or alkalization followed by a return to the pH neutral charge or isoelectric point. The heat-induced gelation process is the result of a sequence of events relating protein transformation from suspension to a semisolid state. The first step is a conformational change from the native state to a pre-gel state by applying heat. This transformation involves dissociation, denaturation, and deployment with exposed protein functional groups, thus making it possible to establish a wide variety of intermacromolecular links (e.g., hydrogen bonds, hydrophobic and electrostatic interactions, disulfide links), which result in an orderly continuous network (Ponce et al., 2000).

If the stage of aggregation is slower than the denaturalization stage, partially deployed polypeptides could be oriented before the aggregation, favoring an orderly, homogeneous, strongly expanded, elastic, transparent, and stable gel of smooth consistency when facing syneresis and exudation. If this stage velocity is faster, gels will be dull, with little elasticity and will clearly be unstable when facing syneresis and exudation (Fennema, 1996). The development of a particular type of link in a gel network depends, among other factors, on the type and concentration of protein, pH, ionic strength, heating rate, and final temperature (Smith, 1988, 1994). If the forces of attraction are weak, proteins will be added and the intersection will be so strong that they can no longer unite water and, consequently, will precipitate (Matsutmura and Mori, 1996; Totosaus and Guerrero-Legarreta, 2006).

Types of Gels

There are two types of gels in food: thermally reversible and nonreversible gels. Many proteins in food are irreversible or thermoformed gels, including muscle proteins (actin and myosin). Such gels are formed during heating and retain a similar structure when cooled (Hill, 1996). The irreversible thermal gels involve denaturation or the partial unfolding of proteins within a certain temperature range to form an increasingly viscous solution, followed by an aggregation of unrolled molecules to form a network cross-linked when it reaches the gelation point. Properties change when the gel cools or when it is left to settle down to reach a balanced network (Matsumura and Mori, 1996). The establishment of covalent unions of disulfide bonds usually leads to the formation of thermally irreversible gels, as with ovalbumin or lactoglobulin gels (Fennema, 1996). On the other hand, reversible gelatin gels are essentially stabilized by hydrogen bonds, and if the gelatin is heated again, the hydrogen bonds are broken and the gel is liquefied, forming a viscous solution (Smith and Culberson, 2000). The relative contribution of each type of bonding to the gel network varies depending on the protein and environmental conditions, temperature, and heating rate, these being the factors that most influence gel properties and appearance (Barbut, 1994).

Factors Affecting Gelation

Temperature, pH, and salts affect the degree of protein binding because they modify the quaternary structure or distribution of the net electric charges, altering the gel nature and structure (Ponce et al., 2000).

Temperature In gels in which the denaturation process is accomplished by heat treatment, it is observed that the higher the temperature, the greater the firmness and the less the water-holding capacity of gels because the gel formed is heterogeneous and protein aggregates present pores filled with aqueous phase that may easily be removed (Ordoñez et al., 2000). Temperature increase favors protein–protein interactions; protein molecules deploy and exposure of hydrophobic groups increases, thus diminishing protein–water interactions, promoting disulfide bridge formation, which reinforces the strength of the intermolecular network, and as a result, the gel tends to be irreversible (Fennema, 1996). In muscle proteins there are two critical transition temperatures for gel formation (43 and 55°C), which implies the participation of two conformational changes in the gelation induced by heat treatment: Myosin heads aggregate at low temperature, and as the temperature rises, cane-shaped segments or tails start to deploy (Wong, 1995). Head aggregation demands disulfide exchanges and possible intermolecular associations of lateral chains (Hoseney, 1984; Wong, 1995). Gel networks are reinforced by means of noncovalent linkages established within available centers through unrolled molecules of myosin. Actin presence improves the characteristics of myosin gels (Wong, 1995). Gelation with myosin starts at 30°C, reaching its maximum state at 60° to 70°C (Ponce et al., 2000).

pH The pH zone in which gelling occurs widens as the protein concentration increases. This would indicate that the many hydrophobic links and disulfides formed with a strong protein concentration could compensate the electrostatic repulsion forces induced by the high protein net charge (away from the isoelectric point). On the other hand, the presence of the isoelectric point and the absence of repulsive forces lead to the formation of a less expanded hydrated and consistent gel. Optimum pH for myosin gelation is pH 6. When myosin concentration is much greater than that of actin, the gelation is higher than pH 6, but when the myosin/actin relationship decreases, the gelation maximum values obtained are below pH 6 (Smith, 1994; Guillen et al., 1996). Lesiów and Xiong (2003) reported that the optimum gelation point in homogenized chicken breast was at pH 6.3, and in chicken thighs pH values were within 5.8 to 6.3; similar results were found in isolated myofibril protein from chicken.

Salt Concentration When NaCl concentration is high (0.6 M, 3%), the fraction of the myosin tail is attached to the myosin filament and cannot contribute effectively to the gelation process (Smith, 1994). Similarly, phosphates may dissociate the actomyosin complex into its components, actin and myosin, and by increasing the concentration of usable myosin, the gelation capacity is enhanced (Flores and Bermell, 1986; Smith, 1994).

Protein Concentration A high protein concentration favors gelation, as it strengthens the contact between proteins, even under unfavorable conditions (Ordoñez et al., 1998a).

EMULSIFICATION

An *emulsion* is described as a set of two immiscible liquid phases scattered one into the other. Proteins act as emulsifying agents, being absorbed at the interface and orienting neutral groups to fat particles and ionic groups toward a watery matrix. Emulsion stability depends on a number of factors, such as geometry and the equipment used, the energy applied to emulsion formation, the speed of fat addition, the mixing temperature, the pH, the ionic strength, and the presence of surface agents (Ponce et al., 2000). Emulsions are thermodynamically unstable as a result of excess free energy associated with the droplet surface. The dispersed droplets therefore try to unite and reduce the surface. To minimize this effect, emulsifiers are added to the system to improve its stability. The stability of an emulsion can therefore be considered as dependent on the balance between the forces of attraction and repulsion (Wong, 1995). Proteins are used widely in the food industry to stabilize emulsions (Kinsella, 1998; Walstra et al., 1999). The protein efficiency as emulsifiers of force depends on the density and structure of protein absorption.

Stability

Among the factors stabilizing emulsions, the presence of a resistant interfacial layer could be mentioned. It consists of a layer of absorbed protein, as well as the presence of electrostatic charges of the same sign on dispersed droplet surfaces due to the ionization or adsorption of ions (Wong, 1995). A small droplet diameter may be obtained through intense agitation and the appropriate concentration of a specific tensoactive agent, also contributing to emulsion stabilization (Totosaus and Guerrero-Legaretta, 2006). Similarly, electrolyte minerals providing electrostatic charges to the dispersed droplets would increase repulsion, a mechanism that is more important in oil-in-water emulsions. Another stabilization mechanism is the formation of a physical barrier with finely divided insoluble materials (basic metal salts) absorbed at the interface level, or tensoactive or surface-active molecules such as proteins which are oriented so that their ends are placed in hydrophobic and hydrophilic surfaces, respectively, on the oil–water interface. This molecule accumulation at the interface reduces the surface tension (Ordoñez et al., 1998b). Stabilization by liquid crystals created by interactions between the emulsifying agent, oil, and water gives way to the formation of a multilayer with a liquid-crystal structure around the droplets. This barrier leads to greater emulsion stability (Fennema, 1996). Stability through macromolecules dissolved in the continuous phase could also be achieved, thus increasing this phase viscosity (polysaccharide thickeners), or they could be adsorbed at the interface (proteins soluble in water), forming a barrier against the coalescence (Ordoñez et al., 1998a).

Destabilization Mechanisms

Destabilization mechanisms of emulsions are coalescence, flocculation, and creaming. During flocculation and creaming there are changes in structure, but drop-size distribution may remain unaltered, in contrast with coalescence, which changes drop-size distribution over time (Tcholakova et al., 2006).

Creaming or Sedimentation The densities of scattered drops are different from those in the scattering phase, and as a result, there may be sedimentation and separation of the phases due to gravity forces. The speed at which this phase occurs is directly proportional to the drop size of the dispersed phase and inversely proportional to the viscosity of the continuous phase or dispersal. It complies with Stokes' law, whose mathematical expression is

$$V = \frac{2r^2 g\rho}{9\eta} \qquad (1)$$

where V = sedimentation rate of the drops in the dispersed phase
r = radius of drops in the dispersed phase
g = gravitational acceleration
ρ = density difference between the dispersing and dispersed phases
η = viscosity coefficient of the dispersing phase

Perlo (2000) mentions that in beef emulsions the factors affecting stability are the size of the fat globule and the viscosity. Therefore, the lower the globule radius and the higher the continuous-phase viscosity, the more stable the emulsion will be (Nawar, 1993).

Flocculation or Aggregation Flocculation is a consequence of the removal of electrical charges, with the subsequent inhibition of electrostatic repulsions. The drops are fused to each other but separated by a layer of fine continuous phase.

With flocculation the droplet size increases, as does the sedimentation speed. The globules move as a whole rather than individually. Flocculation does not imply a break in the interfacial film surrounding the globule, and therefore a change in the size of the original globule should not be expected (Ordoñez et al., 2000).

Coalescence or Fusion Coalescence of emulsion, the fusion of two similarly sized drops forming a larger drop, is manifested by a more or less rapid decrease in the interfacial surface when joining the droplets of the internal surface. The interface area is minimal when the two phases are separated completely and is faster when two immiscible liquids are shaken together. From a thermodynamic viewpoint, this is a spontaneous phenomenon.

Emulsification Agents

Emulsifiers are substances that have two parts, one hydrophilic and the other hydrophobic, separated but being part of the same molecule (Wong, 1995). Monoglycerides, diglycerides, sorbitan fatty acid, lecithin, and some proteins are the principal emulsifiers used in food processing. All of these compounds have hydrophilic or hydrophobic regions (Totosaus and Guerrero-Legarreta, 2006). Emulsifiers are classified according to their ability to form water-in-oil or oil-in-water emulsions, in a system called stock hydrophilic-lipophilic (or HLB, hydrophilic–lipophilic balance), in which they are numbered from 1 to 20. For values around 10, there is a balance between lipophilic and hydrophilic characteristics; values below 10 give tensioactive molecules with lipophilic predominance; values higher than 10 give tensioactive molecules with hydrophilic predominance. Consequently, it is very important to select an emulsifying agent that considers its HLB number, structure, and interaction with other components, such as polysaccharides or proteins (Wong, 1995).

Mono- and Diglycerides Mono- and diglycerides are commonly used as food emulsifiers and are obtained by triglyceride interesterification with excess glycerol. HLB values range from 1 to 10.

Lecithin Lecithin is commonly used as a food emulsifier and is obtained by triglyceride interesterification with excess glycerol. HLB values range from 1 to 10. Lecithin designates a mixture of phospholipids, including phosphatidylcholine, phosphatidylethanolamine, phosphatidylinositol, and phosphatidylserine obtained from oils in the degumming process. Lecithin contains small amounts of triglycerides, fatty acids, pigments, carbohydrates, and sterols. The HLB values range from 2 to 12.

Sorbitan Esters Sorbitan monostearate is the only approved sorbitan ester to be used in processed food. It is produced by the reaction of sorbitol and fatty acid, and its HLB value is 4.7. In meat products, they are used in combination with commercial emulsifiers that have a high HLB value, generally based on pyrophosphate addition and myofibrillar protein participation as an emulsifier (Culberson and Smith, 2000).

Meat Emulsions

Meat emulsion systems are more complex systems than true emulsions, because solid fat is the dispersed phase and the continuous phase is a salty aqueous matrix with soluble and insoluble proteins, particles of muscle fiber, and connective tissue (Perlo, 2000). Because proteins contain both hydrophilic and hydrophobic amino acid lateral chains, they may be used like emulsifiers but must migrate to the interface (Totosaus and Guerrero-Legarreta, 2006). Meat proteins soluble in salt solutions (myofibrillar proteins) are the main emulsifiers, since sarcoplasmic and stromal proteins have little capacity to emulsify fat (Ordoñez et al., 2000). Due to its polar character, a myofibrillar protein such as myosin acts as a union between water and fat, tending to place itself into the water–fat interface with its hydrophobic part oriented toward fat and its hydrophilic part toward water (Perlo, 2007). Myosin and actomyosin are attracted to and concentrate on the fat globule surface, forming a stabilizing membrane (Perlo, 2000), with its viscoelasticity transferring its mechanical strength, related directly to the protein concentration per unit area (Ordoñez et al., 2000). The more soluble proteins present better emulsifying capacity and are good emulsion stabilizers, since they must be dissolved and migrate to the interface so that their surface properties can act. Thus, the ionic strength is crucial to the passing of these proteins into the aqueous phase. On the other hand, muscle proteins are soluble ionic forces from 0.5 to 0.6 M NaCl (Totosaus and Guerrero-Legarreta, 2006). Proteins have their maximum emulsifying capacity when the pH is nearly neutral, which is with pH values between 5 and 7. Emulsion stability and viscosity both decrease as a result of the use of relatively high temperatures. With rising temperatures, fat drops melt and tend to increase in size. High temperatures favor protein denaturation and contribute to a decline in emulsifying capacity. High temperatures favor fat droplet coalescence or reaggregation.

REFERENCES

Barbut S. 1994. Protein gel ultra-structure and functionality. In: Hettiarachchy N, Ziegler GR, eds., *Protein Functionality in Food Systems*. New York: Marcel Dekker.

Barbut S, Findlay CJ 1991. Influence of sodium, potassium and magnesium chlorides on thermal properties of beef muscle. J Food Sci 56:180–182.

Fennema OR. 1996. *Food Chemistry*, 3rd ed. New York: Marcel Dekker.

Flores J, Bermell S. 1985. Capacidad de emulsión de las proteínas miofibrilares. Rev Agroquím Tecnol Aliment 25(4):481–490.

Flores J, Bermell S. 1986. Capacidad de gelificación de las proteinas miofibrilares. Rev Agroquim Tecnol 26(3):151–158.

Foegeding EA. 1988. Molecular properties and functionality of proteins in food gels. In: Kinsella JE, Souci WG, eds., *Food Proteins*. Gaithersburg, MD: Aspen Publishers.

Girard J. 1991. *Tecnología de Carne y los Productos Cárnicos*. Zaragoza, Spain: Acribia.

Guillen C, Solas T, Borderias J, Montero P. 1996. Effect of heating temperature and sodium chloride concentration on ultra-structure and texture of gels made from giant squid. Z Lebensmittel-Untersuchung, 202(3):221–227.

Hill SE. 1996. Emulsions. In: Hall GM, ed., *Methods of Testing Protein Functionality*. New York: Blackie Academic & Professional.

Hoseney RS. 1984. Functional properties of pentonsans in baked foods. Food Technol 38(1):114–116.

Kinsella JE. 1998. Stability of food emulsions: physicol-chemical role of protein and nonprotein emulsifiers. Adv Food Nutr Res 34(81).

Lesiów T, Xiong YL. 2003. Chicken muscle homogenate gelation properties: effect of pH and muscle fiber type. Meat Sci 64:399–403.

Matsumura Y, Mori T. 1996. Gelation. In: Hall GM, ed., *Methods of Testing Protein Functionality*. New York: Blackie Academic & Professional.

Nawar W. 1993. Lípidos. In: Fennema O, ed., *Química de los Alimentos*. Zaragoza, Spain: Acribia.

Ordoñez JA, Cambero MI, Fernández L, García ML, García de Fernando G, de la Hoz L, Selgas MD. 1998a. In: Ordoñez JA, ed., *Tecnología de los Alimentos*, Vol. I. Madrid: Síntesis, pp. 60–74.

Ordoñez JA, Cambero MI, Fernández L, García ML, García de Fernando G, de la Hoz L, Selgas MD. 1998b. In: Ordoñez JA, ed., *Tecnología de los Alimentos*, Vol. II Madrid: Síntesis, pp. 247–249.

Perlo F. 2000. Emulsiones cárnicas. In: Rosmini MR, Pérez-Alvarez JA, Fernández López J, eds., *Nuevas Tendencias en la Tecnología e Higiene de la Industria Cárnica*. Elche, Spain: Universidad Miguel Hernández, pp. 253–274.

Perlo F. 2007. Elaboración de pastas finas. In: Pérez-álvarez JA, Fernández López J, Sayas-Barberá ME, eds., *Industrialización de Productos de Origen Animal*, 3rd ed., Vol. 1. Elche, Spain: Universidad Miguel Hernández, pp. 103–118.

Ponce E, Guerrero-Legarreta I, Pérez L. 2000. Propiedades funcionales de la carne. In: Rosmini MR, Pérez-Alvarez JA, Fernández López J, eds., *Nuevas Tendencias en la Tecnología e Higiene de la Industria Cárnica*. Elche, Spain: Universidad Miguel Hernández, pp. 43–49.

Schmidt RH. 1981. Gelation and coagulation. In: Cherry JP, ed., *Protein Functionality in Foods*. ACS Symp. Ser. 147. Washington, DC: American Chemical Society.

Smith DM. 1988. Meat proteins: functional properties in comminuted meat products. Food Technol 42(5):116–121.

Smith DM. 1994. Protein interactions in gels: protein–protein interactions. In: Hettiarachchy NS, Ziegler GR, eds., *Protein Functionality in Food Systems*. New York: Marcel Dekker, pp. 209–224.

Smith DM, Culberson JD. 2000. Protein: functional properties. In: Scott-Smith J, ed., *Food Chemistry: Principles and Applications*. West Sacramento, CA: Science Technology System.

Tcholakova S, Denkov ND, Ivanov IB. 2006. Coalescence stability of emulsions containing globular milk proteins. Adv Colloid Interface Sci 123–126 (Special Issue): 259–293.

Totosaus A, Guerrero-Legarreta I. 2006. Propiedades funcionales y de textura. In: Hui YH, Guerrero-Legarreta I, Rosmini MR, eds., *Ciencia y Tecnología de Carnes*. Mexico City, Mexico: Limusa, pp. 229–251.

Walstra P, Geurts A, Nomen A, Jellema A, van Boekel AA. 1999. *Dairy Technology*. New York: Marcel Dekker.

Wirth F. 1992. *Tecnología de los Embutidos Escaldados*. Zaragoza, Spain: Acribia.

Wong DWS. 1995. *Química de los Alimentos: Mecanismos y Teorías*. Zaragoza, Spain: Acribia, pp. 53–116.

3

GELATION AND EMULSION: APPLICATIONS

ELVIA HERNÁNDEZ-HERNÁNDEZ AND
ISABEL GUERRERO-LEGARRETA
Departamento de Biotecnología, Universidad Autónoma Metropolitana–Unidad Iztapalapa, México D.F., México

INTRODUCTION

Most finely comminuted meat products are basically gels. During processing, a gel-like matrix is formed which consists of unfolded proteins and carbohydrates that increase batter viscosity. Later, when fat is added, a two-phase system (an

Handbook of Poultry Science and Technology, Volume 2: Secondary Processing, Edited by Isabel Guerrero-Legarreta and Y.H. Hui

emulsion) is formed. After the application of heat, the continuous phase undergoes changes due to protein denaturation and structural changes in other components, such as the added carbohydrates. The solid matrix entraps compounds such as flavoring, antioxidants, and color. Coarsely ground meat product stability is due to the gelling and cohesiveness properties of proteins and other components binding the pieces; meat chunks in products such as cooked hams are bound together by protein and added carbohydrate cohesion properties. Therefore, gel and emulsion systems are found coexisting in meat products.

FOOD GELS

Almost all food gels are a mixture of various components, some of them for individual gels. Several gelling mixtures occur naturally, as in soy protein isolates or myofibrillar proteins, or are due to specific formulations, such as surimi. The main macromolecules of interest in the food industry are proteins and polysaccharides, due to their gelling ability; however, the gelling ability of these two biopolymers is based on different structural characteristics. In polysaccharides, the gelling ability depends on the type of sugar monomers, the length and branching of the polysaccharide, and interactions due to monosaccharide functional groups. Protein gelling ability is due to spatial conformation, but also to amino acid sequence and functional group exposure.

Various polysaccharides have marked differences with respect to solubility and gelling ability, leading to different molecular forms in water systems or after solidifying. The association among polysaccharide chains forming a gel is promoted through various mechanisms, all of them based on chain–chain interaction, competing successfully against chain–solvent interaction (Dea, 1993). A number of polysaccharides have been used in meat products, among them xanthan gum, gellan, carrageenans, microcelluloses, starch, and guar gum, with varying results on gel strength and stability.

Texture is also affected by the type of polysaccharide used for gel formation. Sahin et al. (2005) studied the effect of hydroxypropyl methylcellulose (HPMC), guar gum, xanthan gum, and gum arabic on the quality of deep-fat fried chicken nuggets. The authors reported that HPMC and xanthan gums reduced oil absorption compared to other gums and the control. Pietrasik et al. (2005) studied the combined effect of carrageenan and nonmuscle proteins (e.g., blood plasma, sodium caseinate, soy isolate) on the binding, textural, and color characteristics of meat gels; the authors found that carrageenan favorably affected hydration properties and thermal stability, yielding less cooking loss and higher water-holding capacity for meat gels, but these effects were attenuated by nonmeat proteins.

Protein gelation is probably the main cause of meat gel forming, although in most industrial cases protein gel strength is increased by polysaccharide addition. Protein gelation is achieved by heating or by chemical means (e.g., acid addition, enzyme cross-linking, use of salt and urea), causing modifications in the protein–protein and protein–medium interactions (Totosaus et al., 2002). In the meat industry, gel induction by heating is a widely used process. Protein gelation by heat is an aggregation of denatured molecules with a certain degree of order, resulting in the formation of a continuous network; it is a two-step process: denaturation and aggregation (Matsumura and Mori, 1996).

It is assumed that proteins and polysaccharides are compatible; otherwise, co-gelling cannot occur. Complex gels formed by polysaccharide–protein physical interactions increase gel strength but reduce flexibility (Totosaus, 2001).

FOOD EMULSIONS

An *emulsion* is defined as two immiscible liquids, one of them being droplets or fat globules (the disperse or internal phase) immiscible in another (the continuous phase); the droplet or fat globule diameter in a true emulsion is between 0.1 and 100 μm (McClemens, 1999). In addition to these two basic components, meat emulsions in disperse-phase muscle fibers also contain small connective tissue fractions, carbohydrates, and so on. The diameter of the fat droplets in a meat emulsion's disperse phase is larger than 100 μm; therefore, meat emulsions are not considered true emulsions but rather, batters or pastes. However, batters contain basically the same structural elements (i.e., continuous phase, disperse phase, and an emulsifying element).

The disperse phase in meat emulsions is formed by fat globules, and the continuous phase is a matrix of soluble and myofibrillar proteins, flavorings, salts, additives, and other materials in a gel-like medium. The role of myofibrillar proteins in meat emulsions is mainly as a system stabilizer at the oil–water interphase. Actin, myosin, and actomyosin are responsible for most of the functional properties of meat; they contribute to approximately 95% of the total water-holding capacity of meat tissue and 75 to 90% of the emulsifying capacity (Li-Chan et al., 1985). Although other emulsifying compounds are added at the industrial level, myofibrillar proteins are the main emulsifiers in meat products in addition to forming part of the continuous phase. The ability of myofibrillar proteins to form an interface depends on intrinsic parameters (molecular shape and form, amino acid composition and hydrophobicity, among others) as well as extrinsic conditions (pH, temperature, ionic strength) and processing.

Proteases present in the batter can promote protein unfolding due to mild denaturation; this hydrolysis can increase the emulsifying properties. García-Barrientos et al. (2006) studied the effect of endogenous enzymes and added protease extracts obtained from *Pseudomonas fluorescens* in pork and shark model batters; they found a twofold effect in protease-treated batters. A moderate myofibrillar protein proteolysis in the disperse phase increased batter viscosity, stabilizing the emulsion by preventing fat droplet coalescence or creaming; whereas partially denatured myofibrillar proteins in the oil–water interface improve the emulsifier capacity by unfolding and covering a larger droplet area.

It has also been reported that the functionality of meat protein, including its gelling ability and emulsifying capacity, is species-specific (Jiménez-Colmenero and Borderías, 1983) as well as dependent on whether the proteins originated in smooth or striated muscle (Borderías et al., 1983); striated muscle proteins are more efficient gel-formers than are smooth muscle proteins. Differences seem to be related to polymorphism; the molecular weight of light meromyosin is both species- and muscle-specific (Smith, 1988). This fact was also reported by García-Barrientos et al. (2006) for pork versus shark proteins. In fact, the gelling ability and emulsifying capacity of pork are higher than for beef and much higher than for poultry meat. Fish muscles vary in these properties according to the species (García-Barrientos, 2007).

EMULSION STABILITY

Although several factors involved in emulsion stability are thermodynamic (e.g., free energy, gravitation, electrostatic interactions and repulsions) and chemical (e.g., pH, presence of compounds such as proteases, protein structure) in nature, from an empirical point of view, raw meat emulsion stability can be achieved by several means, such as by increasing the continuous-phase viscosity, by the addition of food-grade emulsifiers, and by adjusting the disperse-phase mean particle size.

Continuous-Phase Viscosity

Increasing the continuous-phase viscosity prevents fat droplet migration, although the disperse-phase migration becomes hindered due to an increase in medium viscosity. This is achieved by the addition of polymers (e.g., carbohydrates, non-meat proteins), which also modifies the texture and mouthfeel of the emulsion, an effect that depends on the molecular size, branching, and flexibility of the thickening agent. However, the disperse phase in a meat batter is a molecular packed structure; it therefore behaves as a gel. Having this structure means that if shear stress is applied, the molecular structure and hence the apparent viscosity decrease with increasing shear stress (this is a pseudoplastic fluid; it behaves as an elastic solid if a shear stress below its yield stress is applied, but behaves as a fluid when the shear stress is above its yield stress) (McClemens, 1999). Therefore, care must be taken when mixing these meat batters, as emulsion may be disrupted if the yield stress is surpassed. Addition of food-grade emulsifiers prevents emulsion disruption.

Emulsifier Addition

Emulsifying agents include amphiphilic compounds such as acylmonoglycerides, diglycerides, polyethylene sorbinate fatty acids (e.g., Tween), sorbitan fatty acid esters (e.g., Spam), and lecithin. These compounds present hydrophilic and hydrophobic regions. Emulsifiers are classified according to their ability to form water-in-oil (w/o) or oil-in-water (o/w) emulsions according to the

HLB (hydrophile–lipophile balance) concept. It is described as a number (1 to 20) that indicates the relative affinity of a surfactant molecule for the oil and aqueous phases, and depends on the emulsifier's chemical structure. Emulsifiers with low numbers are used in w/o emulsions, whereas those with high numbers are used in o/w emulsions. Therefore, an emulsifier is selected according to its HLB value, its structure, and its interaction with other components, such as polysaccharides or proteins. An empirical method to calculate the HLB number is the following equation (Davies, 1994; cited by McClemens, 1999):

$$\text{HLB} = 7 + (\text{hydrophilic group number}) + (\text{lipophilic group number}) \qquad (1)$$

Disperse-Phase Particle Size

Coalesce of two particles in an ideal liquid is described by Stokes' law (McClemens, 2005):

$$\frac{V = 2gr^2(\rho_2 - \rho_1)}{9\eta} \qquad (2)$$

where ρ_1 and ρ_2 = density of continuous and disperse phases
r = particle radius
g = gravitational acceleration
η = medium viscosity

The emulsion stability can be enhanced by reducing the particle size (or droplet size), or by increasing the viscosity of the continuous medium (McClemens, 1999). Therefore, by reducing the particle size, the emulsion becomes more stable. However, excessive particle reduction in meat processing (e.g., when emulsifying in silent choppers) can produce the opposite effect, destabilizing the emulsion. As a result, fat separates from the continuous phase, forming fluid fat packs in, for example, cooked sausage.

Due to the high molecular weight of myofibrillar proteins, they behave as particles in suspension (Li-Chan et al., 1985). Therefore, the stability of a meat batter is based on colloidal factors such as ionic strength, particle shape, particle surface charge, and particle size distribution (Van Ruth et al., 2002).

MEAT POULTRY PRODUCTS

According to their degree of comminution, meat products are divided into complete cuts and restructured, coarsely ground, and finely chopped products. Carballo and López de Torre (1991) divided meat products into those formed by meat pieces and those formed by pastes; the latter are divided into chopped and emulsified. From the point of view of the physical system involved and the stabilizing mechanisms, meat products can be classified as shown in Table 1. However, this is a general classification of meat products; each nation or regional legislation includes particular definitions.

TABLE 1 Physical System and Stabilizing Mechanisms of Fabricated Meat Products

Product Type	Examples	Physical System and Binding Mechanism	Operations During Processing
Complete retail cuts or boneless complete animals	Pork ham and ribs, smoked boneless ducks and chicken, marinated chicken and turkey breasts, lamb cutlets and racks	No binding mechanism is applied	Curing, cooking, (sometimes) smoking
Restructured meat pieces	Cooked ham, head cheese, aspics	Pieces are bound together by gels containing thickening or cohesive agents	Curing, cooking, (sometimes) smoking
Coarsely ground products	Breakfast sausages ("sizzlers"), chorizo (spicy raw sausage), kielbasa, summer sausage	Cohesion is achieved by meat massaging, ripening in some cases as in chorizo, kielbasa, and summer sausage	Curing, cooking previous to consumption, (sometimes) smoking (generally, frying)
Finely chopped products	Emulsified sausages (frankfurters, wieners, etc.), loaves, bologna, pâté	These products are mainly emulsions; have a gel-like in their continuous phase, binding is due to a semisolid-to-solid change when heated	Curing, emulsification, cooking, (sometimes) smoking

LOW-FAT EMULSIONS

Lipids are also a basic component in meat product formulation; however, the meat industry is highly interested in developing low-fat meat emulsions without modifying the sensory and textural attributes of the final product. Protein–polysaccharide co-gelling allows a reduction in lipid content in comminuted sausages. Rogers (2006) comments about the use of κ, ι, and λ carrageenans; gums such as gellan, xanthan, locust beam, sodium and propylene glycol alginates; and pectines to increase viscosity in batters. κ-Carrageenan seems to be the most efficient additive in reducing drip losses, although this effect also depends on salt level and type (Na, Ca, and K chlorides). Other carbohydrates, such as starches (preferably, pregelatized starch), develop instant and stable viscosity; konjac flour also acts by substituting fat particles.

AROMA RELEASE FROM MEAT EMULSIONS

The flavor of a food is determined by the type and concentration of flavor molecules and by their ability to reach the appropriate sensory receptors in the nose and mouth (Van Ruth and Roozen, 2000). Nonvolatile molecules responsible for taste are perceived more intensely when they are present in an aqueous phase than in an oil phase (Smith and Margolskee, 2001). On the other hand, volatile molecules responsible for aroma can be detected by the nose only after they have been released into the vapor phase above an emulsion (Taylor and Linforth, 1996). The aroma molecules may be transported directly into the nose by sniffing a food, or they may be released from a food during mastication and carried into the nose retronasally (Linforth et al., 1996).

Meat aroma comprises a wide variety of chemical compounds. It is the result of a number of factors, such as substrate composition, pH, water activity, processing and storage conditions, and gas atmosphere, among others. It has been reported to be composed of approximately 700 chemical compounds, generated in the fat and lean meat. The contribution of chemical compounds to meat aroma depends on their release from the food matrix, which, in turn, depends on ion strength, temperature, presence and concentration of other compounds, and hydrophobicity. Proteins particularly affect aroma perception, due to interactions with aroma-related compounds. Lipids greatly influence flavor through their effect on perception (mouthfeel, taste, and aroma), flavor generation, and stability, whereas carbohydrates tend to increase retention in the matrix (Seuvre et al., 2000).

Food structure is also associated with the release of aroma compounds. In a two-phase system such as a meat emulsion, the concentration of these compounds in the lipid and aqueous phases, as well as the interface, results in specific contributions due to diffusion to the gas phase. Therefore, each phase contributes in a different way to the overall aroma. The overall flavor profiles depend on the odor compound partitioning coefficient and mass transport throughout the continuous and disperse phases (Brossard et al., 1996). Volatiles must migrate through the continuous and disperse phases to the product surface–vapor interface to be perceived by the consumer (Taylor and Linforth, 1996). However, some of these odor-related compounds are associated with the interfacial region, altering their concentration between the vapor and aqueous phases in the emulsion; therefore it is important to establish the partition coefficient for odor-related compounds of the emulsion (Doyen et al., 2001). In general, odor-related compound liberation is faster from the continuous than from the disperse phase (de Roos, 1997). Herrera-Jiménez et al. (2007) studied the release of five odor-related compounds from meat emulsions. Hexanal, octanal, and nonanal were taken as indicators of lipid odor, whereas 1-ethyl-3,5-dimethylpyrazine and 2-methylpyrazine were Maillard reaction indicators. Low pyrazine release was mainly from the continuous phase; these compounds cannot be considered as important aroma contributors, due to binding to unfolded protein. Conversely, octanal can be considered an important aroma contributor in meat emulsions. In general, aldehydes showed high release indexes. Volatile binding to the interface increases with protein concentration (Fabbiane, 2000); their release is encouraged by large average particle size (Landy et al., 1996).

Triglycerides' low partial vapor pressure increases the perception threshold, mainly in short-chain and unsaturated lipids (Leland, 1997). Interactions between proteins and odor compounds are mainly through adsorption or van del Waals forces or through covalent or electrostatic interactions, including amide and ester formations and aldehyde condensation with amino or sulfhydryl groups or free lysine and histidine (Fischer and Widder, 1997). Emulsion thickeners affect flavor perception by reducing odor compound migration through the vapor phase (Cook et al., 2003). Very efficient thickeners, such as pectin, guar gum, or alginates, can inhibit odor compound volatility (Leland, 1997). Liberation of nonpolar compounds is also affected by particle size, increasing in large average particle-size emulsions (Van Ruth et al., 2002).

REFERENCES

Borderías AJ, Lamua M, Tejada M. 1983. Texture analysis of fish fillets and minced fish by both sensory and instrumental methods. J Food Technol 18:85–95.

Brossard C, Rousseau F, Dumont JP. 1996. Flavour release and flavour perception in oil-in water emulsions: Is the link so close? In: Taylor AJ, Mottram DS, eds., *Flavour Science: Recent Developments*. Cambridge, UK: The Royal Society of Chemistry, pp. 375–379.

Carballo B, López de Torre G. 1991. Tecnología: productos cárnicos. In: *Manual de Bioquímica y Tecnología de la Carne*. Madrid, Spain: A. Madrid Vicente.

Cook DJ, Linforth RST, Taylor AJ. 2003. Effects of hydrocolloid thickeners on the perception of savory flavours. J Agric Food Chem 51:3067.

Davies EA. 1994. Factors determining emulsion type: hydrophilic–lipophilic balance and beyond. Colloid Surf A 91:9.

Dea IC. 1993. Conformational origins of polysaccharide solution and gel properties. In: Whistler R, Bemiller JN, eds., *Polysaccharides and Their Derivatives*, 3rd ed. San Diego, CA: Academic Press, pp. 21–52.

De Roos KB. 1997. How lipids influence food flavor. Food Technol 51(1):60–62.

Doyen K, Carey M, Linforth RST, Marin M, Taylor AS. 2001. Volatile release from an emulsion: headspace and in-mouth studies. J Agric Food Chem 49:804–810.

Fabbiane FV. 2000. Effect of some fat replace on the release of volatile aroma compounds from low-fat meat products. J Agric Food Chem 48:3476.

Fischer N, Widder S. 1997. How proteins influence food flavor. Food Technol 51(1): 68–70.

García-Barrientos R. 2007. Efecto de enzimas proteolíticas endógenas y una bateriana en la formación de emsuliones modelo a partir de músculo estriado de cazón (*Rhizopriondon terranovae*) y cerdo. Ph.D. dissertation, Universidad Autónoma Metropolitana, Mexico City.

García-Barrientos R, Pérez Chabela ML, Montejano JG, Guerrero-Legarreta I. 2006. Changes in pork and shark (*Rhizopriondon terraenovae*) protein emulsions due to exogenous and endogenous proteolytic activity. Food Res Int 39(9):1012–1022.

Herrera-Jiménez M, Escalona-Buendia H, Ponce-Alquicira E, Verde-Calvo R, Guerrero-Legarreta I. 2007. Release of five indicator volatiles from a model meat emulsion to study phase contribution to meat aroma. Int J Food Prop 10(4):807–818.

Jiménez-Colmenero FJ, Borderías AJ. 1983. A study of the effect of frozen storage on certain functional properties of meat and fish protein. J Food Technol 18:731–737.

Leland JV. 1997. Flavor interactions: the greater whole. Food Technol 51(1):75–80.

Li-Chan E, Nakai S, Wood DF. 1985. Relationship between functional (fat binding, emulsifying) and physicochemical properties of muscle proteins: effects of heating, freezing, pH and species. J Food Sci 50(4):1034–1040.

Matsumura Y, Mori T. 1996. Gelation. In: *Methods of Testing Protein Functionality*. Hall GM, ed., Suffolk, UK: Blackie Academic & Professional, pp. 76–109.

McClemens DJ. 2005. Emulsion flavor. In: *Food Emulsions: Principles, Practices, and Techniques*, 2nd ed. Boca Raton, FL: CRC Press, Chap. 9.

Pietrasik Z, Jarmoluk A, Shand P. 2005. Textural and hydration properties of pork meat gels processed with non-muscle proteins and carrageenan. Pol J Food Nutr Sci 14(2):145–150.

Rogers RW. 2006. Manufacturing of reduced-fat, low-fat and fat-free emulsion sausages. In: Hui YH, Nip WK, Rogers RW, Young OA, eds., *Meat Science and Applications*. New York: Marcel Dekker, pp. 443–461.

Sahin S, Sumnu G, Altunakar B. 2005. Effects of batters containing different gum types on the quality of deep-fat fried chicken nuggets. J Sci Food Agric 85(14):2375–2379.

Seuvre AM, Espinoza-Díaz MA, Voilley A. 2002. Transfer aroma compounds through the lipidic–aqueous interface in a complex system. J Agric Food Chem 50:1106–1110.

Smith DM. 1988. Meat proteins: functional properties in comminuted meat products. Food Technol 42(5):116–121.

Smith DV, Margolskee RF. 2001. Making sense of taste. Sci Am 284:32–39.

Taylor AJ, Linforth RS. 1996. Flavour release in the mouth. Trends Food Sci Technol 7:444–448.

Totosaus A. 2001. Estudio del efecto de la adición de polisacáridos a la gelificación de proteínas animales. Ph.D. dissertation, Universidad Autónoma Metropolitana, Mexico City.

Totosaus A, Gault N, Guerrero-Legarreta I. 2000. Dynamic rheological behavior of meat proteins during acid-induced gelation. Int J Food Prop 3(3):465–472.

Totosaus A, Montejano G, Salazar A, Guerrero-Legarreta I. 2002. A review: physical and chemical gel induction. Int J Food Sci Technol 37:589–601.

Van Ruth SM, Roozen JP. 2000. Influence of mastication and saliva on aroma release in a model mouth system. Food Chem 71(3):339–345.

Van Ruth SM, King C, Giannouli P. 2002. Influence of lipid fraction, emulsifier fraction and mean particle diameter of oil-in-water emulsions on the release of 20 aroma compounds. J Agric Food Chem 50:2365–2371.

4

BATTERING AND BREADING: PRINCIPLES AND SYSTEM DEVELOPMENT

Susana Fiszman and Teresa Sanz
Instituto de Agroquímica y Tecnología de Alimentos, Burjassot, Valencia, Spain

INTRODUCTION

Consumers are always looking for safe, tasty products that require less and less preparation time from the moment that they thaw them at home or order them in a restaurant. The battered and breaded chicken product market has been one of the fastest growing in recent decades, as these products are very attractive. Consumers demand to be served pieces of chicken that are hot, have optimum moisture content, excellent flavor and a crisp, savory batter coating, and are served quickly after being ordered. The chicken substrate pieces are first batter coated, then

Handbook of Poultry Science and Technology, Volume 2: Secondary Processing, Edited by Isabel Guerrero-Legarreta and Y.H. Hui

breaded (or not), deep-fried (to prefry them or cook them completely), and finally, frozen. Food-service companies or retail customers buy the frozen products and finish cooking them in an oven or by frying.

The food-processing world is constantly innovating to encounter formulas and processes that will allow all these steps to be followed without producing an undercooked, overcooked, tough, soft, or dry final product in which the external layer has become more of a drawback than an added value. To achieve these objectives, the food-coating industry has developed complex, high-value technologies. Basically, coatings for frying food are made with batters in which flour is the main ingredient. However, an increasing number of other components have gradually been added to the basic formulation, and the list of ingredients is increasingly long and sophisticated. In this chapter we examine the part that each component of a batter plays in achieving a high-quality product.

TYPES OF COATING

Batters need to satisfy a large number of requirements. What is essential is adherence to the substrate, and they must be formulated so that the thickness of the layer they will form can be controlled. The amount of batter clinging to a food piece after application is called *pickup*. The surface to be covered is normally irregular, but an optimum degree of coating needs to be achieved to avoid a peeling tendency or "blowing off." Adhesion depends partly on product composition: The quantity of fat and its proportion relative to the water and proteins present must be taken into account. The internal cohesion of the batter is as important as adhesion. There is a direct relationship between the viscosity of the batter and the thickness of the layer that it can form, whereas the fluidity of the batter determines the uniformity of that layer.

Adhesion Batters

Adhesion or interface batters are a means of binding the crumbed external layer to the food. The bond between the batter and the food is very important, as is the bond between the batter and the crumbs. The complexity of the formulations can vary depending on the type of substrate (whole pieces of chicken meat, their shape, their surface with or without skin, nuggets, etc.), and the type of crumbing. Thickness is an important factor for adhesion; the thickness of the layer that adheres is defined to a large extent by the quantity of dry solids and the final viscosity. In this type of batter, however, the texture it ultimately acquires is not so important, as it will be defined by the nature of the crumbs.

Tempura Batters

Tempura or puff batters must be sufficiently viscous to form the exterior layer of the final product on their own (without adding crumbs). After applying a tempura batter, the product must be fried to set the batter. In products with very smooth surfaces the coatings have difficulty in adhering, so they generally need a predusting treatment with flour to create a rougher, drier surface that will allow the batter to stick. Tempura batters contain baking powder, which gives the layer they form an open, spongy structure. This makes it easier for water to evaporate from the product during frying, preventing the coating from being blown off by a buildup of steam between the product and the batter layer.

ROLE OF INGREDIENTS IN BATTERING SYSTEMS

Predusting

Predust is a dry ingredient, normally flour, which is dusted onto the moist surface of a poultry piece before any other coating is applied. The poultry piece can be frozen, fresh, or steam- or heat-cooked. Predust improves batter adhesion because it absorbs part of the water on the surface of the substrate and creates a rougher surface (Yang and Chen, 1979). If the batter is applied to a surface that is too watery, it will not cover the food piece properly, leaving some areas uncovered or covered with a diluted batter that will then form too thin a layer. If a predust material is applied to very dry product surfaces or deep-frozen substrates, however, the powder ingredient does not adhere to the substrate surface, so it mixes with the liquid batter. In time, as the processing continues, the result is that the mixture becomes too viscous. Predusting also helps to eliminate areas where air can become trapped between the substrate and the batter layer during the coating process. Furthermore, it normally increases yields, as it helps to improve the batter pickup.

In a pioneering work, Baker et al. (1972) evaluated the behavior of a series of ingredients and additives to determine which acted as the best predust. The ingredients belonged to three groups: starches, proteins, and gums. They concluded that the proteins produced crusts with better adhesion, and of these, dried egg albumin produced the best results in terms of yield and visual scores. In some processes, the application of consecutive layers of predust, batter, and breading is recommended to achieve a particular thick crust texture that will provide protection for the substrate it envelops. The most widely used predust is wheat flour, although corn and rice flours are also used. The flours can also be mixed with native and modified starches, those with a high amylase content being used most frequently. In turn, all the materials above can be combined with gums (various cellulose derivatives and other hydrocolloids) or proteins (soy protein isolate, whey protein concentrate, egg white powder, or gluten). As may be gathered from this enumeration, each manufacturer can adopt the solution that it finds works best for its products, as there is no such thing as an ideal predust. For each substrate, depending on its nature, the goal is to create a layer that absorbs moisture and helps to form a barrier against moisture and fat migration from the poultry piece or chicken nugget (Kuntz, 1997; Usawakesmanee et al., 2004).

Batters

Flours Wheat flour is the most usual basic ingredient for formulating a batter. All the main functional properties of the batter (i.e., viscosity, extensibility, degree of adhesion, color after frying, oil absorption behavior, etc.) depend on it. The batter's dry ingredient mixture may contain up to 90% by weight of wheat flour. Its main role is to provide the structure that supports the other components. Essentially, this structure develops because the starch granules gel in the presence of a sufficient quantity of water and heat. The characteristics of the structure are determined by the amylose/amylopectin ratio of the flour in question and its protein content and functionality. Various studies have related characteristics such as oil absorption, texture, crispness, external appearance, and sponginess to the quality of the flour.

Generally speaking, a greater proportion of amylopectin or an excess of proteins causes greater oil absorption when frying a batter-coated food. Another factor that must be borne in mind is the degree of starch damage: Damaged starch grains absorb more water than whole grains, and this influences the functional properties of the batter. Although the functional properties of the flours of most types of wheat in existence are known, the variations are not so large as to justify recommending a specific variety. Normally, a multipurpose wheat flour would be suitable. Sometimes flour types are used that have undergone a specific treatment to make them more suitable for use in batters. The most common treatment is heating: As this usually increases the viscosity that can be developed, it has a positive effect on batter pickup and cohesiveness. Also, heat treatment causes a considerable reduction in the amylase activity, which hydrolyzes starch. Regarding wheat proteins, their capacity to develop cohesive pastes when hydrated is well known. Moreover, wheat proteins are able to stretch, so in the case of tempura-type batters (which are formulated with raising agents), it is they that are responsible for forming an elastic structure that can hold bubbles of gas and give the fried end product its sponginess. This porous appearance contributes to the perception of a light, brittle coating. The elastic properties also help to keep any insoluble solids in suspension.

In recent years, numerous results have been published on the effects of replacing some or all of the wheat flour with the flours of other cereals, mainly rice; they are generally described in terms of improved texture characteristics, bringing greater crunchiness. The effects of soy flour (5%) and rice flour (5%) additions to the batter formulation on the quality of deep-fat fried chicken nuggets have been evaluated in terms of coating pickup, moisture and oil contents, texture, color, and porosity. Soy flour was found to be an effective ingredient for improving color and crispness, and both soy and rice flour provided reduced oil absorption (Dogan et al., 2005).

Replacement of the wheat flour with rice flour, which is less effective than wheat flour as a thickening agent, leads to a batter that requires a greater proportion of solids or the addition of a thickener to achieve a suitable viscosity. Rice flour showed potential as an alternative to wheat flour in a batter formulation used for chicken drumsticks only when appropriate levels of oxidized starch, xanthan gum, and methylcellulose were included in the formulation (Mukprasirt et al., 2000a). However, rice flour-based frying batter has significantly reduced oil uptake during the frying process, due to increased amylase in the amylopectin/amylose ratio of the starch (Shih and Daigle, 1999).

Batter temperature control is also important when replacing part of the wheat flour with rice flour, because this changes the rheological properties of the batter, depending on the replacement ratio (Mukprasirt et al., 2000b). Burge (1990) found that by varying the ratio of wheat–corn blends in the batter composition, different characteristics could be obtained to suit different types of products. Using corn as an ingredient to dilute the tough wheat proteins, the texture of the fried exterior layers could be modified, increasing their crispness and reducing their toughness. Wheat flour has also been replaced by steam jet–cooked barley flour in batter formulations, producing batters with good rheological properties and coatings with low oil uptake and high moisture content, which could be related to the high water-binding capacity of this ingredient (Lee and Inglett, 2006).

Starches The addition of starches from different sources is associated primarily with a reduction in oil absorption and with improvements in the texture of the fried end products. Starches such as those of corn or rice differ from wheat starch in the size, shape, and composition of their granules. These differences are related directly to the quantity and speed of water absorption (swelling) and, consequently, to their gelatinization profile. It is this profile that is responsible for developing viscosity and therefore for the rheological properties that the starch in question contributes to the batter. Generally speaking, native starches confer improved raw batter adhesiveness, better tenderness and crunchiness, a number of textural modifications depending on the starch source, better mechanical stability during the holding time under heat lamps, better moisture retention, and decreased greasiness (Van Beirendonck, 1998). Starch-added batters require good mixing during processing, as the solids tend to settle out easily, leading to changes in batter viscosity over the production period that result in irregular batter pickup (Suderman, 1993). In these cases, gums or thickeners with good suspension properties are usually added to avoid such problems.

Improved brittleness and crispness are generally associated with the use of amylose-rich hybrid starches. These starches have a high gelatinization temperature and create a film in just a few seconds when the battered food is submerged in hot oil; the appearance of the coating is very good and uniform (Bertram, 2001). This film-forming capacity is associated with a reduction in oil absorption during frying (Higgins et al., 1999). A positive correlation between the amylose content and the crispness and hardness of the final fried batters was found when a number of native starches were added to a wheat flour batter (20% starch). The native starches were high-amylose cornstarch (HAC), normal cornstarch (NCS), waxy cornstarch (WCS), rice starch (RS), and waxy rice starch (WRS); their relative amylase contents were HAC > NCS > RS > WRS > WCS (Lee et al., 2004). Lenchin and Bell (1986) formulated a dry mix for coating prefried food products, suitable for cooking in a microwave oven, which are characterized by improved crispness; these authors used flour and cornstarch with a high amylose content for this purpose.

Pregelatinized starch can be added to control batter viscosity. Batters that are continuously pumped and recirculated tend to settle out, causing a lack of homogeneity in the dispersion of the solid matter. The addition of pregelatinized starch to these systems helps to keep the dry ingredients in adequate suspension. The addition of pregelatinized rice flour has been found to impart greater crispness to the fried batter, but also to increase oil absorption (Mohamed et al., 1998). Oil-resistant, rice-based starch products that swell in cold water, such as pregelatinized rice flour, phosphorylated rice starch, and pregelatinized acetylated rice starch, are used in formulations to enhance the viscosity and textural and sensory quality of the fried batter. The effect of amylomaize, corn, waxy maize, and pregelatinized tapioca starches on the texture, moisture content, oil content, color, coating pickup, cooking yield, volume, and porosity of deep-fat fried chicken nuggets was studied by Altunakar et al. (2004). They concluded that adding starch to the formulation increased the crispness of the final product in the last stages of frying; pregelatinized tapioca starch resulted in the lowest oil content and the highest moisture content, coating pickup, and volume, while the highest porosity and oil content were obtained when cornstarch was used.

Starch from various sources can be modified chemically to provide specific characteristics. Modified starches with a reduced starch granule swelling capacity may change the viscosity of the batter when it is heated. Modified starches find many applications in batter development as a result of their wide-ranging functionality. Oxidized starches, for example, have carboxyl groups that bind with proteins in the substrate, and this bonding makes the batter stick (Shinsato et al., 1999). Lee and Lim (2004) found that modified starches (oxidized, acid-treated, cross-linked, hydroxypropylated, and acetylated) provided texture improvements (crispness and hardness) in fried products.

Gums Gums comprise a group of polysaccharides that will hydrate in cold water to form viscous solutions. The best known application of gums in batter-coated products, as a result of their water-binding properties, is related to viscosity control. The advantage of using gums rather than other hydrocolloids such as modified starches is their effectiveness at lower concentrations, thus reducing the dilution of the functional protein in the flour, which plays a critical role in the overall performance of the coating system. At room temperature, at levels of 0.3 to 1% of a total formula, they are effective for maintaining a homogeneous suspension in batters and as viscosity-control agents. Cellulose has been modified to provide several ingredients that are useful for controlling viscosity and maintaining suspensions. Some of these cellulose derivatives (e.g., carboxymethylcellulose) have been shown to reduce fat absorption, while others improve adhesion. Other gums may exhibit similar effects.

The viscosity of the batter is critical to the quality of the coating and is the main factor that determines batter behavior during frying (Loewe, 1990). The degree of coating and the quantity and quality of the batter that adheres to the substrate are dependent on its viscosity (Hsia et al., 1992). Flour, being a natural ingredient, has variable rheological properties; consequently, gums need to be added to the batter to standardize its viscosity (Mukprasirt et al., 2000a).

Gums have the ability to keep solids in suspension. This property has been related particularly to xanthan gum and tragacanth gum, which are able to aid the suspension of heavy particles at low gum concentrations (0.10 to 0.25% w/w) (Hsia et al., 1992). The enhanced viscosity conferred by gums also makes it possible to obtain batters with a greater proportion of water while maintaining appropriate viscosity. This property has been used to increase the quantity of water available for starch gelatinization (Davis, 1983).

The gums most often used have been xanthan gum and the cellulose derivatives carboxymethycellulose (CMC), methylcellulose (MC), and hydroxypropyl methylcellulose (HPMC). In the case of MC and HPMC, temperature control is essential to assure the gum's functional properties. To ensure correct hydration of these cellulose derivatives and their effectiveness as gelling agents during the hot-water-bath coagulation step, the temperature of the water used to make up the batter must be between 10 and 15°C. For the same reason, the batter must be kept at this same temperature during the coating step (Sanz et al., 2004).

In a comparative study of the effects of guar gum, xanthan gum, and CMC on the rheological behavior of batters, xanthan gum produced the batters with the greatest consistency, followed by guar gum and CMC (Hsia et al., 1992). An important factor to consider when choosing the concentration of a hydrocolloid is its effect on the crispness of the final fried crust. Excessive use of gums can cause other problems. They can affect the texture adversely. An increased concentration of certain cellulose derivatives, such as MC, has been associated with greater moisture retention in the product during frying, causing a loss of crispness. A product with too much gum exhibits a chewy texture. With some gums, such as CMC, guar, and xanthan, this can happen at levels above 0.2%. Typically, these thickeners are used at levels below 0.1%. The use of xanthan gum at levels higher than 0.2% imparts a chewy texture (Kuntz, 1995). In the particular case of MC and HPMC, their film-forming and thermal gelation properties also help to maintain batter integrity and structure during frying, which is especially important for batters that use low quality flour or have low solids contents (Meyers, 1990).

Hydrocolloids influence the pickup value of a batter-coated product because of their ability to confer viscosity. The appearance, thickness, and crispness of the fried external crust are critical to the sensory acceptance of fried batter-coated foods and are all closely linked to the pickup value. Pickup is also important for industrial manufacturers of batter-coated products, as the weight of the coagulated batter in relation to the total weight of the batter-coated food determines the process yield, which in turn affects the final cost of the product.

Generally speaking, the higher the viscosity of the batter, the higher the percentage that will adhere to the substrate. In batters formulated with xanthan gum, guar gum, and CMC at concentrations of 0.25%, 0.5%, and 1.0%, a correlation between greater viscosity and greater adhesion of the batter to the food was observed (Hsia et al., 1992). However, this effect is not always so direct, as the increase in viscosity caused by a higher concentration of HPMC does not cause a proportional increase in the quantity of batter on the final product: For HPMC levels between 0.25 and 1.0%, the batter pickup rate was found to increase less than the batter viscosity as the HPMC content increased (Meyers and Conklin, 1990). The presence of HPMC in batter-coated and breaded chicken nuggets has been associated with increased product yields due to moisture loss reduction during frying (Dow Chemical Co., 1996). Also, the body and integrity conferred by gums enhance batter performance before, during, and after the frying operation, leading to improved batter adhesion and, indirectly, to higher yields.

Other Ingredients

Dextrins The use of dextrins in batter formulations is associated with an improvement in the crispness of a fried product (Shinsato et al., 1999). The dextrins used generally have a medium-high viscosity and aid the formation of a continuous, uniform batter. In addition to increasing the crispness of the fried product, they help to maintain crispness when infrared lamps are used to keep the fried products hot. Battered squid rings prepared with a batter that contained dextrins in its formulation have been found to remain crisper for a longer period of time after frying (Baixauli et al., 2003).

Proteins A high proportion of egg albumen in batter formulations for coating chicken nuggets has caused a gummy texture or color problems (Baker and Scott-Kline, 1988), although the coating pickup values and the final fried product yield obtained were slightly better than those of formulations without added egg white; sensory analysis has also given good scores to batters that contained egg. Mohamed et al. (1998) observed that the addition of ovalbumin to a batter formulation improved the crispness and color of the fried product as a result of the proteins that take part in Maillard reactions. They also verified that egg yolk increased the hardness of the coating and the absorption of oil, probably because the proteins for the greater part take the form of lipoproteins and phosphoproteins, which can reduce the surface tension between water and oil. Egg albumen, a protein that coagulates when heated, is useful in binding the batter to the product, and the lecithin in the yolk can act as an emulsifier, contributing to the stability of the batter (Loewe, 1993). Other proteins, such as powdered milk or whey solids, also provide structure and contribute lactose, a reducing sugar that takes part in nonenzymatic browning reactions. Bhardwaj (1990) proposed a batter with a high concentration of a pulse in a particular particle size to obtain fried battered products of high quality with a crispy, chewy crust.

Fiber and Fiber Sources Powdered cellulose with particular fiber lengths (>100 μm) has been described as a batter ingredient that reduces oil absorption and increases moisture retention in the final product after frying (Ang et al., 1991). As the result of a strong interaction due to hydrogen bonding between the cellulose fibers and the water molecules in the batter, the displacement of water by oil during frying is restricted. The nonenzymatic browning properties of powdered cellulose can also be advantageous for controlling the development of color (Ang, 1993). Other dietary fibers, such as oat, soy, pea, or sugar beet fiber, can be used for the same purpose. These substances develop a greater mechanical resistance than in conventional batters, making it easier for the coagulated product to stay intact during handling, and they improve product appearance because they give the coating an even, golden color after frying (Ang, 1991). Microcrystalline cellulose co-dried with whey has demonstrated a similar functionality in tempura-battered food items. Fibers have also been used to promote batter–substrate adhesion. Polydextrose, which is considered a soluble fiber, is another ingredient that has been described as reducing oil absorption in a batter formulation, alone or in combination with other fibers, soy protein, or a cellulose derivative (Kilibwa, 1999). In a recent work Sanz et al. (2008) demonstrated that a resistant starch (type 3) could be incorporated in a batter formulation, up to a concentration of 20%, as a source of dietary fiber without comprising consumer acceptability.

REFERENCES

Altunakar B, Sahin S, Sumnu G. 2004. Functionality of batters containing different starch types for deep-fat frying chicken nuggets. Eur Food Res Technol 218:318–322.

Ang JF. 1991. Water retention capacity and viscosity effect of powdered cellulose. J Food Sci 56:1682–1684.

Ang JF. 1993. Reduction of fat in fried batter coatings with powdered cellulose. J Am Chem Soc 70:619–622.

Ang JF, Miller WB, Blais IM, inventors; James River Corp., assignee. 1991, May 26. Fiber additives for frying batters. U.S. patent 5,019,406.

Baixauli R, Sanz T, Salvador A, Fiszman SM. 2003. Effect of the addition of dextrin or dried egg on the rheological and textural properties of batters for fried foods. Food Hydrocoll 17:305–310.

Baker RC, Scott-Kline D. 1988. Development of high protein coating using egg albumen. Poult Sci 67:557–564.

Baker RC, Darfler JM, Vadehra DV. 1972. Prebrowned fried chicken: II. Evaluation of predust materials. Poult Sci 51:1220–1222.

Bertram A. 2001. Pump up the amylase. Food Process 19 (Feb):19.

Bhardwaj SC, inventor; Bhardwaj Satish C, assignee. 1990, Oct 16. Method for cooking involving high protein frying batter that eliminates the need for breading and produces crispy and chewy crust. U.S. patent 4,963,378.

Burge RM. 1990. Functionality of corn in food coatings. In: Kulp K, Loewe R, eds., *Batters and Breadings in Food Processing*. St Paul, MN: American Association of Cereal Chemists, pp. 29–49.

Davis A. 1983. Batter and breading ingredients. In: Suderman DR, Cunningham FE, eds., *Batter and Breading*. Westport, CT: AVI Publishing, pp. 15–23.

Dogan SF, Sahin S, Sumnu G. 2005. Effects of soy and rice flour addition on batter rheology and quality of deep-fat fried chicken nuggets. J Food Eng 71:127–132.

Dow Chemical Co. 1996. *A Food Technologist's Guide to Methocel Food Grade Gums*Brochure 194–01037-1996 GW.. Midland, MI: Dow.

Higgins C, Qian J, Williams K, inventors; Kerry Inc., assignee. 1999, Nov 2. Water dispersible coating composition for fat-fried foods. U.S. patent 5,976,607.

Hsia HY, Smith DM, Steffe JF. 1992. Rheological properties and adhesion characteristics of flour-based batters for chicken nuggets as affected by three hydrocolloids. J Food Sci 57:16–18, 24.

Kilibwa M, inventor; Cultor Food Science Inc., assignee. 1999, Dec 14. Polydextrose as a fat absorption inhibitor in fried foods. U.S. patent 6,001,399.

Kuntz LA. 1995. Building better fried foods. Food Prod Des 5:129–146.

Kuntz LA. 1997. The great cover-up: batters, breadings, and coatings. Food Prod Des 7:39–57.

Lee S, Inglett G. 2006. Functional characterization of steam jet-cooked–glucan-rich barley flour as an oil barrier in frying batters. J Food Sci 71:308–313.

Lee M-J, Lim S-T. 2004. Utilization of various native and modified starches in deep-fat fried batter. Presented at the Annual Meeting of the Institute of Food Technologists, Las Vegas, NV.

Lenchin JM, Bell H, inventors; National Starch Chem. Co., assignee. 1986, Jun 17. Process for coating foodstuff with batter containing high amylose flour for microwave cooking. U.S. patent 4,529,607.

Loewe R. 1990. Ingredient selection for batter systems. In: Kulp K, Loewe R, eds., *Batters and Breadings in Food Processing*. St Paul, MN: American Association of Cereal Chemists, pp. 11–28.

Loewe R. 1993. Role of ingredients in batter systems. Cereal Foods World 38:673–677.

Meyers MA. 1990. Functionality of hydrocolloids in batter coating systems. In: Kulp K, Loewe R, eds., *Batters and Breadings in Food Processing*. St Paul, MN: American Association of Cereal Chemists, pp. 117–141.

Meyers MA, Conklin JR, inventors; Dow Chemical Co., Inc., assignee. 1990, Feb. 13. Method of inhibiting oil adsorption in coated fried foods using hydroxypropyl methyl cellulose. U.S. patent 4,900,573.

Mohamed S, Hamid NA, Hamid MA. 1998. Food components affecting the oil absorption and crispness of fried batter. J Sci Food Agric 78:39–45.

Mukprasirt A, Herald TJ, Boyle DL, Rausch KD. 2000a. Adhesion of rice flour-based batter to chicken drumsticks evaluated by laser scanning, confocal microscopy, and texture analysis. Poult Sci 79:1356–1363.

Mukprasirt A, Herald TJ, Flores R. 2000b. Rheological characterization of rice flour–based batters. J Food Sci 65:1194–1199.

Sanz T, Fernández MA, Salvador A, Muñoz J, Fiszman SM. 2004. Thermal gelation properties of methylcellulose MC and their effect on a batter formula. Food Hydrocolloids 19:141–147.

Sanz T, Salvador A, Fiszman SM. 2008. Resistant starch (RS) in battered fried products: functionality and high-fibre benefit. Food Hydrocolloids 22:543–549.

Shih F, Daigle K. 1999. Oil uptake properties of fried batters from rice flour. J Agric Food Chem 47:1611–1615.

Shinsato E, Hippleheuser AL, Van Beirendonck K. 1999. Products for batter and coating systems. World Ingred, Jan.–Feb.:38–42.

Suderman DR. 1993. Selecting flavorings and seasonings for batter and breading systems. Cereal Foods World 38:689–694.

Usawakesmanee W, Chinnan MS, Wuttijumnong P, Jangchud A, Raksakulthai N. 2004. Effects of type and level of edible coating ingredient incorporated in predusting mix on fat and moisture content of fried breaded potato. Presented at the Annual Meeting of the Institute of Food Technologists, Las Vegas, NV.

Van Beirendonck K. 1998. Coatings: starch fights the fat. Int Food Ingred 4:43.

Yang CS, Chen TC. 1979. Yields of deep-fat fried chicken parts as affected by preparation frying conditions and shortening. J Food Sci 44:1074–1092.

5

BATTERING AND BREADING: FRYING AND FREEZING

Susana Fiszman and Ana Salvador
Instituto de Agroquímica y Tecnología de Alimentos, Burjassot, Valencia, Spain

INTRODUCTION

The easiest and most common way to have battered or breaded chicken products in a restaurant or at home is to use industrially prepared portions that have been pre-fried or fried and then frozen. Pre-frying is a process whereby pieces of food that have been dipped or coated in a batter mix are submerged in hot oil for a short time (as little as 10 or 20 s is enough in some cases) so that the batter sets around the food, acquiring a solid consistency and a certain degree of light golden coloring. The purpose of this stage is to make the batter solid enough for the pieces to be handled without dripping, sticking to each other, or losing part of their coating. After pre-frying, the pieces are quickly cooled and frozen.

Handbook of Poultry Science and Technology, Volume 2: Secondary Processing, Edited by Isabel Guerrero-Legarreta and Y.H. Hui

However, there is no reason why the products should not be fried for a longer time at this stage, cooking them fully and giving them the typical darker golden color of a fried food and their characteristic final flavor. In this case they will not require final frying but will only need to be heated through before serving.

PRE-FRYING AND FRYING PROCESSES

The raw batter must create a homogeneous layer that covers the food and must adhere to it before and after coagulation, which takes place during the pre-frying step and during final frying. The coagulated external layer needs to acquire a consistency that will withstand freezing temperatures (including possible fluctuations) and normal handling (packaging and transportation) without breaking or cracking and without losing any portion of the external layer. During the final frying performed in the restaurant or in the consumer's home, the batter must create an outer crust with good acceptability in terms of texture (especially crispness or crunchiness), flavor, and color.

Coatings might also need to limit moisture and oil transfer, give freeze–thaw stability to the product they are covering, and extend shelf life if possible. On the other hand, of course, they must also be cost-effective. To develop all these properties, strict control is essential, using methods which ensure that both the raw batter and the crust that is formed through pre-frying or final frying possess appropriate physical characteristics. In recent years, a large amount of research has been published on these subjects, leading to the development of battered foods produced apart from the trial-and-error methods that were the norm at the end of the twentieth century.

Strategies for the Control of Oil Absorption

What are the drawbacks that high oil absorption by foods during industrial pre-frying or frying processes can entail? During these processes there may be a certain misuse of the oil, which is often kept hot for long periods of time. With increasing use, in frying processes that are normally continuous, varying quantities of air and moisture become mixed with the oil and favor certain oxidative and hydrolytic reactions that impair its quality. This misuse gives rise to undesirable rancid flavors—aldehydes and ketones, the breakdown products of hydroperoxides at high temperatures—that the battered food absorbs, together with an excess of oil (Holownia et al., 2000). This type of breakdown is not usual in home frying, as the oil is renewed frequently and the remains of food that may have fallen into the oil are removed by filtering; however, while deep pan frying is a very common cooking method in some countries, such as those of southern Europe, it is less so in others.

Nowadays, low-calorie and low-fat products are attracting considerable interest. Consuming this type of food reduces the risk of developing heart diseases and helps to maintain a healthier lifestyle. As a result, fried foods have somewhat fallen out of favor, owing to their high fat content; nonetheless, they are a food group that has great appeal, as they have a very pleasant flavor and are crisp and juicy, so they are difficult to replace in consumers' decisions and occupy a very large proportion of the market. For this reason, considerable effort is being directed to reducing the quantity of oil absorbed by battered products during frying, and many research projects are addressing the subject. However, one must not lose sight of all the factors that influence the process of oil absorption by the food during frying. Shape, size, porosity, the type and regularity of the surface to be battered and of the final surface, the composition (especially the moisture content), and the weight/surface ratio of the food are all fundamental factors, and the quality of the oil, frying time, and initial temperatures of the oil and food are also very important (Pinthus et al., 1993).

Using Ingredients to Reduce Oil Absorption

A way of reducing the absorption of oil when frying batter-coated products is to incorporate certain substances into the batter. Various ingredients have been proved to reduce the amount of oil absorbed by fried food. Of these, gums are of considerable interest, as their effectiveness is the highest. The gelling ability of hydrocolloids, together with their usually hydrophilic nature, makes them suitable for reducing oil uptake in battered products during frying (Annapure et al., 1999). The hydrocolloids most commonly employed as fat barriers by using them as batter ingredients are the cellulose ethers methylcellulose (MC) and hydroxypropyl methylcellulose (HPMC), which possess a unique ability to gel on heating; this property is reversible and is known as *thermal gelation*. These cellulose derivatives have been more widely investigated and reported than other gums (Lee and Han, 1988; Ang, 1989; Stypula and Buckholz, 1989).

When the product enters the hot oil during pre-frying, the MC or HPMC in the batter gels to form a gelled film. The development of this gel helps the coating to adhere to the food substrate and creates a physical barrier to oil absorption; the final product is healthier and has a less greasy mouthfeel. This structure also protects against moisture loss (Meyers, 1990; Mallikarjunan et al., 1997). The employment of MC and HPMC has been shown to be effective in a wide range of applications: specifically, in chicken nuggets (Meyers and Conklin 1990; Dow Chemical Co., 1991), chicken breast strips (Ang, 1993), chicken balls (Balasubramaniam et al., 1997), and chicken strips (Holownia et al., 2000).

The main mechanism of MC and HPMC gelation is hydrophobic interaction between molecule chains containing methoxyl groups. However, substitution with hydroxypropyl groups significantly alters gelation properties: Given the same degree of methoxyl substitution, an increase in hydroxypropyl substitution raises the gelation temperature and diminishes the strength of the resulting gel. As a result, the gelation temperature of HPMC is higher and the strength of the resulting gels is lower than for MC. As regards the influence of molecular weight, gel strength increases with molecular weight up to a molecular weight of approximately 140,000, at which point it stabilizes. Molecular weight does not affect the gelation temperature (Sarkar, 1979). Although MC and HPMC have different gelling properties, the relationship between these and the gels' oil absorption barrier efficiency is not given in the studies published.

Increasing the MC concentration of a batter from 1% to 2% has been shown to lead to lower oil absorption and a greater moisture reduction in the crust of batter-coated food, both after a 30-s pre-frying step and after the final frying following freezing. For three concentrations studied, the barrier effectiveness was found to be more evident after the pre-frying step (Sanz et al. 2004). Higher barrier effectiveness in the first 30 s of frying has also been found in batter-coated and breaded chicken nuggets containing HPMC levels from 0.25 to 1.0% w/w on a wet batter basis (Dow Chemical Co., 1991).

When incorporating MC and HPMC into a batter, there are two main ways of achieving correct hydrocolloid hydration. One is to disperse the hydrocolloid by blending it with the rest of the dry ingredients; hydration take place when they are mixed with water. It must be taken into account that for correct hydration, cold water must be used; the required temperature decreases as the number of methoxyl substitutes in the anhydroglucose ring rises. The second method is to hydrate the hydrocolloid before mixing it into the batter (Dow Chemical Co., 1996).

These methods of blending the MC or HPMC into the batter have been related to their efficiency in reducing oil uptake. The barrier effect of HPMC in various batter-coated foods, such as pieces of chicken, improved when the prehydrated hydrocolloid method was used. The final viscosity of the batters was also higher, which was linked to a greater level of hydration (Meyers and Conklin, 1990). Another important factor related to barrier efficiency is batter temperature: in batters with differing MC concentrations stored at 5, 15, and 25°C for an hour after preparation (dry-blending technique), lower barrier efficiency was observed as the temperature increased (Sanz et al., 2004).

Another way to reduce oil absorption is to cover the pieces of food in an edible film by dipping them in an MC or HPMC solution. The influence of adding HPMC and MC to the breading formulation or adding them as a film before breading to reduce the amount of oil absorbed by the crust has been evaluated in marinated chicken strips (Holownia et al., 2000); the most efficient method was to add the dry hydrocolloid to the breading formulation. In this work, forming the film before breading reduced its moisture; this was associated with an inhibition of the migration of moisture from the substrate into the crust. Although less efficient than MC and HPMC, sodium carboxymethylcellulose (CMC) has also been used to reduce oil absorption. A study that analyzed different concentrations of CMC (0.5 to 3%), adjusting the proportion of water to obtain an adequate batter viscosity, obtained the greatest barrier efficiency at a concentration of 2% (Priya et al., 1996).

Microcrystalline cellulose co-dried with whey has been found to reduce the oil absorbed and increase the moisture retained by the bread coating of fried batter-coated and breaded food. Annapure et al. (1999) evaluated the oil uptake barrier efficiency of various gums at concentrations of between 0.25 and 2% in model systems based on chickpea flour; they classified HPMC as the most effective at the 0.25% concentration, although its efficiency fell as the concentration increased. For a 2% admixture, gum arabic proved the most effective, followed by carrageenan and karaya gum. They attributed the failures they encountered to the formation of too thick a coating at higher concentrations, resulting in rupture of the film from excessive pressure built up during frying. Other gums studied, such as xanthan gum, ghatti gum, tragacanth gum, and locust bean gum, were not effective.

Other film-forming possibilities that have been investigated include the application of gellan gum, an anionic linear heteropolysaccharide produced by microorganisms. This substance was tested by applying it in a hot solution: The food piece is dipped into the solution and the film forms as it cools. Another possibility is to dip the food into a cold gellan solution: Gelation takes place in the presence of ions such as Na^+, K^+, Ca^{2+}, or Mg^{2+} (Duxbury, 1993). Gellan gum has also been added with calcium chloride to the dry-ingredient mixture of batters for chicken and other substrates, resulting in low oil absorption and the development of appropriate crispness. Gellan solutions gelled in the presence of sodium or calcium ions have also been used to coat the crumbs used for breading; according to Chalupa and Sanderson (1994), use of these rather than conventional breading crumbs achieved a final product with excellent crispness and lower oil absorption.

Pectin is another substance that gels in contact with calcium and forms a film that has proved effective for reducing oil absorption in a batter-coated and breaded product. Ca^{2+} has been added to the breading by dry blending, agglomerating, spray drying, baking the calcium source into the breadcrumb, or any combination of these, followed by treating the batter-coated and breaded food item with a solution of calcium-reactive pectin (Gerrish et al., 1997). The reaction between the Ca^{2+} and the pectin produces a hydrophilic film that reduces the absorption of oil. It is very important to take the level of Ca^{2+} added to the breading into account, because sufficient available calcium for efficient reaction with the pectin should be present. The pectin types selected are those with stronger calcium reactivity: in other words, low-methoxy pectins or amidated low-methoxy pectins.

Method Without Prefrying

An innovative process that suppresses the pre-frying step in manufacturing battered food products has been developed by Fiszman et al. (2003). The invention is an original process that replaces the standard pre-frying step in the industrial preparation of frozen batter-coated food products with an alternative step: setting the batter covering the food substrate by immersion in hot water at 60 to 90°C. This coagulates the batter by gelling the cellulose derivative it contains, and thermoreversion of the reaction is avoided by subsequent heating in a microwave or conventional oven.

The heat of the hot-water bath is sufficient to coagulate the batter in practically the same way as the hot oil does during pre-frying. The cellulose that is used gels per se during this immersion. A quick heating step (microwave, conventional oven, or infrared) is used to "fix" the resulting structure, and the product is then frozen. By eliminating the industrial pre-frying step, this process presents two health benefits: It offers consumers a product that does not contain any frying oil at that point, and it eliminates the health and safety risks related to the use of very hot oil in the factory.

The complete sequence of steps in this process is: predusting the substrate, immersion in the batter, dripping, coagulation of the batter in a hot-water bath, flash heating, and subsequent freezing. The formulation of the batter mix is similar to the commercial wheat flour–based formulations normally available; like them, it allows great flexibility, as in addition to having wheat flour as its main ingredient, it may include starch, salt, leavening, and other ingredients.

One of the key points in this method is temperature control throughout the process, both when the batter is made up (when the dry ingredients are mixed with the water) and during all the battering steps. The water for the batter and the batter itself must be at a temperature between 10 and 15°C. In this temperature range the methylcellulose will become completely hydrated and be capable of forming a strong gel when the battered product is immersed in a hot water bath. Strict temperature control is fundamental because at temperatures below 10°C the batter could become too viscous, impeding proper handling, and at temperatures above 18°C the viscosity could become too low, causing poor adherence.

During immersion in the hot-water bath, the gel formed by the methylcellulose within the batter is thermo-reversible (it would regain its original semifluid consistency on cooling), so after this step the coagulated battered product must be subjected to a heating process, which can be performed in a conventional oven, a microwave, or by infrared radiation. This heating step is another key point in the process, because it blocks or sets the structure formed by thermo-gelation of the methylcellulose, preventing the coating from softening when it cools. After the heating step, the battered product is cooled, frozen, and packaged for storage at temperatures of −18°C or less.

After the final frying, the product prepared by this method has very similar texture, flavor, and color properties to those of a product manufactured using a traditional process that includes pre-frying in oil, with the advantage of being free of the fat absorbed during pre-frying. Sanz et al. (2008) added resistant starch (type 3), replacing up to 20% of the flour, and found a significant increase in the hardness and fragility of the batter crust and a more intense golden-brown color.

The nature of the oil used in the pre-frying step, in terms of its fatty acid composition, for example, or whether it is a mixture of oils of various origins, is something that the consumer normally does not know. A further advantage, perhaps the most important one from the point of view of the manufacturer, is the elimination of industrial fryers, bringing energy savings and cost reductions connected with oil purchasing, testing, quality maintenance, and waste disposal. It also avoids the dangers involved in working at such high temperatures and in the presence of fumes or foam, the former generated by volatile chemical products as a result of overheating the oil and the latter due to the generation of structurally degrading surfactants.

An additional advantage of this alternative method for the preparation of battered foods without pre-frying is that the final product, after frying by the end user, has a significantly lower fat content because the cellulose incorporated into the batter leads to a lower absorption of frying oil, as mentioned earlier. After final frying, a conventional food substrate has an oil content of about 35%, whereas the same product prepared by this novel method contains about 16% oil (Sanz et al., 2004).

CONTROL OF COATING ADHESION: FREEZING AND HANDLING STABILITY

One of the main quality factors in coated products is good coating adhesion. Inspection of the final cooked food pieces makes it possible to determine whether the outer layer is adhering to the substrate or whether a gap has formed between the two. However, it should be borne in mind that good fried coating adhesion to the substrate in the end product is closely related to the adhesion of the raw batter at the moment when it is applied.

Adhesiveness is heavily dependent on the nature of the surfaces. For example, if the raw material is frozen, as in the case of chicken bites, for example, a layer of ice results in poor adhesion; the normal practice is to melt this layer with the help of salt. The same happens if the moisture content of the substrate is too high: It could be a good practice to apply a predust (usually, flour-based) to absorb the surface moisture.

The result of poor adhesion is the appearance of *voids*, as areas that are bare of any coating are called. In zones with poor adhesion, the steam generated during frying causes "pillowing" of the coating, the appearance of air pockets, or "blow-off," with the consequent loss of coating into the frying medium. One way of improving batter adhesion is through the use of different predusts and of gums or proteins in the batter formulations.

The degree of adhesion is very important for manufacturers because it is related to yield: on the one hand, the quantity of batter (greater weight) adhering to a piece of food, and on the other, in the final fried product, the proportion of outer layer to total weight. In tempura-type batters, a batter that is too thin produces a weak, porous coating that is difficult to handle, as well as constituting a poor barrier against oil absorption and the loss of food juices during frying.

In breaded products, the concept of adhesion is slightly different because the breadcrumbs or bread or cereal particles tend to fall off the coated product easily. Obviously, this is related to the yield of the entire process.

Hsia et al. (1992) proposed the following definitions for breaded products:

$$\%\ \text{cooked yield} = CM \times \frac{100}{I} \quad (1)$$

$$\%\ \text{overall yield} = S \times \frac{100}{I} \quad (2)$$

where CM is the mass of a cooked breaded food item, S is the mass of a cooked breaded item after shaking (standardized sieve and shaking process), and I is the initial mass of a raw unbreaded food item. In this case, the percent overall yield accounts for the possible losses from the adhered breading layer during postcooking handling.

A series of factors can influence the greater or lesser adherence of the batter or breading layer; they include the addition of certain ingredients. A higher fat content has been suggested as a possible cause of higher coating loss during shaking in fried, battered, and breaded chicken breasts prepared with different levels of Tween 80 (a surfactant) and batter dry mix/water ratios (Maskat and Kerr, 2004a). The effects of breadcrumb particle size on coating adhesion have also been studied; the results showed that coating adhesion was highest in coatings formed from breadcrumbs with a particle size of less than 250 μm and lowest in those made from larger particles (Maskat and Kerr, 2004b).

The term *pickup* is generally used to quantify the batter that has adhered to the piece of food (substrate). It is a quality index that all manufacturers check because the yield and the final quality of the product depend on it. The batter or coating pickup of the final product is defined as

$$\%\text{ batter pickup} = \frac{B}{B + S} \times 100 \tag{3}$$

where B is the mass of batter coating a food item after final frying and S is the mass of a food item, excluding batter ("peeled"), after final frying (Baixauli et al., 2003).

In this case, the pickup index expresses the proportion of coating in the final weight of the fried food. It must be remembered that these weights include the weight of oil absorbed during frying and that both the substrate food and the external crust will have shed moisture during the entire process. From the point of view of industrial manufacturers, if the factory only pre-fries the battered food, it may also be of interest to calculate this value after pre-frying. This will show what part of the weight of the pre-fried batter-coated food (the product they sell) corresponds to the external crust.

Microscopy has also been used to investigate adhesiveness and has shown that the batter–substrate interaction depends on the ingredients used (Mukprasirt et al., 2000). Cunningham (1983) investigated the influence of the structure of poultry skin on the quality of coated poultry and reached the conclusion that removing the cuticle increased the adhesion of the batters and breadings enormously. The study included other factors that influence adhesion, such as scald temperature, the age of the animal, the cooling and freezing methods (which influence the skin ultrastructure), and the holding temperature (at which the pieces are kept hot just before they are served).

CONTROL OF THE FINAL TEXTURE OF THE FRIED PRODUCT

The texture parameter that is most appreciated in a fried product is crispness. A chewy toughness or a mushy softness in a battered fried product is normally cause for rejection. Ideally, the coating should exhibit a consistency with a little resistance to the initial bite and a crispiness while chewing. The crisp final texture of the fried product can be evaluated instrumentally or by sensory tests.

In breaded and battered poultry products, crispness is among the textural characteristics that are highly valued by consumers (Antonova et al., 2003). As well as differences in composition and process, the final heating method is decisive for retaining crispness. For example, Antonova et al. (2004) found significant differences in the sensory crispness of chicken nuggets: Fried samples were significantly crispier than chicken nuggets cooked in either a convection oven or a microwave oven.

The importance of crispness in the acceptability of crusted foods has created the need to define and measure this perception. Parameters such as crispness or crunchiness, fragility, and tenderness are difficult to quantify using empirical mechanical methods because they are a complex sum of sensations. A puncture test with a plunger is the technique most often used (Fan et al., 1997; Mohamed et al., 1998). Shear-press tests have been used to measure the instrumental texture of the fried crust of battered foods, applied up to the breaking point; parameters such as maximum force/mass and work/mass have been calculated as indices of textural characteristics (Du Ling et al., 1998). However, one-point instrumental parameters are normally not capable of discriminating differences that are sometimes very subtle or complex in nature. Salvador et al. (2002) proposed evaluating the complete penetration curve profile of a fried coating to observe various aspects of texture.

Some simplified methods have measured the fried crust alone (separated from the substrate) by puncturing it with a cylinder or a conical probe (Fan et al., 1997; Mohamed et al., 1998; Matsunaga et al., 2003). However, measuring only the external crust has the limitation of not reflecting reality, as deformation of the outer layer while eating is not only a result of the mechanical behavior of the crust but also depends on the mechanics of the noncrispy core, where shear and compression forces are mixed (Varela et al., 2008). The mechanical, morphological, and compositional differences between the layers of this sandwichlike structure make it difficult to determine the contribution of each part (Luyten et al., 2004), but the mechanics of the layers stacked together are essential for understanding the texture of the food piece as a whole. Other textural parameters, such as greasiness, juiciness, oiliness, or mealiness, are important for fried food coatings; they can be quantified using trained panelists (Prakash and Rajalakshmi, 1999).

Studying the microstructure of the coatings can be a very interesting tool for understanding their physical properties. Battered foods are very complex systems from a structural viewpoint; they normally contain a great number of components, and both the coating and the food substrate change in several ways during frying. Llorca et al. (2001) analyzed the effects of frying on the microstructure of a battered food and pointed out that in the pre-fried products there was a structure that interconnected the batter and the substrate, whereas in the final fried product the two layers separated.

REFERENCES

Ang JF. 1989. The effect of powdered cellulose on oil fat uptake during the frying of battered products. J Am Chem Soc 66:56.

Ang JF. 1993. Reduction of fat in fried batter coatings with powdered cellulose. J Am Chem Soc 70:619–622.

Annapure US, Singhal RS, Kulkarni PR. 1999. Screening of hydrocolloids for reduction in oil uptake of a model deep fat fried product. Fett Lipid 101:217–221.

Antonova I, Mallikarjunan P, Duncan SE. 2003. Correlating objective measurements of crispness in breaded fried chicken nuggets with sensory crispness. J Food Sci 68:1308–1315.

Antonova I, Mallikarjunan P, Duncan SE. 2004. Sensory assessment of crispness in a breaded fried food held under a heat lamp. Foodservice Res Int 14:189–200.

Baixauli R, Sanz T, Salvador A, Fiszman SM. 2003. Effect of the addition of dextrin or dried egg on the rheological and textural properties of batters for fried foods. Food Hydrocoll 17:305–310.

Balasubramaniam VM, Chinnan MS, Mallikarjunan P, Phillips RD. 1997. The effect of edible film on oil uptake and moisture retention of a deep-fat fried poultry product. J Food Process Eng 20:17–29.

Chalupa WF, Sanderson GR, inventors; Merck & Co., Inc., assignee. 1994, Dec 13. Process for preparing low-fat fried food. U.S. patent 5,372,829.

Cunningham FE. 1983. Application of batters and breadings to poultry. In: Suderman DR, Cunningham FE, eds., *Batter and Breading*. Westport, CT: AVI Publishing, pp. 15–23.

Dow Chemical Co. 1991. Fried foods stabilizer keeps moisture in fat out. Prepared Foods (Mar):61.

Dow Chemical Co. 1996. *A Food Technologist's Guide to Methocel Food Grade Gums*. Brochure 194–01037-1996 GW. Midland, MI, Dow.

Du-Ling, Gennadios A, Hanna MA, Cuppet SL. 1998. Quality evaluation of deep-fat fried onion rings. J Food Qual 21:95–105.

Duxbury DD. 1993. Fat reduction without adding fat replacers. Food Process 68(5):70.

Fan J, Singh RP, Pinthus EJ. 1997. Physicochemical changes in starch during deep-fat frying of a model cornstarch patty. J Food Process Preserv 21:443–460.

Fiszman SM, Salvador A, Sanz T, Lluch MA, Castellano JV, Camps JL, Gamero M, inventors; Alimentaria Adin SA, assignee. 2003. Method of preparing a frozen battered food product. WO patent 03/101228-A1.

Gerrish T, Higgins C, Kresl K, inventors. 1997, Feb 11. Method of making battered and breaded food compositions using calcium pectins. U.S. patent 5,601,861.

Holownia KI, Chinnan MS, Erickson MC, Mallikarjunan P. 2000. Quality evaluation of edible film-coated chicken strips and frying oils. J Food Sci 65:1087–1090.

Hsia HY, Smith DM, Steffe JF. 1992. Rheological properties and adhesion characteristics of flour-based batters for chicken nuggets as affected by three hydrocolloids. J Food Sci 57:16–18, 24.

Lee HC, Han I. 1988. Effects of methylcellulose (MC) and microcrystalline cellulose (MCC) on battered deep-fat fried foods. Food Technol 42:244.

Llorca E, Hernando I, Pérez-Munuera I, Quiles A, Fiszman SM, Lluch MA. 2003. Effect of batter formulation on lipid uptake during frying and lipid fraction of frozen battered squid. Eur Food Res Technol 216:297–302.

Luyten A, Plijter JJ, Van Vliet T. 2004. Crispy/crunchy crusts of cellular solid foods: a literature review with discussion. J Texture Stud 35:445–492.

Mallikarjunan P, Chinnan MS, Balasubramaniam VM, Phillips RD. 1997. Edible coatings for deep-fat frying of starchy products. Lebensm-Wiss Technol 30: 709–714.

Maskat MY, Kerr WL. 2004a. Effect of breading particle size on coating adhesion in breaded fried chicken breasts. J Food Qual 27:103–114.

Maskat MY, Kerr WL. 2004b. Effect of surfactant and batter mix ratio on the properties of coated poultry product. J Food Prop 7:341–352.

Matsunaga K, Kawasaki S, Takeda Y. 2003. Influence of physicochemical properties of starch on crispness of tempura fried batter. Cereal Chem 80(3):339–345.

Meyers MA. 1990. Functionality of hydrocolloids in batter coating systems. In: Kulp K, Loewe R, eds., *Batters and Breadings in Food Processing*. St Paul, MN: American Association of Cereal Chemists, pp. 117–141.

Meyers MA, Conklin JR, inventors; Dow Chemical Company, assignee. 1990, Feb. 13. Method of inhibiting oil adsorption in coated fried foods using hydroxypropyl methylcellulose. U.S. patent 4,900,573.

Mohamed S, Hamid NA, Hamid MA. 1998. Food components affecting the oil absorption and crispness of fried batter. J Sci Food Agric 78:39–45.

Mukprasirt A, Herald TJ, Boyle DL, Rausch KD. 2000. Adhesion of rice flour–based batter to chicken drumsticks evaluated by laser scanning confocal microscopy and texture analysis. Poult Sci 79:1356–1363.

Pinthus EJ, Weinberg P, Saguy IS. 1993. Criterion for oil uptake during deep-fat frying. J Food Sci 58:204–205, 222.

Prakash M, Rajalakshmi D. 1999. Effect of steamed wheat flour on the sensory quality of batter coated products. J Food Qual 22:523–533.

Priya R, Singhal RS, Kulkarni PR. 1996. Carboxymethylcellulose and hydroxypropyl methylcellulose as additives in reduction of oil content in batter based deep-fat fried boondis. Carbohydr Polym 29:333–335.

Salvador A, Sanz T, Fiszman SM. 2002. Effect of corn flour salt and leavening on the texture of fried battered squid rings. J Food Sci 67:730–733.

Sanz T, Salvador A, Fiszman SM. 2004. Effect of concentration and temperature on properties of methylcellulose-added batters: application to battered fried seafood. Food Hydrocoll 18:127–131.

Sanz T, Salvador A, Fiszman SM. 2008. Resistant starch RS in battered fried products: functionality and high-fibre benefit. Food Hydrocoll 22:543–549.

Sarkar N. 1979. Thermal gelation properties of methyl and hydroxypropyl methylcellulose. J Appl Polym Sci 24:1073–1087.

Stypula RJ, Buckholz L, inventors; Int. Flavors & Fragrances Inc., assignee. 1989. Process for preparing a coated food product. U.S. patent 4,877,629.

Varela P, Salvador A, Fiszman SM. 2008. Methodological developments in crispness assessment: effects of cooking method on the crispness of crusted foods. LWT 41:1252–1259.

6

MECHANICAL DEBONING: PRINCIPLES AND EQUIPMENT

MANUEL VIUDA-MARTOS, ELENA JOSÉ SÁNCHEZ-ZAPATA, CASILDA NAVARRO-RODRÍGUEZ DE VERA, AND JOSÉ ANGEL PÉREZ-ALVAREZ
Grupo Industrialización de Productos de Origen Animal (IPOA Research Group), Departamento de Tecnología Agroalimentaria, Escuela Politécnica Superior de Orihuela, Universidad Miguel Hernández, Orihuela, Alicante, Spain

Handbook of Poultry Science and Technology, Volume 2: Secondary Processing, Edited by Isabel Guerrero-Legarreta and Y.H. Hui

INTRODUCTION

When high-priced muscles from animals, poultry, or fish are removed, inedible, relatively hard or tough components such as bone, gristle, and tendons remain on the carcass. Thus, *mechanically recovered* or *separated meat* are generic terms used to describe residual meat that has been recovered or separated by the application of pressure or shear force to animal bones or poultry carcasses from which the bulk of the meat has been removed (Crosland et al., 1995). This permits the recovery of most of the residual meat, which would otherwise be difficult or uneconomical to acquire (Day and Brown, 2001).

The consumption of poultry meat and poultry meat products continues to grow and the increased production of cut-up and processed meat has provided considerable quantities of parts suitable for mechanical deboning (Perlo et al., 2006). This process is considered an efficient method of harvesting meat from parts left after hand deboning as well as from poor-quality poultry. The yield of mechanically deboned poultry meat ranges from 55 to 80%, depending on the part deboned and the deboner settings (Mielnik et al., 2002). The composition of mechanically recovered meat is inherently variable, due to the natural variation within and between animal species (chicken, hen, turkey, etc.), previous treatment of the carcasses (trimming, freezing), carcass part used (neck, wing, frames, etc.), and the machine type and operating conditions used in the recovery process (Pizzocaro et al., 1998; Day and Brown, 2001). For example, the meat remaining on chicken carcases and necks constitutes about 24% of the total meat, whereas on turkeys it represents about 12%—not a negligible amount (Sadat and Volle, 2000).

The resulting mechanically recovered poultry meat has the appearance of finely comminuted meat; thus, mechanically recovered poultry meat is used in a wide range of emulsified meat products and, in smaller proportions, in nonemulsified meat products such as frankfurters, various loaf products, fermented sausage, and restructured chicken products (Jones, 1986; Steele et al., 1991; Mielnik et al., 2002; Perlo et al., 2006), opening up additional markets for meat from turkey and chicken frames and necks (Mielnik et al., 2002).

From an engineering point of view, the recovery machines used to separate the residual meat from the bone can essentially be divided into three main types: those that separate the meat from the bones by means of a sieve screen, those that use a stripper disk, and those that use centrifugal force (Crosland et al., 1995).

STRIPPER DISK MECHANICAL DEBONING SYSTEMS

In the processing and preparation of poultry and animals parts such as chicken thighs for sale and consumption in the retail market, grocery stores, and restaurants, it is considered highly desirable to package and serve the meat with the bones removed. Deboned meat can be sliced easily for use in sandwiches, and in other food products it is desirable to have the bones removed prior to cooking and serving. In the past, automated processes have been developed for the removal of meat from the bones of poultry, such as from chicken thighs, by engaging the bone with a scraping tool and scraping along the length of the bone. However, raw meat has a tendency to cling tightly to the bone. Consequently, it is necessary for the scraper blades to surround the thigh bones closely to ensure that will be the meat stripped completely from the bone. One problem that arises with such deboners is that the blades should avoid contact with the bone to avoid gouging or chipping the bones as they strip the meat. This contact between the hard blade material and the bones increases the possibility of bone fragments becoming lodged in the stripped meat, which creates a health risk to the consumer, who expects that a "boneless" product will indeed be completely boneless.

Stripper Disk

To avoid bone fragments during the deboning process, aperture elastic meat stripper disks have been substituted for the scraper blades in some stripping equipment. The bone is pushed longitudinally through the aperture of the disk and the resilient disk retards the movement of the meat, thereby separating the meat from the bone. Species and cuts used in stripper disk debonding machines are listed in Table 1.

Heuvel and Johannes (1989) described the use of an elastic disk for deboning poultry parts. Additionally, Hazenbroek (1992) described an automated deboning apparatus that includes a series of elastic stripping disks, each of which is mounted adjacent and moves with a conveyor tray on which a poultry thigh is received and moved about on a processing path. As each thigh is moved along the path, the thigh bone passes through an opening in a stripper disk, whereupon the meat of the thigh is stripped progressively from its bone.

The meat stripper disk assemblies are mounted on the end plate, which is positioned adjacent to the processing path for the carrier trays. Each stripper disk assembly includes a flexible stripper disk supported within a rigid collar that stabilizes and supports it in each stripper disk. As the thigh bones pass through the apertures in the center of the stripper disks, the meat is stripped from them progressively. Guides aligned with the aperture are mounted behind each stripper disk. The guides receive and stabilize the thigh bones as they pass through and beyond the stripper disks during a meat-stripping operation. The bone pusher assembly of each deboning module is mounted on, and is laterally

TABLE 1 Species and Cuts Used in Continuous-Operation Stripper Disk Deboning Machines

Species	Cuts
Hen	Thigh
Chicken	Frame
Turkey	
Duck	

movable along, a pair of travel rods positioned initially on the opposite side of each carrier tray from the aligned stripper disk. Each bone pusher assembly includes a hollow pusher sleeve with open front and rear ends, which is mounted on a carrier block attached to the travel rods. A cam follower is fixed to the bottom of the carrier block and engages and rolls along a first cam track formed about the circumference of the stationary cam drum of the deboner unit. As the cam follower engages and moves along its cam track, the pusher sleeve is moved and engages with the adjacent knuckle of the bone of the poultry thigh, forcing the thigh toward its aligned meat stripper disk assembly and through the aperture of the stripper disk. The device described by Hazenbroek (1992), however, has the disadvantage that the meat stripper disk, and especially the aperture, show a high degree of wear and need to be replaced regularly.

Pusher Rod

To ensure the proper deboning of poultry parts of widely varying sizes, the pusher rod may have an extendible part that can be displaced to a position within the aperture of the meat stripper disk, or even beyond the stripper. To improve the grip on the bone, the pusher rod has an inwardly tapering front end whose surface is rough, and/or it is provided with one or more pins to engage the bone. In the deboning device of this apparatus, the pusher rod's diameter is larger than that of the aperture of the meat stripper disk, to prevent the rod from being forced through the aperture. In this way, wear in the stripper disk due to frequent contact with the pusher rod is avoided.

Although the pusher rod does not pass through the aperture of the meat stripper disk, complete removal of the meat can still be achieved without the need to grip the bone as soon as it passes through the aperture of the stripper disk. The deboning device comprises a pusher assembly with a pusher rod that is displaceable toward the meat stripper disk until it contacts the disk, and from there on to a position beyond the nondeformed position of the stripper disk, thus deforming the disk and prolonging the distance over which the bone is pushed out of the meat. In a perpendicular direction to the advancing direction of the processing path, a first obstruction is provided to stop the bone extending through the aperture of the meat stripper disk while the deboned meat is still being moved by the conveyor belt. The obstruction causes the bone to be moved or rotated out of the aperture of the stripper disk and to complete the removal of any meat remaining on the bone. This first obstruction is provided at an upstream position of the processing path, preferably beyond the device for advancing the bone pusher assembly toward the meat stripper disk.

The deboning unit preferably comprises means for vertically adjusting the position of the poultry part with respect to the meat stripper disk. In that way, the vertical height of the deboning unit between the poultry part and the aperture of the stripper disk can be varied and adapted to the size of the poultry part that needs to be deboned, to ensure that the bone can be directed toward and through the aperture in the stripper disk regardless of the size of the poultry part. The means of adjusting the vertical position of the poultry part with respect to the meat stripper of the deboning unit includes a socket for carrying at least part of the poultry part in the vicinity of the meat stripper disk.

Socket

The socket is mounted in the carrier in the forward direction of the deboning unit, to allow the deboning unit to be adapted to the size of the poultry part and to allow proper deboning of poultry parts independent of their size. The socket is provided to cooperate with a third guide in the vicinity of the deboning position.

Mechanical Deboning

The means of guiding the bone pusher assembly toward the meat stripper disk assembly provided at the deboning position preferably comprises the first step in guiding the centering tube and pusher rod from their first starting position to their first ending position, and the pusher rod from its first ending position to its second ending position, which corresponds to a position in which the meat stripper disk is deformed elastically. After the second end position, the first guide transfers to a second guide for retracting the pusher rod from the meat stripper disk. Upon the backward retraction of the pusher rod toward its start position away from the meat stripper disk, the flexible disk collapses backward, exerting a forward projecting force on the bone. In this way, the bone is projected through the aperture of the meat stripper disk, and any remaining last traces of meat are stripped from the bone. Thus, a virtually complete removal of meat from the bone can be achieved without having to displace the pusher rod through the aperture disk. Opposite the side facing the stripper disk, the socket may form an upright rim that corresponds to the end position of the pusher sleeve.

The deboning device provides a second means of separating any tendons or ligaments remaining between the bone and the stripped meat: for example, a second obstruction or a cutting process upstream of the moving direction of the conveyor belt. In another position upstream of the deboning position of the processing path, a fourth guiding may be provided for displacing the socket vertically in the deboning unit to force the meat to leave the carrier, which permits automatic removal of the meat from the device. If meat sticks to the carrier, a paddlewheel picks it up from the carrier and transports it toward the meat collector.

Another alternative involves a guiding block to which the pusher rod is connected. The rod is preferably surrounded over at least part of its length, preferably its entire length, by a spring, which is clamped between the guiding block and a rim in the vicinity of the front end of the centering tube. The clamped positioning of the spring ensures, on the one hand, that the bone pusher assembly is advanced toward the poultry part so as to clamp the poultry part between the bone pusher assembly and the meat stripper disk and, on the other hand, that a clamping force is exerted on the bone of the poultry part to hold the bone and direct it toward the aperture in the stripper disk. Upon displacement of the bone pusher assembly, the spring is gradually compressed. The pressure it exerts on the bone at deboning is gradually increased, reducing the risk of breaking the bone.

The guide has a third part connected to the guiding block which cooperates with the first guide to direct the centering tube from its starting position to its first ending position, which, preferably, corresponds to the position where the front end of the centering tube contacts the upright rim of the socket. The third part cooperates further with the first guide to move the pusher rod from this position to its second ending position, which deforms the meat stripper disk elastically in the advancing direction of the pusher rod. This results from the fact that the pusher rod is mounted in the centering tube. In the position at which the first guide stops and the bone has been removed from the poultry part, the first guide transfers to a second guide to direct the centering tube and pusher rod back to their starting positions.

While the centering tube and pusher rod are being displaced toward the meat stripper disk, the spring is compressed. As the centering tube and the rod are returned to their starting position along the second guide, the displacement is controlled by relaxation of the spring and takes place automatically, without having to exert any force on the pusher rod after the meat has been removed completely from the bone. The carrier and/or the socket preferably include a groove to receive the poultry part, and each deboning unit includes at least one travel rod that extends across the chain to guide displacement of the guiding block and to guide the centering tube through the groove of the carrier from its first starting to its first stopping position, and preferably also guiding the pusher rod to its second ending position. Upon displacement of the bone pusher assembly, the front end of the centering tube is located close to the bottom of the groove to pick up the poultry part and redirect it toward the meat stripper disk.

The deboning units may be connected to form an endless chain. In this way there is no need to provide a chain or gear wheel, so that the device can run with less wear. However, if desired, the deboning machine may be mounted on one or more conveyor belts to form an endless chain.

SIEVE SCREEN MECHANICAL DEBONING SYSTEMS

The mechanical separation of various meat and bone combinations to produce a high-quality meat product is a well-established procedure. The apparatus used generally consists of one stage strainer, in which a combination of meat and bone is forced through a perforated screen to separate the meat from the bone. However, it is common for some bone material to pass through the screen and be included in the recovered meat product, often causing the bone calcium content to exceed government and industry standards.

Mechanical Characteristics

The meat deboning machine is characterized by a perforated housing section that includes a rigid perforated cylinder in which a conveying auger rotates at relatively high rotational speeds (500 to 1800 rpm). In normal operation a meat–bone mixture is fed into the in-feed end of the sieve cylinder from an adjoining pressurizing auger section of the machine. Separation of meat from bone solids is effected within the perforated housing section, with the meat, in liquid or fluidized form, extruding or exuding through the foramina, small orifices in the perforated cylinder, while bone solids and sinews leave the discharge end of the perforated cylinder in compacted form through a generally frusto-conical orifice. The size of the orifice is set or adjusted so that it is restricted and back pressure is created on the meat–bone mixture within the perforated housing. The fluids are directed to the outside and remain there until they reach the screen, where they exit first. By the time the material has reached the screen, the root diameter has increased sufficiently to allow the material to begin filling the cavity between the flights if enough material has entered the system.

Flights

The flights and tapered root of the feed and compression augers are constructed as a single piece, which eliminates welds between the auger flights and the auger root and enhances auger sanitation. This device passes the meat radially through a sieve screen and the bony material axially through an outlet (Richburg, 1992). The auger is used to compress and convey the meat and bone material axially through a sieve screen, where the compressed meat passes through the sieve screen and the bony material is conveyed through an outlet at the end of the sieve screen. Species and cuts used in sieve screen deboning machines are listed in Table 2.

The front faces of the auger flights are inclined 15 to 18° toward the entry end of the auger, assisting the outward flow of material toward the sieve screen. This helps retain material behind the flight, as evidenced by the fact that there is far more wear on the trailing edge of the rear face of the flight than there is on the shear edge of the front face. Typically, deboning sieve screen apparatus uses high pressures (1000 to 10,000 psi) to exert compressive forces on a product. Flights are carried by the auger root, which has a depth of about 1.0 to 1.8 in. at the entry end. The flights have a constant outside diameter along the total length of the feed and compression augers. The cross-sectional area ratio change (8 to 1) over the length of the auger, from the entry to the exit ends, is due to the change in root diameters. The cross-sectional area ratio is the ratio of the area of

TABLE 2 Species and Cuts Used in Continuous-Operation Sieve Screen Deboning Machines

Species	Cuts
Hen	Thigh
Chicken	Neck
Turkey	Frame
Duck	Wing

the outside diameter of the flight to the area of the outside diameter of the root at a point along the length of the auger; it is the cross-sectional area ratio at one point on the auger's length divided by the cross-sectional area ratio at a different point on the auger's length. The points referred to are the entry and exit points. Augers typically exhibit a cross-sectional area ratio change (2.2 to 1) over the length of the auger.

The clearance between the outer diameter of the flights on the conveying augers and the interior diameter of the sieve cylinders is substantially uniform from the in-feed end to the discharge end. The root diameter of the conveying augers is either uniform from the in-feed end to the discharge end or increases by a fraction of a degree toward the discharge end with a view to decreasing the space available for the bone-laden material as it approaches the discharge end. This forces the meat material through the perforations or apertures while continuing to agitate and convey the boning components along the helical path.

Auger Root

The feed auger has a first coupling end, and the compression auger has a second coupling end which mates with the first coupling end so that the auger root has a continuous taper and flight. An auger shaft extends through the compression auger and connects to the feed auger, and the auger shaft has a spiral groove on one end. A hollow bore is formed within the feed auger, and a locking pin extends radially within the hollow bore to engage within the spiral groove of the auger shaft, connecting the compression auger, feed auger, and auger shaft. An auger coupling joins together the first coupling end of the feed auger and the second coupling end of the compression auger. The spiral groove of the auger shaft has an extended groove portion which prevents the auger shaft from becoming unlocked in the event that the drive motor is operated in the reverse direction. The auger coupling consists of many coupling pins on the feed auger, with a similar number of corresponding coupling holes formed in the compression auger, which mate to lock the augers together in a drive connection. One coupling pin and hole have a different configuration from the other coupling pins and holes, to align the feed and compression augers.

The taper of the auger root is greater than 5° and is continuous along the length of the feed and compression augers so that the flights of the auger at the entry end are substantially deeper than at the exit end, to allow more meat and bone material to be introduced into the first cavity of the auger flights. Preferably, the auger root has a taper of about 9° relative to the axis of the auger, and the flights of the auger have an outside diameter of about 5 in. The feed can inlet includes a reduced eccentric opening or plate which is oriented to the rear near the entry end of the feed auger to increase its loading capacity by increasing the velocity due to the orifice effect. The feed can inlet of the housing includes an entry for introducing meat and bone material into only one flight of the auger, to prevent subsequent pressure relief and backup of the product in the inlet. A bearing housing includes a housing shaft. A first bearing is carried by the first end of the housing, and a second bearing is carried by the second end of the housing. A thrust bearing is situated in an intermediate portion of the housing, whereby the drive shaft is supported by first and second bearings for rotation and against axial thrust by the thrust bearing.

When material enters the entry end of the auger, it is directed to the root by the sloping front face of the flight. Once it has reached the root, it is directed up the rear face of the flight to the outside diameter of the feed can or is immediately thrown outward by centripetal force. In either case the material goes to the outside. This will begin to take place farther along the length of the auger. If the auger flight is not filled, the particles will stay on the outside and travel in a spiral motion around the inside of the screen as they ride on the top of the front face of the flight. The wedge action of the front face may force material through the screen and then shear some of it off, but little happens until the root starts to compress the material. This can be seen when the deboner is underfed: Only fluids come out of the entry end of the screen and the finished product comes out of the exit end of the screen. The degree of underfeeding will cause a more or less pronounced effect. The pulsing of the power used in this condition serves to further corroborate this observation.

At the point when the material starts to fill the cavity between flights, several things happen. In the area of the auger near the entry end, material is held against the rearward-sloping rear face more effectively than against the forward-sloping rear face, allowing the material to increase its velocity more as it approaches the outside diameter of the flight before being hurled into the screen. The kinetic energy imparted changes as the square of the velocity, so a small velocity change results in higher kinetic energy and a relatively large increase in productivity.

The continually decreasing area ratio causes a gradual increase in root surface velocity at the rear face of the flight, which assists in imparting greater energy to the particles hurled off the flight's rear face. As the material undergoes mild compression, it moves up the front face of the flight in a spiral motion toward the exit end. This motion is forward with respect to the axis but would appear rearward to an observer at the root of the leading edge of the flight beneath the particle being observed. Material on the front face of the flight is directed forward and outward by centripetal force and the wedging action of the rearward-leaning flight face against the screen.

As the particles being processed move farther along the length of the auger toward the exit end, they lose mass through the screen as they are sheared off by the leading face shear edge of the flight and become relatively lighter than particles not as far into the separation process. As a result, the relatively heavier particles being wedged and hurled into the screen push the relatively lighter, more completely separated particles ahead of them forward and toward the root in a spiral path relative to the screen. This happens because centripetal force is proportional to mass, and the relatively lighter particles, although they experience an outward force, literally float to the inside, because they are displaced by heavier particles that experience greater forces. Because inward displacement forces are greater than the outward centripetal force on the relatively lighter particles, the separated bone contained within the screen is forced to congregate near the root of the trailing edge of the flight and follow it to the exit end. The result is a lower bone count and lower temperature rise in the finished product, since the bone near the root acts as a heat sink. This action is verified by observing that the root of the auger near the trailing face of the flight is worn much more than the middle of the trailing face and becomes more severe as the exit end of the auger is approached. Because it takes very little pressure to force meat through an orifice in the screen, because the screen is vented to the atmosphere, and because the system is never filled, the pressure cannot build up excessively as long as the throttling ring valve is adjusted properly. Yarem and Poss (1978) describe the action in a forward-leaning flighted auger. Material is directed forward and down to the root and forward toward the rear face of the flight ahead of it, where the material heads outward to the screen and waits for the leading edge to pick it up again. The wear appears on the front face of the flight, and the shear edge is dulled rapidly. Therefore, it cannot act as a slinger and impart maximum velocity to the product as it heads toward the screen. Meat and bone product is fed to the feed auger in an area that covers more than one auger flight. This allows pressure relief in subsequent flights if the flight is full of product, while centripetal force tends to push the product back up the entry pipe if it opens during a second or third flight. It is common to use a shallow flight at the feed auger entry to reduce its loading capacity. The flight must cut off the product at the entry opening of the feed can more often for the same amount of in-feed product, thus increasing power requirements as well as pulsing and cavitation. Typically, the feed auger is threaded into the motor coupling, which creates assembly and disassembly problems and misalignment due to the threads locking.

Auger Sections

In the typical meat deboning apparatus such as that depicted by Beck et al. (1977), the auger must be divided into two sections, a feed auger and a compression auger, for assembly in the unit. The feed and compression augers include auger roots with different tapers and flights with different configurations, which do not produce the most efficient results. Since the taper changes as the product passes from the feed auger to the compression auger, turbulence and cavitation cause a diminished product flow and increased temperature in the product. Poss (1987) described a compression auger with a root of fairly uniform cross section, although the compression auger was one piece and a separate feed auger was used in another section of the machine. Yarem and Poss (1978) described a compression auger with a relatively uniform taper combined with a feed auger, where the feed and compression auger taper was not constant.

CENTRIFUGAL FORCE MECHANICAL DEBONING SYSTEMS

The use of centrifugal force in food processing is well established. Various forms of apparatus have been proposed to accomplish separation and removal of meat from an associated bone structure; however, such apparatus has not been wholly satisfactory and has not enjoyed any substantial commercial use. In many older-style machines designed to remove meat from bone, animal parts were caused to impact a wall or other structure, the force of the impact causing the meat to separate from the bone. Zartman (1962) describes the use of centrifugal force as an aid in delivering fragments of meat and bone after they have been separated by conventional impact methods. Note also that devices of the type described involve a wet process, with meat and bone fragments floating in a container of water. Kaplan (1974) describes a carcass mounted axially on a rod. Rotational motion is imparted to the carcass by high-speed fluid jets. The meat separates from the skeleton due to the combined effects of the impact force of the fluid jets and whatever centrifugal forces are developed from the spinning. Note, however, that centrifugal force plays only a partial role in this separation apparatus. Much of the separation is due to impact. This process also involves wet components.

Although centrifugal force has been used in the past in preparing poultry, the devices usually required substantial preparation of the carcass before processing. Some devices require that the carcass be cooked prior to separation. Other devices, although they use centrifugal force to a limited extent, rely principally on various types of cutting edges and the like to achieve actual separation of the meat.

As mentioned above, many of the older processes involved the use of water, which creates its own problems. When water is used, valuable protein and fat tend to be washed away, and some of the nutritional value of the meat is lost. Also, applicable processing standards set a limit on the amount of moisture that meat may contain, and if too much water is absorbed by a piece of meat, it is necessary to employ a means to remove that water. This often requires the use of an oven or a blower, or even a blast freezer, to dry the meat sufficiently.

Mechanical Basis

Helmer and Small (1980) describe a mechanical means of separating poultry meat from bone in which separation is accomplished by centrifugal force alone, without the need for numerous and substantial preparatory steps. Poultry sections are attached continuously to holders mounted on a number of moving trolleys. The trolleys hold one or more poultry sections, and when the trolleys move into the proper position, the holders rotate at high speed, up to 400 rpm, and the resulting centrifugal force causes the meat to separate from the bone. The separated meat chunks are collected by a conveyor belt and removed from the apparatus.

Poultry sections are processed in batches rather than continuously. A poultry section is mounted on a rotatable holder, and direct-drive rotation of the holder at high speeds causes the meat to separate from the bone. The motor used to rotate the holder is then disengaged and the meat and skeleton are removed automatically. This equipment can be used with either cooked or raw poultry. Greater centrifugal force is required to separate the meat of raw poultry from the

TABLE 3 Species and Cuts Used in Batch-Operation Centrifugal Force Deboning Machines

Species	Cuts
Hen	Thigh
Chicken	
Turkey	
Duck	

bone. The centrifugal force experienced by each poultry section is a uniform force felt substantially equally at all locations on the poultry section. Consequently, separation occurs in neat chunks, without substantial shredding of the meat itself. Species and cuts used in centrifugal force deboning machines are listed in Table 3.

Mechanical Characteristics

In most apparatus in current use, a motor is mounted on a frame, connected by pulleys and a belt to a rod caused to rotate by the motor. Attached to the rod is a piece holder, to which poultry pieces are attached. An enclosure prevents pieces of meat from flying out in all directions. When the motor is energized, the rod and piece holder begin to spin with a controlled acceleration. For previously cooked poultry pieces, the thigh meat will separate from the bone when the speed of rotation is about 400 rpm; the leg meat will separate when the speed of rotation is about 900 rpm; any residual meat will separate at about 1100 rpm. Separation occurs almost instantaneously; there is no need to maintain a high speed of rotation for an appreciable length of time.

A variable resistance can be used to control the speed and acceleration of the motor, thereby regulating the amount and rate of increase of the centrifugal force. It is desirable to accelerate the holder at rates in the range 1 to 40 rad/s. Controlled acceleration tends to prevent portions of a poultry piece from snapping off or shattering before having been deboned by the apparatus. If the poultry pieces have not been cooked, higher speeds of rotation are necessary to accomplish the same result. In the latter case it would be necessary to provide a motor that can generate speeds of rotation of up to about 4000 rpm. Whether or not the meat is precooked, separation occurs without the use of water or other fluids, and the meat separates in neat chunks. Separation is due solely to centrifugal forces, which pull substantially uniformly on all portions of each piece. Thus, there is little or no shredding of the meat. The enclosure prevents pieces of meat from being scattered and permits the separated chunks to drop into a pan or other collection device below. The bones remaining on the holder would then be removed, a new set of poultry pieces attached, and the process repeated. For best results, the tendons of leg–thigh poultry piece are cut and the skin is removed before the piece is placed in the apparatus. If a breast section is to be deboned, it is desirable to cut the section along the keel bone and the wishbone and remove the skin.

This apparatus has the disadvantage that only one set of poultry pieces can be processed at a time. However, operation is continuous, as many of the necessary steps are performed automatically.

REFERENCES

Beck NR, Leonard GC, Prince JA, inventors; Meat Separator Corp., assignee. 1977, Aug. 16. Deboning apparatus and method. U.S. patent 4,042,176.

Crosland AR, Patterson RLS, Rosmary CH. 1995. Investigation of methods to detect mechanically recovered meat in meat products: I. Chemical composition. Meat Sci 40:289–302.

Day L, Brown H. 2001. Detection of mechanically recovered chicken meat using capillary gel electrophoresis. Meat Sci 57:31–37.

Hazenbroek JE, inventor; Hazenbroek JE, assignee. 1992, Dec. 22. Thigh deboner with tray conveyor. U.S. patent 5,173,076.

Helmer WD, Small RE, inventors; Campbell Soup, assignee. 1980, Jul. 22. Mechanical deboning of poultry. U.S. patent 4,213,229.

Heuvel V, Johannes D, inventors; Mayn Maschf., assignee. 1989, Mar. 14. Apparatus for separating bone and flesh of the legs of poultry or part therof. U.S. patent 4,811,456.

Jones JM. 1986. Review: application of science and technology to poultry meat processing. J Food Technol 21:663–681.

Kaplan JJ, Richert WH, Trumblee JW, inventors, Jo Bi Farms Inc., assignee. 1974, Oct. 17. Poultry leg boning machine. U.S. patent 3,965,535.

Mielnik MB, Aaby K, Rolfen K, Ellekjaer M, Nilsson A. 2002. Quality of comminute sausage formulated from mechanically deboned poultry meat. Meat Sci 61:73–84.

Navarro C. 2008. Carne de ave mecánicamente recuperada. In: Fernández-López J, Pérez-Álvarez JA, Sayas ME, eds., *Industrialziación de Productos de Origen Animal*, vol. 1. Elche, Spain: Universidad Miguel Hernandez, pp. 268–275.

Perlo F, Bonato P, Teira G, Fabre R, Kueider S. 2006. Physicochemical and sensory properties of chicken nuggets with washed mechanically deboned chicken meat: research note. Meat Sci 72:785–788.

Pizzocaro F, Senesi E, Veronese P, Gasparoli A. 1998. Mechanically deboned poultry meat hamburgers: Note 1. Effect of processing and frozen storage on lipid stability. Ind Aliment 37(369):449–454.

Poss W, inventor; Poss Desing Ltd., assignee. 1987, Jan. 1. Apparatus for the separation of mixtures of materials of different consistencies, such as meat and bone. U.S. patent 4,638,954.

Richburg JB, inventor; Richburg JB, assignee. 1992, Mar. 11. Meat deboning apparatus and method. U.S. patent 5,160,290.

Sadat T, Volle C. 2000. Integration of a linear accelerator into a production line of mechanically deboned separated poultry meat. Radiat Phys Chem 57:613–617.

Steele FM, Huber CS, Orme L, Pike O. 1991. Textural qualities of turkey frankfurters incorporating fish and turkey based surimi. Poult Sci 70:1434–1437.

Yarem J, Poss W, inventors; Chemtrom Corp., assignee. 1978, Jan. 24. Process and apparatus for mechanical separation of a combination of meat and bone into useful fractions. U.S. patent 4,069,980.

Zartman WA, inventor. Meat Separator Corp., assignee. 1962, Jan. 23. Method for separation of meat and bone. U.S. patent 3,017,661.

7

MECHANICAL DEBONING: APPLICATIONS AND PRODUCT TYPES

CASILDA NAVARRO-RODRÍGUEZ DE VERA, ELENA JOSÉ SÁNCHEZ-ZAPATA, MANUEL VIUDA-MARTOS, AND JOSÉ ANGEL PÉREZ-ALVAREZ
Grupo Industrialización de Productos de Origen Animal (IPOA Research Group), Departamento de Tecnología Agroalimentaria, Escuela Politécnica Superior de Orihuela, Universidad Miguel Hernández, Orihuela, Alicante, Spain

Handbook of Poultry Science and Technology, Volume 2: Secondary Processing, Edited by Isabel Guerrero-Legarreta and Y.H. Hui

INTRODUCTION

The cost of producing meat products has encouraged the industry to optimize the use of all available protein sources, including mechanically deboned chicken meat and carcass trimmings from duck, turkey, and ostriches (Abdullah and Al-Najdawi, 2005). Yields of mechanically deboned poultry meat range from 55 to 80%, depending on the part deboned and the deboner settings (Serdaroglu et al., 2005). This technology allows meat packers and processors to recover fragments of meat that are left on the bones of carcasses after hand trimming (McNeil, 1980; Navarro, 2007). Recovery of meat from bones of filleted fish began in Japan in the late 1940s and increased as the amount of filleted fish increased. Mechanical recovery of poultry from necks, backs, and other bones with meat attached began in the late 1950s (Field, 2004).

Definition

Mechanically deboned meat (MDPM) and *mechanically separated meat* are terms used to describe meat that remains on the bones of a carcass after hand trimming, that is removed using deboning machines (Serdaroglu et al., 2005). In the mechanical deboning process, bones and attached meat are ground up and fed into special deboning machines. The bone bits are screened out while the meat passes through (McNeil, 1980). This permits the recovery of most of the residual meat, which would otherwise be difficult or uneconomical to acquire (Serdaroglu et al., 2005). Inevitably, crushing of the material leads to changes in the chemical, physical, sensory, and functional properties of the meat (Abdullah and Al-Najdawi, 2005). The resulting MDPM has the appearance of finally comminuted meat, a raw material with excellent nutritional and functional properties used for the formulation of a wide variety of emulsified meat products (Serdaroglu et al., 2005).

The deboning process can be applied to whole poultry carcasses, parts of carcasses, necks, backs, and in particular, to residual meat left on the bones after the completion of manual deboning operations. It is also a way to use whole spent layer hens (Barbut, 2002; Abdullah and Al-Najdawi, 2005). The recovery machines used to separate the residual meat from the bone can be divided into two main types: those that exert pressure to force the meat to flow from bones by means of a hydraulically powered piston and those that use an auger feed (Crosland et al., 1995). The most popular mechanical deboner for poultry is the auger type. In this system, bone is retained on the inside and augured out of the end of the cylinder to separate it from meat (Barbut, 2002; Field, 2004).

Regulations

The U.S. Department of Agriculture (USDA) began formulating a policy on the use of MDPM in November 1974 (McNeil, 1980). In 1995, the Food Safety and Inspection Service (USDA-FSIS) amended the regulations to prescribe a definition and standard of identity and composition for poultry products that result from mechanical separation of carcasses and parts (Field, 2004). In Europe, several council directives are related to mechanical deboned meat: Council Directives

71/118/CEE, 77/99/EEC, 91/495/CEE and 94/65/EC. One statement is of particular importance: "Member States shall ensure that mechanically recovered meat may be traded only if it has previously undergone heat treatment" (EC, 1994).

PROXIMATE COMPOSITION OF MECHANICALLY DEBONED POULTRY MEAT

Proteins, Bone, and Skin Content

The chemical composition of MDPM is subjected to an extremely wide variability and depends on species, breed, and age of the animal as well as the carcass parts used, the proportion of bone and fat in the material being deboned, the degree of trimming, and the machine type and setting (Henckel et al., 2004; Field, 2004; Navarro, 2005). The protein content in MDPM varies from 11.4 to 20.4% depending on the material being deboned (e.g., backs, necks, spent hens). As the protein content increases, the fat decreases from 24.7% to 7.5%. Skin-on poultry always contains more fat before and after mechanical recovery than poultry with the skin removed because most fat in poultry is associated with the skin (Field, 2004).

Effects of Deboning Methods on Chemical Composition

Mechanical deboning results in cellular disruption, protein denaturation, and increased lipid and heme oxidation. Storage conditions are therefore very important in this type of meat (Serdaroglu et al., 2005). The proximate composition of deboned meat showed that MDPM had higher fat and lower moisture and protein contents than did hand deboned meat (HDPM). Ash content was significantly higher in MDPM also. Collagen content was significantly higher in MDPM than in HDPM (Al-Najdawi and Abdullah, 2002).

Calcium content is an indicator of the amount of bone in meat. The higher calcium content of mechanically deboned samples arises from the higher content of bone particles in MDPM (Serdaroglu et al., 2005). By USDA regulation, calcium must not exceed 0.235% in mechanically recovered meat from mature poultry, or 0.175% when it is made from other poultry. These calcium values are intended to limit bone content to 1%. Iron in MDPM is higher than in HDPM, reflecting the incorporation of bone marrow (Field, 2004).

As a result of the inclusion of bone marrow in MDPM, there is a great variation in the fatty acid content and a higher percentage of cholesterol and phospholipid in MDPM. Mechanical deboning resulted in increased lipid concentration because of the high fat content of bone marrow. Therefore, higher pH values for mechanically deboned meat than for hand deboned meat can be attributed to the presence of marrow in the product. Bone marrow, fat, and skin are the factors that affect the cholesterol level in MDPM (Al-Najdawi and Abdullah, 2002; Serdaroglu et al., 2005). Sodium, aluminum, potassium, and magnesium contents show no differences between MDPM and HDPM, whereas significantly more calcium, manganese, and zinc are found in MDPM than in HDPM (Al-Najdawi and Abdullah, 2002).

Microbiological Aspects

The microbiology of MDPM depends on the microflora that is on the raw meat, the type of preservation process used, and the storage conditions (Guerrero and Pérez-Chabela, 2004). Contamination of meats with pathogenic microorganisms continues to be a major health problem. MDPM is usually contaminated with microorganisms, which are introduced during HDPM and MDPM processing, This makes MDPM highly perishable. The small particle size, the release of cellular fluids due to tissue maceration, and the heat generated during mechanical deboning all enhance bacterial growth Therefore, MDPM has a short shelf life even under refrigeration (Yuste et al., 1998). Frozen storage is commonly used, for these reasons.

New technologies such as high hydrostatic pressure (HHP) may also help. This treatment inactivates microorganisms and certain enzymes and makes it possible to increase the shelf life of foods. Yuste et al., (1998) reported on the microbiological quality of recovered poultry meat treated with HHP and nisin. The nonthermal character of this technology provides a unique opportunity for the development of novel foods of superior nutritional and sensory quality. However, pressurization can induce lipid oxidation, resulting in certain deterioration of stored meat and meat products with MDPM as an ingredient (Tuboly et al., 2003).

Because there are microbiological concerns related to the production and use of mechanically recovered poultry, some countries have adopted specific regulations relating to the source of bones, the anatomical regions from which they come, and the temperature and time under which they can be held prior to mechanical recovery of lean meat. Handling mechanically recovered products, including room temperature, chilling, and freezing, and their use have also been specified (Field, 2004).

FUNCTIONAL AND SENSORY PROPERTIES OF MECHANICALLY DEBONED POULTRY MEAT

MDPM is widely used in the manufacture of emulsioned meat products. An important property of the raw material used in this type of product is the emulsifying capacity (EC). Another important property of meat used for product manufacture is the water-holding capacity (WHC). The skin content of deboned meat is a factor affecting the EC and gelling properties (Abdullah and Al-Najdawi, 2005; Navarro, 2005).

Meat color, one of the most important meat quality characteristics, has a strong influence on consumer acceptance of retail products. Mechanical deboning results in the release of significant amounts of heme pigments from the bone marrow, which become a part of the MDPM. Higher pigment levels in MDPM can be attributed to the release of bone marrow during the deboning process. Any meat product containing bone marrow varies in WHC, color, flavor, pH, and EC (Field, 1999; Abdullah and Al-Najdawi, 2005). The texture is finely comminuted or pasty, the color is redder, reflecting the incorporation of heme pigments from blood and marrow, and the flavour changes relate to increased oxidation of fat, resulting in a shorter storage life (Field, 2004).

STORAGE CONDITIONS OF MECHANICALLY DEBONED POULTRY MEAT

MDPM is a raw material with limited processability due to its disadvantages: chemical composition, lowered functionality, highly comminuted structure, and limited storage and microbiological stability. Through this, frozen storage is frequently used as a preservation method for this type of meat. To maintain the highest possible processability, cryoprotectants are added to MDPM prior to freezing (Stangieriski and Kijowski, 2003). The cryoprotectants include all compounds that aid in preventing changes induced in foods by freezing, frozen storage, or thawing. Cryoprotectants stabilize the myofibrillar proteins. Sugars, amino acids, polyls, methyl amines, polymers, and even inorganic salts are considered cryoprotectants (Park, 2000).

Testing results show that it is possible to reduce the effect of freezing and frozen storage on the functional properties of myofibrillar protein by the addition of cryoprotectants. Cryoprotection prevents protein dehydration. In meat and meat products, the undesirable freezing changes may be associated with changes in the physicochemical properties of myofibrillar proteins, whose principal representative is myosin.

Although frozen storage prevents many undesirable changes in meat, significant deterioration of some functional properties and some oxidative reactions that affect product quality adversely still occur (Stangieriski and Kijowski, 2003). Lipid oxidation is considered a major problem in the meat industry, due to the resulting deterioration in flavor and loss of nutritional value. MDPM is highly susceptible to oxidative deterioration, due to extensive stress and aeration during the machine deboning process and the compositional nature (bone marrow, heme, and lipids) of the product (Mielnik et al., 2003; Pettersen et al., 2004).

The major strategies for preventing lipid oxidation in MDPM are the use of antioxidants (Püssa et al., 2008) and restricting the access of oxygen during storage (Pettersen et al., 2004). The storage life of mechanically deboned meat can be extended through the use of antioxidants, compounds capable of delaying, retarding, or preventing autooxidation processes of MDPM. However, antioxidants cannot reverse the oxidation process or suppress the development of hydrolytic rancidity (Mielnik et al., 2003).

APPLICATIONS OF MECHANICALLY DEBONED POULTRY MEAT

Mechanically recovered poultry meat is a processing material with rather limited technological processability. In most countries, meat recovered from bones or carcass parts by mechanical procedures is generally considered to be of poor quality and has therefore been subjected to strict regulations concerning use of the product as a binding agent or as a source of meat in minced meat products The implications of a general ban would be that huge amounts of potentially high-quality meat would have to be discarded or processed into pet food (Rivera et al., 2000; Henckel et al., 2004).

MDPM is frequently used in the formulation of comminuted meat products, due to its fine consistency. The main applications of MDPM are in products such as sausage and salami, which do not require a fibrous texture but demand emulsion stability and also benefit from MDPM's natural color and relatively low cost (Barbut, 2002; Guerra Daros et al., 2005). The use of mechanically deboned poultry meat in the formulation of sausage is considered recent in the food industry since it did not begin until the 1960s, when there was a strong tendency to replace red meat with healthier white meat in Western industrialized countries and due to the lower price of the latter compared with other types of meat (Guerra Daros et al., 2005).

The composition and storage stability of the final product is affected by the raw materials and conditions used in mechanical separation (Serdaroglu et al., 2005). Mechanically separated poultry is used in sausages such as frankfurters and bologna, chicken patties, nuggets, baby foods, chilli, hamburgers, and poultry rolls. In some countries, sausages have mechanically separated chicken or turkey along with beef and pork as ingredients (Schnell et al., 1973; Pizzocaro et al., 1998; Kolsarici and Candogan, 2002; Mielnik et al., 2002; Field, 2004).

To provide new alternatives for the poultry industry in its use of spent layer hens, the sensory acceptance and storage stability of mortadella made with different levels of MDPM were evaluated. Another end use for spent layer hens is the production of mechanically separated poultry for the sausage industry. Sausages formulated with up to 50% mechanically separated chicken meat presented good consistency and texture, but higher percentages of MDPM resulted in excessively soft products (Trindade et al., 2005). Trindade et al. (2006) recommended previous blending of nitrite and erythorbate in MDPM soon after extraction when the raw material will eventually go through prolonged frozen storage before its use in mortadella sausage processing, because the chemical and structural alterations during mechanical separation cause MDPM have low stability during storage.

Surimi technology is also used in MDPM processing. Meat washing produced an extraction of lipids and pigments of the MDPM (Dawson et al., 1988; Lin and Chen, 1989; Smyth and O'Neil, 1997; Navarro, 2005). Washed MDPM could be incorporated into nugget formulation as a substitute for hand deboned chicken meat without affecting sensory attributes of the product (Perlo et al., 2006). Production of protein preparations using the technology of fish surimi in MDPM is also employed (Stangierski and Kijowski, 2007).

REFERENCES

Abdullah B, Al-Najdawi R. 2005. Functional and sensory properties of chicken meat from spent-hen carcasses deboned manually or mechanically in Jordan. Int J Food Sci Technol 40:537–543.

Al-Najdawi R, Abdullah B. 2002. Proximate composition, selected minerals, cholesterol content and lipid oxidation of mechanically and hand deboned chickens from the Jordanian market. Meat Sci 61:243–247.

Barbut S. 2002. *Poultry Products Processing: An Industry Guide*. Boca Raton, FL: CRC Press.

Crosland AR, Patterson RLS, Higman RC, Stewart CA, Hargin KD. 1995. Investigation of methods to detect mechanically recovered meat in meat products: I. Chemical composition. Meat Sci 40:289–302.

Dawson PL, Sheldon BW, Ball JR. 1988. Extraction of lipid and pigment component from mechanically deboned chicken meat. J Food Sci 53(6):1615–1617.

EC (European Commission). 1994. Directive 1994/65/EC of Dec. 14. Laying down the requirements for the production and placing on the market of minced meat and meat preparations.

Field RA. 1999. Bone marrow measurement for mechanically deboned recovered products from machines that press bones. Meat Sci 51:205–214.

Field RA. 2004. Mechanically recovered meat. In: *Encyclopedia of Meat Sciences*. San Diego, CA: Academic Press.

Guerra Daros F, Masson ML, Campos Amico S. 2005. The influence of the addition of mechanically deboned poultry meat on the rheological properties of sausage. J Food Eng 68:185–189.

Guerrero I, Pérez-Chabela L. 2004. Spoilage of cooked meats and meat products. In: *Encyclopedia of Meat Sciences*. San Diego, CA: Academic Press.

Henckel P, Vyberg M, Thodec S, Hermansen S. 2004. Assessing the quality of mechanically and manually recovered chicken meat. Lebensm-Wiss Technol 37:593–601.

Kolsarici N, Candogan K. 2002. Quality characteristics and uses of mechanically deboned meat. Gida 27(4):277–283.

Li CT, Wick M. 2001. Improvement of the physicochemical properties of pale soft and exudative (PSE) pork meat products with an extract from mechanically deboned turkey meat, (MDTM). Meat Sci 58:189–195.

Lin SW, Chen TC. 1989. Yields, colour and composition of washed, kneaded and heated mechanically debonded poultry meat. J Food Sci 54:561–563.

MacDonald GA, Lanier TC, Carvajal PA. 2000. Stabilization of proteins in surimi. In: Park JW, ed., *Surimi and Surimi Seafood*. New York: Marcel Dekker, pp. 91–126.

McNeil DW. 1980. Economic welfare and food safety regulation: the case of mechanically deboned meat. Am J Agric Econ 62(1):1–9.

Mielnik MB, Aaby K, Rolfsen K, Ellerkjaer MR, Nilsson A. 2002. Quality of comminuted sausages formulated from mechanically deboned poultry meat. Meat Sci 61:73–84.

Mielnik MB, Aaby K, Skrede G. 2003 Commercial antioxidants control lipid oxidation in mechanically deboned turkey meat. Meat Sci 65:1147–1155.

Navarro C. 2005. Optimización del procesos de obtención de geles cárnicos partir de carne de ave mecánicamente recuperada. Ph.D. dissertation, Universidad Miguel Hernández, Elche, Spain.

Navarro C. 2007. Carne de ave mecánicamente recuperada. In: Fernández-López J, Pérez-Álvarez JA, Sayas ME, eds., *Industrialización de Productos de Origen Animal*. Elche, Spain: Universidad Miguel Hernández, pp. 87–96.

Park JW. 2000. Ingredient technology and formulation develpment. In: Park JW, ed., *Surimi and Surimi Seafood*. New York: Marcel Dekker, pp. 343–392.

Perlo F, Bonato P, Teira G, Fabre R, Kueider S. 2006. Physicochemical and sensory properties of chicken nuggets with washed mechanically deboned chicken meat: research note. Meat Sci 72:785–788.

Pettersen MK, Mielnik MB, Eie T, Skrede G, Nilsson A. 2004. Lipid oxidation in frozen, mechanically deboned turkey meat as affected by packaging parameters and storage conditions. Poult Sci. 83:1240–1248.

Pizzocaro F, Senesi E, Veronese P, Gasparoli A. 1998. Mechanically deboned poultry meat hamburgers: II. Protective and antioxidant effect of the carrot and spinach tissues during frozen storage. Ind Aliment 37(371):710–720.

Püssa T, Pällin R, Raudsepp P, Soidla R, Rei M. 2008. Inhibition of lipid oxidation and dynamics of polyphenol content in mechanically deboned meat supplemented with sea buckthorn (*Hippophae rhamnoides*) berry residues. Food Chem 107:714–721.

Rivera JA, Sebranek JG, Rust RE. 2000. Functional properties of meat by-products and mechanically separated chicken (MSC) in high-moisture model petfood system. Meat Sci 55:61–66.

Schnell PG, Nath KR, Darfler JM, Vadehra DV, Baker RC. 1973. Physical and functional properties of mechanically deboned poultry meat as used in the manufacture of frankfurters. Poult Sci 52:1363–1369.

Serdaroglu M, Yildiz Turp GL, Baúdatlioglu N. 2005. Effects of deboning methods on chemical composition and some properties of beef and turkey meat. Turk J Vet Anim Sci 29:797–802.

Smyth AB, O'Neil E. 1997. Heat-induced gelation properties of surimi from mechanically separated chicken. J Food Sci 62(2):326–330.

Stangierski J, Kijowski J. 2003. Effect of selected commercial substances with cryoprotective activity on the quality of mechanically recovered, washed and frozen stored poultry meat. Nahrung Food 47(1):49–53.

Stangierski J, Kijowski J. 2007. Effect of selected substances on the properties of frozen myofibril preparation obtained from mechanically recovered poultry meat. Eur Food Res Technol. (doi 10.1007/s00217-007-0672-2).

Trindade MA, Contreras Castillo CJ, De Felíci PE. 2005. Mortadella sausage formulations with partial and total replacement of beef and pork backfat with mechanically separated meat from spent layer hens. J Food Sci 70(3):S236–S241.

Trindade MA, Contreras Castillo CJ, De Felíci PE. 2006. Mortadella sausage formulations with mechanically separated layer hen meat preblended with antioxidants. Sci Agric 63(3):240–245.

Tuboly E, Lebovics VK, Gaal O, Meszaros L, Farkas J. 2003. Microbiological and lipid oxidation studies on mechanically deboned turkey meat treated by high hydrostatic pressure. J Food Eng 56:241–244.

Wimmer MP, Sebranek JG, McKeith FK. 1993. Washed mechanically separated pork as surimi-like meat product ingredient. J Food Sci 58(2):254–258.

Yuste J, Mor-Mur M, Capellas M, Guamis B, Pla R. 1998. Microbiological quality of mechanically recovered poultry meat treated with high hydrostatic pressure and nisin. Food Microbiol 15:407–414.

8

MARINATION, COOKING, AND CURING: PRINCIPLES

FRANCISCO ALFREDO NUÑEZ-GONZÁLEZ

Facultad de Zootecnia, Universidad Autónoma de Chihuahua, Chihuahua, México

INTRODUCTION

The market trend today is toward increased demand for novelty poultry products with excellent sensory characteristics, particularly tenderness and juiciness; however, these characteristics are the most variable in poultry or meat. It must be kept in mind that tenderness is a sensory characteristic of immense economic value

Handbook of Poultry Science and Technology, Volume 2: Secondary Processing, Edited by Isabel Guerrero-Legarreta and Y.H. Hui

for the meat industry. This is because consumers usually regard tender meat to be of the top quality. As meat tenderness depends on many stress factors, ranging from animal transport, genetics, nutrition, and pre- and postslaughter handling, marination has been developed to reduce meat toughness, improving tenderness as well as other meat attributes, producing improved or standardized meat. Therefore, poultry processors have been steadily incorporating this technique in their plants to increase product yield and increase product uniformity. Marinating is a traditional process in which meat is immersed in a solution that extends its shelf life and imparts a specific flavor characteristic. More recently, marination has been used in meat and poultry to improve tenderness, juiciness, flavor, color, and cooking yield.

PHYSICOCHEMICAL BASIS FOR POULTRY MARINATION

The basis of marination is the ability of animal muscles to bind water when transformed into meat. Myofibrillar, or structural, proteins forming the actomyosin complex are mainly responsible for water retention; this is achieved through the *steric effect*, caused by the opening of spaces existing between the myosin and actin filaments of the muscular structure and by the degree of reactivity of these proteins (Aberle et al., 2001; Lawrie and Ledward, 2006). Not all muscle proteins retain water to the same extent; globular sarcoplasmic proteins are soluble in water and have a very low water retention capacity, whereas myofibrillar proteins, actin and myosin and the actomyosin complex, are fibrous proteins, soluble in salt solutions and with a high water-holding capacity (WHC). This attribute has an important effect on meat texture; on the other hand, muscle water retention is affected by rigor mortis development and final meat pH and temperature (Offer and Knight, 1988). Texture, color, juiciness, and flavor are meat physicochemical characteristics highly dependent on WHC (Aberle et al., 2001). After an animal's death, the muscle produces lactic acid as a residual anaerobic glycolysis to maintain homeostasis; free hydrogen ions accumulate, resulting in a progressive pH reduction until reaching the ultimate pH. Thus, among other factors, meat quality is directly related to final pH. Reactive muscle proteins reach their isoelectric point, only a few charged groups are available for water binding (net charge effect) (Lawrie and Ledward, 2006). As meat pH shifts out of the protein isoelectric point, the amount of immobilized water increases, due to an increase in reactive groups, binding free water. Water retention is also affected by the reduction in spaces between myofibrils during rigor mortis onset and development, causing a steric effect (Offer and Trinick, 1983).

TYPES OF MARINATION

In the meat industry, *marination* is defined as incorporation and retention of water by meat and poultry, to improve tenderness, juiciness, color, and flavor, by adding an alkaline solution, pH 7.5 or above, containing mainly sodium chloride and phosphates, although other additives are also used. Cuisine marination is mainly acidic and is used to improve or change flavor. This marinating process is accomplished by the immersion of meat and poultry in an acid solution. Traditionally, this type of marinade was used to preserve meat; later, it was also used to improve juiciness, odor, and flavor or to develop distinctive sensory characteristics pleasant for a given culture. These traditional marinades are solutions containing oils, vinegar, wine, fruit juice (e.g., lime, orange, grapefruit, lemon), vegetal enzymes, herbs, and spices. Acid and enzymes in the marinating solution cause the meat structure to loosen, allowing water absorption and producing a tenderer and juicier product, although an unpleasant texture can also develop.

MARINADE ABSORPTION SYSTEMS

Industrial marination processing systems consist of brine injection and tumbling; cuisine marinating is carried out only by meat immersion in the marinade solution (Toledo, 2001). Immersion marinating is not commonly used at the industrial level, as diffusion of marinade components by meat is a time-consuming operation; distribution time is extended considerably, and microbial contamination risks increases (Xargayó et al., 2001). Marination by injection is a quick process, achieved by forcing marinade solution into meat by means of one or a number of needle devices. The amount and concentration of solution is regulated by government agencies, such as the U.S. Department of Agriculture's Food Safety and Inspection Service (USDA, 1999a). The method used most commonly is multiple-needle marination; it provides an exact amount of solution to the meat and ensures the replication of processing conditions (Sams, 2003). The marinade can be as simple as water, salt, and phosphates; a typical formulation consists of 92% water, 7% salt, and 1% phosphate. Breast fillets injected with a 10% marinade solution contain 0.3% salt, 0.35% phosphate, and 0.05% herb extract (Brashears et al., 2007).

Tumbling or massaging marination is a method usually carried out in processing plants or retail facilities. The process can be operated at ambient, positive, or vacuum pressure, although optimum conditions are not fully documented (Young et al., 2005). According to Smith and Young (2005), the pressure applied during tumbling has no effect on drip loss, cook yield, or color, and only a minor effect on marinade retention. Vacuum tumble marination is the most widely practiced process for poultry marinating in the United States. Ten percent marinade contains 4% sodium tripolyphosphate and 15% NaCl. Use of this type of marination has greatly increased in the United States in the past few years, especially in the poultry industry, due to its ability to enhance breast fillet tenderness, yield, and product acceptability.

MAIN FUNCTIONAL INGREDIENTS FOR MARINATION AT THE INDUSTRIAL LEVEL

The main objective of marination is to disperse the functional ingredients in the marinade uniformly throughout the muscle (Toledo, 2001). The basic marinade ingredients are sodium chloride and phosphates.

Sodium Chloride

Sodium chloride has long been used by humans in a variety of products to modify shelf life and enhance flavor. In solution, it dissociates into Na^+ and Cl^- ions, increasing the ionic strength (Medynsky et al., 2000). The molecule contains 40% sodium and when added to commercial poultry marinades (Gillette, 1985), enhances flavor, extracts salt-soluble proteins, increases water-holding capacity, and reduces drip and cooking losses (Juncá et al., 2004). Offer and Trinick (1983) reported that water uptake by muscles in the presence of high-ionic-strength salt solutions is caused by expansion of the actin and myosin filament network due to an increase of the negative charges in myofibril components, increasing their repulsive force. NaCl then depolymerizes the myosin filament by weakening the interaction between the heavy meromyosin tails and exposing charged group, binding the water molecules (Offer and Knight, 1988). Xargayó et al. (2001) explained that meat proteins swell to double their size in the presence of salt concentrations used in meat processing, the ion exchange in the meat results in improved water binding Sodium chloride action results in protein isoelectric point reduction. Meat binding properties are improved by increasing the myofibrillar or salt-soluble protein solubility (Babji et al., 1982). As NaCl increases myofibril ionic strength, meat is tenderized, due to muscle protein solubilization (Wu and Smith, 1987).

Phosphates

Consumer demands for NaCl reduction have increased the use of phosphates in meat products (Trout and Schmidt, 1983). Food-grade phosphates perform a number of functions in processed poultry products, including pH and ionic strength increase and myofibrillar protein modification to improve the water-holding capacity and to reduce drip and cooking loss. Phosphates also protect lipids from oxidizing by binding metal ions that accelerate oxidation; they protect and preserve the natural flavor and color of poultry, and they solubilize the proteins necessary to fabricate formed and restructured products. The *Code of Federal Regulations* (USDA, 1999b) limits the use of sodium phosphates in poultry products to a maximum of 0.50% of the finished product weight. Sodium tripolyphosphate (STPP) is the form of phosphate used most commonly in the poultry industry (Zheng et al., 2000). The addition of STPP and sodium chloride (NaCl) in poultry marinade formulations has several advantages. Polyphosphates and NaCl are synergists in increasing ionic strength and extracting myofibrillar proteins; they improve the texture properties of poultry products; and they increase pH and complex with magnesium and calcium bound to the proteins, allowing actomyosin dissociation and exposing water bonding sites, hence increasing the water-holding capacity (Froning and Sackett, 1985; Wong, 1989). STPP also retards lipid oxidation, maintaining flavor characteristics. By the effect of phosphate salts on meat proteins, muscle fibers swell and water is retained within the muscle protein network (Offer and Trinick, 1983). It has been reported that poultry toughness is reduced significantly by STPP injection (Zheng et al., 2000). Tetrasodium pyrophosphate (TSPP) and sodium acid pyrophosphate (SAPP) are the most rapidly acting phosphates in solubilizing myofibrillar protein (Ebert, 2007). STPP addition to industrial marinades increases the poultry isoelectric point 0.1 to 0.3 pH unit (approximately 5.2) (Xiong, 1999; Xiong and Kupski, 1999). Therefore, alkaline phosphates are used in poultry marinades to control pH, increase water-holding capacity and tenderness, and decrease lipid oxidation (Dziezak, 1990). In addition, the use of phosphates in combination with NaCl results in an NaCl level of 1.0 to 0.5% (Ebert, 2007). Production yields can be improved by phosphate addition to the marinade. The quality of PSE (pale, soft, and exudative) chicken breast fillets was improved when marinated prerigor with phosphate solutions; prerigor marination with NaCl and sodium phosphates (pH 9) reduced cooking losses compared to those in normal breast fillets (Alvarado and Sams, 2003). The use of high-pH phosphate marinades and sodium bicarbonate increases pH and improves the water-holding capacity in PSE meat (Woelfel and Sams, 2001).

WATER QUALITY

Water quality can alter marinade performance due to hardness (Ebert, 2007). The solubility of alkaline pyrophosphates in water is limited; therefore, these salts are usually mixed with more-soluble long-chain phosphates, such as STPP and SHMP, which are then hydrolyzed in the meat to the active pyrophosphate form. This is a strong metal ion sequestrant, binding calcium and magnesium, the minerals responsible for water hardness.

MEAT AND POULTRY MARINATION AND HUMAN HEALTH

Salt or sodium chloride in poultry marinades can increase the sodium content of meat and poultry products. An intake of sodium in excess leads to an increase in blood pressure in salt-sensitive persons. Blood pressure increases to allow the kidneys to excrete excess sodium to maintain homeostasis (Preuss, 1997). High blood pressure, or hypertension, is the most common chronic disease in the United States, leading to heart conditions and kidney failure. The average adult requires less than 1.0 g of salt, or approximately 400 mg of sodium, to maintain a proper sodium balance (Haddy and Pamnani, 1995). *Dietary Guidelines for Americans* (DHHS, 2005) recommends decreasing salt consumption to prevent hypertension; 2.3 g of salt is recommended as the level of daily consumption. Thus, if not labeled properly (USDA, 1999a,b), "enhanced" poultry products may lead to excessive sodium intake. The labeling term *marinated* can be used only with specific solution amounts. "Marinated" meats can contain no more than 10% solution; boneless poultry no more than 8% solution; and bone-in poultry no more than 3% solution. For this reason, there are concerns of increased unsafe sodium intake in consumer diets, although it is possibly to marinate with lower salt concentration while still improving meat quality characteristics and keeping ingredients cost low (Owens, 2007). To produce enhanced poultry products with reduced sodium, sodium potassium tripolyphosphate (SKTP), potassium tripolyphosphate (KTPP), or tetrapotassium pyrophosphate (TKPP), marinades must be tested due to the astringent flavor caused by the potassium ion (Ebert, 2007).

Another problem associated with industrial marination is possible pathogen microorganism contamination. As marination procedures are applied to improve marinade penetration into muscle, they can also increase bacterial penetration (Ryser, 2003). Finally, several research works have shown that a reaction between creatine and amino acids caused by flame-cooking at high temperature, such as grilling or roasting, produces heterocyclic amines (HCAs), possible carcinogenic substances (Knize et al., 1995). Conversely, Salmon et al. (1997) reported that marinades can reduce the formation of certain HCAs. Marination by immersion in an acid-based marinade resulted in a decrease of 92 to 99% of heterocyclic amines (HCAs) in grilled meat. Further work is necessary on the effect of various marinade types in HCA production in grilled poultry. Acidic marinade, including olive oil, brown sugar, vinegar, lemon juice, garlic, salt, and mustard, caused an HCA reduction of $\frac{1}{10}$ or less (Nerurkar et al., 1999). Marinating is currently the best known method of reducing HCAs, although the mechanisms of marinade effects are not known.

CONCLUSIONS

The homogeneous quality of poultry and meat marination is troublesome, due to the number of variables involved in the processes, such as deboning times, time between deboning and marination, temperatures of solution and meat, water quality, and marination procedure, type, and concentration. It is possibly to marinate with lower concentrations of salt while still improving meat characteristics, keeping the ingredient cost low and reducing consumer concerns regarding sodium consumption. It is also possible to combine industrial marinades with antibacterial phosphates or organic acids to produce a safer poultry product.

REFERENCES

Aberle ED, Forrest JC, Gerrard DE, Mills W. 2001. *Principles of Meat Science*, 4th ed. Dubuque, IA: Kendall/Hunt.

Alvarado C, McKee S. 2007. Marination to improve functional properties and safety of poultry meat. J Appl Poult Res 16:113–120.

Alvarado CZ, Sams AR. 2003. Injection marination strategies for remediation of pale, soft, exudative broiler breast meat. Poult Sci 82:1332–1336.

Babji AS, Froning GW, Ngoka DA. 1982. The effect of short-term tumbling and salting on the quality of turkey breast muscle. Poult Sci 61:300–303.

Brashears MM, Brooks JC, Miller MF. 2007. Characterizing the safety and quality of fresh beef cuts subjected to deep muscle marination. Poult Sci 86(Suppl 1):200 (abstract).

DHHS (U.S. Department of Health and Human Services). 2005. *Dietary Guidelines for Americans*. Washington, DC: DHHS.

Dziezak JD. 1990. Phosphates improve many foods. Food Technol 44:80–92.

Ebert AG. 2007. Phosphates function in poultry products. http://www.foodproductdesign.com. Copyright 2008 by Virgo Publishing. Accessed Jan. 25, 2008.

Froning GW, Sackett B. 1985. Effect of salt and phosphates during tumbling of turkey breast muscle on meat characteristics. Poult Sci 64:1328–1333.

Gillette M. 1985. Flavor effects of sodium chloride. Food Technol 39:47–52, 56.

Haddy FJ, Pamnani MB. 1995. Role of dietary salt in hypertension. J Am Coll Nutr 14:428–438.

Juncá G, Borrell D, Fernandez E, Lagares J, Zargayó M. 2004. Mejoramiento de la textura de la carne en productos marinados empleando inyectora multiagujas espreadoras. *Mundo Lact Carn*, Sept., pp. 3–9. Accessed at Alimentaria online.com Jan. 16, 2008.

Knize MG, Salmon CP, Mehta SS, Felton JS. 1995. Heterocyclic amine content in fast food meat products. Food Chem Toxicol 33:545–541.

Lawrie RA, Ledward DA. 2006. *Lawrie's Meat Science*, 7th Engl. ed. Boca Raton, FL: CRC Press.

Medinsky A, Pospiech E, Keniat R. 2000. Effect of various concentrations of lactic acid and sodium chloride on selected physiochemical meat traits. Meat Sci 55:285–290.

Nerurkar PV, Le Merchand L, Conney RV. 1999. Effect of marinating with Asian marinades or Western barbecue sauce on PhIP and MeIQx formation in barbecued beef. Nutr Cancer 34:147–152.

Offer G, Knight DL. 1988. The structural basis of water holding in meat. In: Lawrie RA, ed., *Developments in Meat Science*, vol. 4. London: Elsevier Applied Science.

Offer G, Trinick J. 1983. On the mechanism of water holding in meat: the swelling and shrinking of myofibrils. Meat Sci 8:245–281.

Owens CM. 2007. Impact of marination and deboning time on poultry meat tenderness. Poult Sci 86(Suppl 1):200 (abstract).

Post RC, Heath JL. 1983. Marinating broiler parts: the use of a viscous type marinade. Poult Sci 62:977–984.

Preuss HG. 1997. Diet, genetics and hypertension. J Am Coll Nutr 16:296–305.

Ryser ET. 2003. Examine *Salmonella* penetration into turkey during vacuum marination. Microbial Update International, format online. Accessed Jan. 25, 2008.

Salmon CP, Knize MG, Felton JS. 1997. Effects of marinating on heterocyclic amine carcinogen formation in grilled chicken. Food Chem Toxicol 35:443–451.

Sams AR. 2003. Dr. marinade: marination is not just for flavor anymore: remedial marination can correct problems. WATT Poult USA, pp. 18–25.

Smith DP, Young LL. 2005. The effect of pressure and phosphates on yield, shear, and color of marinated broiler breast meat. In: *Proceedings of the 17th European Symposium on Quality of Poultry Meat*, Doorwerth, The Netherlands, pp. 139–144.

Toledo RT. 2001. Marinating techniques. Presented at Session 36, Industry Needs, New Ingredients Technology for Further Processed Meat Products, IFT 2001, New Orleans, LA.

Trout GR, Schmidt GR. 1983. Utilization of phosphates in meat products. In: *Proceedings of the 36th Annual Reciprocal Meat Conference*, American Meat Science Association, Fargo, ND, pp. 24–27.

USDA (U.S. Department of Agriculture). 1999a. FSIS Label Policy Memorandum 042: Ready-to-cook poultry products to which solutions are added. *Code of Federal Regulations*, Title 9, Chapter III, Part 381, Subpart P, § 381.169. Washington, DC: U.S. Government Printing Office.

USDA. 1999b. Restrictions on the use of substances in poultry products. *Code of Federal Regulations*, Title 9, Chapter III, Part 381, Subpart O, § 381.147. Washington, DC: U.S. Government Printing Office.

Woelfel RL, Sams AR. 2001. Marination performance of pale broiler breast meat. Poult Sci 80:1519–1522.

Wong DWS. 1989. Additives. In: *Mechanism and Theory in Food Chemistry*. New York: Van Nostrand Reinhold, p. 314.

Wu FY, Smith SB. 1987. Ionic strength and myofibrillar protein solubilization. J Anim Sci 65:597–608.

Xargayó M, Lagares J, Fernandez E, Ruiz D, Borrell D. 2001. Marination of fresh meats by means of spray effect: influence of spray on the quality of marinated products. http://www.metalquimia.com. Accessed Nov. 2007.

Xiong YL. 1999. Phosphate-mediated water uptake, swelling, and functionality of the myofibril architecture. In: *Quality Attributes of Muscle Foods*. New York: Springer. pp. 319–334.

Xiong YL, Kupski DR. 1999. Time-dependent marinade absorption and retention, cooking yield, and palatability of chicken fillets marinated with various phosphate solutions. Poult Sci 78:103–105.

Zheng M, Detienne NA, Barnes BW, Wicker L. 2000. Tenderness and yields of poultry breast are influenced by phosphate type and concentration of marinade. J Sci Food Agric 81:82–87.

9

MARINATION, COOKING, AND CURING: APPLICATIONS

ALMA DELIA ALARCÓN-ROJO

Facultad de Zootecnia, Universidad Autónoma de Chihuahua, Chihuahua, México

INTRODUCTION

Poultry meat consumption around the world is increasing steadily, due to its cost-effectiveness and nutritional quality. The increased demand for convenience foods has resulted in expansion of the processed meat and poultry industry. This has led to the development of further-processed poultry products that include such items as cut and deboned portions, battered pieces, breaded and precooked cold cuts, nuggets, or marinated portions (Baker and Bruce, 1989). The development of marination and injection technologies has contributed to product diversification at the retail level and will continue to grow. In today's market, meat is often sold free of bones, frozen, and premarinated to improve color, flavor, texture, storage characteristics, or other attributes (Young and Buhr, 2000). Most of these value-added products, formulated primarily to suit the local palate, not only target the

Handbook of Poultry Science and Technology, Volume 2: Secondary Processing, Edited by Isabel Guerrero-Legarreta and Y.H. Hui

changing needs of consumers (i.e., convenience, nutrition, health, quality, variety, shelf life), but also allow a marketing edge over imports (Bilgili, 2001). Today, almost all poultry products sold into food service are marinated.

Marination of meat has been an active topic of research because of its potential to extend the versatility of processed products (Baker and Bruce, 1989). Moreover, marination brings about changes in the structure of meat and can affect many properties, such as color, flavor, texture, storage characteristics, and other attributes (Young and Buhr, 2000). Marination originated in the Mediterranean region (*mare* is Latin for "sea"). The first marinades were probably salt solutions, or brines. A marinade is generally a seasoned liquid in which meat, fish, and poultry are soaked to become tenderized and/or to absorb flavor. Today, most marinades are composed of brines, acids, oils, and aromatic seasonings. Marination consists of the incorporation of ingredients such as salt, phosphates, and proteins in a water solution or applied dry. Whether they are injected, massaged, or vacuum tumbled, marinated products are becoming more prevalent on meat shelves everywhere. Application of a marinade to poultry enriches the flavor, increases moisture retention, improves tenderness (Aog and Young, 1987), inhibits warmed-over flavor (Mahon, 1963), and preserves color and flavor (Cassidy, 1977). According to the American Institute for Cancer Research, marination reduces the formation of cancer-causing heterocyclic amines in grilled meat (Castaldo, 2007).

INGREDIENTS

Marinades are generally included in 12 to 15% of breast fillets. Marination involves synergistic combinations of flavorful and functional ingredients that go far beyond salty soaking solutions. Primary ingredients of a typical marinade include salt, phosphate, and water, while other ingredients, such as flavors, seasonings, starches, vegetable or dairy proteins, acids, antimicrobials, and antioxidants, may also be added. Salt and phosphate help to improve water-holding capacity and the tenderness of pre- and postrigor deboned meat. Partially prepared or cooked products can be excessively dry, so food manufacturers often include moisture-retention aids such as sodium tripolyphosphate (STPP) in product formulation to improve the juiciness of partially prepared or cooked products and to improve water-holding capacity and yield (Hamm, 1960; Froning and Sackett, 1985; Young and Lyon, 1986; Lemos et al., 1999) due to their synergistic effect in increasing water binding by rising pH and ionic strength and dissociating actomyosin to expose more water-binding sites (Wong, 1989).

The extent of meat quality improvements depends on the type of phosphates used. Young et al. (1992) showed that the effects of STPP on the water-holding properties of poultry meat results primarily from the salt effect on the ionic strength of meat fluids, but pH affects the STPP efficacy. Phosphates vary in their solubility and effect on pH. Although diphosphates may be somewhat more difficult to solubilize, they react with the muscle enzymes quickly to help hold water in the system.

During rigor mortis, ATP breaks down into adenosine diphosphate (ADP) and monophosphate. All phosphates must break down to diphosphate to promote marinade retention and water binding. For example, even though both phosphates promote myofibril swelling and myosin extraction, pyrophosphates are a more soluble form of diphosphates and therefore are easier to use. Tetrasodium phosphates produce good binding ability because of their high pH (approximately 11), whereas sodium acid pyrophosphate decreases pH and therefore reduces water-holding capacity and yield. Short-chain phosphates such as orthosphoshates and pyrophosphates present the best buffering capacity (Van Wazer, 1971).

Actomyosin is a muscle protein formed by actin and myosin polymerization when the muscle contracts or when it goes into rigor mortis. Prevention of actomyosin formation occurs in the presence of salts and phosphates, which increase meat water-holding capacity (WHC). Meat fibers swell due to their marinade uptake, resulting in increased juiciness and tenderness and a reduction in cooking losses. In most countries, phosphate concentrations are limited to 0.5%.

Several reports indicate that some textural and color problems in finished cooked products can be attributed to the addition of polyphosphates. An excess in phosphate concentration produces a soapy-tasting marinade; another effect is a perceived dry mouthfeel, especially with tripolyphosphate (Brandt, 2001). A functional marination system includes adding ingredients that promote the capability of muscle to bind water, such as salt and phosphates, and those that actually bind water themselves, such as soluble proteins and starches.

Sodium chloride or salt, an important component of the meat marinade solution, helps to solubilize proteins to increase water-holding capacity, improves tenderness, and enhances flavor. Although salt is part of a traditional marination systems, salt levels in meat products are restricted, due to consumer concerns. Normal sodium chloride levels in marinated meat are 1.5 to 2%, and lower levels result in problems related to spoilage and shelf-life reduction.

Ionic strength is a key factor in marinade retention, and it is dependent on the molar concentration of the various ions in solution and their charges. Adding various ionic components, such as salts of phosphates, diacetates, and organic acids, can boost the total ionic strength and help maintain low levels of sodium chloride.

Flavor ingredients for marinades can include oleoresins, spices, hydrolyzed vegetable proteins, flavors, and yeast extracts. Spices or spice extracts can be incorporated into the marinate solution, depending on the application. Marinating chicken breasts with solutions containing 0.5% clove oleoresin (CLO) and peppermint leaf oil (PLO) by soaking or immersion for 1 h and by marinade injection inhibited pseudomona populations and reduced yeast counts significantly (with 0.2% PLO) even after one week of storage at 4°C, but none of these treatments showed significant activity against *Listeria monocytogenes* (Carlos and Harrison, 1999). As cooking temperature increases, flavor is more susceptible to degradation. Flavors can be added in microcapsules or as prepared bases in a variety of flavors, such as prime rib, roasted tomato, onion blend, carrots, and celery. Generally, about 1 to 3% base is added to an injected marinade. Popular ethnic cuisines such as Latin American, Asian, and Caribbean incorporate a variety of flavors, including citrus types.

Typically, highly acidic marinades (pH below 5.0) can denature proteins, causing the meat to become too soft, resulting in a mushy texture when cooked and making possible negative characteristics such as moisture release, toughness, and chewiness (Brandt, 2001). Lower salt concentrations are commonly used to enhance boneless breast fillets, including portioned fillets, to improve water-holding capacity, tenderness, yield, and overall quality and uniformity of products. The size or thickness of a fillet determines the time required for appropriate marination (Owens, 2007).

In recent years, the poultry industry has seen a significant increase in the incidence of pale, soft, and exudative (PSE) meat exhibiting characteristics unsuitable for further processing, due to excessive variation in color, poor binding, and decreased water-holding capacity. Incorporating different nonmeat binders into whole muscle products may be an effective way to improve the water-holding capacity of PSE meat. Cavitt and Owens (2001) suggested that using modified food starch in commercial marinade solutions can enhance the water-holding capacity in broiler breast fillets, including those that are PSE. Previous research indicated that the use of high-pH phosphate marinades and sodium bicarbonate increases pH and improves the water-holding capacity in PSE meat to those of normal meat (Van Laack et al., 1998; Woelfel and Sams, 2001). Marination with sodium bicarbonate has been studied to a lesser extent (Kauffman et al., 1993; Van Laack et al., 1996, 1998).

Flavor caused by lipid oxidation of irradiated poultry has been a problem for the industry; research has been focused on a study of this problem. Rababah et al. (2004) investigated the effectiveness of synthetic and natural antioxidants to avoid flavor modification of poultry lipids. They used additives such as green tea, commercial grape seed extract, and tertiary butylhydroquinone (TBHQ), with varying concentrations of lipid oxidation of nonirradiated and irradiated breasts stored at 5°C for 12 days, and observed that the mixture of meat with selected plant extracts is an effective method of minimizing lipid oxidation and volatile developments caused by irradiation. In another work with natural ingredients in poultry meat processing, Busquets et al. (2006) tested the effect of red wine addition on genotoxic heterocyclic amine formation in fried chicken breast; the authors found that red wine marinades reduce the formation of some of the heterocyclic amines formed during the frying process of chicken.

Sodium lactate is frequently added to meat and poultry products; it is recommended as a flavor enhancer in fresh and cooked meat and poultry products and as a pH-control agent. The antimicrobial effects of organic acids, such as propionic and lactic acids, are due both to pH reductions below microbial growth conditions and to metabolic inhibition by the undissociated acid molecules. High levels of organic acid salt can inhibit or inactivate *L. monocytogenes*, even at neutral pH. Some authors concluded that the inactivation rate depends not only on the environmental pH but also on the type and concentration of the acid used (Buchanan et al., 1993).

APPLICATIONS

The marination process consists of adding to meat, salt and a small amount of phosphate dissolved in water. The meat and solution are put in a vessel, the air is removed to create a vacuum, and the vessel is rolled to tumble the contents for a few minutes. This process helps the solution to penetrate the meat (Smith and Young, 2007). The three main marinating methods are immersion, injection, and vacuum tumbling (Xargayó et al., 2001; Sams, 2003). Immersion, the oldest method, consists of submerging the meat in the marinade and allowing the ingredients to diffuse into the meat; the process takes from hours to several days.

Direct contact is the important point, since it is necessary for a chemical reaction to occur. This means that soaking a piece of meat in a marinade will penetrate only a few millimeters into the meat surface. Thus, this method does not provide a homogeneous distribution of ingredients; conversely, it increases the risk of bacterial contamination. It is difficult to maintain the uniformity of marinating ingredients in large pieces; in bone-in cuts, the bones can be damaged or separate from the meat (Xargayó et al., 2001). Moreover, this method requires a long time and a storage space to be practical in most commercial operations (Owens, 2007).

Multineedle injection marinating is the most widely used method because it allows marinade dosification and ensures product homogeneity. Another important factor is dripping taking place subsequent to injection; it must be minimized to avoid appearance deterioration (Xargayó et al., 2001). To inject a marinade, needles are inserted and the liquid forces though the needles into the meat, spreading the marinade throughout the entire piece, without the time loss occurring in immersion marinating (Sams, 2003).

Spray injection has been used with optimum results in cooked meat products. It has the advantage of not forming marinade pockets around the needle; this is achieved by forcing the marinade through needles of lesser diameter (0.6 mm) at a higher flow rate, causing dispersion of the liquid into thousands of atomized microdrops that are introduced deeply between the meat fibers without damaging the muscle structure. Portioning meat for marination has the advantage of improving the yield by increasing marinade incorporation and retention and reducing drip and cooking losses (Owens, 2007). Portioned fillets can potentially incorporate more marinade in the tumbler, due to the increased surface area of exposed muscle fibers, especially in fillets cut lengthwise of the muscle fibers. Additionally, because fillets are substantially thinner than nonportioned meat, marination is carried out faster and is more evenly distributed. Fillet thickness has a significant effect on increased marinade incorporation and retention and reduction in drip and cooking loss (Owens, 2007).

A procedure is used to accelerate the marination process, such as injecting the marination solution under pressure directly into the meat through needles.

This method represents improvements in the marination process by enhancing its efficiency and improving the finished meat product quality. Today's processors are also using a combination of injection and vacuum tumbling to develop new products (Castaldo, 2007).

Vacuum tumbling has been used to develop ready-to-cook meat, poultry, and seafood. Vacuum causes the product to incorporate larger marinade volumes uniformly into deeper meat sections. Protein is extracted during this process, promoting meat chunk cohesion when a thermal process is used (Xargayó et al., 2001; Barbut, 2002). Tumbling with salt solution reduces dripping significantly (Froning and Sackett, 1985), as well as cooking loss and shear values (Maki and Froning, 1987). Myofibrillar protein extraction from the meat surface has two objectives: (1) protein coagulates upon heating, improving binding properties, and (2) the extracted protein acts as a seal when thermally processed, facilitating water retention (Maki and Froning, 1987).

Barbut et al. (2005) discussed the fact that results of marinating and tumbling chicken breast meat fillets are often variable, whereas Woelfel and Sams (2001) reported no significant differences in marinade incorporation and drip loss between PSE and normal vacuum-tumbled chicken fillets: higher cooking loss in PSE meat at pH 9 but not at pH 11. Qiao et al. (2002) studied vacuum-tumbled broiler breast meat in brine and reported significant differences in marinade absorption, cooking yield, and shear values among PSE, normal, and DFD (dark, firm, dry) fillets. Allen et al. (1998) studied vacuum-tumbled broiler breast fillets and found that dark fillets had significantly higher marinade pickup, a larger amount of bound water, and lower drip and cooking losses than did PSE fillets. Conversely, Smith and Young (2007) reported that vacuum pressure is unnecessary to improve meat yield, tenderness, or color; phosphates have little or no effect on tenderness or color but do substantially improve yields. Low-temperature meat mixing, tumbling, and massaging facilitates tenderization by muscle fiber disintegration and myofibril stretching. However, Young and Smith (2004) reported that the use of vacuum during marination increased moisture absorption, but after cooking, yields were similar to marination at atmospheric pressure. The authors concluded that the use of vacuum during marination have no significant advantages over marination at atmospheric pressure.

To improve ingredient penetration, marinating meat under pressure had been studied. Marination under pressure reduced muscle pH 2 to 5%, increases titrable acidity 4 to 19%, increases salt content 15 to 45%, improves water-holding capacity 10 to 21%, and reduces shear values 24 to 48%, depending on the muscle and the pressure applied. Marination under pressure between 0.25 and 0.75 kg/cm^2 enhances salt and acid penetration and improves texture quality (Yashoda et al., 2005).

Traditionally, marination additives were incorporated in postrigor breast meat during ripening. However, a reduction in aging time can be carried out by marinating prior to completion of the rigor process. Heath and Owens (1992) marinated broiler pieces in sealed polyester bags and observed that in-bag marinating process retained 2.1% more marinade in the chicken pieces and caused less variation in the amount of marinade retained. They also observed an improvement in yield and shear values of raw and cooked breast meat (Young and Lyon, 1997).

Electrical stimulation (ES) has proven as efficient and commercially feasible for poultry processing. Young et al. (2004) studied how ES reduces the pH of nonaged muscle sufficiently to alter the efficacy of sodium tripolyphosphate (STPP) as a moisture-binding additive. They observed that ES improved marinade absorption but did not affect cooking loss; conversely, polyphosphates did not affect marinade absorption but reduced cooking losses significantly. Shear values of ES-treated meat are reduced 50% compared to a control (non-ES meat); ES-treated meat injected with a phosphate solution was as tender as meat aged for 8 h postmortem (Hirschler and Sams, 1998). Both electrical stimulation and marination result in lower shear values (more tender meat) compared to nonstimulated and nonmarinated controls. Combining both treatments, tender cooked breast meat was obtained without additional aging after 1 h of chilling (Lyon et al.,1998).

Several negative effects of polyphosphate-containing marinades on incompletely conditioned meat can be avoided if the decline in meat pH is accelerated by applying ES to the carcass during slaughtering (Young and Lyon, 1994; Young et al., 1999). It is known that stimulated carcasses produce lower-pH broiler breast fillets compared to nonstimulated carcasses (6.1 ± 0.1 and 6.5 ± 0, respectively) as well as lower shear values (6.4 ± 0.3 and 15.5 ± 0.3 kg, respectively). Polyphosphate addition increases nonstimulated fillet shear values in approximately 1 kg; this was not observed in stimulated carcasses treated with polyphosphate (Young et al., 2005). Electrical stimulation during bleeding, together with marination, improves poultry breast texture; marination alone improved all sensory attributes (Lyon and Lyon, 2000).

PRODUCT QUALITY

Marination is a popular technique used to tenderize and improve the flavor and succulence of meat (Lemos et al., 1999). It has two functions, as a tenderizer and flavor enhancer; broiler marination has been used as a means to improve taste, tenderness, and protein functionality. The cooking process turns connective tissues into gelatin to various degrees; some plant and fungus enzymes, as well as several acids, also break down muscle and connective protein. As far back as pre-Columbian Mexico, it was found that wrapping meat in papaya leaves before cooking produced more tender meat. The active enzyme in papaya leaves is papain, now refined into a commercial additive. On the other hand, connective tissue in direct contact with protein-digesting enzymes is also degraded to a certain extent.

The addition of phosphate mixtures, salt, and spices affects meat functional properties (water-holding capacity) and textural properties (tenderness). As a general rule, the addition of phosphates increases water-holding capacity and decreases the force required to cut cooked meat. Optimum textural benefits due to an increased protein charge and improved hydration have also been reported when phosphates are used in combination with salt. For example, mechanically tenderized and tumbled breast meat, added with a commercial phosphate–salt blend, exhibited improved tenderness, juiciness, flavor, and overall sensory acceptability compared to muscles tumbled with salt (Lyon and Lyon, 2000). Results also indicated that meat tenderization using a citrus juice marinade could be attributed to marinade uptake by muscle proteins and collagen solubilization (Burk and Monahan, 2003). The effectiveness of marination and the level of salt concentration to improve broiler breast fillet tenderness, juiciness, and flavor in deboned fillets at varying postmortem times was studied by Owens (2001). The author reported that marination of prerigor deboned meat is effective in producing meat similar to deboned meat marinated postrigor. Although meat marinated with high salt concentration give more tender meat, a negative impact on flavor and saltiness can occur. Therefore, it is possible to marinate with low salt concentrations while still improving meat characteristics and keeping low ingredient costs.

Yield increase can be controlled, increasing the value added to the final product (Xargayó et al., 2001). Consumer benefits of marinated products range from the tender, moist characteristics associated with enhanced meat quality, to ready-to-eat marinated products, easy to prepare and consume. Consumers benefit from this process, as marinades increase perceived juiciness, tenderness, and flavor, with possible color improvement (Smith and Young, 2007). Tenderization depends on various factors, however, such as deboning, time elapsed between deboning and marination, product type, ingredient concentration, and marination method (e.g., tumbling, injection) (Owens, 2007).

Traditionally, meat has been marinated to improve and diversify flavors, to promote tenderization in tough muscles, and to extend shelf life. Marination affects meat palatability, particularly by reducing toughness and increasing juiciness and reducing water loss during cooking. Flavor is enhanced by the addition of spices, fruit extracts, herb extracts, and essential oils (Xargayó et al., 2001). Marination helps to tenderize tough poultry breast meat; the best results are obtained if meat destined for marination is aged at least 3 h prior to deboning (Young et al., 2004).

Regarding color, Qiao et al. (2002) studied the effect of raw broiler breast meat color (light: L^* (lightness) > 53; normal: $48 < L^* < 51$; dark: $L^* < 46$) on marinated and cooked meat quality. The results showed that pH variation, associated with extreme raw breast meat color, affected breast meat marination and cooked meat quality. Pink meat or bone-darkening problems were studied by Smith and Northcutt (2004) in a model system made by chicken breast meat, blood, and bone marrow, the components found in processed carcasses, and the effect of blood and marrow in inducing red discoloration was analyzed. Blood, marrow, and blood–marrow combinations produced darker and redder raw and cooked breast meat; marination did not improve this color condition. However, bone marrow was the main component for inducing red discoloration (Smith and Northcutt, 2004). Experiments conducted to determine the effect of raw broiler breast meat color variation on marinated and cooked meat quality showed that breast meat color is affected by pH, marination, marinade retention, cooking yield, and texture, but is not related to cooked meat moisture content. These results indicated that pH variation associated with different raw breast color affects marination and cooked meat quality (Qiao et al., 2001). Smith et al. (2001) studied the effects of selected chemicals on red discoloration in uncured, fully cooked broiler breast meat and found that citric acid and ethylenediamenetetraacetic acid (EDTA) reduced redness significantly, whereas ascorbic acid and nonfat dry milk did not affect redness.

The effect of marination on raw, skinless, and cooked broiler breast fillets was studied by Northcutt et al. (2000). These authors reported that marination has no effect on the lightness of bruised, raw, or cooked fillets; however, marination decreased yellowness in cooked bruised fillets. They also reported that after 1 h of processing, broiler fillet marination resulted in increased redness compared to nonmarinated fillets.

Postmortem time elapsed until marination affects several color attributes (Young and Lyon, 1994, 1997; Young et al., 1996a,b, 1999), cooking loss (Young and Buhr, 1997; Young et al., 1999), and texture (Young and Lyon, 1994; Young et al., 1996b, 1999). These effects are related to interactions between times postmortem at which marinade was applied as well as to marinade components, especially polyphosphates, which affect meat pH. In general, polyphosphate-containing marinades applied before rigor mortis resolution tend to increase meat pH more than similar marinades applied after rigor resolution. An answer to this problem is to "condition" the meat under refrigeration for 6 to 8 h prior to further processing, although this is an expensive method (Owens, 2001). Alvarado and Sams (2003) studied prerigor injection of broiler breast fillets with sodium phosphate (STPP) and sodium bicarbonate, and reported that PSE meat quality was improved by prerigor marination with high-pH phosphate marinades.

Barbut et al. (2005) investigated the relationships of microstructure, protein extraction, water retention, and tumbling marination of PSE, normal, and DFD chicken meat. PSE meat showed significantly lower salt-soluble protein extraction with less heavy myosin chains than did DFD meat. In addition, PSE meat showed larger intercellular spaces among muscle fibers and bundles than did normal and DFD meat. Further cooking resulted in lower yield and higher shear force values for PSE than for normal and DFD fillets.

REFERENCES

Allen CD, Fletcher DL, Northcutt JK, Russell SM. 1998. The relationship of broiler breast color to meat quality and shelf life. Poult Sci 77:361–366.

Alvarado CZ, Sams AR. 2003. Injection marination strategies for remediation of pale, exudative broiler breast meat. Poult Sci 82:1332–1336.

American Meat Institute. 1991. *Annual Report*. Washington, DC: AMI.

Aog CYW, Young LL. 1987. Effect of marination with sodium pyrophosphate solution on oxidative stability of frozen cooked broiler leg meat. Poult Sci 66:676–678.

Baker RC, Bruce CK. 1989. Further processing of poultry. In: *Processing of Poultry*. Essex, UK: Science Publishers, pp. 251–282.

Barbut S. (2002). Poultry products processing. New York: CRC Press. 548 pp.

Barbut S, Zhang L, and Marcone M. 2005. Effects of pale, normal, and dark chicken breast meat on microstructure, extractable proteins, and cooking of marinated fillets. Poult Sci 84:797–802.

Bilgili SF. 2001. Poultry products and processing in the international market place. http://www.fass.org/fass01/pdfs/Bilgili.pdf. Accessed Nov. 2007.

Brandt LA. 2001. Marinades 'meat' challenges: brief article. *Prepared Foods*, Jan.

Buchanan RL, Golden MH, Whiting RC. 1993. Differentiation of the effects of pH and lactic or acetic acid concentration on the kinetics of Listeria monocytogenes inactivation. J Food Protect 56:474–478.

Burk RM, Monahan FJ. 2003. The tenderisation of shin beef using a citrus juice marinade. Meat Sci 63(2):161–168.

Busquets R, Puignou L, Galceran MT, Skog K. 2006. Effect of red wine marinades on the formation of heterocyclic amines in fried chicken breast. J Agric Food Chem. 54(21):8376–8384.

Carlos AM, Harrison MA. 1999. Inhibition of selected microorganisms in marinated chicken by pimento leaf oil and clove oleoresin. J Appl Poult Res 8:100–109.

Carroll CD. 2005. Marination of turkey breast fillets using organic acids to control the growth of *Listeria monocytogenes* and improve meat quality in deli loaves. M.Sc. thesis, Texas Tech University, Lubbock, TX, p. 90.

Cassidy JP. 1977. Phosphates in meat processing. Food Prod Dev 11:74–77.

Castaldo DJ. 2007. Getting to the point. Equipment update. http://www.kochequipment.com/site/news/articles_getting_point.php. Accessed Nov. 2007.

Cavitt LC, Owens CM. 2001. Marination of PSE broiler meat using non-meat binders. J Anim Sci 79(Suppl 1)/ J Dairy Sci 84(Suppl 1)/ Poult Sci 80(Suppl 1)/54th Annu Rec Meat Conf II:137.

Chen TC. 1982. Studies on the marination of chicken parts for deep-fat frying. J Food Sci 47:1016–1019.

Farr AJ, May KN. 1970. The effect of polyphosphates and sodium chloride on cooking yields and oxidative stability of chicken. Poult Sci 49:268–275.

Froning GW, Sackett B. 1985. Effect of salt and phosphates during tumbling of turkey breast muscle on meat characteristics. Poult Sci 64:1328–1333.

Heath JL, Owens SL. 1992. Effect of storage on yield and shear valued of in-bag marinated chicken breasts. J Appl Poult Res 1:325–330.

Hirschler EM, Sams AR. 1998. The influence of commercial-scale electrical stimulation on tenderness breast meat yield and production costs. Appl Poult Res 7:99–103.

Kauffman RG, Sybesma W, Smulders FMJ, Eikelenboom G, Engel B, Van Laack RLJM, Hoving-Bolink AH, Sterrenburg P, Nordheim EV, Walstra P, van der Wal PG. 1993. The effectiveness of examing early postmortem musculature to predict ultimate pork quality. Meat Sci 34:283–300.

Lemos ALSC, Nunes DRM, Viana AG. 1999. Optimization of the still-marinating process of chicken parts. Meat Science 52(2):227–234.

Lyon CE, Lyon BG. 2000. Sensory differences in broiler breast meat due to electrical stimulation, deboning time, and marination. J Appl Poult Res 9:234–241.

Lyon CE, Lyon BG, Dickens JA. 1998. Effects of carcass stimulation, deboning time and marination on color and texture of broiler breast meat. J Appl Poult Res 7:53–60.

Mahon JA. 1962. Phosphates improve poultry. Food Eng 34:108, 113.

Mahon JH. 1963. U.S. patent 3,104,170. Food Technol 18:101–102.

Maki AA, Froning GW. 1987. Effect on the quality characteristics of turkey breast muscle of tumbling whole carcasses in the presence of salt and phosphate. Poult Sci 66:1180–1183.

Northcutt JK, Smith DI, Buhr RJ. 2000. Effects of bruising and marination on broiler breast fillet surface appearance and cookyield. J Appl Poult Res 9:21–28.

Owens CM. 2001. Impact of marination and deboning time on poultry meat tenderness. J Anim Sci 79(Suppl 1)/ J Dairy Sci 84(Suppl 1)/ Poult Sci 80(Suppl 1)/54th Annu Rec Meat Conf II:200.

Owens C. 2007. Targeting tenderness in portioned breast fillets. Poult Int, Nov. 11.

Qiao M, Northcutt JK, Fletcher DL, Smith DP. 2001. The effects of raw broiler breast meat color variation on marination and cooked meat quality. J Anim Sci 79(Suppl 1)/ J Dairy Sci 84(Suppl 1)/ Poult Sci 80(Suppl 1)/54th Annu Rec Meat Conf II:138.

Qiao M, Fletcher DL, Smith DP, Northcutt JK. 2002. Effects of raw broiler breast meat color variation on marination and cooked meat quality. Poult Sci 81(2):276–280.

Rababah TM, Hettiarachchy NS, Horex R. 2004. Total phenolics and antioxidant activities of fenugreek, green tea, black tea, grape seed, ginger, rosemary, gotu kola and ginkgo extracts, vitamin E and *tert*. Butylhydroquinone. J Agri Food Chem 52:5183–5086.

Sams AR. 2003. Dr. Marinade. Marination is not just for flavor anymore: remedial marination can correct problems. Watt Poultry USA 4(11):18–25.

Smith DP, Northcutt JK. 2004. Induced red discoloration of broiler breast meat: effect of blood, bone marrow, and marination. Int J Poult Sci 3:248–252.

Smith DP, Young LL. 2007. Marination pressure and phosphate effects on broiler breast fillet yield, tenderness, and color. Poult Sci 86:2666–2670.

Smith DP, Northcutt JK, Claus JR. 2001. Effects of selected chemicals on red discoloration in fully cooked broiler breast meat. J Anim Sci 79(Suppl 1)/ J Dairy Sci 84(Suppl 1)/ Poult Sci 80(Suppl 1)/54th Annu Rec Meat Conf II:139.

Trout GR, Schmidt GR. 1984. The effect of phosphate type, salt concentration and processing conditions on binding in restructured beef rolls. J Food Sci 49:687–694.

USDA (U.S. Department of Agriculture). 1994. *The Livestock and Poultry Meat Situation*. Washington, DC: USDA.

Van Laack RLJM, Kauffman RG, Lee S, Pospeich E, Greaser ML. 1996. The effect of pre-rigor application of sodium bicarbonate on the quality of pork. 42nd Int Congr Meat Sci Technol. pp 386–387.

Van Laack RLJM, Kauffman RG, Pospeich E, Greaser ML, Lee S, Solomon MB. 1998. Sodium bicarbonate perfusion affects quality of porcine M semimembranosus. J. Muscle Foods 9:185–191.

Van Wazer JR. 1971. Chemistry of phosphates and condensed phosphates. In: *Symposium: Phosphates in Food Processing*. Deman JM, Melnychyn P., eds Westport, CT: AVI, pp. 1–19.

Woelfel RL, Sams AR. 2001. Marination performance of pale broiler breast meat. Poult Sci 80:1519–1522.

Xargayó M, Lagares J, Fernández E, Ruiz D, Borrell D. 2001. Marination of fresh meats by means of spray effect: influence of spray injection on the quality of marinated products. http://metalquimia.com/images/doctecnologic/art13.pdf. Metalquimia, pp. 118–126. Accessed Nov. 2007.

Yashoda KP, Rao RJ, Mahendrakar NS, Rao DN. 2005. Marination of sheep muscles under pressure and its effect on meat texture quality. J Muscle Foods 16(3):184–191.

Youm GW, Meullenet J-F, Owens C. 2004. Consumer acceptance of marinated chicken breast meat (pectoralis major) deboned at various times post-mortem and its relationship to sensory profile. Presented at the 2004 IFT Annual Meeting, July 12–16, Las Vegas, NV, paper 112-10.

Young LL, Buhr RJ. 1997. Effects of stunning duration on quality characteristics of early deboned chicken fillets. Poult Sci 76:1052–1055.

Young LL, Buhr RJ. 2000. Effect of Electrical stimulation and polyphosphate marination on drip from early-harvested, individually quick-frozen chicken breast fillets. Poult Sci 79:925–927.

Young LL, Lyon CE. 1994. Effects of rigor state and addition of polyphosphates on the color of cooked turkey meat. Poult Sci 73:1149–1152.

Young LL, Lyon CE. 1997. Effect of postchill aging and sodium tripolyphosphate on moisture binding properties, color, and Warner–Bratzler shear values of chicken breast meat. Poult Sci 76:1587–1590.

Young L, Smith DP. 2004. Effect of vacuum on moisture absorption and retention by marinated broiler fillets. Poult Sci 83:129–131.

Young LL, Papa CM, Lyon CE, Wilson RL. 1992. Moisture retention and textural properties of ground chicken meat as affected by sodium tripolyphosphate, ionic strength and pH. J Food Sci 57:1291–1293.

Young LL, Lyon CE, Northcutt JK, Dickens JA. 1996a. Effect of time post-mortem on development of pink discoloration in cooked turkey breast meat. Poult Sci 75:140–143.

Young LL, Northcutt JK, Lyon CE. 1996b. Effect of stunning time and polyphosphates on quality of cooked chicken breast meat. Poult Sci 75:677–681.

Young LL, Buhr RJ, Lyon CE. 1999. Effect of polyphosphate treatment and electrical stimulation on postchill changes of broiler breast meat. Poult Sci 78:267–271.

Young LL, Cason JA, Buhr RJ, Lyon CE. 2001. Effects of vacuum marination on moisture absorption and retention by boneless broiler chicken breasts. http://ift.confex.com/ift/2001/techprogram/paper_7445.htm

Young LL, Smith DP, Cason JA, Walker JM. 2004. Effects of intact carcass electrical stimulation on moisture retention characteristics of polyphosphate-treated non-aged boneless broiler breast fillets. Int J Poult Sci 3(12):796–798.

Young LL, Smith DP, Cason JA, Walker JM. 2005. Effects of pre-evisceration electrical stimulation and polyphosphate marination on color and texture of early harvested chicken broiler breast fillets. Int J Poult Sci 4(2):52–54.

10

NONMEAT INGREDIENTS

Elena José Sánchez-Zapata, Manuel Viuda-Martos, Casilda Navarro-Rodríguez de Vera, and José Angel Pérez-Alvarez
Grupo Industrialización de Productos de Origen Animal (IPOA Research Group), Departamento de Tecnología Agroalimentaria, Escuela Politécnica Superior de Orihuela, Universidad Miguel Hernández, Orihuela, Alicante, Spain

INTRODUCTION

The consumption of poultry products in Europe and North America is significantly higher than in other parts of the world. This is despite of the fact that over the last few years the meat industry has been facing a growing number of difficulties in a range of contexts:

Handbook of Poultry Science and Technology, Volume 2: Secondary Processing, Edited by Isabel Guerrero-Legarreta and Y.H. Hui

- *Health:* pressure from politicians and dieticians concerned about cardiovascular disease, who wish to reduce the consumption of animal fats and hence meat (poultry) products
- *Snacking:* changes in eating habits that are reducing the consumption of meat prepared and cooked at home
- *Food safety:* recurrent food scandals, such as dioxins in Belgian poultry meat

Challenges and problems such as these have tarnished the image of poultry products, even though they offer nutritional benefits such as supplying essential amino acids, iron, and vitamins. Now it is necessary to improve meat safety to win back and maintain the confidence of consumers. Farmers and food-processing companies must project an image of a dynamic modern industry that meets consumer expectations by making products that deliver energy and promote health and well-being. Functional and nutritional nonmeat ingredients can offer promising ways to address these issues. Such ingredients should be as natural as possible, neither additives nor allergens, and they may not be obtained from genetically modified organisms, especially in Europe. Their origin should also be specified clearly and their manufacture should meet environmental protection standards. In addition, they should possess excellent sensorial properties and be easily identifiable by the consumer.

Scientific–technological advances made in the poultry meat industry are very important. Some of them are related to improved technological properties and others with health, the last one having given rise to a new sector in the world of food technology known as *functional foods*, not a totally new concept. In countries of the Far East influenced by Chinese culture, food and drugs have been thought of by many as materials from the same source. Factors contributing to this reshaping of the food supply include the following: an aging population; increased health care costs; self-efficacy, autonomy in health care, and an awareness and desire to enhance personal health; and advancing scientific evidence that diet can alter disease (Pérez-Alvarez, 2008a).

From a technological and scientific point of view, many ingredients have been shown to play a beneficial role in health (Won et al., 2007; Englyst et al., 2007), improving the physiological functions of the human organism and permitting the design and optimization of foods that prevent or diminish the risk of certain chronic diseases (Ohr, 2007). It has been estimated that 5% of the foods consumed in Europe can be classified as functional (Pérez-Alvarez, 2008b).

Nutrition and health claims regulations may be stifling innovation; its resolution in Europe is quite recent (EC, 1924/2006). Moreover, a certain saturation of the healthy ingredients sector has occurred, as over two decades of intensive research and development has raised innovations in the field. Functional foods have been reported as the top trend facing the food industry (Sloan, 2000, 2006; Dixon et al., 2006) in general and the poultry meat industry in particular. With emerging scientific evidence of their effectiveness and a wide diversity of products offered for sale, a broader range of consumers are likely to become interested in dietary interventions to promote health (Teratanavat and Hooker, 2006).

The poultry meat industry cannot afford to be left behind in this respect, both as regards meat itself and poultry meat products, since consumers are increasingly concerned about the food–health relationship, while the new technologies available for this sector mean that there is a huge potential market (Kern, 2006). Thus, dietary recommendations aimed at an elderly population or women should support a dietary pattern characterized by a high consumption of olive oil, raw vegetables, and poultry (Esmaillzadeh et al., 2007; Massala et al., 2007).

The role played by the poultry meat industry worldwide is important, as regards both the number of people it employs and the volume of sales. This industry can be favored if consumers gain in understanding and faith regarding nutrition and health claims surrounding poultry meat and poultry products. Thus, consumer attitudes toward meat influence meat consumption intention most markedly with regard to poultry meat, due to health concerns, eating enjoyment, safety, and price (McCarthy et al., 2004). The importance of poultry meat products can be evidenced by the fact that consumption trends for beef in the United States were affected negatively by consumer attitudes toward health claims as to chicken (Koizumi et al., 2001). Thus, cholesterol information has a significant impact on the market, with a positive effect on demand for chicken meat (Adhikari et al., 2006).

During the last two decades, some health authorities have indicated that beef and pork meat and meat products should be avoided because, they are all unhealthy (Adhikari et al., 2006). But fortunately, poultry is not viewed in the same way by most consumers. Animal fat (including poultry fat) was prosecuted in most foods. Food manufacturers are now attempting to encourage the concept of healthy eating among consumers, often with limited success. Much consumer perception about healthy foods is that "sensory pleasure" must be sacrificed to achieve the goal of a healthy diet (Urala and Lahteenmaki, 2003, 2006; Hanley and Angus, 2004; Ollila et al., 2004).

The poultry meat industry is already assuming as its role offering enriched meat products, primarily with nonmeat ingredients (water, carbohydrates, nonmeat proteins, other additives) (Jiménez-Colmenero, 1996), which is just the beginning of the industry's goal to improve the quality and properties of processed products, thereby retaining even the most demanding consumers in terms of *functional foods* (Jiménez-Colmenero et al., 2005). This reflects the general move toward functional foods, which can be defined simply as foods that contain a component (nutrient or nonnutrient) with a selective activity related to one or several functions of the human organism and having a physiological effect (Pérez-Alvarez et al., 2003) over and above its nutritional value and whose positive action justifies its being termed as functional (physiological) or even healthy. But scientifically proved substantiation of the health effect is required, and legal authorizations are required to establish "healthiness" for specific functional meat products developed by manufacturers (Fernández-López et al., 2006). This is one reason why the development and marketing of functional foods is expensive and exceptionally risky. It is necessary to invest more in finding out what functional foods can contribute to individual and public health in relation to the promises made by manufacturers.

DEVELOPMENT OF FUNCTIONAL POULTRY MEAT PRODUCTS

A variety of strategies are possible for developing healthier meat and meat products, including functional foods. Jiménez-Colmenero et al. (2001) have suggested modification of carcass composition, manipulation of meat raw materials, reformulation of meat products, reduction of fat content, modification of the fatty acid profile, reduction of cholesterol, reduction of calories, reduction of sodium content, reduction of nitrites, and the incorporation of functional ingredients. Functional foods are also viewed as being members of the particular food category to which they belong rather than being considered as a specific homogeneous group of products (Urala and Lahteenmaki, 2007).

New research lines are opening up and companies are investing in R&D (Fernández-López et al., 2006; Grashorn, 2006) and looking into the use of components or substances (macro- and micronutrients and bioactive compounds) that will benefit consumer health, making this an area of huge potential importance for large food companies. In addition to being an excellent poultry meat product as such, a functional food has to offer a specific health effect. Compared to conventional foods, the development of poultry functional food requires a strategy of its own by tightly linking research and business. Development of new functional components and technological solutions for functional poultry meat products can be very challenging and expensive. To avoid major failures in investments, manufacturers also have to utilize new methods of consumer research in each country.

Meat-based functional foods are seen by a manufacturer as an opportunity to improve their image and address the needs of consumers as well as to update nutrient dietary goals. Poultry meat products can be processed with a wide variety of functional ingredients: dietary fiber, oligosaccharides, sugar alcohols, peptides and proteins, prebiotics and probiotics, phytochemicals and antioxidants, and polyunsaturated fatty acids (Roberfroid, 2002). Roberfroid (1999) classified substances that can lead to the production of functional foods: an essential macronutrient having specific physiological effects, such as resistant starch or omega-3 fatty acids; an essential micronutrient if it confers a special benefit through intake over and above the recommended daily intake (RDI); and a nonessential nutrient such as some oligosaccharides and phytochemicals that provide specific physiological effects (de Jong et al., 2007).

Functional poultry meat products differ from conventional meat products in several ways:

1. Conventional "healthy" poultry meat products are typically presented as types of foods contributing to a healthy diet [e.g., olive oil meat products

(Serrano-Pérez, 2005), high-fiber meat products (Fernández-López et al., 2004)] without emphasizing the role of any single product.
2. Functionality adds a novelty aspect to meat functional food without necessarily changing the sensory quality of the product.
3. Functional meat product processing often requires knowledge of modern food technology practices since a constituent needs to be added, removed, or modified, or adapted to new processing technologies such as nanotechnologies (Kern, 2006) and vacuum osmotic dehydration (Barrera et al., 2007; Galaverna et al., 2008).

Poultry meat and meat products are fundamental components in the human diet (Barroeta, 2007) and play an especially important role in developed countries, in which the food technologist must preserve all healthy components. Besides water, their principal components are proteins and fats, with a substantial proportion of highly bioavailable vitamins and minerals. Now, the foods showed a new function (a tertiary function) in which they play an important role in preventing diseases by modulating physiological systems. Examples of such functions of foods are anticarcinogenicity, antimutagenicity, antioxidative activity, and anti-aging activity (Parrado et al., 2006; Pszczola, 2007). Consumers have to trust information concerning the functional effect, as the functional and conventional products can appear to be identical when used. A meat-based product in which a health effect is added can also affect the credibility. The manufacturer has to offer accurate information in a credible way to the right consumers. This means that there is a risk that functional poultry meat products are perceived as being less natural than conventional products and are thus avoided by those who value naturalness in food choices. For example, Pegg et al., (2006) reported a potential for an emu (*Dromaius novaehollandiae*) meat snack to be considered as a functional food for athletes looking for performance enhancement who are interested in consuming greater quantities of creatine from a natural food source. Factors contributing to this reshaping of the food supply include an aging population (Hughes, 2006); increased health care costs; self-efficacy, autonomy in health care, and an awareness and desire to enhance personal health; and advancing scientific evidence that diet can alter disease (Wahlqvist, 2008).

One thing that the entire poultry meat industry must avoid is a "food phobia" because of its negative effect on the willingness to buy and use functional meat products. Food safety is another critical aspect of food quality; efforts should also be directed to ensuring that new functional poultry meat products are safe (Burdock et al., 2006). Without proof of product safety, most consumers would hesitate to adopt new foods in their diet (Niva, 2006). There is little understanding of the circumstances under which functional foods, in general, and poultry meat and meat products, in particular, are eaten, whether target groups are reached, and if targeted education programs or health policies should be recommended. Very little is known about exposure, long term or otherwise, and safety under free conditions of use, and whether and how functional foods interfere with drugs designed for the same target (de Jong et al., 2007).

From the consumer's point of view, it is very important for functional poultry meat products to show almost the same sensorial characteristics as are shown by similar or traditional meat products. Also in the specific case of functional foods, taste expectations and experiences are extremely critical factors when selecting this food category. Functional poultry meat products, cooked or dry-cured, must also be perceived as natural, tasty (mainly for dry-cured, as many bioactive compounds or functional ingredients are bitter and aversive to the consumer) (Drewnowski and Gomez-Carneros, 2000; Anon., 2006), safe, and healthy, the last especially as related to perceptions about turkey meat products. Consumers in general are not very willing to compromise on the taste of poultry meat functional foods for health benefits. In recent years it is possible to see a variety of behaviors between the United States and Europe regarding the acceptance of these new types of meat products, particularly between cooked and dry-cured functional poultry meat products. In the Spanish market, several poultry meat prototypes are now available in which low-salt poultry meat products are available to consumers. U.S. consumers accept all applied new technologies better, than Europeans, who are far more critical and largely reject the use of new technologies for meat products.

Jaeger (2006) described a methodology for researching functional food sensory topics that is build on three premises: (1) that multimethod and interdisciplinary approaches are required to research people's relationships with food; (2) that tools and techniques are needed that are tailored to food-related research; and (3) that taking the context into account threatens the validity of food-related research.

Healthy images are projected in the sale of chicken and turkey, dry-cured salchichón, and chorizo. The same aspects are related to the processing of ostrich and emu meat (Sayas-Barberá and Fernández-López, 2000; Pegg et al., 2006). Consumers who purchase functional foods do so for one or all of the following reasons: to prevent diseases or ailments directly (e.g., to prevent heart disease or colds), as diet therapy (to promote health and prevent disease), or for performance in activities based on energy or athletic pursuits (Brugarolas et al., 2006; Hart, 2007).

Acording to Roberfroid (1999), functional foods are generally produced using one or more of four approaches:

1. Eliminating a component known or identified as having a deleterious effect to the consumer (e.g., an allergenic protein).
2. Increasing the concentration of a component, naturally present in food, for its beneficial effect.
3. Adding a component that is not normally present in most foods, for a beneficial effect.
4. Replacing a component, usually a macronutrient whose intake is often excessive, by a component whose beneficial effects have been demonstrated.

Meat and meat products can be modified by including ingredients considered beneficial or by eliminating or reducing those components considered harmful to health. The meat itself can be modified by manipulating animal feed (Kennedy et al., 2005; Assi and King, 2007) or by postmortem manipulation of the carcass. As regards meat products, efforts are directed principally toward modifying the lipid content (fatty acids) (Grashorn, 2006), fortifying poultry meat with organic selenium (Sevcikova et al., 2007), or incorporating a series of functional ingredients [fruit or cereal fiber, vegetal proteins (Shaw, 2008), monounsaturated or polyunsaturated fatty acids (omega-3), vitamins, calcium, inulin, etc. (Farrell, 1995; Kennedy et al., 2005)].

Functional ingredients may be added to poultry products using various methods, depending on the properties of the functional ingredient and the product. For comminuted or heavily macerated meats (turkey ham), all types of ingredients are easily dispersed by mixing or massaging. Mixing is a rapid process that requires only minutes to achieve uniform distribution of ingredients. The addition of ingredients to intact muscle (e.g., dry-cured duck breast or turkey leg) is more complex. Low-molecular-weight easily soluble ingredients such as salt or nitrite may be added by surface application of the dry ingredients to "adobo" with the addition of paprika, black or white pepper, vinegar, or white wine (Pérez-Alvarez, 2006), by injection of a water solution into the meat or by immersion of the meat in a water solution. Injection is quite rapid, whereas surface application or immersion requires days or weeks for ingredients to diffuse throughout the intact muscle. A high-molecular-weight ingredient (e.g., starch) or insoluble ingredients may be added by injecting a suspension of the ingredients. Continuous agitation is needed to keep the ingredients in suspension during injection. Also, the suspended materials tend to remain in the path opened by the injection needle and become visible when the finished product is sliced (Mills, 2004). The main nonmeat ingredient used in poultry meat products are listed in Table 1.

The use of proteins, peptides, and free amino acids obtained by enzyme biotransformation are currently under study as functional ingredients. Recently, the trypsin hydrolysate from duck meat has been proved to be a functional ingredient (So et al., 2003). From an industrial and consumer point of view, then, the development of functional poultry meat products with a reduced or modified fat content, or whose formulation has been modified by incorporating a functional ingredient, are of great interest. When food research developers formulate a reduced-fat healthy poultry meat product, nonmeat ingredients are added to a variety of low-fat meat products to counteract changes in product characteristics associated with fat reduction (Hughes et al., 1998). Technicians must take into account that emulsifiers play a central role in this process. When fat is removed, the stability of the formulation is reduced, so these ingredients, additives that can bind water and oil together in an emulsion, are very important.

However, one of the inconveniences associated with the incorporation of a new ingredient in any food (and poultry meat products are not the exception) is the effect that it might have on the corresponding technological, nutritional, and sensorial properties. Some ingredients or bioactive compound additions can produce bitter, acrid, or astringent off-tastes (Aleson-Carbonell et al., 2004). That

TABLE 1 Main Nonmeat Ingredients Used in the Poultry Meat Product Elaboration Process

Ingredient	Technological Properties	Poultry Meat Products	References
Dextrose	Water retention	Brines, "adobos," dry-cured sausage	Elvira-García (1992)
Lactose	Fermentation substrate	Dry-cured sausage	Elvira-García (1992), Sayas-Barberá et al. (2003)
Starch	Water retention, gelling capacity	Burgers, fresh and dry-cured sausages, cooked sausages, surimi	Lee et al. (1992), Park (2000), Tabilo-Munizaga and Barbosa-Cánovas (2004)
Dietetic fiber	Prebiotic, antioxidant, water and oil binding, gelling	Cooked products, restructured products, dry-cured poultry meat products	García et al. (2002), Redondo (2002), Muguerza et al. (2004), Sendra et al. (2008)
Sorbitol	Emulsifying, antifreeze agent	Cooked sausages, poultry surimi	Elvira-García (1992)
Glycerine	Water retention and fermentation substrate	Brine	Elvira-García (1992)
Omega-3 fatty acids	Enriched with EPA and DHA	Cooked and dry-cured meat products	García et al. (2002), Mugerza et al. (2004)
Whole soy flour	Emulsifier, water and oil retention	Cooked and comminuted poultry meat products	Endres (2001), López-Vargas et al. (2007)
Soy protein isolate	Emulsifier, water and oil retention, gelling action	Cooked and comminuted poultry meat products	Sipos (1999), López-Vargas et al. (2007)
Soy-textured protein	Meat analog, water and oil retention, gelling action	Cooked and comminuted poultry meat products, meat analog	Sipos (1999), López-Vargas et al. (2007)
Quinoa	Meat analog	Meat analog	Jiménez-Colmenero (2004)
Pea	Water/lipid linkage, emulsifying, texture improvement, increased cooking yields	Sausages, pâtés, marinades, meat-substituted reconstituted poultry products	Shaw (2008)

TABLE 1 ***(Continued)***

Ingredient	Technological Properties	Poultry Meat Products	References
Caseinate	Water retention, emulsifying, and gelling	Dry-cured poultry meat sausages	Jiménez-Colmenero (2004), Fernádez-Sevilla (2005)
Lactoalbumin and lactoglobulin	Water retention	Cooked meat products	Jiménez-Colmenero (2004), Fernádez-Sevilla (2005)
Egg white	Gelling agent	Cooked meat products	Lee et al. (1992), Park (2000), Fernández-López et al. (2007b)
Egg yolk	Emulsifying	Cooked meat product	Fernández-López et al. (2007b)

underscores the dilemma of trading off taste for health benefits. These characteristics are particularly problematic for dry-cured sausages. In these types of meat product, botanicals and herbals are common (depending on the region or country). Botanical or herbal incorporation, increased protein levels, and vitamin and mineral fortification can lead to unacceptable flavors and marked alteration in a product's color (Pérez-Alvarez, 2006). It must also be taken into account that a food's concentration can disrupt the normal evolution (chemical, biochemical, enzymatic, etc.) of these types of meat products, mainly in dry-cured products (Sayas-Barberá et al., 2003). These ingredients, as polyphenols, can act as antioxidants (in low doses) (Vuorela et al., 2005) or prooxidants (in high doses) (Segura-Ferrero, 2007). The presence of off-flavors (bitter/salty in this case) can dramatically affect both consumer liking for a product and consumers' likelihood of consuming it.

Many nonmeat ingredients are used in the manufacture of comminuted meat products; of these, protein-based ingredients are among the most important. They are very widely used, and in many cases there are specific regulations of the amounts that can be used and in what products they can be used. The range of applicable nonmeat proteins from animal and plant sources is very large (Jiménez-Colmenero, 2004). Nonmeat ingredients containing proteins from other plant sources have been used as binders and extenders in comminuted meat products. Pea proteins are a co-product of starch extraction from peas that have a good potential as an additives for emulsions and other meat products, thanks to its promising functional properties (water and fat binding, emulsifying, whippability, foam stability). Pea proteins in the form of flours, concentrates, and isolates (up to 90% protein) have been used in both normal-fat and low-fat sausages. Different oilseed (other than soy) protein ingredients (sunflower, rapeseed, peanut, and cottonseed, in the form of flours, protein concentrates, protein isolates, and extrusions) have been used experimentally in minced meats, patties, and sausages. These generally improve cooking yield and emulsion stability and retard oxidative rancidity. Rice products (flours and protein isolates), oat proteins, and bean products (flours) have been used in sausage meat.

FUNCTIONAL POULTRY MEAT PRODUCT ADDITIVES

There is limited information on the physiological functions of meat, but most recent efforts made in meat science and technology are focused on improving the health image of meat and meat products (Arihara, 2006). Consumers often associate meat and meat products with the negative image that meat has a high fat content, and red meat is regarded as a cancer-promoting food (Valsta, 2005) (the exception is ostrich meat). Fernández-Ginés et al. (2005) state that the functional properties of meat products can be improved by adding ingredients considered beneficial for health or by eliminating components that are considered harmful. Functional modifications of meat and meat products (poultry meat products are included) include the modification of fatty acid and cholesterol levels in meat, the addition of vegetal oils to meat products, the addition of soy, the addition of natural extracts with antioxidant properties, sodium chloride control, the addition of fish oils, the addition of vegetal products, and the addition of fiber (Monreal-Revuelta et al., 2002). There are still some hurdles to be overcome in developing and marketing novel functional poultry meat products since such products are unconventional and consumers in many countries recognize meat and meat products to be bad for health (Arihara, 2006).

Poultry meat and meat products are important sources of proteins, although other components, such as fat, play an important role in their composition. In meat products, the content of this macronutrient is between 40 and 50% (Ordoñez et al., 1999). Fat fulfills a primordial role in all meat products, in sensorial aspects such as taste and juiciness (Lucca and Tepper, 1994; Hughes et al., 1997; Cofrades et al., 2000), so that any reduction in its content will have an important effect on the acceptability of meat products (Giese, 1996). In addition, one must bear in mind the conditions under which meat products are processed, because the type of processing may affect the development of functional poultry meat products and the content, formation, or bioavailability of some compounds (Fernández-López et al., 2007a), considering the technological challenge that a change in the composition of a given product represents (variation of lipid content or presence of unusual ingredients) and which may affect the properties of the same (sensorial, water-holding capacity, fat, etc.). Furthermore, the new formula must fulfill the same quality requisites as any other type of derived product (Jimenez-Colmenero, 1996). The effect of treatment on the content, formation, or bioavailability of some of the compounds contained in the product must also be taken into consideration, in some cases necessitating modifications in the processing conditions (Purchas et al., 2004). Many processes used in the elaboration of these products (e.g., mincing, drying, maturing) reduce or increase the presence of some constituents (e.g., protein, fat, salt) (Monreal-Revuelta et al., 2002). They may even lead to the formation of many compounds, some of which may be harmful to health (e.g., nitrosamines, lipid oxidation products) (Maijala and Eerola, 1993; Papadima and Bloukas, 1999). Therefore, such approaches tend to concentrate on aspects related with treatments directed at obtaining adequate raw materials, the formulation of products that induce changes in composition, and the development of new technologies for elaborating poultry meat products (Jimenez-Colmenero et al., 2001).

Oil

Vegetable oils with a high polyunsaturated fat (PUFA) content, such as corn, sunflower, cottonseed oil, partially hydrogenated vegetable fat, palm fat, and soybean oil have been mainly used to substitute for the fat content in cooked sausages, in some cases producing problems related to a lipid oxidation process (Fernández-López et al., 2003a; Muguerza et al., 2003). Olive oil has the highest content of monounsaturated acids of all vegetal oils, and its consumption has been related to a diminution of some cardiovascular diseases and breast cancer (Gil-Conejero, 2001).

Poultry meat products are formulated with poultry meat and pork fatty cuts. Thus, a fresh poultry sausage was produced with soy protein (at 2 to 10% of the final product), olive oil (at 0.1 to 1% of the final product), pork, poultry meat, and additives, including salt. The appearance and flavor of these sausages were identical to those of conventional fresh sausages but had a healthier composition (Serrano-Pérez, 2005). Some dry-cured sausages are formulated with poultry meat and pork fatty cuts. Some of the main considerations related to this type of product can be adapted to poultry-dry-cured meat products. Thus, when Muguerza et al. (2008) made a traditional Spanish sausage replacing pork backfat (0 to 30%) with preemulsified olive oil, the oleic and linoleic acid levels increased and the cholesterol content was reduced. Sensory characteristics are comparable to those of the traditionally made product. This result clearly points to the possibility of replacing up to 25% of pork backfat by olive oil to increase the nutritional value. Other studies (Muguerza et al., 2002) with fermented meat products showed that using olive oil to replace 20% of pork backfat did not affect the weight loss. The product had an acceptable flavor and odor, and the product was paler and yellower, although the general appearance was unacceptable because of intense wrinkling of the surface and overhardening. Muguerza and co-workers in another study (Muguerza et al., 2003a) showed that replacing 20% of porkback fat by olive oil in the manufacture of Greek meat products lessened oxidation processes and increased the monounsaturated fat (MUFA) content significantly. Vacuum packing of olive oil meat products prevents lipid oxidation during storage and increases the MUFA content at the same time (Ansorena and Astiasarán, 2004a,b). The addition of extra virgin olive oil to salami as a partial replacer of pork fat did not substantially affect the chemical, physical, and sensorial characteristics of the products, except for the water activity and firmness. The addition of the oil did not reduce the shelf life in terms of lipid oxidation, probably due to the antioxidant effect of the polyphenols and tocoferols that it contains. The replacement of pork backfat by linseed oil decreased the omega-6/omega-3 ratio (from 14.1 to 2.1) as a consequence of the increase in α-linoleic acid. This change increases the nutritional quality of the product and was not accompanied by any great change in taste or oxidation (Ansorena and Astiasarán, 2004a).

Carotenes

Carotenes are responsible for the color of poultry meat, fat, and skin and are of great economic importance (Pérez-Alvarez and Fernández-López, 2006). The consumers accepted yellow poultry meat products, and consumers in some countries considered the color as natural. Carotenes have the ability to quench singlet oxygen and trap peroxyl radicals; carotenoids act as excellent antioxidants. In meat products such as minced meat, the addition of lycopene from natural sources could lead to a meat product with a different taste, better color, and a well-documented health benefit (Osterlie and Lerfall, 2005). New functional meat products would be developed and natural lycopene might replace nitrite added to minced meat. Some colorants interact with nitrites, which could reduce the quantity of nitrite added for color development and stability in meat products (Bloukas et al., 1999).

Soy

Processed poultry products represent a major growth area for meat product manufacturers. Creation of acceptable products from poultry meat pieces now includes the use of mechanically deboned meat (involves control of microbiology status and pH value) and the use of ingredients such as soy proteins to obtain the sensory characteristics demanded from the end product (Barrett Dixon Bell, 1995). Soy proteins are a traditional ingredient of fresh and cooked meat products, where it may replace up to 30% of the fat. Soy proteins (flour, concentrates, or isolates) are used for their functional properties and relatively low cost. They bind to the water and fat, improving the stability of the emulsion and increasing yield (Chin et al., 2000). Unlike animal protein, soy protein diminishes blood lipid levels. But now the most important aspect of soy are the isoflavones (Marín et al., 2005), Clinical trials identified the potential efficacy of soy isoflavones in the prevention of coronary heart disease, osteoporosis, and breast and prostate cancer. The consumption of up to 45 g/day of soy protein can lower serum cholesterol levels by 10%: Also, isoflavones can reduce low-density lipids (LDLs) and the total concentration of cholesterol more readily than can soy extract (Marín et al., 2005). Soy oil can also be added to meat products (directly or preemulsified) as a fat replacer. When preemulsified soy oil is added, cholesterol levels changed very little, and lipid oxidation of meat products was not altered. Saturated and monounsaturated fatty acids decreased and polyunsaturated fatty acids increased, due to the significant increase in linoleic and α-linolenic acids (Marín et al., 2005).

Soy and wheat proteins are used to replace the functional and nutritional properties of poultry meat in lower-priced poultry meat products (Mott, 2007). Aloida-Guerra et al. (2005) produced poultry burgers with soy flour and texturized soy protein (65%) without an effect on product quality. Similar results were obtained by Hoffman and Mellet (2003) when a fat substitute (70% water, 20% modified starch, and 10% soy protein isolate) was used in a low-fat ostrich meat patties. Chia et al. (2001) produced restructured pressed smoked duck steak with soy protein and carrageenan. From a technological point of view, both ingredients reduced drip losses, and the flavor was improved by their addition.

Sodium Chloride

Because of the role played by salt in hypertension in sodium-sensitive persons, health authorities have recommended that the intake of sodium chloride be reduced, although consumption still exceeds recommended levels (Ruusunen et al., 2003). Taking into account estimations of consumer habits, this suggests that 20 to 30% of salt consumption comes from meat and meat products (Jimenez-Colmenero et al., 2001). Salt reduction is an important handicap, because the technological and sensorial properties of salt can be lost, and meat products can be lost. Salt contributes to several functions (water retention and fat binding) in meat products and any reduction would have an adverse effect on these parameters, increasing losses during cooking and weakening the texture (Ruusunen et al., 2003).

Gimeno et al. (2001), in a study of the use of calcium ascorbate as a partial substitute for NaCl in dry-cured meat products, the substitution led to increased acidification following the development of lactic acid bacteria, probably as a result of the presence of calcium. The partial replacement of NaCl by calcium ascorbate seems to be a feasible way to reduce the sodium content of dry-cured meat products, at the same time enriching the diet in ascorbate and calcium. In general, reduction in the salt level affects the color and textural characteristics of meat products.

Fish Oils (Omega-3)

Omega-3 polyunsaturated fatty acids are mainly extracted from several marine sources, especially cold-water fish such as mackerel (*Scomber scombrus*), halibut (*Hippoglossus hipoglossus*), and rainbow trout (*Oncorhynchus mykiss*); omega-6 PUFAs are found mainly in plants (Halsted, 2003). The omega-6/omega-3 ratio is strongly correlated with disease-related fatalities in developed countries, so health authorities recommend the consumption of food rich in omega-3, bringing the omega-6/omega-3 ratio into the range 5:1 to 10:1 (Fernández-López et al., 2003a). Epidemiological, clinical, and biochemical studies have provided much evidence concerning the beneficial effect of omega-3 fatty acids in the prevention and treatment of several chronic diseases, such as some common types of cancer (lung and colon), mental illnesses (e.g., Alzheimer disease, depression, migraine), rheumatoid arthritis, asthma, psoriasis, inflammatory diseases of the intestine, and cardiovascular diseases (Fernández-López et al., 2003a). Dietary levels of fish oil and antioxidants significantly influence the omega-3 fatty acid and cholesterol content of meat (Jeun-Horng et al., 2002).

As mentioned above, dry-cured sausage technology using poultry meat products is basically the same. But when this type of healthy ingredient is used, the technician must take into account the way in which ingredient is incorporated. Thus, Hoz et al. (2004) elaborated a traditional Spanish sausage using omega-3 PUFAs and α-tocoferol and concluded that it is possible to produce a dry-cured meat product enriched with omega-3 with no adverse effects on the composition, lipid stability, texture, or sensory characteristics. Valencia and co-workers obtained similar characteristics (Valencia et al., 2006). Omega-3 fatty acids were also added to a chorizo (Muguerza et al., 2004) in which the nutritional quality was improved while the omega-6/omega-3 ratio was reduced significantly, reaching values very close to those considered optimal. These authors also showed that it is possible to manufacture a dry-cured meat product enriched in omega-3 fatty acids by the correct choice of oil and storage conditions. Schneiderova et al., (2007) reported that the inclusion of linseed oil with a high content of α-linolenic acid in chicken feed could allow poultry meat to be produced as a functional food with a very narrow omega-6/omega-3 ratio.

Fiber

Epidemiological studies have shown that the consumption of fruit and vegetables is beneficial to health, reducing, for example, the risk of cardiovascular disease, heart attack, and some types of cancer. Apart from dietary fiber, fruit and vegetables contain organic micronutrients such as carotenoids, polyphenols, tocoferols, vitamin C, and others (Schieber et al., 2001). Market studies suggest that in Europe foods containing fiber have great potential and that many products are suitable for its addition, meat products being among the most promising (Aleson-Carbonell et al., 2004).

Fiber is defined as the edible part of plants (Svensson, 1991) and carbohydrate (Klein-Essink, 1991) analogs that resist digestion and absorption in the human small intestine (Prosky, 1999). In Table 2 dietary fiber is expressed according to its solubility. Epidemiological studies have pointed to a close relationship between a diet containing an excess of fat- and sugar-rich energetic foods, which are poor in fiber, and the appearance of chronic diseases such as cancer of the colon, obesity, and cardiovascular disease (Beecher, 1999), for which reason an increase in the amount of fiber in the diet is recommended (Eastwood, 1992). Fiber is desirable not only for its nutritional value but also for its functional and technological properties (Thebaudin et al., 1997), its presence in foods also lowering the calorie content.

Fiber has been added to cooked meat products to increase cooking yield due to its water- and fat-fixing properties, which improve texture (Cofrades et al., 2000).

TABLE 2 Dietary Fiber Classification and Types Used in Poultry Meat Products

Classification	Solubility	Types
Fermentable	Soluble	Pectin, guar gum, arabic gum, tragacant, inulin, fructooligosaccharides
	Insoluble	Resistant starch
Partially fermentable	Soluble	Agar-agar
	Insoluble	Cellulose
Nonfermentable	Soluble	Carrageenan, methylcellulose, carboxymethylcellulose
	Insoluble	Lignin

Several fiber types are used in meat products (i.e., fruit, cereal, pulses, etc.). Some advantages are found when fruit fibers are added, as most of these fibers showed antioxidant activity, while others reduce the residual nitrite level (Fernández-Ginés et al., 2003). It must also be taken into account that these fibers can act as a source of active biocompounds. The main advantage of dietary fiber obtained from citrus is its high proportion of soluble fiber and its better quality than that of other sources, due to the bioactive compounds with antioxidant properties that it contains (Fernández-López et al., 2004). Several types of fiber have been studied alone or in combination with other ingredients in the elaboration of low-fat minced and restructured meat products (Grigelmo et al., 1999) in meat emulsions (Mansour and Khalil, 1999) and dry-cured meat products (Ruiz-Cano et al., 2008).

Inulin is another source of soluble dietary fiber, which can be used to replace fat because it enhances the flavor, bringing out the taste and diminishing the caloric value (1.0 kcal/g). Mendoza et al. (2001) discussed a dry-cured meat product with inulin with a smoother texture, while textural characteristics were very close to those of traditional products. A low-calorie product (30% of original) can be obtained with approximately 10% inulin.

Another source of fiber is that of co-products generated in the industrial processing of citrus; for example, lemon albedo and orange fiber extract have been added in different concentrations to cooked meat products (Fernández-Ginés et al., 2004), nonfermented dry-cured embutidos (Alesón-Carbonell et al., 2004), and a typical Spanish dry-cured meat product (salchichón) (Ruiz-Cano et al., 2008), with excellent results. Also, Tsujita and Takaku (2007) reported that the segment wall of satsuma mandarin oranges might be useful as a functional ingredient, especially as a fat-reducing material.

Cereal fiber added to meat products improves their nutritional properties and is accompanied by an acceptable sensory profile, but meat products do not incorporate higher fiber concentrations because textural characteristics are not accepted by consumers. Viuda et al. (2008a,b) reported that the use of citrus fiber wastewaters can reduce the nitrite residual level in meat products. This was also observed when citrus fiber (orange and lemon) and lemon albedo are incorporated in meat products (Sendra et al., 2008).

CONCLUSIONS

Healthy poultry meat and meat products can be obtained by the addition of several ingredients that have demonstrated beneficial properties by eliminating or reducing components considered harmful to health. The use of these ingredients offers the possibility of improving the nutritional value and offering potential health benefits, although as we have seen, the inappropriate use of some ingredients may well result in a product of low sensory quality.

REFERENCES

Adhikari M, Paudel L, Houston JE, Paudel KP, Bukenya J. 2006. The impact of cholesterol information on meat demand: application of an updated cholesterol index. J Food Distrib Res 37(2):60–69.

Aleson-Carbonell L, Fernández-López J, Sendra E, Sayas-Barberá E, Pérez-Alvarez JA. 2004. Quality characteristics of a non-fermented dry-cured sausage formulated with lemon albedo. J Sci Food Agric 84(15):2077–2084.

Anon. 2006. Taste masking functional foods. Nutraceut Business Technol 2(1):26–29.

Ansorena D, Astiasarán I. 2004a. The use of linseed oil improves nutritional quality of the lipid fraction of dry-fermented sausages. Food Chem 87:69–74.

Ansorena D, Astiasarán I. 2004b. Effect of storage and packaging on fatty acid composition and oxidation in dry fermented sausages made with added olive oil and antioxidants. Meat Sci 67:237–244.

Arihara K. 2006. Strategies for designing novel functional meat products (52nd International Congress of Meat Science and Technology, Aug. 13–18, Dublin, Ireland). Meat Sci 74(1):219–229.

Assi JA, King AJ. 2007. Assessment of selected antioxidants in tomato pomace subsequent to treatment with the edible oyster mushroom, *Pleurotus ostreatus*, under solid-state fermentation. J Agric Food Chem 55(22):9095–9098.

Barrett Dixon Bell. 1995. Perfecting your processed poultry products. Asia Middle East Food Trade 12(3):22–23.

Barrera C, Betoret N, Heredia A, Fito P. 2007. Application of SAFES (systematic approach to food engineering systems) methodology to apple (EFFoST 2005 Annual Meeting: Innovations in Traditional Foods). J Food Eng 83(2):193–200.

Barroeta AC. 2007. Nutritive value of poultry meat: relationship between vitamin E and PUFA. World's Poult Sci J 63(2):277–284.

Beecher GR. 1999. Phytonutrients' role in metabolism: effects on resistance to degenerative processes. Nutr Rev 57:3–6.

Bloukas JG, Arvanitoyannis IS, Siopi AA. 1999. Effect of natural colorants and nitrites on colour attributes of frankfurters. Meat Sci 52:257–265.

Brugarolas M, Martinez-Carrasco L, Martinez-Poveda A, Llorca L, Gamero N. 2006. Consumer opinions of functional foods. Alim Equip Tecnol 25(215):71–74.

Burdock GA, Carabin IG, Griffiths JC. 2006. The importance of GRAS to the functional food and nutraceutical industries. Toxicology 221(1):17–27.

Chia CH, Tzu YW, Jeng-Fang-Huang A, Chai-Ching-Lin S. 2001. Studies on the quality of restructured pressed smoked duck steak. Australas J Anim Sci 14(9):1316–1320.

Chin KB, Keeton JT, Miller RK, Longnecker MT, Lamkey JW. 2000. Evaluation of konjac blends and soy protein isolate as fat replacements in low-fat bologna. J Food Sci 65:756–763.

Cofrades S, Guerra MA, Carballo J, Fernández-Martín F, Jiménez-Colmenero F. 2000. Plasma protein and soy fiber content effect on bologna sausage properties as influenced by fat level. J Food Sci 65:281–287.

de Jong N, Klungel OH, Verhagen H, Wolfs MCJ, Ocke MC, Leufkens HHM. 2007. Functional foods: the case for closer evaluation. Br Med J 334(7602):1037–1039.

Dixon JM, Hinde SJ, Banwell CL. 2006. Obesity, convenience and 'phood.' Br Food J 108(8):634–645.

Drewnowski A, Gomez-Carneros C. 2000. Bitter taste, phytonutrients, and the consumer: a review. Am J Clin Nutr 72(6):1424–1435.

Eastwood MA. 1992. The physiological effect of dietary fiber: an update. Annu Rev Nutr 12:19–35.

Elvira-García S. 1992. Aditivos. In: Martín-Bejarano S, ed., *Manual Práctico de la Carne*. Madrid: Martin & Macias, pp. 331–382.

Englyst KN, Liu S, Englyst HN. 2007. Nutritional characterization and measurement of dietary carbohydrates. Eur J Clin Nutr 61(1):19–34.

Esmaillzadeh A, Kimiagar M, Mehrabi Y, Azadbakht L, Hu FB, Willett WC. 2007. Dietary patterns and markers of systemic inflammation among Iranian women. J Nutr 137(4):992–998.

Farrell DJ. 1995. The enrichment of poultry products with the omega (ω)-3 polyunsaturated fatty acids: a selected review. Proc Aust Poult Sci Symp 7:16–22.

Fernández-Ginés JM, Fernández-López J, Sayas-Barberá ME, Sendra E, Pérez-Alvarez JA. 2003. Effect of storage conditions on quality characteristics of bologna sausages made with citrus fiber. J Food Sci 68:710.

Fernández-Ginés JM, Fernández-López J, Sayas-Barberá ME, Sendra E, Pérez-Alvarez JA. 2004. Lemon albedo as a new source of dietary fiber: application to bologna sausages. Meat Sci 67:7–13.

Fernández-Ginés JM, Fernández-López J, Sayas-Barberá E, Pérez-Alvarez JA. 2005. Meat products as functional foods: a review. J Food Sci 70:37–43.

Fernández-López J, Pérez-Alvarez JA, Sayas-Barberá ME. 2003a. Aplicación de los ácidos grasos omega 3 procedentes del pescado en el desarrollo de alimentos funcionales. In: Pérez-Alvarez JA, Fernández-López J, Sayas-Barberá ME, eds., *Alimentos Funcionales y Dieta Mediterránea*. Elche, Spain: Universidad Miguel Hernández, p. 79.

Fernández-López J, Sevilla L, Sayas-Barberá E, Navarro C, Marin F, Pérez-Alvarez JA. 2003b. Evaluation of the antioxidant potential of hyssop (*Hyssopus officinalis* L.) and rosemary (*Rosmarinus officinalis* L.) extracts in cooked pork meat. J Food Sci 68(2):660–664.

Fernández-López J, Fernandez-Ginés JM, Aleson-Carbonell L, Sendra E, Sayas-Barberá E, Pérez-Alvarez JA. 2004. Application of functional citrus by-products to meat products. Trends Food Sci Technol 15(3–4):176–185.

Fernández-López J, Sayas E, Sendra E, Navarro C, Pérez-Alvarez JA. 2006. Utilization of citrus fruit fibre in development of functional meat products. Alim Equip Tecnol 25(211):63–69.

Fernández-López J, Viuda M, Sendra E, Sayas-Barberá E, Navarro C, Pérez-Alvarez JA. 2007a. Orange fibre as potential functional ingredient for dry-cured sausages. Eur Food Res Technol 226(1–2):1–6.

Fernández-López J, Pérez-Alvarez JA, Sayas-Barberá E. 2007b. Procesamiento del huevo: ovoproductos. In: Pérez-Alvarez JA, Fernández-López J, Sayas-Barberá E, eds., *Industrialización de Productos de Origen Animal*, 3rd ed. Elche, Spain: Universidad Miguel Hernández, pp. 205–229.

Fernández-Sevilla JM. 2005. Estructura y función de las proteínas. In: *Ampliación de Tecnología de Alimentos*. http://www.ual.es/docencia/jfernandez/ATA/Tema3/Tema3-EstructuraFuncionProteinas.pdf.

Galaverna G, diSilvestro G, Cassano A, Sforza S, Dossena A, Drioli E, Marchelli R. 2008. A new integrated membrane process for the production of concentrated blood orange juice: effect on bioactive compounds and antioxidant activity. Food Chem 106(3):1021–1030.

García ML, Domínguez R, Gálvez MD, Casas C, Selgas MD. 2002. Utilization of cereal and fruit fibres in low fat dry fermented sausages. Meat Sci 60:227–236.

Giese J. 1996. Fats and fat replacers: balancing the health benefits. Food Technol 50(9):78.

Gil-Conejero A. 2001. Bs.Sci. thesis, Miguel Hernández University, Orihuela, Spain.

Gimeno O, Astiasarán I, Bello J. 2001. Calcium ascorbate as a potential partial substitute for NaCl in dry fermented sausages: effect on colour, texture and hygienic quality at different concentrations. Meat Sci 57:23–29.

Grashorn M. 2006. Poultry meat as functional food: enrichment with conjugated linoleic acid, omega-3 fatty acids and selenium and impact on meat quality. Fleischwirtschaft 86(2):100–103.

Griguelmo-Miguel N, Abadías-Seros M, Martín-Belloso O. 1999. Characterisation of low-fat high-dietary fibre frankfurters. Meat Sci 52:247–256.

Halsted CH. 2003. Dietary supplements and functional foods: two sides of a coin? Am J Clin Nutr 77:1001–1007.

Hanley B, Angus F. 2004. Functional foods: a taste for success. Food Technol Int 28:30.

Hart P. 2007. Heart of the matter. Food Ing Health Nutr (Nov.–Dec.):39–43.

Hoffman LC, Mellett FD. 2003. Quality characteristics of low fat ostrich meat patties formulated with either pork lard or modified cornstarch, soya isolate and water. Meat Sci 65(2):869–875.

Hoz L, D'Arrigo M, Cambero I, Ordóñez JA. 2004. Development of an *n*-3 fatty acid and alpha-tocopherol enriched dry fermented sausage. Meat Sci 67:485–495.

Hughes K. 2006. A cornucopia of anti-aging nutritionals. Prep Foods 175(6):9–15.

Hughes F, Cofrades S, Troy D. 1997. Effects of fat level, oat fibre and carrageenan on frankfurters formulated with 5, 12 or 30% fat. Meat Sci 45(3):273–281.

Hughes E, Mullen AM, Troy DJ. 1998. Effects of fat level, tapioca starch and whey protein on frankfurters formulated with 5% and 12% fat. Meat Sci 48(1–2):169–180.

Jaeger Sr. 2006. Non-sensory factors in sensory science research (The First European Conference on Sensory Science of Food and Beverages: A Sense of Identity). Food Qual Pref 17(1–2):132–144.

Jeun-Horng L, Yuan-Hui L, Chun-Chin K. 2002. Effect of dietary fish oil on fatty acid composition, lipid oxidation and sensory property of chicken frankfurters during storage. Meat Sci 60:161–167.

Jiménez-Colmenero F. 1996. Low fat products manufacturing technologies. Eurocarne 6(43):49–101.

Jiménez-Colmenero F. 2004. Chemistry and physics of comminuted meat products: non-meat proteins. In: Jensen WK, Devine C, Dikeman M, eds., *Encyclopedia of Meat Sciences*. London: Academic Press–Elsevier Science, pp. 271–278.

Jiménez-Colmenero F, Carballo J, Cofrades S. 2001. Healthier meat and meat products: their role as functional foods. Meat Sci 59(1):5–13.

Jiménez-Colmenero F, Ayo MJ, Carballo J. 2005. Physicochemical properties of low sodium frankfurter with added walnut: effect of transglutaminase combined with caseinate, KCl and dietary fibre as salt replacers. Meat Sci 69(4):781–788.

Kennedy OB, Stewart-Knox BJ, Mitchell PC, Thurnham DI. 2005. Vitamin E supplementation, cereal feed type and consumer sensory perceptions of poultry meat quality. Br J Nutr 93(3):333–338.

Kern M. 2006. 2025: global trends to improve human health. From basic food via functional food, pharma-food to pharma-farming and pharmaceuticals: Part 1. Basic food, functional food. AgroFood Ind High Technol 17(6):39–42.

Klein-Essink G. 1991. Potex: a functional food ingredient. Eur Food Drink Rev 4:87.

Koizumi S, Jussaume RA, Kobayashi S, In Jen Pan, Takaku S, Nishino M, Saito H, Baba M, Nagano M. 2001. Study on consumer behavior for meat consumption in the US. Anim Sci J 72(4):329–343.

Lee C, Wu MC, Okada M. 1992. Ingredients and formulation technology for surimi based products. In: *Surimi Technology*. New York: Marcel Dekker, pp. 273–302.

López-Vargas JH, Restrepo-Molina DA, Navarro C, Fernández-López J, Sayas-Barberá E, Pérez-Alvarez JA. 2007. Caracterización química y fisicoquímica de un análogo cárnico a base de proteína de soja: la seguridad y la calidad de los alimentos. *VI Congreso Nacional de Ciencia y Tecnología de los Alimentos*, Tenerife, Spain, pp. 389–390.

Lucca PA, Tepper BJ. 1994. Fat replacers and the functionality of fat in foods. Trends Food Sci Technol 5(1):2–9.

Maijala RL, Eerola S. 1993. Contaminant lactic acid bacteria of dry sausages produce histamine and tyramine. Meat Sci 35:387–395.

Mansour EH, Khalil AH. 1999. Characteristics of low-fat beefburgers as influenced by various types of wheat fibres. J Sci Food Agric 79:493–498.

Marín FR, Pérez-Alvarez JA, Soler-Rivas C. 2005. Isoflavones as functional foods. In: Rahman AU, ed. *Studies in Natural Product Chemistry*, vol. 32. Amsterdam: Elsevier Science, p. 1177.

Massala G, Ceroti M, Pala V, Krogh V, Vineis P, Sacerdote C, Saieva C, Salvini S, Sieri S, Berrino F, et al. 2007. A dietary pattern rich in olive oil and raw vegetables is associated with lower mortality in Italian elderly subjects. Br J Nutr 98(2):406–415.

McCarthy M, O'Reilly S, Cotter L, de Boer M. 2004. Factors influencing consumption of pork and poultry in the Irish market. Appetite 43(1):19–28.

Mendoza E, Garcia ML, Casas C, Selgas MD. 2001. Inulin as fat substitute in low fat, dry fermented sausages. Meat Sci 57(4):387–393.

Mills E. 2004. Additives/functional. In: Jensen WK, Devine C, Dikeman M, eds., *Encyclopedia of Meat Sciences*. London: Academic Press–Elsevier Science, pp. 1–6.

Monreal-Revuelta S, Fernández-Ginés JM, Fernández-López J, Sayas-Barberá ME, Pérez-Alvarez JA. 2002. Physiological and nutritional aspects of functional foods. Aliment Equip Tecnol 21(165):132–139.

Mott S. 2007. Replacing egg and poultry meat. Wellness Foods Eur 3:12–14.

Muguerza E, Gimeno O, Ansorena D, Bloukas JG, Astiasarán I. 2008. Effect of replacing pork backfat with pre-emulsified olive oil on lipid fraction and sensory quality of Chorizo de Pamplona—a traditional Spanish fermented sausage. Meat Sci 59(3):251–258.

Muguerza E, Fista G, Ansorena D, Astiasarán I, Bloukas JG. 2002. Effect of fat level and partial replacement of pork backfat with olive oil on processing and quality characteristics of fermented sausages. Meat Sci 61:397–404.

Muguerza E, Ansorena D, Bloukas JG, Astiasarán I. 2003a. Effect of fat level and partial replacement of pork backfat with olive oil on the lipid oxidation and volatile compounds of Greek dry fermented sausages. J Food Sci 68:1531–1536.

Muguerza E, Ansorena D, Astiasarán I. 2003b. Improvement of nutritional properties of chorizo de Pamplona by replacement of pork backfat with soy oil. Meat Sci 65:1361–1367.

Muguerza E, Gimeno O, Ansorena D, Astiasarán I. 2004. New formulations for healthier dry fermented sausages: a review. Trends Food Sci Technol 15:452–457.

Niva M. 2006. Can we predict who adopts health-promoting foods? Users of functional foods in Finland. Scand J Food Nutr 50(1):13–24.

Ohr LM. 2007. Facets of ageing. Food Technol 61(12):79–84.

Ollila S, Tuomi-Nurmi S, Immonen H. 2004. *Finnish Consumers' Willingness to Buy Functional Foods*. VTT Research Notes 2241.

Ordoñez JA, Hierro EM, Bruna JM, De la Hoz L. 1999. Changes in the components of dry-fermented sausages during ripening. Crit Rev Food Sci Nutr 39:329–334.

Osterlie M, Lerfall J. 2005. Lycopene from tomato products added minced meat: effect on storage quality and colour. Food Res Int 38:925–929.

Papadima SN, Bloukas JG. 1999. Effect of fat level and storage conditions on quality characteristics of traditional Greek sausages. Meat Sci 51:103–113.

Park JW. 2000. Ingredients technology and formulation development. In: Park JW, ed., *Surimi and Surimi Seafood*. New York: Marcel Dekker, pp. 343–391.

Parrado J, Miramontes E, Jover M, Gutierrez JF, Collantes-de-Teran L, Bautista J. 2006. Preparation of a rice bran enzymatic extract with potential use as functional food. Food Chem 98(4):742–748.

Pegg RB, Amarowicz R, Code WE. 2006. Nutritional characteristics of emu (*Dromaius novaehollandiae*) meat and its value-added products. Food Chem 97(2):193–202.

Pérez-Alvarez JA. 2006. Aspectos tecnológicos de los productos crudo-curados. In: Hui YH, Guerrero I, Rosmini MR, eds., *Ciencia y Tecnología de Carnes*. Mexico City: Limusa, pp. 463–492.

Pérez-Alvarez JA. 2008a. Overview of meat products as functional foods. In: Fernández-López J, Pérez-Alvarez JA, eds., *Technological Strategies for Functional Meat Products Development*. Kerala, India: Transworld Research Network, pp. 2–17.

Pérez-Alvarez JA. 2008b. Los alimentos del bienestar: la alimentación del siglo XXI. Ind Aliment 2(3):54.

Pérez-Alvarez JA, Sayas-Barberá ME, Fernández-López J. 2003. La alimentación en las sociedades occidentales. In: Pérez-Alvarez JA, Fernández-López J, Sayas-Barberá ME, eds., *Alimentos Funcionales y Dieta Mediterránea*. Elche, Spain: Universidad Miguel Hernández, pp. 27–38.

Pérez-Alvarez JA, Fernández-López J 2006. Chemistry and biochemistry of color in muscle foods. In: Hui YH, Nip W-K, Leo ML, Nollet, Paliyath G, Simpson BK, eds., *Food Biochemistry and Food Processing*. Ames, IA: Blackwell Publishing, pp. 337–350.

Prosky L. 1999. What is fibre? Current controversies. Trends Food Sci Technol 10: 271–275.

Pszczola DE. 2007. Bringing home the bacon. Food Technol 61(12):66–76.

Purchas RW, Rutherfurd SM, Pearce PD, Vather R, Wilkinson BHP. 2004. Concentrations in beef and lamb of taurine, carnosine, coenzyme Q_{10}, and creatine. Meat Sci 66:629–637.

Redondo L. 2002. *La Fibra Dietética*, 2nd ed. Barcelona, Spain: Glosa.

Roberfroid MB. 1999. Concepts in functional foods: a European perspective. Nutr Today 34(4):162–165.

Roberfroid MB. 2002. Global view on functional foods: European perspectives. Br J Nutr 88(2):133–138.

Ruiz-Cano D, López-Vargas JH, Pérez-Alvarez JA. 2008. Addition of fibre-rich orange extract to raw cured meat products. Alim Equip Tecnol 229:50–54.

Ruusunen M, Vainionpaa J, Poulanne E, Lyly M, Lähteenmäki L, Niemistö M, Ahvenainen R. 2003a. Physical and sensory properties of low-salt phosphate-free frankfurters composed with various ingredients. Meat Sci 63:9–16.

Ruusunen M, Vainionpaa J, Poulanne E, Lyly M, Lähteenmäki L, Niemistö M, Ahvenainen R. 2003b. Effect of sodium citrate, carboxymethyl cellulose and carrageenan levels on quality characteristics of low-salt and low-fat bologna type sausages. Meat Sci 64:371–381.

Sayas-Barberá E, Fernández-López J. 2000. El avestruz como animal de abasto en el siglo XXI. In: Rosmini MR, Pérez-Alvarez JA, Fernández-López J, eds., *Nuevas Tendencias en la Tecnología e Higiene de la Industria Cárnica*. Elche, Spain: Universidad Miguel Hernández, pp. 218–236.

Sayas-Barberá E, Fernández-López J, Pérez-Álvarez JA. 2003. Embutidos crudo-curados. In: Pérez-Alvarez JA, Fernández-López J, Sayas-Barberá E, eds., *Industrialziación de Productos de Origen Animal*. Elche, Spain: Universidad Miguel Hernandez, p. 56

Schieber A, Stintzing FC, Carle R. 2001. By-products of plant food processing as a source of functional compounds: recent developments. Trends Food Sci Technol 12:401–413.

Schneiderova D, Zelenka J, Mrkvicova E. 2007. Poultry meat production as a functional food with a voluntary ω-6 and ω-3 polyunsaturated fatty acids ratio. Czech J Anim Sci 52(7):203–213.

Segura-Ferrero M. 2007. Utilizanción de subproductos de la industria de zumos cítricos en el proceso de elaboración de pate de higado de cerdo. Bs.Sci. thesis, Miguel Hernández University, Orihuela, Spain.

Sendra E, Lario Y, Sayas-Barberá E, Pérez-Alvarez JA, Alesón-Carbonell L, Fernández-Ginés JM, Fernández-López J. 2008. Obtention of lemon fibre from lemon juice by-products. AgroFood Ind High Technol 19s(2):22–24.

Serrano-Pérez C. 2005. Process for manufacture of a fresh sausage free from animal fat made with olive oil and enriched with soy protein, and the products so obtained. Spanish patent application. ES 2 238 927 AI.

Sevcikova S, Skrivan M, Koucky M. 2007. The influence of selenium addition on slaughterhouse evaluation and meat quality of broiler chickens. Maso 18(2):14–16.

Shaw L. 2008. Could pea protein be the next valuable functional ingredient? Food Rev 35(2):19–21.

Sipos EF. 1999. *Usos Comestibles de la Proteína de Soya*. Fort Wayne, IN: Sipos & Associates, Inc. http://www.aces.uiuc.edu/asamex/.

Sloan AE. 2000. The top 10 functional food trends. Food Technol 54(4):33–60.

Sloan AE. 2006. Top 10 functional food trends. Food Technol 60(4):22–40.

So YK, Sun HK, Kyung BS. 2003. Purification of an ACE inhibitory peptide from hydrolysates of duck meat protein. Nutr Food 8(1):66–69.

Svensson S. 1991. A functional food fibre: Fibrex. Eur Food Drink Rev 4:93–94.

Tabilo-Munizaga G, Barbosa-Cánovas GV. 2004. Color and textural parameters of pressurized and heat-treated surimi gels as affected by potato starch and egg white. Food Res Int 37:767–775.

Teratanavat R, Hooker NH. 2006. Consumer valuations and preference heterogeneity for a novel functional food. J Food Sci 71(7):533–541.

Thebaudin JY, Lefebvre AC, Harrington M, Bourgeois CM. 1997. Dietary fibres: nutritional and technological interest. Trends Food Sci Technol 8:41–48.

Tsujita T, Takaku T. 2007. Lipolysis induced by segment wall extract from satsuma mandarin orange (*Citrus unshu* Mark). J Nutr Sci Vitaminol 53(6):547–551.

Urala N, Lähteenmäki L. 2003. Reasons behind consumers' functional food choices. Nutr Food Sci 33(4):148–158.

Urala N, Lähteenmäki L. 2006. Hedonic ratings and perceived healthiness in experimental functional food choices. Appetite 47(3):302–314.

Urala N, Lähteenmäki L. 2007. Consumers' changing attitudes towards functional foods. Food Qual Pref 18(1):1–12.

Valencia I, Ansorena D, Astiasaran I. 2006. Nutritional and sensory properties of dry fermented sausages enriched with n - 3 PUFAs. Meat Sci 72(4):727–733.

Valsta LM, Tapanainen H, Mannisto S. 2005. Meat fats in nutrition. Meat Sci. 70(3): 525–530.

Viuda-Martos M, Fernández-López J, Sendra E, Sayas ME, Navarro C, Sánchez-Zapata E, Pérez-Alvarez JA. 2008a. Physical and chemical properties of cooked meat products added with orange juice wastewater and thyme essential oil. *International Functional Foods Conference*, Pôrto, Portugal, p. 79.

Viuda-Martos M, Fernández-López J, Sendra E, Sayas ME, Navarro C, Sánchez-Zapata E, Pérez-Alvarez JA. 2008b. Physical, chemical and sensory properties of bologna sausages added with orange fibre and oregano essential oil. *International Functional Foods Conference*, Pôrto, Portugal, p. 79.

Vuorela S, Salminen H, Makela M, Kivikari R, Karonen M, Heinonen M. 2005. Effect of plant phenolics on protein and lipid oxidation in cooked pork meat patties. J Agric Food Chem 53(22):8492–8497.

Wahlqvist ML. 2008. New nutrition science in practice. Asia Pacific J Clin Nutr 17(1): 5–11.

Won OS, Ock KC, Inkyeong H, Han SS, Bong GK, Kun SK, Sang YL, Dayeon S, Sung GL. 2007. Soy isoflavones as safe functional ingredients. J Med Food 10(4):571–580.

PART III

PRODUCT MANUFACTURING

11

OVERVIEW OF PROCESSED POULTRY PRODUCTS

Y. H. Hui
Science Technology System, West Sacramento, California

Isabel Guerrero-Legarreta
Departamento de Biotecnología, Universidad Autónoma Metropolitana, Mexico D.F., México

INTRODUCTION

Most of us like to eat muscle foods such as meat, poultry, and seafood. Consumers tend to prefer poultry and processed poultry products to meat, mostly because of

Handbook of Poultry Science and Technology, Volume 2: Secondary Processing, Edited by Isabel Guerrero-Legarreta and Y.H. Hui

their price. In the United States and most Western countries, the type of processed poultry products in the market is determined by two factors: legal requirements and commercial demands and preferences. In the United States, the primary job of the Food Safety and Inspection Service (FSIS) of the U.S. Department of Agriculture (USDA) is the regulation of meat and poultry products to achieve two goals: safety for public consumption, and absence of economic fraud. This means that the products sold in the market are wholesome and that a consumer will not get sick by eating them, and that they are labeled truthfully so that the consumer gets what he or she pays for.

For a producer or manufacturer, a processed poultry product must satisfy the following premises:

1. The products can be mass-produced scientifically and technically and backed up by the normal requirements, such as safety, cost, labor, and distribution.
2. There is an adequate demand or preference by consumers.
3. The products must comply with government regulations for safety and labeling.

In this chapter we review the legal requirements and commercial production as related to processed poultry products. The information is a general overview and should be interpreted accordingly.

1. There are thousands of printed pages of discussion, legal, scientific, and technical, distributed by the government, industry, and academia. There is no attempt here to be comprehensive in coverage.
2. The legal requirements are those promulgated in the United States, and their subsequent implications in the science and technology in the production of processed poultry products apply only in this country. The practices in other countries are not involved.
3. Other chapters in this book may include discussions of the science and technology of the processing of specific poultry products; all safety factors included in those discussions do not necessarily reflect compliance to regulations in the United States.

It must be emphasized that the information in this chapter is for educational purposes and should be used only after due considerations of the following two guidelines, especially for small poultry processing establishments:

1. It should not be used as an official source of legal references. All legal documents and citations should be obtained from the official FSIS Web site, www.FSIS.USDA.gov.
2. It should not be the sole source of information for manufacturing purposes. For the processing of any poultry product for public consumption, recognized and accepted professional reference materials in the printed and

electronic media should always be consulted. Further, the manufacturing recommendations of the applicable poultry trade associations for a specific poultry product should always be consulted.

CATEGORIES OF PRODUCTS ACCORDING TO LEGAL REQUIREMENTS

In the United States, according to government regulations promulgated by the USDA/FSIS, processed poultry products are divided into two categories: shelf stable (SS) and not shelf stable (NSS). Shelf-stable processed poultry products can be prepared by heating, drying, salting, and other techniques (sugar, acid, chemicals, etc.), Non-shelf-stable processed poultry products include those fully cooked, heat treated but not fully cooked, and those using secondary inhibitors. Both categories can be:

1. Not ready-to-eat (NRTE)
2. Ready-to-eat (RTE)
3. Not ready-to-eat (NRTE) or ready-to-eat (RTE)

Table 1 shows shelf-stable processed poultry products and Table 2 shows non-shelf-stable processed poultry products.

CATEGORIES OF PRODUCTS ACCORDING TO COMMERCIAL PRACTICES AND PREFERENCES

For the manufacturer or producer, processed poultry products in general can be one of three types: (1) not ready to eat (NRTE, or "raw"); (2) ready to eat

TABLE 1 Shelf-Stable Poultry Products

	Preparation	Products
Thermal processing (sterile)	Heated in container (metal, glass, pouches)	Canned ham, canned soups with poultry, canned poultry stew, ravioli, lasagna
	pH (not heated in container)	Pickling
Drying		
	Dry and semidry sausages	Poultry bologna sausage
	Dry or dry-cured whole-muscle poultry	Turkey ham
	Dry or dry-cured whole-muscle poultry or poultry snacks	Turkey jerky, turkey jerky bites, turkey jerky nuggets
	Freeze-dried	Poultry gravy mix
	Dehydrated	Poultry gravy mix
Salt	Poultry base	

TABLE 2 Non-Shelf-Stable Poultry Products

Preparation	Readiness for Eating	Examples	Labels
Fully cooked	Ready	General poultry (turkey) products: ham, loaves, sliced luncheon poultry meat, poultry rolls, baked chicken, poultry salads Cooked and smoked turkey sausages: bologna, hot dogs Fresh or frozen entrees: precooked chicken pieces, barbecued chicken, prepared poultry dinners with rice, pasta, sauce, vegetables, etc.	Product must be kept refrigerated or kept frozen
Heat treated but not fully cooked	Not ready	Bacon, char-marked patties, partially cooked nuggets, frozen entrees or dinners, and low-temperature rendered products (partially defatted chopped poultry, partially defatted poultry fatty tissue) Partially cooked battered and breaded poultry products	Product must be kept refrigerated or kept frozen "Cook and serve" "Cook before eating"
Prepared with secondary inhibitors with heat or no heat	Ready or not ready	Sliced country-style ham, salted turkey, semidry fermented sausage	Product must be kept refrigerated or kept frozen
Uncooked, cured, fermented, dried, salted, or brine treated	Ready on not ready	Fermented sausage, acidulated sausage, country ham	Special handling instructions if not ready to eat

(RTE), and (3) products that can be NRTE or RTE. They categorize products from the following perspectives instead of as being shelf stable or not shelf stable.

Raw Products

1. *Raw-salted and salted-cured products* (salt pork, dry-cured bacon, country ham): shelf stable, ready to eat, not ready to eat
2. *Perishable raw-salted and salted-cured products* (fresh sausage, chorizo, bratwurst, and Polish and Italian sausage): shelf unstable, not ready to eat

3. *Marinated products*: shelf unstable, not ready to eat
4. *Raw breaded products*: shelf unstable, not ready to eat

Ready-to-Eat Products Ready-to-eat products include the following:

1. *Perishable cooked uncured products* (cooked roast beef, cooked pork, and cooked turkey)
2. *Perishable cooked cured products* (franks, bologna, ham, and a variety of luncheon meats)
3. *Canned shelf-stable cured products* [Vienna sausages, corned beef, meat spreads, small canned hams, canned sausages with oil and water activity (a_w) < 0.92, dried beef, and prefried bacon]
4. *Perishable canned cured products* (ham and other cured meats)
5. *Shelf-stable canned uncured products* (roast beef with gravy, meat stew, chili, chicken, and spaghetti sauce with meat)
6. *Fermented and acidulated sausages* (German and Italian salamis, pepperoni, Lebanon bologna, and summer sausage)
7. *Dried meat products* (jerky, beef sticks, basturma, and other dried meats)

A ready-to-eat product:

1. Is a meat or poultry product that is in edible form and needs no additional preparation to achieve food safety
2. May receive additional preparation to make a product taste and/or look better
3. Can include frozen meat and poultry products
4. Is exemplified by deli and hotdog products

There are five types of RTE poultry products:

1. Cooked or otherwise processed whole or comminuted products
2. Fermented meat and poultry products
3. Salt-cured products
4. Dried products
5. Thermally processed commercially sterile products

Examples of RTE cooked or otherwise-processed whole or comminuted poultry products include such poultry products as:

- Cooked/cured sausages (e.g., bologna, hotdogs, weiners, turkey franks, poultry roll)
- Cooked/smoked sausages (e.g., turkey cheese smokies)

- Cooked sausages (e.g., brown-and-serve turkey sausages and sausage patties)
- Cooked roast turkey, cooked ham, fried chicken, cooked/breaded chicken nuggets
- Others:
  - Poultry loaf
  - Cooked poultry chili, stew, ravioli
  - Cooked chicken/turkey BBQ
  - Chicken burritos, poultry egg rolls
  - Poultry entrées/dinners

There many RTE fermented meat products: for example, Lebanon bologna, pepperoni, cervelat, chorizo, Genoa or Italian salami, summer sausage, and cacciatore (a dry sausage). Recently, many sausage manufacturers have produced turkey versions of these sausages, using the formula for beef with some scientific and technical adjustments. Examples of salt-cured poultry products include country ham and dry-cured chicken, duck, and turkey. Examples of dried poultry products include turkey sticks, turkey jerky, and dried turkey meat.

RTE products manufactured in the United States according to the safety requirements of the federal government then face additional considerations regarding contamination:

1. Is there direct exposure of the RTE product to a food-contact surface or processing environment after the lethality treatment?
2. Examples of routes of exposure to a food-contact surface in a processing environment:
   - Slicing
   - Peeling
   - Re-bagging
   - Cooling a semipermeable encased product using a brine solution
3. Environmental routes of contamination
   - *Direct contact:* direct exposure of an RTE product to a food-contact surface
   - *Indirect contact:* potential contact of an exposed RTE product
     - Handling a mop handle and then touching an RTE product
     - A soiled apron touching a product
   - Floors, drains, and overhead structures

Of all RTE products, deli items and hotdog-type varieties are of major concern because they are popular and easily contaminated. Deli poultry RTE products are those typically sliced, either in an official establishment or after distribution from an establishment, and typically assembled in a sandwich for consumption [e.g., poultry hams (boiled or baked), turkey breast, chicken roll]. Typical hotdogs made from poultry are RTE franks, frankfurters, weiners, and similar items. They do not include such products as bratwurst, Polish sausage, other cooked sausages made with poultry ingredients.

EXAMPLES OF PROCESSED POULTRY PRODUCTS

Several chapters in this book discuss in detail the following commercially processed products:

- Canned and uncured products
- Luncheon meat, including bologna
- Turkey sausages
- Turkey bacon
- Breaded product (nuggets)
- Paste product (paté)
- Poultry ham
- Special dietary products for the aged, diseased, children, and infants

In this chapter we describe several other processed poultry products:

1. Dry whole-muscle poultry and poultry snacks
2. Fermented sausage

Dry or Dry-Cured Whole-Muscle Poultry or Poultry Snacks

Chemical Deterioration of Dried Poultry Meats Chemical deterioration in dried poultry meats usually refers to lipid or fat oxidation. Oxidation is usually begun by the action of a catalyst: heat, light, or oxygen being most common. Many different compounds form during oxidation, imparting an atypical flavor and odor known as rancidity. The oxidation of fatty acids develops spontaneously. This chemical reaction is slowed at low temperatures but is not stopped. Poultry meat stored in a freezer will turn rancid more slowly than poultry meat stored in a cooler. Rancid poultry meat is often associated more with frozen poultry meat than chilled poultry meat because bacteria can spoil poultry meat in a cooler before rancidity develops. Rancidity can develop rapidly in fermented products if fermentation bacteria do not produce acid fast enough. This is one reason that lactic acid starter cultures are recommended for fermented products rather than relying on viable bacteria occurring naturally or being added from previously fermented products. To prolong shelf life and slow fatty acid oxidation, antioxidants (e.g., butylated hydroxyanisole, butylated hydroxytoluene, tertiary butylhydroquinone, rosemary extracts) are commonly used. Smoke components can have an antioxidant effect. Antioxidants are not used in some processed poultry meats that contain nitrite, since nitrite is a potent antioxidant. Vacuum packaging of poultry meats removes oxygen, reducing rancidity. Cured items have a longer shelf life than noncured items, due to the antioxidant effect of the nitrite (e.g., ham vs. pork roast). Rancidity is a quality issue, not a safety issue.

Dried Whole-Muscle Poultry Meat Snacks This product category consists primarily of sliced whole-muscle beef jerky and similar whole-muscle poultry meat snacks. A variety of other poultry meat snacks, such as poultry meat sticks and assorted chopped and formed or ground and formed products, are manufactured using methods similar to those used for sausage-type products. These can be considered a type of dried fermented or nonfermented poultry sausage with differences of particle size and final product form (round, flat, etc.).

Most dried poultry meat snacks have a much lower moisture content (22 to 24%), water activity (a_w; < 0.80), and moisture protein ratio (0.75:1.0 or less) than other products that we have discussed. (*Note:* The jerky compliance guidelines recommend that a_w be below 0.80; other levels may be acceptable with proper validation.) Typically, the final yield is only about 45 to 55% from the original marinated or injected product.

Whole-muscle jerky products are manufactured by pickle curing the poultry meat in either the entire original piece or the first slicing, and then curing or marinating the individual slices. For the latter case, a raw poultry piece is first sliced, then the slices are tumbled with the brine, cure, and spices (optionally, marinated slices are held overnight) and placed on screens for subsequent smoking, cooking, and drying. This process is used more often by smaller jerky producers. Larger producers of whole-muscle jerky generally go with the former process in which whole-muscle pieces are injected with a complete brine containing salt, cure, spices, any preservatives, and often some percentage of ground poultry meat. This injection of ground poultry meat is permitted up to a certain percentage and requires special equipment to inject the ground poultry meat. After injection, the poultry meats (often 35 to 50% injection) are tumbled and held for a defined period before slicing and placing on screens for smokehouse smoking, cooking, and drying.

A typical jerky process would consist of first preparing the brine solution with salt, ground meat, seasoning, and cure, followed by injection, tumbling, slicing, and smokehouse processing (with 90% relative humidity) as follows:

135°F (57.2°C)	5 hours
150°F (65.6°C)	5 hours
170°F (76.7°C)	Time to reach a 160°F (71.1°C) internal temperature and water activity (a_w) of 0.80 (or other validated a_w value)

Processing and Safety Factors Poultry jerky is a ready-to-eat dried product that is generally considered to be shelf-stable (i.e., it does not require refrigeration after proper processing). Two safety concerns are noted.

1. Jerky may not be adequately heat-treated to meet lethality standards if the requirement for moist cooking is not achieved. Some processors use dry heat to both heat and dry their products and thus do not achieve adequate lethality during the heating process because the product dries prematurely and the lethality process stops.
2. Some manufacturers rely on the maximum moisture–protein ratio (MPR) rather than water activity to determine whether their process dries the jerky adequately to produce a shelf-stable product. Although an MPR of 0.75:1 or less remains part of the standard of identity for jerky, water activity, as measured by laboratory analysis, should be used to verify that the jerky is dried properly. Water activity is a better measure than MPR of water available for microbial growth. Minimizing available water (e.g., achieving an a_w value of 0.80 or less) is critical for controlling the growth of pathogens.

In general, jerky processing includes slicing or forming the poultry, then marinating heating, and drying the strips. The purpose of the heating step is to apply a lethality treatment to kill or reduce the numbers of microorganisms so that the jerky is safe for human consumption. Drying the jerky stabilizes the final product and prevents the growth of toxigenic microorganisms such as *Staphylococcus aureus*. Some processors combine the heating and drying procedures into a single step. However, it is critical that heating accompanied by adequate humidity precedes the drying. If the times and temperatures in the lethality compliance guidelines are used, it is critical that the humidity criteria be followed rigorously during the cooking/heating (lethality) steps.

The following are general or common processing steps used in jerky production. Although a particular establishment's process may not include all these steps, the lethality treatment and drying are required to produce a safe product. The intervention step may be required for those processes that do not achieve adequate lethality. The seven steps listed as heating and drying are consecutive steps. Drying should closely follow heating. Heating is used to achieve lethality of harmful microorganisms, and drying is used to stabilize the product.

Step 1: Strip Preparation Whole muscle is sliced or ground; the ground product is then formed into strips. (Some jerky is formed).

Step 2: Marination The strips are marinated in a solution that often contains salt, sugar, and flavoring ingredients.

Step 3: Interventions Antimicrobial interventions before and after marinating strips of raw product have been shown to increase the level of pathogen reduction above that achieved by heating alone. Some processes may not deliver an adequate lethality and therefore may require an additional intervention step to ensure product safety. Examples of such interventions are:

- Preheating the meat or poultry jerky strips in the marinade to achieve a minimum internal temperature of 160°F will provide an immediate reduction of *Salmonella*. Because heating in the marinade may produce an unacceptable flavor for some products, other liquids, such as water, could be used. The times and temperatures in the lethality compliance guidelines could be used for preheating in the liquid.
- Dipping the product in 5% acetic acid for 10 minutes before placing it in the marinade can augment the log reduction effects of drying but not enough to eliminate pathogens. This intervention may also result in an undesirable flavor.

Step 4: Lethality Treatment The establishment must apply a treatment to control, reduce, or eliminate the biological hazards identified in the hazard analysis. For meat and poultry jerky, these hazards will probably include the microbiological hazards from *Salmonella* spp., *Listeria monocytogenes*, and *Staphylococcus aureus*. For beef jerky, *Escherichia coli* O157:H7 may also be a hazard reasonably likely to occur. In recent years, several jerky products have been found to be adulterated with *Salmonella* and *E. coli* O157:H7.

For meat jerky, use of the time–temperature combinations provided in the lethality compliance guidelines should help to ensure the safety of the product. These combinations are based on experiments that were done with ground beef without added salt or sugar. Added salt, sugar, or other substances that reduce water activity will increase the heat resistance of bacteria in a product. However, time and experience have shown that the time–temperature combinations in the lethality compliance guidelines have been sufficient to produce safe products even with both salt and sugar additives, but the humidity during heating is a critical factor.

To produce a safe poultry jerky product, producers can use the minimum internal temperatures listed in the lethality compliance guidelines: 160°F for uncured poultry and 155°F for cured poultry. They can also use the time–temperature combinations listed in the poultry tables provided at the Web site of the U.S. Department of Agriculture's Food Safety and Inspection Service. However, humidity during heating is a critical factor regardless of which USDA compliance guideline is used. The time–temperature combinations would be sufficient to produce safe products with both salt and sugar additives if the processor uses the humidity parameters recommended by the USDA.

The heating temperature and humidity (e.g., steam) are critical for achieving adequate lethality. As the water activity is reduced, the heat resistance (D value) of the bacteria increases. Therefore, if adequate humidity is not maintained during heating, the time required to eliminate *Salmonella* at a particular temperature will be greatly increased. It is crucial that the processor prevent drying of the product until a lethal time–temperature combination is attained. The humidity requirement must be applied during the first part of the heating process, before any drying and thus increase in solute concentration occur.

Some simple and practical measures can be used to meet the humidity parameters in the lethality prescribed:

- *Seal the oven*. Close the oven dampers to provide a closed system and prevent moisture loss. Steam may be observed venting when the dampers are closed, similar to venting that occurs in a steam retort during canning.
- *Add humidity*. Place a wide, shallow pan of hot water in the oven to provide humidity to the system. Conduct a test run to determine whether the water evaporates. Injecting steam or a fine water mist into the oven can also add humidity. Use of a wet bulb thermometer in addition to a dry bulb thermometer would also enable the operator to determine if adequate humidity is being applied.
- *Monitor humidity*. Use a wet bulb thermometer in combination with a dry bulb thermometer. A basic wet bulb thermometer can be prepared by fitting a wet, moisture-wicking cloth around a dry bulb thermometer. To maintain a wet cloth during the process, submerse an end of the cloth in a water supply. The cloth must remain wet during the entire cooking step and should be changed daily, especially if smoke is applied. The use of a wet bulb thermometer is especially important for production at high altitudes or areas of low humidity where evaporation is facilitated.

Step 5: Drying After the lethality treatment, dry the product should to meet the MPR product standard and to stabilize the finished product for food safety purposes. If the product is insufficiently dried, *S. aureus* and mold are potential hazards. These organisms are not expected to grow in products dried properly. A suggested a_w critical limit for stabilization of jerky is 0.80 or lower. This range of water activity should control the growth of all bacterial pathogens of concern.

The establishment should verify the water activity value to demonstrate that the product has attained the critical limit for shelf stability. Water activity, which is the key to determining the proper level of drying, can vary greatly at any given MPR (as a result of the presence and level of various solutes, such as sugar and salt). Therefore, a laboratory test for water activity should be used to verify proper drying.

Step 6: Postdrying Heat Heat the dried product in a 275°F oven for 10 min. This has the potential of reducing *Salmonella* levels by approximately 2 logs from the level of reduction achieved during the initial heat step. This step may be needed for processes that do not result in an adequate reduction of *Salmonella* through the heating process.

Step 7: Handling The establishment's sanitation standard operating procedures should ensure that products are handled properly to prevent recontamination or cross-contamination of the meat and poultry products by bacterial pathogens of concern.

Figure 1 shows a hypothetical process flow for the manufacture of poultry jerky. Professional consultation is needed for actual manufacture.

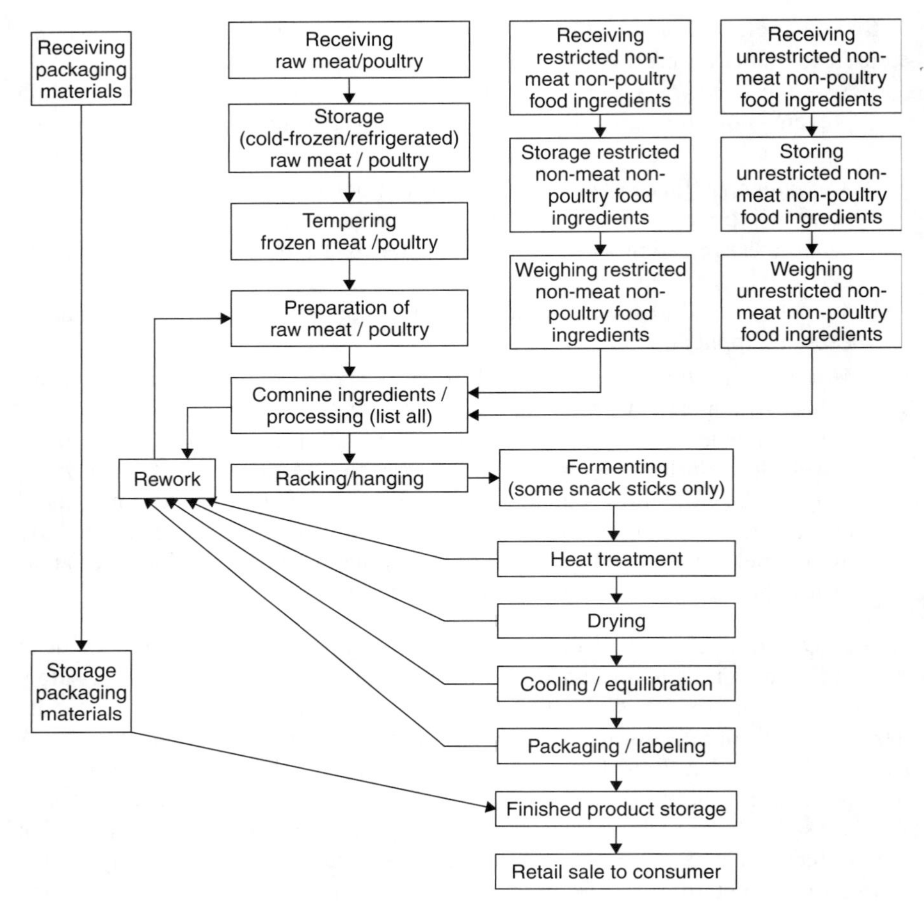

FIGURE 1 Process flow for meat/poultry jerky or snacks.

Fermented Poultry Sausages

Most fermented sausages are dry or semidry sausages utilizing controlled, bacterially induced fermentation to preserve the meat and impart flavor. The most common examples of dry sausages are salami and pepperoni. The dry category also includes shelf-stable nonfermented products such as beef jerky. Again, there are many variations of process steps and ingredients, resulting in a vast array of products available to the consumer. We concentrate on common examples of these types of products.

See Chapter 14, in which the manufacture of fresh and cooked turkey sausage is discussed. The steps in the manufacture of sausages are:

1. Formulation and blending
2. Grinding meat ingredients

3. Adding nonmeat ingredients
4. Blending
5. Stuffing
6. Packaging

The stages involving in processing product components are:

- Smoking
- Cooking
- Showering
- Chilling
- Peeling

In addition to the standard production stages identified above, dry sausages have two additional production steps:

- Blending special curing ingredients
- A drying process

Let's briefly review these additional special production steps used to produce dry and semidry sausages (Figure 2).

Blending Special Curing Ingredients The meats used in dry sausages are typically ground or chopped at low meat temperatures (20 to 250°F) to maintain the well-defined fat and lean particles that are desired in this type of sausage. The

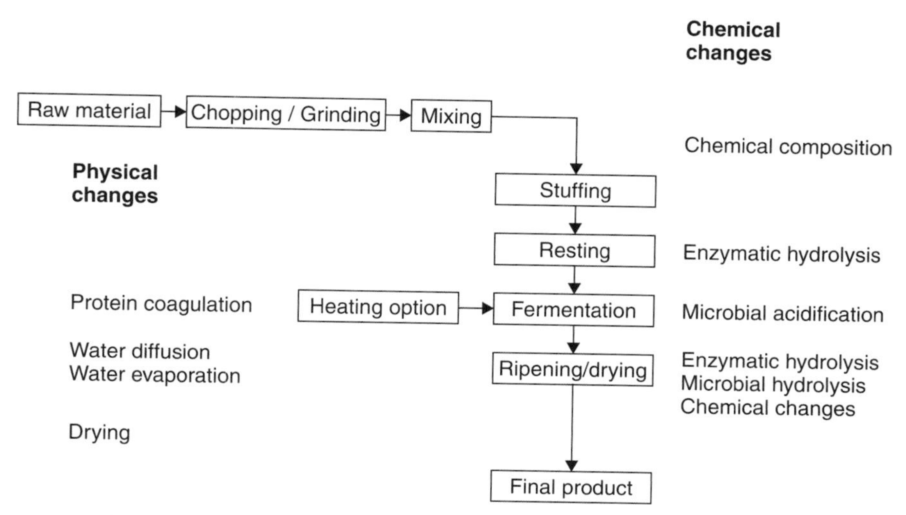

FIGURE 2 Changes during the processing of fermented sausages.

ground meats and spices are then mixed with curing ingredients such as nitrates, nitrites, antioxidants, and bacterial starter cultures. Salts have traditionally been used to help preserve sausages. Eventually, producers learned that nitrates and nitrites in the salts were essential to the curing process. Manufacturers of dry and semidry sausages use a curing agent consisting of salt and nitrates and/or nitrites. Approved antioxidants may also be added to protect flavor and prevent rancidity, and are limited to 0.003% individually or 0.006% in combination with other antioxidants. These ingredients must be distributed uniformly throughout the mixture to achieve the maximum microbiological stability.

Bacterial fermentation is then used to produce a lower pH (4.7 to 5.4), which results in the tangy flavor associated with this type of sausage. In the earlier days of sausage making, the bacteria growth was uncontrolled, resulting from bacterial contamination of the meat or production equipment, producing unreliable results. Traditionally, sausage makers held the salted meat at low temperatures for a week or more to allow plenty of time for the lactic acid bacteria in the environment to reproduce in the meat mixture. Unfortunately, this traditional method was not always reliable and was subject to error, such as cutting the time short, adding too much or too little salt, or growth of the wrong lactic acid bacteria.

Modern producers of dry sausage use a commercial lactic bacteria starter culture and simple sugars, such as dextrose or corn sugar, that promote lactic acid bacterial growth by serving as food to fuel the bacteria during fermentation. The bacteria starters are harmless and are limited to 0.5% in both dry and semidry sausage formulations. Commercial bacteria starter cultures typically consist of a blend of microorganisms such as *Pediococcus, Micrococcus*, and *Lactobacillus*, using specific species such as *P. cerevisiae, P. acidilactici, M. aurantiacus*, and *L. plantarum*. The bacterial fermentation lowers the sausage pH by producing lactic acid. This lower pH also causes the proteins to give up water, resulting in a drying effect that creates an environment that is unfavorable to spoilage organisms, which helps to preserve the product. However, mold may grow, and could become a problem.

Some small producers may still allow the mixture to age in a refrigerator for several days to encourage the fermentation process, even though they are using starter cultures. However, most modern processors stuff the mixture directly into casings, which then undergo a fermentation process at about 70 to 110°F (depending on the starter type). This fermentation process is designed to allow the bacteria to continue to incubate. The fermentation occurs during a one- to three-day process that takes place in a "greening room," which provides a carefully controlled environment designed to obtain specific fermentation results. Temperatures are typically maintained at approximately 750°F and at a relative humidity at 80%. Semidry sausages are usually fermented for shorter periods at slightly higher temperatures.

Cooking/Smoking What happens after fermentation depends on the type of product being produced. Semidry sausages such as summer sausage are almost always smoked and cooked before drying. Dry sausages such as pepperoni are rarely smoked, and may or may not be cooked. Today, some establishments choose to heat-treat dry sausages as a critical step designed to eliminate *E.coli*. A moist-heat process may be used for some products. This process utilizes a sealed oven or steam injection to raise the heat and relative humidity to meet a specific temperature–time requirement sufficient to eliminate pathogens (e.g., 1300°F minimum internal temperature for 121 min or 1410°F minimum internal temperature for 10 min). (Processors should consult FSIS regulation 9 CFR 318.10 and the *Cattlemen's Blue Ribbon* book for processes designed to control *E. coli*.) The product is then sent on to the drying stage of the process.

Drying Process The drying process is a critical step in ensuring product safety. The FSIS requires that these products undergo a carefully controlled and monitored air-drying process that cures the product by removing moisture. Pork-containing products must be treated to destroy trichinae (see 9 CFR 318.10). Sausages not containing pork have no such requirement. Manufacturers are required to control the ratio of moisture to protein in the final product. The moisture–protein ratio (MPR) is controlled by varying the amount of water added based on the overall product formulation, and primarily by the drying procedure. In some products the MPR can affect the final microbiological stability of the product. In other products the MPR is important to ensure elements of the overall product quality, such as the texture. The minimum requirement for all products produced in FSIS-inspected facilities is that they must meet the FSIS policy standards for MPR. However, these prescribed treatments have proved to be insufficiently lethal for some bacterial pathogens. Thus, most of the industry has volunteered to implement more rigorous treatment.

The drying process consists of placing a product in a drying room under a relative humidity of 55 to 65% in a process that can last from 10 days to as long as 120 days, depending on the product diameter, size, and type. The drying process is designed to produce a final product with approximately 30 to 40% moisture and an MPR generally of 1.9:1 or less, to ensure proper drying and a safe product. Facilities must keep accurate records of the temperature and the number of days in the drying room for each product manufacturing run, to help ensure product safety and consistency.

The drying environment is controlled to ensure that the drying rate is slightly higher than the rate required to remove moisture from the sausage surface as it migrates from the sausage center. Drying too quickly will produce a product with a hard and dry casing. Drying too slowly results in excessive mold and yeast growth and in excessive bacterial slime on the product surface. The controlled drying process is designed to reduce moisture to the point where the final product has a specific MPR. This process will vary depending on the specific facility operation and the specifications desired for the final product. For example, pepperoni is required to maintain an MPR of 1.6:1, indicating a requirement of 1.6 parts moisture to 1 part protein. Genoa salami should have an MPR of 2.3:1, and all other dry sausages an MPR of 1.9:1. Inspectors may sample the products to determine compliance with the MPR specified.

Semidry sausages are prepared in a similar manner but undergo a shorter drying period, producing a product with a moisture level of about 50%. These products are often fermented and finished by cooking in a smokehouse, at first at a temperature of approximately 1000°F and a relative humidity of 80%. The temperature is later increased to approximately 140 to 1550°F to ensure that microbiological activity is halted. Since the moisture level of the final product is about 50%, semidry sausages must be refrigerated to prevent spoilage. Examples of semidry sausages include summer sausage, cervelat, chorizo, Lebanon bologna, and thuringer.

FSIS MPR policy standards for some dry and semidry sausage products are not covered in the regulations; they are contained in the FSIS Food Standards and Labeling Policy Book. The FSIS policy standards call for a for dry and semidry sausage MPR of 3.1:1 for summer sausage, and it must be shelf-stable with a pH of 5.0 or higher. To ensure that the fermentation and drying processes are sufficient to reduce or eliminate any pathogens present in the product, the procedures must be validated to demonstrate that they achieve a specific reduction in organisms (e.g., a 5-log reduction in *E. coli* O157:H7). Professional consultation is needed in the actual manufacturing process.

12

CANNED POULTRY MEAT

ALICIA GRAJALES-LAGUNES AND ALICIA DE ANDA-SALAZAR
Universidad Autánoma de San Luis Potosí, Facultad de Ciencias Quimicas, Centro de Investigación de Estudios de Posgrado, San Luis Potosí, México

UNCURED POULTRY PRODUCTS

Currently, the food industry is concerned about consumer preference regarding uncured products or products cured using natural ingredients. By definition, all organic and natural products are considered uncured products, but not all uncured products are organic or natural (Bacus, 2007). Normally, when nitrites or nitrates are not added to meat products during their fabrication, they are considered uncured. Meat products are those in which ingredients such as nitrites, nitrates, and salts are added for processing or preservation improvement (Pegg

Handbook of Poultry Science and Technology, Volume 2: Secondary Processing, Edited by Isabel Guerrero-Legarreta and Y.H. Hui

and Shahidi, 2000). However, there are several natural meat products in the marketplace that do not have added nitrites, nitrates, or other ingredients.

A pink color is a problem presented by some uncured white meat products. It is the appearance of undercooking in uncured meat products, even though these have been well cooked; this defect is due primarily to myoglobin denaturation. Several endogenous factors may affect the development of pink coloring in poultry white meat (Holownia et al., 2003). In raw meat, this defect is observed primarily on the surface; after cutting, meat exposure to air causes pink coloration (Cornforth et al., 1986). In some products this defect appears when the meat is undercooked, which affects the meat industry widely, as consumers regard the meat as not being safe for consumption (Holownia et al., 2003).

INFLUENCE OF pH REDUCTION RATE ON MYOGLOBIN DENATURALIZATION

Several researchers have demonstrated that meat pH is correlated with its color. The range and rate at which pH decreases during rigor mortis onset depends on intrinsic factors such as species, variability among animals, muscle type and feeding, as well as on extrinsic factors such as like drug intake before slaughter, premortem handling, and ambient temperature (Lawrie, 1998). Conditions at which slaughter is carried out also affect the pH reduction rate; if conditions are not those necessary to ensure humane slaughter, stress occurs that affect final muscle pH, which in turn affects the color.

Color is the factor that most affects meat appearance, and it has a marked influence on consumer preference (Pérez-Dubé and Andújar-Robles 2000). In the presence of air, natural meat color is bright red, due to the predominance of oxymyoglobin, the reduced form of myoglobin in the presence of oxygen. Meat color depends to a large extent on myoglobin content as well as the oxidation–reduction state. Muscle pH has a strong influence on myoglobin chemical reactions. At a pH close to neutral, heme group affinity for protein is very high. At this pH condition, the heme group protects the protein from oxidation (Livingston and Brown, 1981). At pH < 5.4, sarcoplasmic proteins such as myoglobin become denatured, increasing light reflection and resulting in a pale and unpleasant aspect. At high pH levels, breast muscle is dark in color due to high light absorption and penetration.

Myoglobin oxygen absorption remains unchanged within a wide pH range, causing pigment stability. This protein is less sensitive to heat when it is near its isoelectric point (5.5). At a pH below the isoelectric point, myoglobin thermal stability decreases; at pH < 6.0, myoglobin is more susceptible to denaturation (Janky and Froning, 1973). Therefore, myoglobin heat denaturation has an inverse relationship to pH. At 76°C or below and pH > 6.0, myoglobin denaturation is limited, causing an increase in myoglobin redness, which, in turn, causes a pink effect similar to that of an undercooked product (Trout, 1989). Girard et al. (1989) demonstrated that even when turkey muscle is cooked at 85°C, pink coloration remains.

In addition of changes in myoglobin, there are other mechanisms associated with pink coloration. Nitrites and nitrates can be reduced to nitric oxide, which reacts with myoglobin during heating, forming nitrosylhemochrome, the cured meat pigment (Fox, 1987). Cytochrome *c* has also been identified as a factor related to the pink defect, keeping its natural color during cooking up to 105°C (Izumi et al., 1982); on the other hand, there is a higher cytochrome *c* concentration at high pH (Pikul et al., 1986). The redox potential also affects pink color formation by affecting the oxidation state of globin hemochrome (Fe^{2+}, Fe^{3+}); it is the main factor determining the pigment's ability to combine with other molecules (Holownia et al., 2003). Nitrogen-containing ligands (pyrimidine, nicotinamide, and amino acids, among others), in addition to nitrites and nitrates, also interact with meat pigments to generate hemochrome (Ahn and Maurer, 1990). Free amino acid generation by protein denaturation during cooking also contribute to the development of a pink color, due to the formation of hemochromes (Schwarz et al., 1997).

In addition to endogenous factors, a variety of exogenous factors contribute to pink color development, such as premortem handling and meat storage and processing (Froning, 1995). Carbon monoxide and nitric oxide generated by a gas flame during cooking can combine with myoglobin to generate a pink color. These gases, generated by motor vehicles and inhaled by poultry during transportation to the processing plant, also combine with myoglobin in vivo. Ligands related to pink color production, such as nitric oxide, nicotinamide, carbon monoxide, and oxygen, link at position 6 on the heme-iron site (Schwarz et al., 1999). The ingredients used during marination change the pH and redox potential, inducing pink color in meat after cooking (Trout, 1989).

Poultry feed and water may contain nitrites and cause a pink color in the final product (Froning and Hartung, 1967; Froning et al., 1969). Water used for carcass washing may also be a source of nitrogen-containing compounds, as well as the equipment used during the curing process (Mugler et al., 1970). It has been demonstrated that nitrites in concentrations as low as 1 ppm cause pink color in poultry after cooking (Nash et al., 1985; Heaton et al., 2000; Holownia et al., 2004).

Radiation is used to reduce microbial populations, including pathogens, in poultry meat. Ionization generates free radicals in the meat; these radicals can initiate various reactions, such as discoloring, lipid oxidation, and off-odors. Radiation up to 5 kGy produces pink color in raw and cooked turkey meat (Nam and Ahn, 2002).

PREVENTION OF PINK COLORATION

Consumers associate pink coloration with undercooked meat, even though it is cooked at temperatures that ensure microbial destruction (Schwarz et al., 1999; Bagorogoza et al., 2001). Inhibiting this defect is very important since product acceptance by consumers depends on it. According to Schwarz et al. (1999), to inhibit pink color development, ingredients or ligands with affinity to the heme-iron linking site should be identified and avoided. Among these ingredients are milk whey protein concentrates (Schwarz et al., 1997; Slesinski et al., 2000), nonfat dry milk products (Schwarz et al., 1999), chlorine and calcium triphosphate (Sammel and Claus, 2007), and several metal chelates, such as citric acid, sodium citrate, *trans*-1,2-diaminocyclohexane-N,N,N',N'-tetraacetic acid monohydrate (CDTA), diethylenetriaminepentaacetic acid (DTPA), and ethylenediaminetetraacetic acid (EDTA) (Kieffer et al., 2000; Sammel and Claus, 2003).

Schwarz et al. (1997) used nonfat dry milk (NFDM) and 13 other additives to inhibit pink coloring. These authors assessed these additives in uncured cooked turkey products using NFDM in the presence or absence of ingredients known to generate pink coloring (i.e., nitrite, nicotinamide). The additive efficiency studied was reported as CIE L^* (lightness), a^* (redness), and b^* (yellowness) values, and spectrophotometrically by concentration of nitrosylhemochrome and nicotinamide-hemochrome. Of the 14 additives assessed, the most efficient in reducing the pink color produced by added nicotinamide or nitrite were NFMD, CDTA, DTPA, and EDTA. CDTA, DTPA, and EDTA were able to bind heme iron, whereas NFDM did not. The mechanism was not clarified where it was due to color masking by hemochrome generation, or inhibition compounds responsible for pink color development by hemochromes. In general, the authors observed that a marked decrease in pink color occurred with a combination of nicotinamide with the additives studied.

The effect of these additive concentration (CDTA, DTPA, EDTA, and NFDM) on reducing pink color was assessed on turkey breasts by Schwarz et al. (1999). Pink color development was promoted by the addition of nicotinamide and sodium nitrite. Various concentrations of CDTA, DTPA, EDTA, and NFDM were added. The samples were stored for a variety of time periods and exposed to air and light. The results show that EDTA, CDTA, and DTPA at 50 ppm was enough to reduce pink color. Additives were observed to be more efficient when mixed with nicotinamide than with sodium nitrite; all additives delayed a pink color onset, associated with prolonged storage. DTPA proved to be more efficient than EDTA and CDTA, whereas NFDM was efficient in pink color reduction and is approved as a nonmeat ingredient.

Several compounds have been used for their ability in diminishing both pink color defect and pink cooked color promoted by the presence of sodium nitrite and nicotinamide in ground turkey: β-lactoglobulin (at 1.8%), α-lactalbumin (0.8%), bovine serum albumin (0.15 to 0.3%), lactose (1.0 to 3.0%), potassium chloride (500 to 1500 ppm), and ferrous chloride (0.3 to 30 ppm). Lactoferrin efficiency with regard to pink coloration depends on the concentration. Annatto reduced the pink color defect when added at 1 ppm concentration, whereas magnesium chloride (22 to 88 ppm) and ferric chloride (0.3 to 30 ppm) increased pink coloration in samples treated with nicotinamide; calcium chloride (160 to 480 ppm) reduced the pink color in cooked samples with and without nitrate and nicotinamide. In general, it was demonstrated that whey proteins do not reduce pink coloration; in the presence of these proteins, addition of minerals such as calcium, phosphate, and citrate is necessary to reduce this defect (Sammel et al., 2007).

Citric acid and sodium citrate have also been used to reduce pink color. These compounds are widely used in the food industry, their main functions being to controll pH, trap metallic ions, and probably to increase myoglobin denaturation while trapping heme iron in the same manner as ligands generating the pink color (nicotinamide and sodium nitrate) (Kieffer et al., 2000). In ground turkey, citric acid (at 0.2 and 0.3%) and sodium citrate (at 1%) reduce pink color caused by the presence of nicotinamide or sodium nitrate (Kieffer et al., 2000; Sammel and Claus, 2003).

Sammel and Claus (2006) studied the effect of citric acid (0.15 and 0.3%) and sodium citrate (0.5 and 1%) on pink color in ground turkey meat, irradiated before and after cooking. The results showed that these ingredients were more effective on samples cooked before irradiation. The efficiency of these ingredients could be due to the pigment type in the irradiated meat before and after cooking. When samples were irradiated after cooking, the predominant pigment was oxymyoglobin, whereas deoxymyoblobin was the predominant pigment in samples irradiated before cooking. Oxymyoglobin denaturizes faster than deoxymyoglobin; therefore, the pigment in irradiated meat after cooking was denatured to a greater extent.

COLOR CHANGES IN CHICKEN AND TURKEY BREASTS

Poultry meat color sets the guideline for purchase and final acceptance by consumers (Fletcher, 1999). Several studies have demonstrated that when postmortem glycolysis is accelerated, the muscle acidifies and a pale, soft, exudative condition (PSE) is produced in pork (Briskey and Wismer-Pedersen, 1961; Bendall et al., 1963; Fernández et al., 1994; Jossell et al., 2003). Ferket (1995) found that a large proportion of turkey breast meat showed characteristics similar to those of PSE pork meat. Stress before and after slaughter may accelerate postmortem glycolysis in pork (Jossell et al., 2003) as well as in turkey (McKee and Sams, 1997). A rapid decrease in pH in turkey breast also causes protein extraction reduction, water-holding capacity, and reduced cooking yield (Pietrzak et al., 1997). Boulianne and King (1995, 1998) studied the biochemical properties associated with PSE and dark, firm, and dry (DFD) poultry breast fillets. PSE fillets had higher lightness (L) and yellowness (b) and lower redness (a) values; they also had less myoglobin iron content and higher pH than DFD fillets, which had higher total pigment and myoglobin iron concentration and higher pH as well as higher redness (a) values, and lower lightness (L) and yellowness (b) values. The authors explained that the pale color was caused by leakage of heme pigments during cooling. Temperature during cooling can also produce PSE meat resulting from fast pH drop postmortem while the carcass temperature is still high. At low cooling rates, low-pH meat is obtained that shows high L^* values (pale meat), high cooking loss, and reduced gelling ability of proteins (Sams and Alvarado, 2004).

Fletcher (1999) carried out experiments to determine color variation in broiler breast meat and the possible correlation with pH and texture. Meat color was evaluated visually and by CIE parameters (L^*, a^*, b^*). Despite pH and color variations, the authors conclude that there was a high correlation between muscle color and pH. Other authors reported that L^* values increase with meat aging (Le Bihan-Duval et al., 1999; Alvarado and Sams, 2000; Owen and Sams, 2000; Qiao et al., 2001). Petracci and Fletcher (2002) also reported that meat color changed drastically during the first 4 h postmortem; after this time, color still changes but in a lesser extent up to 24 h postmortem. Color changes during storage (1 to 8 days postmortem) varies depending on handling conditions (Petracci and Fletcher, 2002). Santé et al. (1996) developed two statistical models to predict color modification in turkey breast meat. These authors observed that the redness (a^*) increased during the first two days of storage, after which a decrease in this parameter was observed.

To diminish color problems and increase poultry meat quality, rapid cooling is necessary to avoid a fast glycolysis. In a study carried out by Rathgeber et al. (1999) it was demonstrated that carcasses held at high temperature postmortem (accelerated glycolysis) had a reduced sarchoplasmic and myofibril protein extractability compared to carcasses subjected to rapid postmortem cooling; the delay in cooling increased L^*, a^*, and b^* values and also decreased the cook yield. The results reported by Rathgeber et al. (1999) agreed with those obtained by Bravo-Sierra et al. (2005) in pork muscles (semimembranosus, semitendinosus, and longissimus dorsi); the authors reported that PSE occurrence was reduced with rapid cooling.

CANNED POULTRY PRODUCTS

Canning was developed originally to extend the shelf life of foods for French soldiers. Napoleon was engaged in territorial expansion; his soldiers were decimated by scurvy when the emperor offered a prize of 12,000 francs to anyone who would develop a preservation method to transport food to the battle fields. Nicolas-François Appert (1750–1841), a French candy maker, got the prize in 1809, after 14 years of observation and experimentation. His process was the basis for food heat treatment; he placed glass containers, sealed with a cork and wax, in a boiling-water bath. In 1810, Peter Durand filed a patent, sold later to a company that developed iron cans covered with tin. A great amount of work has since been done by many researchers to describe the importance of the thermophile bacteria (aerobic and anaerobic) that are fully developed in the canning process and in learning the biological and toxicological characteristics of *Clostridium botulinum* and other indicator microorganisms.

FACTORS AFFECTING GROWTH OF MICROORGANISMS IN POULTRY PRODUCTS

Meat obtained from of healthy birds generally presents a low microbial population propagated during the various killing and handling operations. Poultry skin is colonized by a diversity of microorganisms; specific populations depend on the hygienic conditions under which an animal is slaughtered, stored, transported, and merchandized (Carroll et al., 2007). Feces are the main source of contamination; pathogens from the genera *Campylobacter, Salmonella, Clostridium, Escherichia, Yersinia*, and *Staphylococcus*, and some spoilage microorganisms of the genera *Alcaligenes, Citrobacter, Enterobacter, Lactobacillus, Proteus, Streptococcus, Pseudomonas, Acinetobacter, Aeromonas*, and *Flavobacterium*, as well as yeasts are present. *Salmonella* spp. and *Campylobacter jejuni* are the most important pathogens contaminating the carcasses; these strains generally come from the fecal material and are associated with a number of food-poisoning outbreaks.

In 2000, 4640 cases of campylobacteriosis, 4237 cases of salmonellosis, 2324 cases of shigellosis, 631 cases of *E. coli* O157, 484 cases of cryptosporidosis, 131 cases of yersinosis, 101 cases of listeriosis, 61 cases of *Vibrio* infections, and 22 cases of cyclosporiasis were confirmed in the United States. Of the salmonellosis cases, 23% corresponds to *S. typhimurium*, 15% to *S. enteritidis*, 11% to *S. newport*, and 7% to *S. heidelberg*. For *Campylobacter*, the highest incidence was for *C. jejuni*; and for *Shigella*, 85% was due to *S. sonnei* and 13% to *S. flexneri* (CDC, 2001). In 2006, the European Food Safety Authority highlighted campylobacteriosis as the most often reported zoonosis in the European Union (EFSA, 2006). Foods of animal origin, including poultry, are the most often reported vector for *Campylobacter, Salmonella*, and *E. coli* O157 (CDC, 2001). The growth of microorganisms in poultry meat depends especially on the factors discussed below.

pH Due to its composition and a pH slightly above the pH of red meat, poultry meat is more susceptible to microbial spoilage. In addition, poultry carcasses generally retain the skin, increasing the risk of spoilage (Fernández Escartín, 2000). The optimum pH range for bacterial growth is 6.0 to 8.0, for yeasts is 2.8 to 8.5, and for molds is 1.5 to 9.0. Gram-positive bacteria grow between 4.0 and 8.5 and gram-negative between 4.5 and 9.0 (Ray 2001). Poultry meat also has a buffering ability due to its protein content; this effect must be considered, as it may increase the bacterial growth rate (Carroll et al., 2007).

Temperature Temperature is the most influential factor in microbial growth, and is considered a critical control point throughout the slaughtering, eviscerating, cleaning, and storage steps, as well as during transport and merchandising. If the meat is kept at 4°C or below, the shelf life can be up to one week. The predominant microorganisms are *Pseudomonas, Alteromonas, Acinetobacter*, and *Moraxella* (Fernández Escartín, 2000). Eviscerated turkeys show signs of decomposition, such as off-odor, after 7 days at 5°C; at 0°C, off-odors occur after 14 days of storage, whereas at −2°C turkeys can be preserved up to 38 days (Barnes, 1976). At 5°C, chicken carcasses can be preserved for 7 days; the decomposition rate doubles at 10°C and triples at 15°C (Daud et al., 1978).

Water Activity (a_w) The water activity is defined as the vapor pressure of water in the food divided by the vapor pressure of pure water at the same temperature

(Jay et al., 2005). The optimum water activity for the growth of most bacteria is above 0.92; the lowest water activity at which most food spoilage bacteria can grow is around 0.90 (Fontana, 2000). The minimum water activity for the growth of gram-negative bacteria is 0.97, for gram-positive bacteria is 0.90, for molds and yeast is 0.88, for halophilic bacteria is 0.75, and for xerophilic fungi is 0.61 (Adams and Moss, 2000).

Redox Potential (Eh) The redox potential measures the ability of a substrate to gain or lose electrons; it is defined as the ratio of the total oxidizing power to the total reducing power of a food (FDA, 2001) expressed in milivolts (mV) (Jay et al., 2005). Aerobic bacteria can tolerate higher Eh values than can anaerobes. In general, aerobic bacteria grow at +300 to +500 mV, facultative anaerobes at −100 to +300 mV, and anaerobes at −250 to +100 mV (Ray, 2001).

Low- and High-Acidity Foods Canned foods can undergo heat treatment to the point of commercial sterilization, which means that:

- The food is free of microorganisms that have the ability to grow in the food if it is stored under normal conditions; refrigeration is not required.
- The food contains no viable microorganisms (including spores) of public health importance.

To reach this condition, a time–temperature combination is calculated specifically for a given canned food. The heat treatment severity is influenced by food composition, pH, heat transfer mechanism, type of retort used in the process (static or agitating retort), and heating medium (vapor or hot water). On the other hand, food is divided according to its pH and a_w value:

- *Low-acidity foods* (except alcoholic beverages) at $a_w > 0.85$ and pH > 4.6. At these conditions *Clostridium botulinum* thermophilic spores can grow; therefore, processing must be severe to ensure commercial sterilization.
- *High-acidity foods* at pH < 4.6. These conditions are not favorable for *C. botulinum* spores. Heat treatments are designed to destroy all vegetative cells and some spores, being less severe than those for high-pH foods.

MICROBIAL SPOILAGE

Canned food spoilage is approached from the points of view of enzymatic reaction, chemical deterioration, or microbial spoilage. The enzymatic and chemical reaction points of view consider that canned foods can present gas production, such as CO_2 or H_2, food darkening caused by several chemical reactions, and can corrosion, all of which result in chemical deterioration. The principal growth-promoting condition is insufficient processing time, particularly for *C. botulinum* if the temperature–time processing combination is not the suitable for a given canned food. If the spoilage includes carbohydrate or protein decomposition by microorganisms such as *C. butyricum* and *C. pasteurianum*, the fermented carbohydrates produce H_2, CO_2 and volatile acids, causing can blowing. Proteolytic species such as *C. sporogenes, C. putrefacience*, and *C. botulinum* metabolize proteins and produce H_2S, CO_2, H_2, mercaptanes, indol, skatol, and ammonia, also causing can blowing. Spores of aerobic bacteria of *Bacillus* spp. survive insufficient heat treatment but do not germinate within the cans. However, spores of some facultative anaerobes, such as *Bacillus subtilis* and *B. coagulans*, can germinate and produce acid and CO_2. If inadequate can cooling after heat thermal treatment occurs, including high-temperature storage, it allows the growth of thermophilic spore-producing bacteria, mainly *C. botulinum* and *Bacilllus* spp.

Thermophilic bacteria can cause spoilage of low-acidity foods when stored at temperatures over 43°C, even for short periods. The main spoilage condition promoted by these bacteria is flat sourin. In this case, cans do not blow, but the food acidifies due to the growth of facultative anaerobic microorganisms such as *Baciliilus stearothemophilus* and *B. coagulans*. They can grow at 30 to 45°C; these microorganisms ferment carbohydrates and produce acids without gas.

The most important thermophile anaerobic bacteria are *Clostridium thermosaccharolyticum* and *Bacillus polymyxa*, which produce large quantities of H_2, CO_2, a sour and cheesy odor, and can blowing. The latter can be hard or soft; in *hard blowing*, can bulges due to gas and frequently explode. There are two types of soft blowing; in both, one end bulges and explodes. In *flipper blowing*, when pressing one end, the other pops out; in *springer blowing*, when pressing the bulged end, the other end does not pop out.

Anaerobic sporeforming gram-negative bacteria are responsible for a spoilage condition known as *sulfide stinker*, caused by bacteria from the genus *Desulfotomaculum*, such as *D. nigrificans*. The spoilage is characterized by product darkening and a strong egg-like odor caused by H_2S production. This acid, a reaction product from sulfur-containing amino acids, dissolves in water and reacts with iron, forming iron sulfide, the black color in the can.

Damaged cans allow microorganisms to get inside after thermal treatment, microorganisms that come from the ambient (air or bad-quality cooling water); the type of spoilage depends on the microorganisms present (Ray, 2004). Leakages caused by a deficient hermetic seam, even if they are microscopic, cause contamination after the cooling step.

THERMAL PROCESSING OF CANNED FOODS

Thermal processing is used successfully for treatment of foods; it is an efficient preservation method for several foods for further commercialization. The scientific basis for thermal process calculation was developed by Bigelow and Ball in 1920. Heat application to canned foods involves the utilization of a high temperature for enough time to destroy microorganisms of importance in public health or that cause food deterioration. Cans are also designed to keep a "commercial sterilization" condition within the can; double seams prevent recontamination during storage (Awuah et al., 2007).

The equipment used in canning operations has been improved to minimize the damage caused to the food nutrients and to prevent sensory changes; to optimize time–temperature processes, and therefore to increase heating rates. In addition, the packaging industry has developed better or more suitable material, such as pouches or colaminates that resist commercial sterility treatments.

In food-processing operations, several methods are used to calculate microbial reduction as a function of the time–temperature relationship. Microbial thermal inactivation kinetics is obtained initially by calculating a survivor curve, a logarithmic plot of microorganisms surviving a given heat treatment severity versus heating time, assuming that microbial destruction is a first-order reaction described by the equation

$$\frac{dN}{dt} = -kN \tag{1}$$

where k is the reaction rate constant (first order), the line slope resulting from plotting the natural logarithm of survivors vs. time. Integrating equation (1) yields

$$\ln \frac{N}{N_0} = -kt \tag{2}$$

The relationship between the decimal reduction time and a first-order reaction rate constant is

$$k = \frac{2.303}{D} \tag{3}$$

To establish processing conditions, D and z values are necessary. The D value is the heating time (in minutes or seconds) that results in a 90% reduction of the existing microbial population. The mathematical expression is

$$D = \frac{t_2 - t_1}{\log A - \log B} \tag{4}$$

where A and B are the survivor counts after heating at times t_1 and t_2. The temperature sensitivity is called the *z-value:*

$$z = \frac{T_2 - T_1}{\log D_1 - \log D_2} \tag{5}$$

D_1 and D_2 correspond to T_1 and T_2, respectively.

Deviations from a first-order reaction rate have been reported, however Because applying a log-linear model for a nonlinear survival curve will have serious implications and potential health-related risks, as the D and z values can be underestimated (Akterian et al., 1999; Awuah et al., 2007). Kinetic data using a classical first-order equation is linked to the time–temperature profile of a given food at a predefined position; the sterilization value or process lethality is calculated as

$$F_o = \int 10^{(T-T_o)/z} dt \tag{6}$$

where F_o = the overall lethality of the process
t = the time (min)
z = the temperature sensitivity of the target microorganism
T = the temperature at any given time
T_o = the reference temperature, generally 121.1°C (250°F) for low-acid foods

For low-acid foods, the minimum lethality is $F_0 = 3$ min, that is, the time–temperature condition necessary to reduce 12 log cycles of a *C. botulinum* population (Brown et al., 1991).

Another mathematical model used to calculate the processing conditions is the Ball equation:

$$B_B = f_h \log \frac{j_{ch}(T_r - T_i)}{g} \tag{7}$$

where B_B = the processing time (min)
f_h = the heating rate (min), related to the time–temperature food profile
j_{ch} = the lag factor
T_r = the retort temperature
T_i = the initial temperature of the food
g = the number of degrees below the retort temperature at the slowest heating point of the container

To ensure the safety of certain foods, the USDA has established a 6.5D reduction for *Salmonella* spp. in ready-to-eat beef products and a 7D reduction in fully cooked poultry products (USDA, 1999). However, kinetics data on *Salmonella* destruction is necessary to calculate specific D-values (Murphy et al., 2001, 2002).

Other process calculation methods use simulated conditions to estimate processing times, mainly if it is necessary to obtain data on the physical properties of foods, such as like density, c_p, thermal diffusivity, and thermal conductivity. These types of properties usually change depending on temperature and food composition, and strongly influence thermal processes. At present, networks use thermal sterilization processes as a tool to computerize procedures for thermal process calculations (Barker et al., 2002).

REFERENCES

Adams MR, Moss MO. 2000. *Food Microbiology*, 2nd ed. Royal Society of Chemistry, Cambridge, U.K.

Ahn DU, Maurer AJ. 1990. Poultry meat color: kinds of heme pigments and concentrations of the ligands. Poult Sci 69(1):157–165.

Akterian SG, Fernández PS, Hendrickx ME, Tobback PP, Periago PM, Martínez A. 1999. Risk analysis of the thermal sterilization process. Int J Food Microbiol 47:51–57.

Alvarado CZ, Sams AR. 2000. Rigor mortis development in turkey breast muscle and the effect of electrical stunning. Poult Sci 79(11):1694–1698.

Awuah GB, Ramaswamy HS, Economides A. 2007. Thermal processing and quality: principles and overview. J Chem Eng Process 46(6):584–602.

Bacus JN. 2007. Navigating the processed meats labeling maze. Food Technol 61(11): 28–32.

Bagorogoza K, Bowers J, Okot-Kotber M. 2001. The effect of irradiation and modified atmosphere packaging on the quality of intact chill-stored turkey breast. J Food Sci 66(2):367–372.

Barker GC, Talbot NLC, Peck MW. 2002 Risk assessment for *Clostridium botulinum*: a network approach. Int Biodeter Biodegrad 50(3):167–175.

Barnes EM. 1976. Microbiological Problems of poultry at refrigerator temperatures: a review. J Food Sci Agric 27(8):776–782.

Bendall JR, Hallund O, Wismer PJ. 1963. Post mortem changes in the muscles of landrace pigs. J Food Sci 28(2):156–162.

Boulianne M, King AJ. 1995. Biochemical and color characteristics of skinless boneless pale chicken breast. Poult Sci 74(12):1693–1698.

Boulianne M, King AJ. 1998. Meat color and biochemical characteristics of unacceptable dark-colored broiler chicken carcasses. J Food Sci 63(5):759–762.

Bravo-Sierra AP, Ruiz-Cabrera MA, González-Garcia R, Grajales-Lagunes A. 2005. Influencia de la temperatura de refriegeración (pre rigor) sobre la incidencia de carne PSE en cerdo. Rev Mex Ing Quim 4(2):181–189.

Briskey E, Wismer-Pedersen PJ. 1961. Biochemistry of pork muscle structure: I. Rate of anaerobic glycolysis and temperature changes versus the apparent structure of muscle tissue. J Food Sci 26(3):297–305.

Brown KL, Rees JAG, Bettison J. 1991. *Principles of Heat Preservation, Processing and Packaging of Heat Preserved Foods*, New York: Van Nostrand Reinhold.

Carroll CD, Alvarado CZ, Brashears MM, Thompson LD, Boyce J. 2007. Marination of turkey breast fillets to control the gowth of *Listeria monocytogenes* and improve meat quality in deli loaves. Poult Sci 86:150–155.

Cason JA, Bailey JS, Stern NJ, Whittemore AD, Cox NA. 1997. Relationship between aerobic bacteria, salmonellae, and *Campylobacter* on broiler carcasses. Poult Sci 76(7):1037–1041.

CDC (Centers for Disease Control). 2001. Incidence of foodborne Illnesses: FoodNet Data 2000. MMWR 50(13):241–246.

CFR (Code of Federal Regulations). 2002. Part 113.2002, title 21, vol. 2, 21 CFR 113.3, pp. 224–226.

Cornforth DP, Vahabzadeh F, Carpenter CE, Bartholomew DT. 1986. Role of reduced hemochromes in pink color defect of cooked turkey rolls. J Food Sci 51(5):1132–1135.

Daud HB, McKeen TA, Olley J. 1978. Temperature function integration and the development and metabolism of poultry spoilage bacteria. Appl Environ Microbiol 36(5): 650–654.

E.C. TO, Robach MC. 1980. Inhibition of potential food poisoning microorganisms by sorbic acid in cooked, uncured, vacuum packaged turkey products. J Food Technol 15(5):543–547.

EFSA (European Food Safety Authority). 2006. The community summary report on trends and sources of zoonoses, zoonotic agents, antimicrobial resistance and foodborne outbreaks in the European Union in 2005. http://www.efsa.europa.eu/enscience/monitoring_zoonoses/reports/zoonoses_report 2005.

FDA (U.S. Food and Drug Administration). 2000. Kinetics of Microbial Inactivation for Alternative Food Processing Technologies. Overarching Principles: Kinetics and Pathogens of Concern for All Technologies. Washington, DC: FDA Center for Food Safety and Applied Nutrition. http://vm.cfsan.fda.gov/~comm/ift-toc.htlm.

FDA (U.S. Food and Drug Administration). 2001. Factors that Influence Microbiol Growth. Evaluation and Definition of Potentially Hazardous Foods. http://www.cfsan.fda.gov/~comm/ift4-3.html.

Fernández Escartín E. 2000. *Microbiología e Inocuidad de los Alimentos*. Querétaro, Mexico: Universidad Autónoma de Querétaro, pp. 501–509.

Fernández X, Forslid A, Tornberg E. 1994. The effect of high postmortem temperature on the development of pale, soft and exudative pork: interaction with ultimate pH. Meat Sci 37(1):133–147.

Fletcher DL. 1999. Broiler breast meat color variation, pH, and texture. Poult Sci 78(9):1323–1327.

Fontana AJ. 2000. Water activity's role in food safety and quality. 2nd NSF International Conference on Food Safety. Savannah, GA.

Fox JB Jr. 1987. The pigments of meat. In: Price JF, Schweigert, eds., *The Science of Meat Products*, 3rd ed. Westport, CT: Food and Nutrition Press, pp. 193–216.

Froning GW. 1995. Color of poultry meat. Poult Avian Biol Rev 6(1):83–93.

Froning GW, Hartung TE. 1967. Effect of age, sex, and strain on color and texture of turkey meat. Poult Sci 46(5):1261.

Froning GW, Daddario J, Hartung TE, Sullivan TW, Hill RM. 1969. Color of poultry meat as influenced by dietary nitrates and nitrites. Poult Sci 48(2):668–674.

Gaurav T, Vijay KJ. 2007. *Advances in Thermal and Non-thermal Food Preservation*. Ames, IA: Blackwell Publishing, pp. 4–16.

Girard B, Vanderstoep J, Richards JF. 1989. Residual pinkness in cooked turkey and pork muscle. Can Inst Food Sci Technol J 22(4):372–377.

Heaton KM, Cornforth DP, Moiseev IV, Egbert WR, Carpenter CE. 2000. Minimum sodium nitrite levels for pinking of various cooked meats as related to use of direct or indirect dried soy isolates in poultry rolls. Meat Sci 55(3):321–329.

Holownia K, Chinnan MS, Reynolds AE. 2003. Pink color in poultry white meat as affected by endogenous conditions. J Food Sci 68(3):742–747.

Holownia K, Chinnan MS, Reynolds AE, Davis JW. 2004. Relating induced in situ conditions of raw chicken breast meat to pinking. Poult Sci 83(1):109–118.

Holdsworth SD. 1997. *Thermal Processing of Packaged Foods*. London: Blackie Academic & Professional, pp. 139–161.

Izumi K, Cassens RG, Greaser ML. 1982. Reaction of nitrite and cytochrome *c* in the presence or absence of ascorbate. J Food Sci 47(5):1419–1422.

Janky DM, Froning GW. 1973. The effect of pH and certain additives on heat denaturation of turkey meat myoglobin. Poult Sci 52(1):152–159.

Jay JM, Loessner MJ, Golden DA. 2005. *Modern Food Microbiology*. 7th ed. Springer, New York.

Jossell A, Seth GV, Tornberg E. 2003. Sensory quality and the incidence of PSE of pork in relation to crossbreed in an *RN* phenotype. Meat Sci 65(1):651–660.

Kieffer KJ, Claus JR, Wang H. 2000. Inhibition of pink color development in cooked uncured ground turkey by the addition of citric acid. J Muscle Sci 11(3):235–243.

Lawrie RA. 1998. Constitución química y bioquímica del músculo. In: *Ciencia de la Carne*. Zaragoza, Spain: Editorial Acribia, Tercera Edición, pp. 67–109.

Le Bihan-Duval E, Millet N, Remignon H. 1999. Broiler meat quality: effect of selection for increased carcass quality and estimates of genetic parameters. Poult Sci 78(6):822–826.

Livingston DJ, Brown WD. 1981. The chemistry of myoglobin and its reactions. Food Technol 35(5):244–252.

McKee SR, Sams AR. 1997. The effect of seasonal heat stress on rigor development and the incidence of pale, exudative turkey meat. Poult Sci 76(11):1616–1620.

Mugler DJ, Mitchell JD, Adams AW. 1970. Factors affecting turkey meat color. Poult Sci 49(6):1510–1513.

Murphy RY, Johnson ER, Marks BP, Johnson MG, Marcy JA. 2001. Thermal inactivation of *Salmonella senftenberg* and *Listeria innocua* in ground chicken breast patties processed in an air convection Oven. Poult Sci 80(4):515–521.

Murphy RY, Duncan LK, Johnson ER, Davis MD, Marcy JA. 2002. Thermal inactivation of *Salmonella senftenberg* and *Listeria innocua* in beef/turkey blended patties cooked via fryer and/or air convection oven. J Food Sci 67(5):1879–1885.

Nam KC, Ahn DU. 2002. Mechanisms of pink color formation in irradiated precooked turkey breast meat. J Food Sci 67(2):600–607.

Nash DM, Proudfoot FG, Hulan HW. 1985. Pink discoloration in cooked broiler chicken. Poult Sci 64(5):917–919.

Owens CM, Sams AR. 2000. The influence of transportation on turkey meat quality. Poult Sci 79(8):1204–1207.

Pegg RB, Shahidi F. 2000. *Nitrite Curing of Meat. The N-Nitrosamine Problem and Nitrite Alternatives*. Trumbull, CT: Food and Nutrition Press, Inc.

Pérez-Dubé D, Andújar-Robles G. 2000. Cambios de coloración de los productos cárnicos. Rev Cub Aliment Nutri 14(2):114–123.

Petracci M, Fletcher DL. 2002. Broiler skin and color changes during storage. Poult Sci 81(10):1589–1597.

Pietrzak M, Greaser ML, Sosnicki AA. 1997. Effect of rapid rigor mortis processes on protein functionality in pectoralis major muscle of domestic turkeys. J Anim Sci 75(8):2106–2116.

Pikul J, Niewiarowicz A, Kupijaj H. 1986. The cytochrome *c* content of various poultry meats. J Sci Food Agric 37(12):1236–1240.

Prandhan AK, Li Y, Swem BL, Mauromoustakos A. 2005. Predictive model for the survival, death, and growth of *Salmonella typhimurium* in a broiler hatchery. Poult Sci 84(12):1959–1966.

Qiao M, Fletcher DL, Smith DP, Northcutt JT. 2001. The effect of broiler breast meat color on pH, moisture, water-holding capacity, and emulsification capacity. Poult Sci 80(5):676–680.

Rathgeber BM, Boles JA, Shand PJ. 1999. Rapid postmortem pH decline and delayed chilling reduce quality of turkey breast meat. Poult Sci 78(3):477–484.

Ray B. 2001. *Fundamental Food Microbiology*. Boca Raton, FL: CRC Press.

Ray B. 2004. *Fundamental Food Microbiology*. Boca Raton, FL: CRC Press.

Sammel LM, Claus JR. 2003. Citric acid and sodium citrate effects on reducing pink color defect of cooked intact turkey breast and ground turkey rolls. J Food Sci 68(3):874–878.

Sammel LM, Claus JR. 2006. Citric acid and sodium citrate effects on pink color development of cooked ground turkey irradiated pre- and post-cooking. Meat Sci 72(3):567–573.

Sammel LM, Claus JR. 2007. Calcium chloride and tricalcium phosphate effects on the pink color defect in cooked ground and intact turkey breast. Meat Sci 77(4):492–498.

Sammel LM, Claus JR, Greaser ML, Lucey LA. 2007. Identifying constituents of whey protein concentrates that reduce the pink color defect in cooked ground turkey. Meat Sci 77(4):529–539.

Sams AR, Alvarado CZ. 2004. Turkey carcass chilling and protein denaturation in the development of pale, soft, and exudative meat. Poult Sci 83(6):1039–1046.

Santé VS, Lebert A, Le Pottire G, Quali A. 1996. Comparison between two statistical models for prediction of turkey breast meat colour. Meat Sci 43(3):283–290.

Schwarz SJ, Claus JR, Wang H, Marriott NG, Graham PP, Fernandes CF. 1997. Inhibition of pink color development in cooked, uncured ground turkey through the binding of non-pink generating ligands to muscle pigments. Poult Sci 76(10):1450–1456.

Schwarz SJ, Claus JR, Wang H, Marriott NG, Graham PP, Fernandes CF. 1999. Inhibition of pink color development in cooked, uncured turkey breast through ingredient incorporation. Poult Sci 78(2):255–266.

Slesinski AJ, Claus JR, Anderson-Cook CM, Eigel WE, Graham PP, Lenz GE, Noble RB. 2000. Ability of various dairy proteins to reduce pink color development in cooked ground turkey breast. J Food Sci 65(3):417–420.

Stumbo CR. 1973. *Thermobacteriology in Food Processing*. New York: Academic Press, pp. 93–188.

Trout GR. 1989. Variation in myoglobin denaturation and color of cooked beef, pork, and turkey meat as influenced by pH, sodium chloride, sodium tripolyphosphate, and cooking temperature. J Food Sci 54(3):536–540, 544.

USDA (U.S. Department of Agriculture). 1999. Performance Standards for the Production of Certain Meat and Poultry Products. Fed Reg. 64(3):732–749. Washington, D.C. Food Safety and Inspection Service, United States Department of Agriculture.

13

TURKEY BACON

EDITH PONCE-ALQUICIRA
Departamento de Biotecnología, Universidad Autónoma Metropolitana–Unidad Iztapalapa, México D.F., México

OCTAVIO DUBLÁN-GARCÍA
Facultad de Química, Universidad Autónoma del Estado de México, Toluca, México

Handbook of Poultry Science and Technology, Volume 2: Secondary Processing, Edited by Isabel Guerrero-Legarreta and Y.H. Hui

INTRODUCTION

For decades poultry meat consumption has been increasing considerably all around the world. There are several reasons behind this trend, such as the relatively low production costs and the fast animal growth, together with a rapid inversion recovery and easy handling and adaptability. In addition, poultry meat is recognized as being low in fat and cholesterol, high in protein, and a good source of vitamins and minerals; therefore, poultry is an alternative for a healthy low-fat diet. Additionally, poultry meat possesses an excellent texture and great versatility, as it is easily adapted to many cuisine styles, due to its neutral taste and flavor. All these characteristics explain its increased acceptability.

Consumers demand tasty poultry products that offer convenience, variety, and a low-fat content. The poultry industry has understood these demands by changing from fresh, minimally processed birds to products further processed, as shown in Figure 1. Many new processed poultry products have been introduced and accepted widely (Rogers, 2002). In many cases these new poultry products originate from traditional meat products, with beef or pork partially or fully substituted by turkey or chicken meat as an alternative low-fat, value-enhanced meat product. Examples include frankfurters, turkey ham, smoked turkey breast, luncheon meats, chicken nuggets, and other restructured meat products where poultry bacon is included, including microwavable precooked entrées and ready-to-cook main dishes. All these products cover new niches and offer alternatives for poultry processing. Further processing permits product differentiation and competitiveness however, it is necessary to ensure the continued growth of the poultry sector by

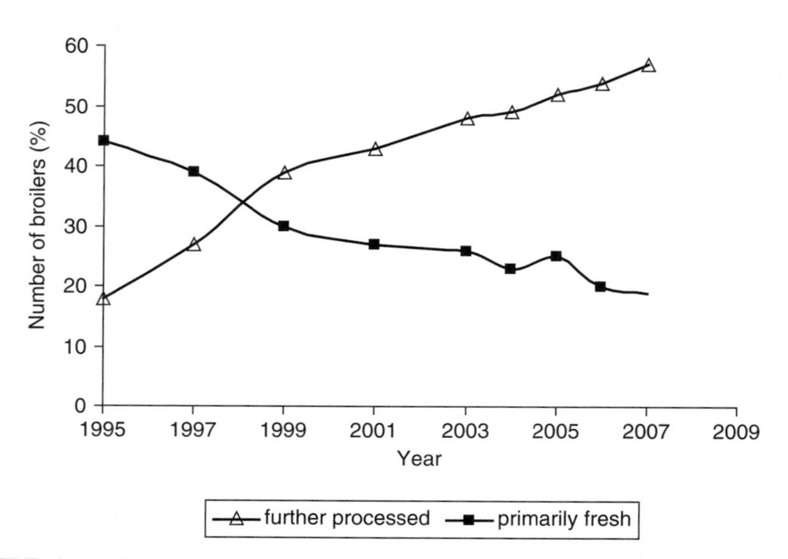

FIGURE 1 Major broiler market destinations as a percentage. (From the National Chicken Council, February 2008.)

continuous improvement in meat quality and safety in a vertical integrated chain from the production sector up to the consumer's table.

Traditional bacon is a cured and smoked meat product obtained from the pork belly or pancetta. It is usually served thinly sliced, uncooked, fried, baked, or grilled; otherwise, it is used as a condiment or topping in dishes. This meat product is highly appreciated because of its characteristic flavor and texture, but it is also used as a source of fat. However, bacon consumption is related to a high cholesterol diet as it contains 40% total fat, from which 14% is saturated, and has 70 mg cholesterol. Several bacon analogues are made from other meat cuts, as well as from other meat sources with the purpose of reduce fat content. Among those products, turkey bacon is gaining popularity as an alternative for the low fat diet, without diminishing flavor and texture characteristics. Fat operates as a flavor carrier, helps flavor perception, juiciness and adds mouthfeel, the correct application of meat technology makes possible replace fat by "structured water" as in an emulsion meat systems, taking into consideration that consumers wants low fat and healthier products without diminishing flavor and taste (Vandendriessche, 2008). Therefore, the aim of the present chapter is to present an outline for the manufacture turkey bacon process, focusing on the selection of raw materials and on those process parameters that may affect meat functionality in order to ensure the quality of the final product.

DEFINITION AND PROCESS

According to the Food Safety and Inspection Service of the U.S. Department of Agriculture (USDA) (USDA–FSIS, 2008a), *bacon* is the cured and smoked belly of a swine carcass. This product is characterized by presenting visible fat and muscle bands. If meat from other cuts is used, the product name must be qualified to identify the specific cut, such as "Pork Shoulder Bacon." Meat from other species, such as beef, lamb, chicken, or turkey, can be processed to resemble bacon, but products are designed as analogs, such as turkey bacon. Several works have been reported in relation to the manufacture of bacon analogs involving the use of turkey, such as the methods reported by Shanbhag et al. (1978), Roth (1984), Gundlach et al. (1999), Gruis (2005), and more recently, Couttolenc-Echeverria (2007). Turkey bacon can be described as a multiphase meat product. Most processes include the selection of materials and the elaboration of two phases: a lean, dark turkey meat emulsion or meat slurry in combination with a white, high-fat emulsion, where subsequent layers of the high fat and lean comminuted mixtures are molded into a rectangular shape and subjected to cooking, smoking, cooling, slicing, and packaging, as shown in Figure 2.

Darker and white emulsions are extruded from two separated stuffing machines to the desired ratio of appearance of dark and white portions (see Figure 3). The product obtained resembles the traditional pork bacon in appearance and flavor but with a lower fat content. Table 1 compares the composition of the pork and turkey bacon. However, it is claimed that emulsion-type bacon is rather dry and has a rubbery texture, which can be overcome by incorporation of entire meat

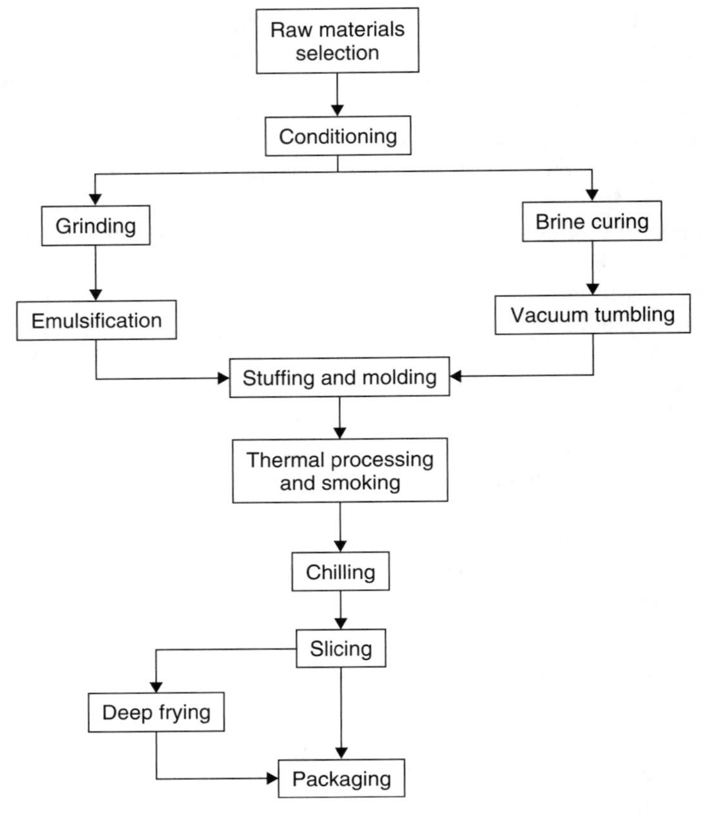

FIGURE 2 Flowchart for turkey bacon processing. (Adapted from Shanbhag et al., 1978; Roth, 1984; Gundlach et al., 1999; Gruis, 2005; Couttolenc-Echeverria, 2007.)

portions (Gruis, 2005; Couttolenc-Echeverria, 2007). In contrast to the majority of turkey bacon processes, Gruis 2005 describes ready-to-eat bacon; the cooked product is sliced and fried in oil, then a flavoring is added and it is packaged. As a result of frying, the product develops a crispy texture very similar to that of fried bacon.

INGREDIENTS

Selection of Raw Materials

Turkey bacon is prepared from lean and fatty cuts derived from boneless breast, legs, thighs, desinewed drumsticks, and dark mechanically deboned turkey (MDT) with or without skin, and fat. Turkey fat is preferable, but other fat sources can be used. These raw materials may be chilled or frozen, but with no deterioration, off-color, off-odor, or apparent microbial growth. The internal

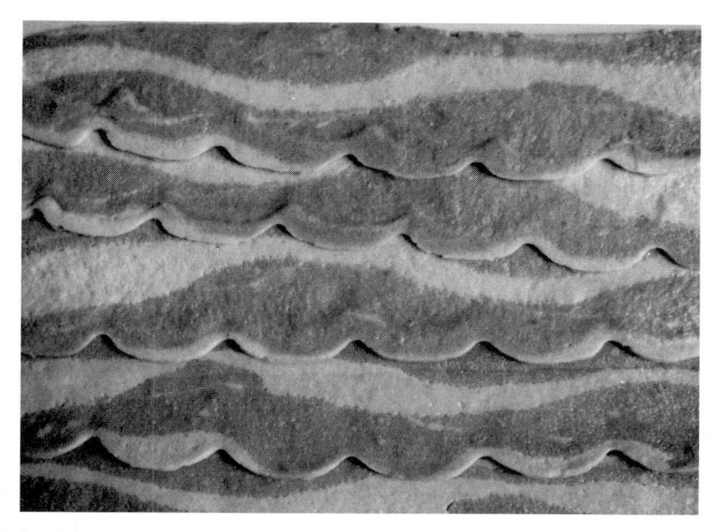

FIGURE 3 Example of a commercial turkey bacon analog showing the dark and white meat bands that resemble pork bacon appearance.

TABLE 1 Chemical Composition of Pork and Turkey Bacon

	Raw Pork bacon	Turkey Bacon
Carbohydrates (%)	0	0
Protein (%)	31	15
Total fat (%)	41	18
Saturated fat (%)	14	9
Trans fat (%)	0	0
Cholesterol (mg/100 g)	68	90
Sodium (%)	2.1	0.7
Calories (kcal/100 g)	457	220

Source: Data from http://www.nutritiondata.com/facts-C00001-01c20Zc.html, and from commercial turkey bacon brands distributed in Mexico City.

temperature of fresh cuts should not be above 4.4°C, while frozen materials should be −18°C when received. Frozen cuts must be packaged during thawing to prevent dehydration and to avoid the risk of microbial contamination (Keeton, 2002; Smith, 2001; Smith and Acton, 2001). Elaboration of any poultry product including turkey bacon requires postmortem inspection of all carcasses, where one or two inspectors on each eviscerating line examine the whole carcass and viscera according to the US-NTI System regulations (9 CFR 381.68).

Turkey meat contains 75% moisture, 23% protein, 1.2% lipids, and 1% minerals. Among them, proteins are the most important meat components in the bacon manufacture process. Myoglobin is a water-soluble protein responsible for meat color, the concentration varies with age, feed, and muscle fiber distribution;

for instance, thigh dark muscles contain more myoglobin (0.6 to 2 mg/g) than white breast muscles (0.1 to 0.4 mg/g); MDT contains some bone marrow and will have higher pigment levels than those of manually deboned meat; therefore, MDT serves is a source of dark meat to prepare turkey bacon (Froning and McKee, 2001; Smith, 2001). On the other hand, myofibrillar proteins comprise 50% of the total meat proteins; among them, myosin and actin from the actomyosin complex in postrigor, and are responsible of the physicochemical properties of meat involved in the bacon manufacture, such as water holding capacity (WHC), binding, and emulsion capacity. WHC refers to the ability of meat to retain or absorb added water in response to an external force such as cooking. WHC is also responsible for sensory quality, juiciness, and product yield. Binding refers to the ability to join meat pieces in a continuous structure upon heating. Emulsion capacity is based on protein–fat interactions, where fat is dispersed in fine droplets and dispersed into the continuous phase, composed of water, protein, and salt and forming a meat batter; proteins unfold and form a protein film, in the polar regions orient themselves toward the water, while nonpolar regions move toward the surface of fat droplets. Thus, proteins reduce interfacial tension between fat and water, preventing fat coalescence (Smith, 2001). Those properties are affected by intrinsic and extrinsic factors such as the amount of myofibrillar proteins, pre- and postmortem condition, muscle integrity, pH, and the addition of salts and other nonmeat constituents. Thus, turkey bacon processing must encourage the development of the functional properties above.

Curing Ingredients

The basic curing ingredients are listed in Table 2, and include salt, phosphates, sodium nitrite, reducing agents, and sugars (Pearson and Gillett, 1999). These can be incorporated directly during emulsification or by immersion into curing brine. Salt improves flavor, and in conjunction with phosphates extracts myofibrillar proteins for emulsification and binding. Sodium chloride increases protein negative charge as well as protein repulsion, allowing more water to bind within the muscle fibers, leading to the solubilization of myofibrilar proteins. On the other hand, alkaline phosphates increase pH and ionic strength, inducing protein to uncoil, exposing hydrophilic and hydrophobic sites needed to form a stable meat emulsion; therefore, phosphates act in a synergistic way with sodium chloride to increase meat functionality.

Sodium nitrite is a multifunctional ingredient which is essential to prevent the growth of *Clostridium botulinum*. Nitrite is also responsible for the development of the distinctive color and flavor of cured products such as bacon; the nitric oxide derived from sodium nitrite reacts with the heme iron of myoglobin and metmyoglobin to form nitrosylmyoglobin. Nitrite also contributes to the flavor stability and consumer acceptance by preventing lipid oxidation and warmed-over flavors by complexing the heme iron, which could promote lipid oxidation reactions (Van Laack, 1994; Claus et al., 1994; Sindelar et al., 2007). Legal limits of initial nitrite levels vary from 120 to 200 ppm for injected and immersed cured

TABLE 2 Basic Curing Ingredients for the Manufacture of Turkey Bacon

Component	Percent
Water	65–80
Salt	3.5–5
Phosphate	0.6–2
Sweeteners	0.5–3
Antioxidants	0.15–0.3
Antimicrobials	0.2–0.5
Nonmeat proteins	0–7.5
Hydrocolloids	0–3
Herbs and spices	0–0.7
Liquid smoke and bacon flavor	1
Sodium nitrite	156 ppm
Sodium eritorbate	450 ppm
Colorants	0–0.3

Source: Adapted from Pearson and Gillett (1999); Couttolenc-Echeverria (2007).

bacon products, with residual levels in the final product varying from 40 to 120 $NaNO_2$ ppm (the latter always in combination with 550 ppm of sodium erythorbate or ascorbate). However, elevated residual nitrite is a well-known risk factor in the potential formation of carcinogenic nitrosamines; thus, ingoing and residual nitrite concentrations must be controlled carefully to ensure product safety. Addition of reducing agents such as sodium erythorbate accelerates curing, promotes formation of nitrosylhemochrome, contributes to flavor and color stability, and prevents the formation of carcinogenic nitrosamines. Nitrite is generally used as a curing salt consisting of 6.25% sodium or potassium nitrite and 93.75% salt; it is usually colored pink or yellow to distinguish it from salt or any other white crystalline ingredient and to avoid the risk of intoxication.

Finally, sugars may be added at a low level, as they play a minor role in flavor and color development; the addition of sugars softens the hardening effects of salt, reduces moisture loss, and interacts with amino groups during heating, producing brown colors and flavor compounds. Common sweeteners include dextrose, sugarcane, molasses, sorbitol, and corn syrup solids (Pearson and Gillett, 1999).

Auxiliary Additives

Several auxiliary ingredients (antioxidants, colorants, emulsifiers, flavorings, nonmeat proteins, preservatives, seasonings, and spices) may be incorporated to improve the stability or to improve binding, texture, and water-retention properties, as well as fat replacers.

One of the main problems of turkey bacon, as of other additionally-processed products, is the high susceptibility of flavor and color deterioration brought about by oxidation. Antioxidants can retard oxidative deterioration by forming derivatives that block the oxidation pathway. Common antioxidant compounds include butylated hydroxyanisole (BHA), butylated hydroxytoluene (BHT), tertiary butylhydroquinone (TBHQ), and propyl gallate (PG). Recently, natural and innovative antioxidants have been introduced as an option to increase product stability. Metal chelators (e.g., citric acid, ascorbic acid, phosphoric acid), polyphenols (e.g., green tea and apple extracts, epicatechin, chlorogenic acid, quercetin glucosides), α-tocopherols, and spice extracts (e.g., garlic, rosemary, sage), and others have been reported to be suitable for oil-in-water emulsions such as turkey bacon (Liu and Yang, 2008). However, antioxidants must be used in combination with a low-oxygen packaging such as vacuum packaging.

Nonmeat proteins used for meat and poultry products include soy proteins (flours, concentrates, and isolates), milk proteins (whey protein concentrates and isolates, nonfat dry milk, and sodium caseinates), egg white, and gluten. Protein concentrates and isolates contain 70 and 90% dry-weight protein, respectively; they have a rather light flavor profile, with the advantage that they can improve emulsion properties and contribute to water and fat binding within the meat matrix. In particular, soy proteins and whey proteins (WPC and WPI) form a gel matrix comparable to that of meat in appearance and texture. Thus, the addition of nonmeat proteins increase binding, reduce water losses, increase brine retention for entire cuts, improve emulsion stability, and maintain a meatlike texture. Concentrates can be incorporated at levels up to 11%, while isolates are restricted to 2%. Other nonmeat proteins include gelatin and protein hydrolyzates. In particular, gelatin possesses high water binding and gelling properties, improving texture when added at levels from 0.5 to 3%. However, gelatin has a low nutritional value compared with soy and milk proteins. Finally, hydrolyzed plant and animal proteins that contain peptides and free amino acids are incorporated primarily as meat flavor enhancers.

Hydrocoloids such as starches, carrageenan, konjac, and alginates may be included, due to their low cost and their ability to absorb water. These ingredients are used as fat replacers, to improve yield, emulsion, and water binding, as well as to enhance product sliceability and juiciness, and to protect the product from excessive water loss during freezing and thawing. In most cases added levels of hydrocolloids are regulated; for example, carrageenan is used at levels below 1%, while starches may be used up to 10% according to the product and regulatory restrictions (Keeton, 2001).

It is also common to use flavor enhancers such as monosodium glutamate (MSG), inosine 5′-monophosphate (IMP), and guanosine 5-monophosphate. These compounds are accepted as GRAS food additives, but enhancing mechanisms are not well understood; they are responsible for the umami flavor perception and also increase the time of residence of active flavorings with the gustative receptors. The amount of MSG added varies from 0.1 to 1%; levels for IMP and GMP range from 0.002 to 0.03% (Hettiarachchy and Kalapathy, 2000; Marcus, 2005).

EMULSIFICATION

White and dark meats are first ground and analyzed for fat content, then, separately, the meats are placed into a bowl chopper with fat, salt, phosphates, and ice to form a homogeneous meat batter. The dark batter includes approximately 50 to 65% MDT, 10% fat, and 20% white turkey meat; while the white batter includes 60% white meat and 30% fat. Both white and dark turkey meat batters are complex emulsions that consist of fine fat droplets dispersed into a continuous phase comprised of water, meat, and nonmeat proteins and salts that form a stable gel matrix when heated. Emulsifying involve three stages: protein extraction, fat encapsulation, and formation of a stable heat-set gel.

The first emulsifying stage is initiated by chopping lean meats with curing ingredients; water is added in the form of ice to maintain a temperature below 4°C. Sarcoplasmic and myofibrillar proteins solubilize and swell as a result of partial unfolding due to the combined effect of ionic interactions and pH increase induced by the addition of salt and alkaline phosphates. Collagen and other stroma proteins are insoluble and exert a limited emulsifying activity under these conditions; therefore, connective tissue must be limited to 15% of the total meat content. After protein extraction, fat tissues and ice are incorporated and homogenized, preferably under vacuum. During this second homogenizing stage, fat tissues are dispersed into small particles within the aqueous phase; at the same time, myofibrillar proteins undergo conformational changes. Myosin forms a monomolecular protein film around fat particles by exposing the hydrophobic myosin heads that dip into the fat droplets; the hydrophilic tails shift toward the water phase, interacting with the actomyosin complex present in the continuous phase. Therefore, myofibrillar proteins, mainly myosin, are absorbed within the water–fat interface entrapping fat particles, thus forming an emulsion. Temperature must be kept below 10°C during homogenization to maintain the fat droplets in a plastic or semisolid state. If the temperature increases above the fat melting point, the liquid fat will be expelled and the emulsion will break down. Once the desired texture is obtained, meat batters are extruded into a rectangular frame and heated to an internal temperature of 68 to 73°C. The heat treatment denatures myofibrillar proteins, causing formation of a gel where fat and water are trapped into a stable matrix. Coagulation of meat proteins is initiated at 57°C and continues up to 90°C; however, temperatures above 75°C causes excessive fiber shrinkage, water loss, and fat melting; thus, the maximum cooking temperature for a stable meat emulsion is around 73°C (Belitz and Grosch, 1999; Keeton, 2001).

BRINE CURING AND TUMBLING

In addition to the dark batter, turkey cuts can be incorporated to give a meatlike texture to turkey bacon. In this case, meat portions are cured by immersion in brine followed by tumbling. Injection of brine under pressure using a multineedle system may also be used, as it facilitates and accelerates incorporation of curing salts. Injection followed by noncontinuous tumbling cycles allows a uniform distribution and rapid absorption of curing ingredients; in addition, extracted salt-soluble proteins improve binding and product texture. The temperature should be kept between 4 and 8°C during this stage and the rotation rate between 3 and 15 rpm; higher speeds can cause cell breakdown and increased temperature, reducing the quality of the final product. Vacuum tumbling has the advantage of speeding up the brine uptake, thus avoiding the formation of air bubbles within the product.

MOLDING AND THERMAL PROCESSING

Once meat emulsions or turkey meat cuts are ready, those portions are molded in alternative layers to resemble pork bacon appearance. Several procedures are available to create turkey bacon. In general, the dark and white portions are stuffed separately and extruded concurrently into a silicone-lined board pan or cook rack, or stuffed into a plastic or cellulose heating bag, placed in an oven for cooking, and finally are smoked (Gruis, 2005). Cheney (1980) invented an apparatus to form bacon analogs, in which the two meat portions are pumped thorough separate manifolds, forming a die that includes several partitions, according to the desired lean and fat configuration.

Thermal processing is usually performed in a steam oven with a gradual temperature increase from 50°C until the internal temperature reaches 73 to 74°C; the total cycle can last about 5 h. However, the USDA allows an internal temperature of 60°C and is then cooled to 26°C and kept under refrigeration (<4°C) for poultry products such as turkey bacon that will be cooked prior to consumption. During cooking, as proteins denature and form a meat gel entrapping water and fat particles within in a solid matrix, the product develops a firm texture that can be sliced. Turkey bacon is then cooled down and chilled overnight to reach 4°C prior to being taken apart for smoking and slicing. Heating also destroys microorganisms, but special care needs to be taken during slicing and packaging to prevent recontamination. (Keeton, 2001; Gruis, 2005; Couttolenc-Echeverria, 2007). Even though turkey bacon receives a heat treatment, it is not fully cooked; thus, this product still needs to be cooked thoroughly before consumption, as must be indicated on the label (USDA–FSIS, 2008b).

SMOKING

Usually, turkey bacon is smoked by direct exposure to wood combustion products, or by addition of liquid smoke to enhance and modify flavor; smoking also aids in the development of surface color and extends shelf life by deposition of several antibacterial substances. Smoke components include phenols, organic acids, hydrocarbons, alcohols, and carbonyl compounds derived from the oxidation and controlled pyrolysis of lignin and cellulose. Among them, the phenolic compound syringol, which is derived from lignin, is the main compound responsible for the smoke and hickory flavor. Moreover, phenolics possess high antioxidant activity, reducing lipid oxidation. During smoking a brown color is also developed, derived from the Maillard reaction between free amino groups of proteins or amino acids and the carbonyls in smoke. Hardwoods produce high-quality smoke, whereas softwoods (e.g., pine, fir, cedar, spruce) produce a turpentine or resin flavor and black spots. Force-conditioned and continuous smoke systems offer more advantages, as they require less space and lower labor, and allow better control than does the traditional natural air smokehouse. Liquid smoke preparations are used by several processors because use of liquid smoke does not require a smoke generator, thus saving time and money, allowing flavor homogeneity between batches, and being free of carcinogenic compounds. Liquid smoke can be added directly to a meat emulsion or applied by spraying onto the product surface (Pearson and Gillet, 1999).

PACKAGING

Packaging protects and preserves meat products during storage and distribution, as it operates as a selective barrier to the passage of oxygen, carbon dioxide, water, steam, and other volatile contaminants (Nychas et al., 2008). Packaging materials also serve as a barrier against light and biological contamination caused by particles, insects, rodents, microorganisms, and other biological agents present in the environment. Turkey bacon is usually vacuum packaged to reduce lipid oxidation and to preserve color and flavor. Vacuum packaging also inhibits the growth of aerobic microorganism, thus increasing shelf life and quality (Nychas et al., 2008). Common polymers used for food packaging include polyethylene (PE), polypropylene (PP), ionomers, poly(vinyl chloride) (PVC), poly(vinylidene chloride) (PVdC), ethylene vinyl alcohol (EVOH), ethylene vinyl acetate (EVA), polystyrene (PS), polyamides (nylons), polyesters, polycarbonates (PCs), and others. These plastic materials have good heat-seal and excellent moisture-barrier properties. However, during recent years, several packaging materials based on biopolymers are under development for commercial use. Biodegradable packaging materials have captured people's interest, due to the elevated costs of petroleum and the reduction in proven reserves of this resource, and also because of the increasing preoccupation with the destructive environmental impact involved in the use of nonbiodegradable synthetic materials (Quintero Salazar and Ponce Alquicira, 2007). The U.S. Environmental Protection Agency (EPA) states that food packaging generates two-thirds of the total municipal residues during 2006. Nevertheless, vacuum packaging is now the best option for turkey bacon distribution and commercialization, as it retards lipid oxidation and microbial growth (Dawson, 2001, Marsh and Bugusu, 2007).

Recently, the industrial application of high-pressure processing (HPP) has been applied to meat and meat products as a means of reducing the microbial population and extending shelf life without the use of thermal treatment, particularly in products that are sold sliced, such as turkey bacon. HPP-treated meat products are first vacuum packaged and placed into a pressurizing vessel, where a pressure liquid (usually, water) is introduced to generate a pressure between 200 and 800 MPa (1 MPa = 9.86 atm). Under these conditions intra- and intermolecular weak energy bonds are affected, inducing microbial and enzyme inactivation as in a pasteurization process but without the use of an extra heat treatment (Lamballerie-Anton, 2002).

CONCLUSIONS

Turkey bacon is a multiphase meat product. Its manufacture includes the selection of materials, elaboration of lean dark meat emulsions and a white (high-fat) emulsion, molding in alternate layers of the high-fat and lean comminuted mixtures into a rectangular shape to resemble bacon. The product is then cooked, smoked, cooled down, and sliced. It is then ready to be packed.

REFERENCES

Belitz HD, Grosch W. 1999. *Food Chemistry*, 2nd ed. New York: Springer.

Cheney EJ. 1980. Apparatus for forming bacon product analogues and the like. U.S. patent 4,200,959.

Claus JR, Colby JW, Flick GJ. 1994. Processed meats/poultry/seafood. In: Kinsman DM, Kotula AW, Breidenstein BC, eds., *Muscle Foods: Meat, Poultry, and Seafood Technology*. New York: Chapman & Hall, Chap. 5.

Couttolenc-Echeverria JG. 2007. Method for preparing bacon analogues. MXPA05007331-A Instituto Mexicano de la Propiedad Industrial.

Dawson PL. 2001. Packaging. In: Sams AR, ed., *Poultry Meat Processing*. Raton, FL: CRC Press, pp. 73–95.

Froning GW, McKee SR. 2001. Mechanical separation of poultry meat and its use in products. In: Sams AR, ed., *Poultry Meat Processing*, Boca Raton, FL: CRC Press, Chap. 14.

Gruis DJ. 2005. Method for making turkey bacon. U.S. patent 20050175749. http://www.freepatentsonline.com/4104415.html. Accessed Feb. 2008.

Gundlach LC, Wauters RP, Raap TA, Selz ME. 1999. Process for making low fat bacon. U.S. patent 5925400. http://www.freepatentsonline.com/5925400.html. Accessed Feb. 2008.

Hettiarachchy N, Gnanasambandam R. 2000. Poultry products. In: Christen GL, Smith JS, eds., *Food Chemistry: Principles and Applications*. West Sacramento, CA: Science Technology System, pp. 387–398.

Hettiarachchy N, Kalapathy U. 2000. Food additives. In: Christen GL, Smith JS, eds., *Food Chemistry: Principles and Applications*. West Sacramento, CA: Science Technology System, pp. 261–273.

Keeton JT. 2001. Formed and emulsion products. In: Sams AR, ed., *Poultry Meat Processing*. Boca Raton, FL: CRC Press, Chap. 12.

Lamballerie-Anton M. 2002 High pressure processing. In: Kerry JF, Kerry JP, Ledward D, eds., *Meat Processing*. London: CRC Press, pp. 313–326.

Liu TT, Yang TS. 2008. Effects of water soluble natural antioxidants on photosensitized oxidation of conjugated linoleic acid in an oil in water emulsion system. J Food Sci 73(4):C256–C261.

Marcus JB. 2005. Culinary applications of umami. Food Technol 598(5):24–29.

Marsh K, Bugusu B. 2007. Food packaging: roles, materials, and enviromental issues. J Food Sci 72(3):R39–R55.

Nychas GJ, Skandamis PN, Tassou CC, Koutsoumanis KP. 2008. Meat spolilage during distribution. Meat Sci 78:77–89.

Pearson AM, Gillett TA. 1999. *Processed Meats*, 3rd ed. Gaithersburg, MD: Aspen Publishers, pp. 53–78, 372–413.

Quintero Salazar B, Ponce Alquicira E. 2007. Edible packaging for poultry and poultry products. In: Hui YH, ed., *Handbook of Food Products Manufacturing*, vol. 2. Hoboken, NJ: Wiley, pp. 796–815.

Rogers RT. 2002. Broilers: differentiating a commodity. In: Deutsch LL, ed., *Industry Studies*, 3rd ed. New York: M.E Sharpe, p. 59.

Roth EN. 1984. Method for forming bacon-like products. U.S. patent 4,446,159. http://www.freepatentsonline.com/4446159.html. Accessed Feb. 2008.

Shanbhag SP, Liggett LG, Mikovits AC. 1978. Process for preparing bacon analog. U.S. patent 4,104,415. http://www.freepatentsonline.com/4104415.html. Accessed Feb. 2008.

Sindelar J, Cordray JC, Olsin DG, Sebranek JG, Love LA. 2007. Investigating quality attributes and consumer acceptance of uncured, no-nitrate/nitrite-added commercial hams, bacons, and frankfurters. J Food Sci 72(8):S551–S559.

Smith DM. 2001. Functional properties of muscle proteins in processed poultry products. In: Sams AR, ed., *Poultry Meat Processing*. Boca Raton, FL: CRC Press, Chap. 11.

Smith DP, Acton JC. 2001. Marination, cooking and curing of poultry products. In: Sams AR, ed., *Poultry Meat Processing*. Boca Raton, FL: CRC Press, Chap. 15.

USDA–FSIS (U.S. Department of Agriculture–Food Safety and Inspection Service). 2008a. http://www.fsis.usda.gov/Help/glossary-B/index.asp. Accessed May 2008.

USDA–FSIS. 2008b. The regulated industries: characteristics and manufacturing. http://www.fsis.usda.gov/PDF/PHVt-Regulated_Industries.pdf Processes. Accessed May 2008.

Vandendriessche F. 2008. Meat products in the past, today and in the future. Meat Sci 78:104–113.

Van Laack RLJM. 1994. Spoilage and preservation of muscle foods. In: Kinsman DM, Kotula AW, Breidenstein BC, eds., *Muscle Foods: Meat, Poultry, and Seafood Technology*, New York: Chapman & Hall, Chap. 14.

Walsh HM, Kerry JP. 2002. Meat packaging. In: Kerry JF, Kerry JP, Ledward D, eds., *Meat Processing*. London: CRC Press, pp. 416–451.

14

TURKEY SAUSAGES

Alfonso Totosaus-Sánchez and Juan Francisco Hernández Chávez
Tecnológico de Estudios Superiores de Ecatepec, Laboratorio de Alimentos, Ecatepec de Morelos Estado de México, México

INTRODUCTION

Turkey sausages are emulsified cooked products generally processed using mechanically deboned meat, since the main cuts, breasts and thighs, are commercialized frozen or fresh. As in other poultry species, mechanically deboned turkey meat represents an important functional, nutritional, and cheap

Handbook of Poultry Science and Technology, Volume 2: Secondary Processing, Edited by Isabel Guerrero-Legarreta and Y.H. Hui

protein source for emulsified meat products. Since turkey sausages are often the subject of religious concern, control of the species employed is also important.

MECHANICALLY DEBONED TURKEY MEAT

After slaughtering, the first processing stage, a second processing of poultry involves a considerable amount of manual labor to perform intricate cutting, trimming, and portioning. However, the second processing is where much of the value is added to the profit and therefore where most of the plant's profit is derived. Production of parts and boneless breast meat are the major functions of modern processing plants and are excellent examples of adding value to a processed carcass. Portion control is a growing segment of turkey processing because turkey carcasses are more easily made into consumer-ready portions of parts and boneless fillets than are other meat types (Sams, 2001). This involves a considerable quantity of leftover parts, since the popularity of cut-up and processed poultry meat produces considerable quantities of backs, necks, and frames with meat recoverable by mechanical separators, enabling an efficient use of whole-turkey carcasses that cannot economically be deboned by hand, and providing raw matter for the manufacture of sausages (Totosaus and Pérez-Chabela, 2007). Figure 1 is a scheme of the complete turkey processing, where the first processing involves slaughter and related operations (e.g., reception, inspection, carcass

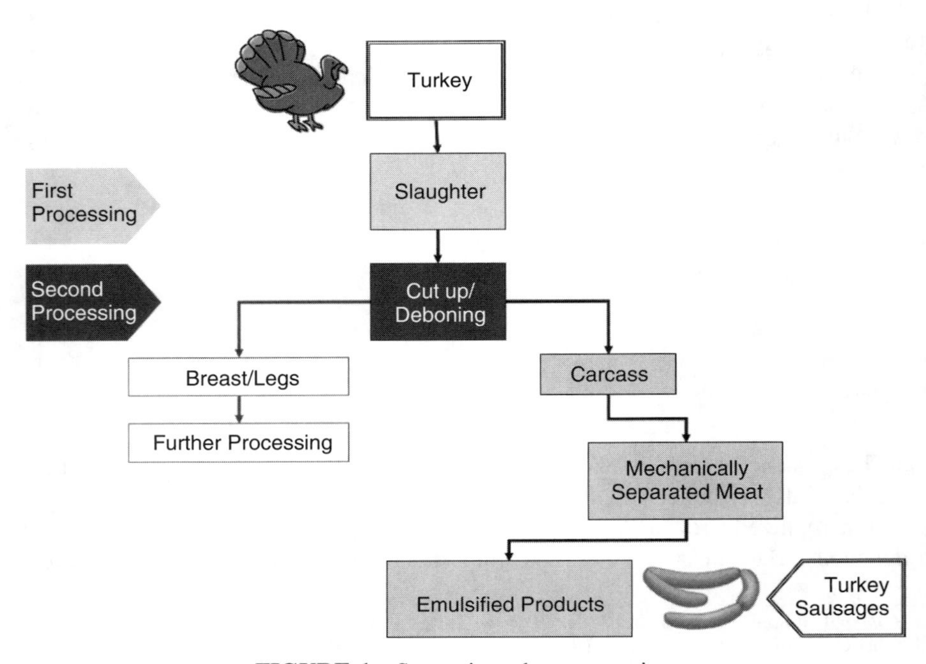

FIGURE 1 Stages in turkey processing.

evaluation). For more details, see Pérez-Chabela and Totosaus (2005). In this view, second processing is cutting, trimming, and portioning with concomitant production of carcass leftovers used in mechanical deboning. This raw material is employed in emulsified products such as turkey sausages.

Mechanical deboning or separation provides a means of harvesting functional proteins which can be used in the preparation of a variety of further-processed meat products. Mechanically separated or deboned poultry meat has been widely utilized in further-processed poultry meat products, such as bologna, salami, frankfurters, turkey rolls, restructured meat products, and soup mixes. This low-cost meat source has led to poultry meat products being more cost-effective in the marketplace. As the growth of further-processed turkey products has grown, more parts have become available for mechanical separation. Turkeys are hand deboned and processed further. After hand deboning of turkey carcasses, the frame, drumsticks, backs, and necks are also separated mechanically (Froning and McKee, 2001). Much of the mechanically separated turkey and poultry has been used in emulsified products. The mechanical separation process on salt-soluble proteins and fat content has been found to influence functional properties. McMahon and Dawson (1976) reported that mechanically separated turkey meat has been found to have fewer salt-soluble proteins, with superior emulsion capacity in hand-deboned turkey compared to mechanically separated turkey. Although water-holding capacity appeared to be higher in mechanically separated turkey, a combination of 0.5% polyphosphate and 3% sodium chloride improved extractable protein, water-holding capacity, and the emulsifying capacity of mechanically separated turkey meat. On the other hand, Lampila et al. (1985) texturized mechanically separated turkey meat by extrusion and heat setting, proposing their utilization in restructured meat products.

FUNCTIONAL PROPERTIES AND SAUSAGE PROCESSING

Proteins are required to fulfill a variety of functions in turkey products, where the typical characteristics of many poultry products are dependent on the successful manipulation of protein functional properties during processing (i.e., yield, quality, and sensory attributes). Functional properties of the muscle proteins are influenced by processing conditions and ingredients used in a formulation. Any changes in a product formulation or process require an appreciation of the effect of that change on the muscle protein structure. Formulation changes can alter the pH, salt concentration, and protein concentration of a product, among other factors, all of which have an effect on the biochemical properties, and subsequently the functional properties, of poultry or turkey muscle proteins. Changes in processing conditions, especially those that alter product temperature or the extent of comminution, can also affect the biochemical properties of muscle proteins. All of these changes, which affect protein structure, affect the quality of the final product (Smith, 2001). In this view, turkey sausages properties such as texture, sensory properties (e.g., color, flavor, mouthfeel), and cooking

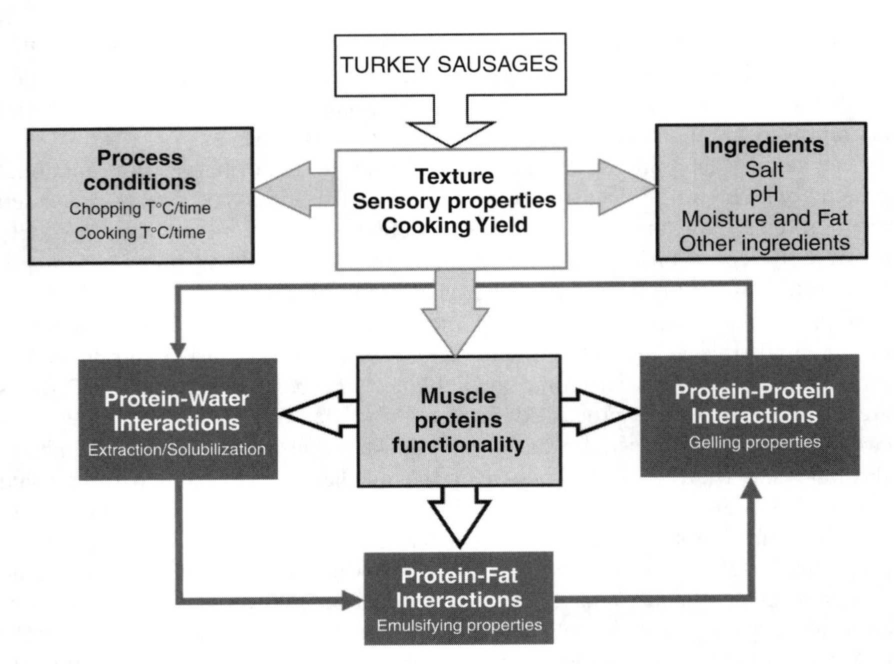

FIGURE 2 Functional properties implicated in turkey sausage elaboration.

yield (related to water-holding properties) are interrelated to muscle proteins functionality (Figure 2).

In general, emulsified poultry products are processed by homogenizing turkey meat in a bowl chopper with iced water, salt, cure, alkaline phosphates, sodium erythorbate, and other additives to an endpoint temperature of approximately 10°C. To avoid overheating the fat that would result in processing defects during thermal processing, batter temperatures should not exceed 12°C. Sausages are then encased in a natural casing (frankfurters) or moistureproof fibrous casing (bologna) and fully cooked (Keeton, 2001). Main ingredients employed are listed in Table 1.

During sausage manufacture, four main stages can be identified in emulsified product manufacturing: (1) protein extraction, (2) protein hydratation activation, (3) fat emulsification, and (4) protein matrix gelification (Guerrero and Totosaus 2006). Muscle protein functionality affects the final product properties, especially process conditions such as the temperature and time of chopping in the cutter and cooking rate and the use of other ingredients (listed in Table 2), since protein functional properties are governed by the hydrodynamic and relative surface properties. These properties are related to the way in which proteins behave in solution (conformation and hydrophobic–hydrophilic balance). When proteins are functional (soluble or active at suitable pH and ionic strength), three types of interactions can be achieved during sausage processing: protein–water, protein–fat, and protein–protein interactions (Figure 2).

TABLE 1 Common Nonmeat Ingredients Used in Further-Processed Products

Additive	Examples
Salt	Sodium chloride
	Potassium, magnesium, or calcium chloride (in sodium-reduced products)
Alkaline phosphates	Sodium tripolyphosphate
	Sodium hexametaphosphate
	Sodium tetrapyrophosphate
Sweeteners	Dextrose
	Corn syrup
	Sorbitol
Curing salt and cure accelerators	Sodium or potassium nitrite
	Sodium or potassium erythorbate
	Sodium or potassium ascorbate
Antimicrobials	Sodium or potassium lactate
	Sodium acetate or diacetate
Antioxidants	Butylated hydroxyanisole (BHA)
	Butlyated hydroxytoluene (BHT)
	Spice extracts
Seasonings, spices, and flavorants	Black pepper, basil, cardamon, cloves, ginger, fennel, nutmeg, mustard, paprika, pimento, cayenne, pepper, white pepper, caraway, coriander, celery seed, cumin, marjoram, thyme, savory, sage, anise, cinnamon, capsicum, onion, garlic, sesame
	Liquid smoke

Source: Adapted from Keeton (2001).

Protein Extraction and Protein Hydration Activation

During sausage processing, process conditions and ingredients added are responsible for the functional performance of muscle proteins. Skeletal muscle can be employed to extract myofibrillar proteins, and mechanically deboned meat is a good option to reduce formulation costs. The first stage of the process includes the extraction and/or activation of muscle proteins to enhance their functionality. When muscle proteins are in situ, their function is merely one of support and/or contraction–relaxation, depending on muscle type. At physiological conditions (before slaughter and aging, i.e., pH close to 7.0 and very low ionic strength), muscle proteins, particularly myofribrillar proteins, are insoluble and hence have minimal functionality. Cell disruption to liberate the proteins is not sufficient to activate them since they need a stronger ionic force to be soluble. The addition of 2.5 to 3.0% salt corresponds approximately to 0.5 to 0.6 M of NaCl, enough to make myofibrillar proteins soluble and to activate them. In the same way, alkaline phosphates (a mixture of alkaline sodium phosphate salts and sodium pyrophosphate) could increase the meat's final pH

TABLE 2 Main Analytical Techniques Employed in Turkey and Other Species Determination in Meat Products

Analytical Technique	Description	Reference
Lateral flow immunoassay	Applied to detect low levels of beef and sheep meat in a wide range of meat products, as beef-in-chicken, beef-in-turkey, and lamb-in-pork products	Rao and Hsieh (2007)
HPLC with electrochemical detection	Detect adulteration and degradative changes of meat proteins in fresh or cooked meats	Chou et al. (2007)
Polymerase chain reaction (PCR)	Primers that exploit intron variability in α-cardiac actin	Lockley and Ronald (2002)
Random amplified polymerphic DNA (RAPD)-PCR	Discrete and reproducible bands which allowed discrimination using visual inspection compared with arbitrary primer PCR of more complex patterns and low-intensity bands	Saez et al. (2004)
PCR-restriction fragment length polymorphism	Amplification of conserved areas of the vertebrate mitochondrial cytochrome *b* gene yielding a 359-bp fragment including a variable 307-bp region	Meyer et al. (1995)
	Oligonucleotide primer pair designed to amplify partial sequences within the 12S rRNA gene of mitochondrial DNA; cooking and autoclaving of meats did not influence the generation of the PCR-RFLP profiles or the analytical accuracy	Sun and Lin (2003)
PCR-amplicon visualization with vista green	PCR-amplicons can be detected in less than 5 min using vista green and a fluorescent plate reader, suggested as a routine procedure in the detection of turkey and chicken in processed meat products	Hird et al. (2003)
DNA detection	Primers amplified at a single actin gene locus, giving a positive band with DNA	Hopwood et al. (1999)
Satellite DNA probes	DNA hybridization to a conjugate of a specific oligonucleotide and alkaline phosphatase for differentiation from closely related species such as turkey and chicken	Buntzer et al. (1995)
RAPD	Generation of fingerprint patterns for pork, chicken, turkey, goose, and duck meats	Calvo et al. (2001)

[reduced from ca. 7.0 (physiological value) to close 6.0 to 5.5 during anaerobic muscle cell metabolism after slaughter and aging] in order to move the myofibrillar proteins away from their isoelectric point, where the solubility is minimum. Added ice controls the batter temperature and allows protein solubilization. Protein–water interactions are responsible for this stage functionality. When mechanically deboned meat is employed, the material has a higher fat and mineral content by the incorporation of lipids and heme pigments from marrow bone and the subcutaneous fat layer, in addition to calcium and phosphorus from bone particles, and the high heme pigment content results in a darker-colored product more susceptible to lipid oxidation (Totosaus and Pérez-Chabela, 2007), depending on packaging and storage conditions (Pettersen et al., 2004).

Curing

The curing process involves many complex reactions, but three reactions can be considered central. The first reaction occurs when curing salts (sodium nitrate) react with a reducing substance (sodium ascorbate or erythorbate). Since sodium nitrite is a strong oxidant, reaction with sodium ascorbate or erythorbate accelerates the conversion of nitrite to nitric oxide. The reverse reaction is suppressed, resulting in a more complete conversion of the muscle pigment to the cured pigment form. In the second stage, nitric oxide is transferred to myoglobin to yield nitrosylmyoglobin, and finally, nitrosohemochrome is formed during cooking, giving the particular pink coloration. An important aspect of the curing process is the immobilization of iron in the heme complex, retarding the catalytic activity (Totosaus, 2008). Turkey batters with sodium nitrite became less fluid in the initial stages of heating and expanded more with heating than those with water only. Sodium chloride increased peak heights and increased the temperatures at which the batters reached those peaks. Sodium chloride alone and in combination with other salts increased batter expansion and increased the temperature at which the batters reached maximum expansion. Batters with sodium chloride and sodium tripolyphosphate were the least fluid during heating and began to expand at the lowest temperatures (Prusa et al., 1985). On the other hand, the nitrite content can determine a consumer's acceptance of turkey frankfurters. Turkey frankfurters containing nitrite were preferred to those containing no nitrite by a consumer panel, but there were no significant differences between frankfurters containing 40 ppm nitrite and those containing 100 ppm nitrite, with no significant difference in the flavor and aroma of the frankfurters (Sales et al., 1980). Emulsions with additives such as sodium nitrite and sodium ascorbate contained less malonaldehyde than those with sodium chloride or no additive, and raw turkey emulsions generally contain less malonaldehyde than did cooked emulsions. Sodium nitrite, ascorbate, or chloride affect aroma as well. Emulsions containing both sodium nitrite and sodium ascorbate had the most meaty aroma and the least stale aroma, whereas emulsions with sodium chloride tended to have a higher level of stale aroma (Tellefson et al., 1982).

Fat Emulsification

When muscle proteins are activated (i.e., at a pH distant form their isoelectric point in a relatively higher ionic strength medium due to NaCl addition), fat can be added to be emulsified by the protein matrix. After the addition of salt, the viscosity of the meat batters changes drastically, increasing the viscosity due to the effect of ions (positive sodium and negative chloride) on protein solubilization. This highly charged protein system entraps fat globules, forming an emulsion. Protein–fat interactions are responsible for this stage's functionality. Lard or pork backfat areas are added frozen, to control temperature and the melting of fat during chopping. Fat dispersion is important in obtaining a homogeneous texture after cooking. The protein–water–salt–fat network formed prevents the coalescence of fat and arranges for fixation or structural enforcement of all supporting compositional ingredients, such as meat and nonmeat proteins (Hoogenkamp, 1995).

There are two theories about fat emulsion in meat systems. One concerns the stabilization of fat in a meat batter by the formation of an interfacial protein film around the small fat globules, the film serving as an interface between the aqueous and the fat phases that prevents coalescence. The other is related to physical entrapment, with emphasis on the fact that fat is entrapped within the protein matrix before and during cooking by an ordered protein spongelike structure formed by the interconnected protein strands forming a coherent three-dimensional structure. This protein structure is formed before thermal processing at a stage in which the batter still flows, retained throughout the cooking process by the embedded fat globules (Barbut, 1995).

Protein Matrix Gelification

In the final step of sausage processing, thermal treatment has two main objectives: to destroy vegetative spoilage microflora to ensure the microbiological safety of the foodstuff, and to form an ordered protein structure responsible for the chemical and physical retention of water and fat in the cooked product. This could be the most important functional factor in emulsified meat products, since conversion from a raw solubilized protein–salt–water–fat batter to a self-supporting gel structure defines most of the textural, sensory, and yield characteristics of this type of product (Figure 2). Raw batters are enclosed in tubular casings to provide physical constraints during thermal processing,

During cooking, the heat being lost from the surface must be transferred to the surface from the interior of the cylinder by conduction. The heat transfer from the interior to the surface is difficult to determine, but as an approximation it may be assumed that all the heat is being transferred from the center of the cylinder. Evaluating the temperature drop required producing the same rate of heat flow from the center to the surface as it passed from the surface to the air. Unsteady-state heat transfer arises from the heating of solid bodies made of good thermal conductors (e.g., meat sausage), and assuming that all the heat flows from the center of the cylinder to the outside. The thermal conductivity is described as (Earle and Earle, 2003)

$$\frac{dQ}{dt} = \frac{k}{L} A(T_C - T_S) \quad (1)$$

where k is the thermal conductivity of the material, L is the radius of the cylinder, A is the area of the cylinder, and T_c and T_s are, respectively, the temperature at the center and surface of the cylinder.

Modified starches vary in their ability to improve firmness and other textural characteristics of reduced-fat high-water-added turkey batters, where cross-linked starches serve to reduce cooking and reheating losses and improve textural attributes with higher smokehouse yields and enhanced product quality (Hachmeister and Herald, 1998). In the gelation of turkey breast meat, protein fraction analysis of extracted salt-soluble proteins by SDS-PAGE identified α-tropomyosin as a salt-soluble protein positively associated with viscoelastic properties at 80°C, and the proteins pyruvate kinase and triose phosphate isomerase as proteins associated negatively with viscoelastic properties, demonstrating that proteins other than myosin and actin participate in the mechanisms that form thermally induced meat gels (Updike et al., 2006).

Endpoint temperature is an important factor in foodborne disease outbreaks. In poultry products it has been suggested that the concentration of lactate dehydrogenase, immunoglobulin, and serum albumin can be employed as an approach to verify internal product temperature after a poultry product such as bologna, pastrami, smoked sausage, and frankfurters has been cooked (to target temperatures between 65.5 and 73.8°C). In same manner, turkey thigh muscle contains higher concentrations of serum albumin and immunoglobulin and lower concentrations of lactate dehydrogenase than breast muscle does at the same endpoint temperature. The concentrations of the three indicators decreased as the endpoint temperature of four turkey thigh products were increased (Veeramuthu et al., 1997).

TURKEY SAUSAGE PRESERVATION

Packaging

With a risk of *Listeria monocytogenes* in processed meats, addition of another safeguard via a package may reduce the risk of foodborne illness to consumers. Biocides incorporated into the matrix of films will release their antimicrobial activity to the surrounding environment. This inhibition is somewhat time released in that continued inhibition has been observed up to 48 h after exposure to liquid media and 21 days when exposed to refrigerated meat surfaces, such as when presliced low-fat turkey bologna with 2 mm thick was cut into 6-cm squares using a sterile knife. Antimicrobial packaging films open the possibility of increasing the shelf-life safety and quality of many food products (Dawson et al., 2002).

Irradiation

Du and Ahn (2002) studied the effect of antioxidants on irradiated turkey sausages. When oxygen is available, irradiation can induce fatty acid degradation through a mechanism similar to lipid oxidation. The effects were studied of antioxidants on the flavor and color of electron-beam-irradiated turkey sausages prepared from turkey thigh meat with one of five antioxidant treatments (none, vitamin E, sesamol, rosemary extract, or gallic acid at 0.02%), packaged and divided randomly into three groups and irradiated at 0, 1.5, or 3.0 kGy. The antioxidant effect of sesamol was the highest, followed by vitamin E and gallic acid; rosemary extract had the weakest antioxidant effect. Irradiation induced red color in sausages, but the addition of gallic acid, rosemary extract, or sesamol reduced it. This study indicated that gallic acid at 0.02% dramatically lowered the redness of irradiated and nonirradiated turkey sausages, and sesamol and rosemary were also effective in reducing irradiation-induced redness.

ANIMAL SPECIES IDENTIFICATION

The fraudulent substitution of meat species in meat products has developed a need for specific and reliable methods pertaining to meat determination of species. Analyses of proteins or immunoassays predominate in the current analytical techniques employed. Nevertheless, the majority of proteins become less stable in cooked meat products, modifying their electrophoretical mobility or specific antigenicity of the protein, resulting in misinterpretation of the results (Hernández-Chávez et al., 2007). Species identification is very important for many ethnic groups, especially those following kosher or halal regulations. Che Man et al. (2007) suggested that species-specific polymerase chain reaction (PCR) detection of a conserved region in the mitochondrial 12S rRNA gene can be employed for halal certification.

Methods employed for meat species authentication include chromatographic, electrophoretic, immunological, enzymatic, and near-infrared spectroscopic assays. Most of these methods are lipid- or protein-based methods; however, for highly processed and heated meat products, DNA-based methods are recommended since DNA is more thermostable than many proteins. Furthermore, DNA is present in the majority of the cells of an organism, potentially enabling identical information to be obtained from an appropriate sample from the same source. Therefore, numerous DNA-based methods dealing with meat species identification have been reported using the polymerase chain reaction (PCR) technique (Vallejo-Cordóba et al., 2005). The substitution or addition of ingredients in meat products can be of two types: protein and/or fat substitution and animal and/or vegetable protein substitution. The animal-origin protein from mechanically boned poultry meat is used most in the processing industry (Hernández-Chávez et al., 2007). The principal deficiency of analytical procedures for the purpose of food labeling is their semiquantitative character. This is due to the kinetics of the PCR amplification per se and to the lack of precision in the quantification of DNA by traditional electrophoretic techniques such as densitometry, resulting in false negatives or positives. PCR techniques, based in the use of competitive amplification targets, have been developed that allow carrying out more reliable quantitative analysis for transgenic food. In this view, the combination of two procedures could ensure a better approach to species identification. Hernández-Chávez (2006), reported on the combined use of quantitative competitive PCR and gel capillary electrophoresis to determine the quantity of turkey in processing products. The combined use of these methodologies can be employed for the determination and quantification of turkey meat in processing meat products. Table 2 lists most of the analytical techniques employed for turkey and others species determination in meat products.

REFERENCES

Barbut S. 1995. Importance of fat emulsification and protein matrix characteristics in meat batter stability. J Muscle Foods 6:161–177.

Buntzer JB, Lenstra JA, Haagsma N. 1995. Rapid species identification in meat by using satellite DNA probes. Z Lebensm-Unters-Forsch 201:577–582.

Calvo JH, Zaragoza P, Osta R. 2001. Random amplified polymorphic DNA fingerprints for identification of species in poultry paté. Poult Sci 80:522–524.

Che Man YB, Aida AA, Raha AR, Son R. 2007. Identification of pork derivatives in food products by species-specific polymerase chain reaction (PCR) for halal verification. Food Control 18:885–889.

Chou C-C, Lin S-P, Lee K-M, Hsu CT, Vickroy TW, Zen JM. 2007. Fast differentiation of meats from fifteen animal species by liquid chromatography with electrochemical detection using copper nanoparticle plated electrodes. J Chromatogr B 846:230–239.

Dawson PL, Carl GD, Acton JC, Han IY. 2002. Effect of lauric acid and nisin-impregnated soy-based films on the growth of *Listeria monocytogenes* on turkey bologna. Poult Sci 81:721–726.

Du M, Ahn DU. 2002. Effect of antioxidants on the quality of irradiated sausages prepared with turkey thigh meat. Poult Sci 81:1251–1256.

Earle RL, Earle MD. 2003. *Unit Operations in Food Processing*, Web edition. New Zealand Institute of Food Science and Technology. http://www.nzifst.org.nz/unitoperations/index.htm.

Froning GW, McKee SR. 2001. Mechanical separation of poultry meat and its use in products. In: Sams AR, ed., *Poultry Meat Processing*. Boca Raton, FL: CRC Press, Chap. 14.

Guerrero I, Totosaus A. 2006. Propiedades funcionales y textura. In: Hui YH, Rosmini M, Guerrero I, eds., *Ciencia y Tecnología de Carnes*. Mexico City: Limusa, pp. 205–227.

Hachmeister KA, Herald TJ. 1998. Thermal and rheological properties and textural attributes of reduced-fat turkey batters. Poult Sci. 77:632–638.

Hernández-Chávez JF. 2006. Identificación y cuantificación de especies en productos cárnicos procesados térmicamente mediante el uso combinación de electroforesis capilar y técnicas moleculares. Ph.D. dissertation, Centro de Investigación en Alimentación y Desarrollo, A.C. México.

Hernández-Chávez JF, González-Córdova AF, Sánchez-Esacalante A, Torrescano G, Camou JP, Vallejo-Córdoba B. 2007. Técnicas analíticas para la determinación de la autenticidad de la carne y de los productos cárnicos procesados térmicamente. Nacameh 1:97–109. http://www.geocities.com/nacameh_carnes/index.html.

Hird H, Goodier R, Hill M. 2003. Rapid detection of chicken and turkey in heated meat products using the polymerase chain reaction followed by amplicon visualization with vista green. Meat Sci 65:1117–1123.

Hoogenkamp HW. 1995. Meat emulsion variables. Fleischwirtschaft Int 1:29–32.

Hopwood AJ, Fairbrother KS, Lockley AK, Bardsley RG. 1999. An actin gene-related polymerase chain reaction test for identification of chicken in meat mixtures. Meat Sci 53(4):227–231.

Keeton JT. 2001. Formed and emulsion products. In: Sams AR, ed., *Poultry Meat Processing*. Boca Raton, FL: CRC Press, Chap. 12.

Lampila LE, Froning GW, Acton JC. 1985. Restructured turkey products from texturized mechanically deboned turkey. Poult Sci 64:653–660.

Lockley AK, Ronald GB. 2002. Intron variability in an actin gene can be used to discriminate between chicken and turkey DNA. Meat Sci 61:163–168.

McMahon EF, Dawson LE. 1976. Effects of salt and phosphates on some functional characteristics of hand and mechanically deboned turkey meat. Poult Sci 55:573–578.

Meyer R, Hofelein C, Luthy J, Candrian U. 1995. Polymerase chain reaction restriction fragment length polymorphism analysis: a simple method for species identification in food. J AOAC Int 78:1542–1551.

Pérez-Chabela ML, Totosaus A. 2005. Poultry carcass evaluation. In: Hui YH, ed., *Handbook of Food Technology and Food Engineering*, vol. 4, *Food Technology and Food Processing*. Boca Raton, FL: CRC Press, pp. 165-1 to 165-6.

Pettersen MK, Mielnik MB, Eie T, Skrede G, Nilsson A. 2004. Lipid oxidation in frozen, mechanically deboned turkey meat as affected by packaging parameters and storage conditions. Poult Sci 83:1240–1248.

Prusa KJ, Bowers JA, Craig JA. 1985. Rates of heating and Instron measurements for turkey batters with sodium salts of nitrite, chloride and phosphate. J Food Sci 50:573–576.

Rao Q, Hsieh Y-HP. 2007. Evaluation of a commercial lateral flow feed test for rapid detection of beef and sheep content in raw and cooked meats. Meat Sci 76:489–494.

Saez R, Sanz Y, Toldrá F. 2004. PCR-based fingerprinting technique for rapid detection of animal species in meat products. Meat Sci 66:659–665.

Sales CA, Bowers JA, Kropf D. 1980. Consumer acceptability of turkey frankfurters with 0, 40, and 100ppm nitrite. J Food Sci 45:1060–1061.

Sams AR. 2001. Introduction to poultry meat processing. In: Sams AR, ed., *Poultry Meat Processing*. Boca Raton, FL: CRC Press, pp. 1–3.

Smith DM. 2001. Functional properties of muscle proteins in processed poultry products. In: Sams AR, ed., *Poultry Meat Processing*. Boca Raton, FL: CRC Press, Chap. 11.

Sun YL, Lin CS. 2003. Establishment and application of a fluorescent polymerase chain reaction–restriction fragment length polymorphism method for identifying porcine, caprine, and bovine meats. J Agric Food Chem 51:1771–1776.

Tellefson CS, Bowers JA, Marshall C, Dayton AD. 1982. Aroma, color, and lipid oxidation of turkey muscle emulsions. J Food Sci 47:393–396.

Totosaus A. 2008. Colorants. In: Nollet LML, Toldrá F, eds., *Handbook of Processed Meats and Poultry Analysis*. Boca Raton, FL: CRC Press, Chap. 9.

Totosaus A, Pérez-Chabela ML. 2007. Poultry sausages. In: Hui YH, ed., *Handbook of Food Products Manufacturing*, vol. II. Hoboken, NJ: Wiley, pp. 775–781.

Updike MS, Zerby H, Utrata KL, Lilburn M, Kaletunc G, Wick M. 2006. Proteins associated with thermally induced gelation of turkey breast meat. J Food Sci 71:E398–E402.

Vallejo-Córdoba B, González-Córdova AF, Mazorra-Manzano MA, Rodríguez-Ramírez R. 2005. Capillary electrophoresis for the analysis of meat authenticity. Sep Sci Food Anal 28:826–836.

Veeramuthu GJ, Booren AM, Smith DM. 1997. Species, muscle type, and formulation influence the residual concentrations of three endpoint temperature indicators in poultry products. Poult Sci 76:642–648.

15

BREADED PRODUCTS (NUGGETS)

MARÍA DE LOURDES PÉREZ-CHABELA
Departamento de Biotecnología, Universidad Autónoma Metropolitana–Unidad Iztapalapa, México D.F., México

ALFONSO TOTOSAUS-SÁNCHEZ
Tecnológico de Estudios Superiores de Ecatepec, Laboratorio de Alimentos, Ecatepec de Morelos Estado de México, México

Handbook of Poultry Science and Technology, Volume 2: Secondary Processing, Edited by Isabel Guerrero-Legarreta and Y.H. Hui

INTRODUCTION

Breaded products represent a convenient food item that is tasty and can be prepared to fit most consumers' needs. One of the best success stories is the chicken nugget product introduced to the North American market by fast-food chains in the 1970s. Originally, the product was made from a whole breast muscle strip that was battered and breaded and then fried. Today, chicken nuggets are made from whole-muscle breast meat, ground breast meat, chunks of dark meat, ground dark meat with and without the inclusion of mechanically deboned poultry meat (MDPM), and skin. The traditional oval shape of the nugget has also been modified in some markets (e.g., a dinosaur-shaped nugget is a common shape used to attract young children). The breading and spices used also vary, depending on the market (Barbut, 2002).

Breaded fried foods are favored by consumers, due to unique and desirable characteristic of crispness provided by a soft and moist interior with an outer crispy crust. Crispness is one of the most distinctive characteristics of fried foods. Loss of crispness in fried foods is due primarily to the absorption of moisture by the crust, causing it to become soggy, and the breaded fried products become less desirable to consumers (Antonova, 2001).

POULTRY MEAT

Nuggets can be made from a variety of meat sources. They are usually made from whole-muscle trimmings and usually reflect the preference in the locality. The most common formulation for poultry nuggets is breast meat and skin, since breast meat has a uniformly soft texture and light color. However, other meat sources can be incorporated, as thigh or drums or MDPM. Processors must label the meat composition according to U.S. Department of Agriculture (USDA) standards (Figure 1) (Owens, 2001).

Other Ingredients

There are many ingredients that can be added to nuggets during production, for various reasons. One of the most important added ingredients is salt. Salt has two main functions in the production of nuggets; it adds flavor and aids in myofibrillar protein extraction, a necessary step for meat particle binding in forming a nugget. It is added at a concentration of less than 2% of the formula. Sodium tripolyphosphate is used to increase the water-holding capacity by increasing the pH and unfolding muscle proteins to allow for more water-binding sites. A variety of species and seasonings can also be added, depending on product specifications (Owens, 2001).

Extension of meat products with legume and oil seed derivatives has the potential to reduce product cost while maintaining nutritional and sensory qualities. Soybean proteins are predominant additives for use as fillers, binders, and

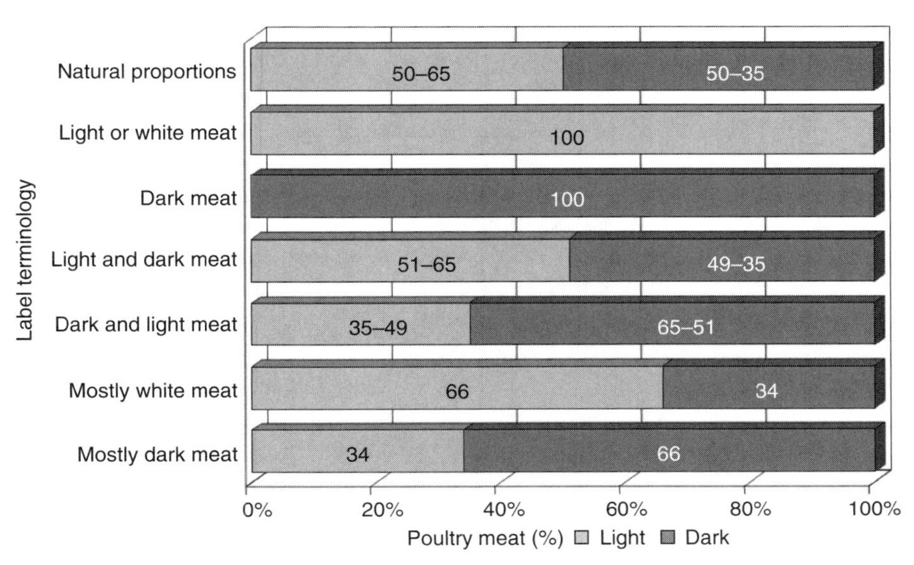

FIGURE 1 Poultry meat content standards for certain poultry products according to the USDA (From *Code of Federal Regulations*, Title 9, § 381.156.)

extenders in meat systems (McWatters, 1977). Nonmeat protein additives derived from cowpeas and peanuts are less common. Prinyawiwatkul et al. (1997) studied the incorporation of flours from cowpeas and fermented partially defatted peanuts to produce acceptable extended chicken nuggets. Flavor and texture more than appearance and color influenced the overall acceptance of nuggets. Nuggets containing a mixture of 2.5% fermented cowpeas and 2.5% fermented partially defatted peanuts were as acceptable as the control, with a sweet, chicken flavor. This suggests a market potential for such poultry products. Jackson et al. (2006) studied the utilization of rice starch in the formulation of low-fat, wheat-free chicken nuggets and demonstrated that the majority of consumers found the rice flour, dry batter, baked treatment acceptable. However, rice could be substituted for wheat in fried products with added health benefits and without decreasing product acceptability in some consumer groups. Kumar et al. (2007) studied the use of liquid whey to replace water added to pork nuggets and its effect on storage stability of the product at refrigeration temperatures. The study revealed that up to 20% of liquid whey can be utilized to replace the added water without affecting the quality of pork nuggets. This formulation with liquid whey is more beneficial for the meat product industry as well as the dairy industry.

PROCESSING

Nugget processing can be divided broadly into four principal steps: product forming, coating, frying, and packaging and storage (Figure 2). Coating and frying are

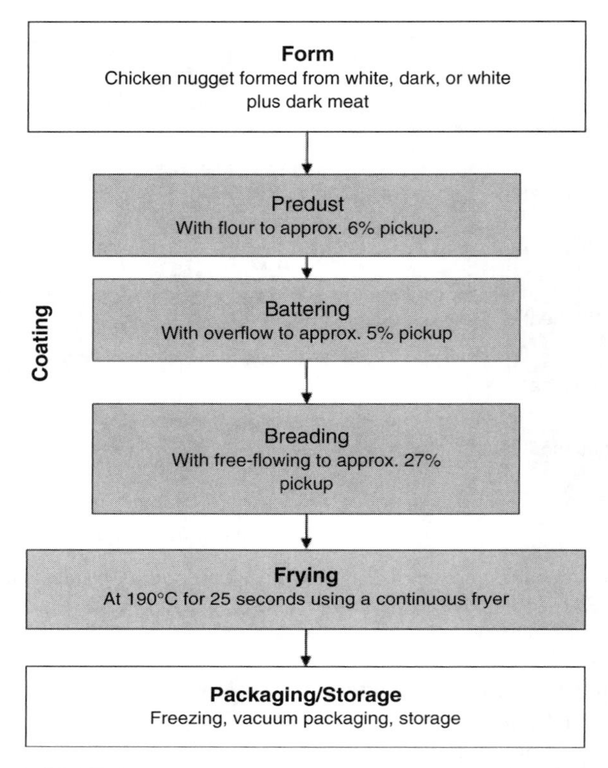

FIGURE 2 Main steps in poultry nugget manufacturing.

important since besides contributing functional ingredients and developing texture and flavor characteristics, respectively, they are responsible for an increase in weight. Coating represents a way to add water that can be retained during frying by the additives, with no weight losses. Added ingredients (proteins and carbohydrates) are also responsible for color and flavor development during thermal processing, caused by Maillard reactions between proteins and reducing sugars.

Coating

Coating poultry products with a batter and/or breading before cooking is a common practice of homemakers, food processors, and commercial fast-food outlets. *Batter* is a liquid mixture comprising water, flour, starch, and seasonings, into which food products are dipped prior to cooking. *Breading* is a dry mixture of flour, starch, and seasonings, coarse in nature, applied to moistened or battered food products prior to cooking. *Coating* is batter and/or breading adhering to a food product after cooking by interface/adhesion. Batters are used with breading, serving primarily as an adhesive layer between the product's surface and the breading, and chemical leavening is not normally used (Antonova, 2001).

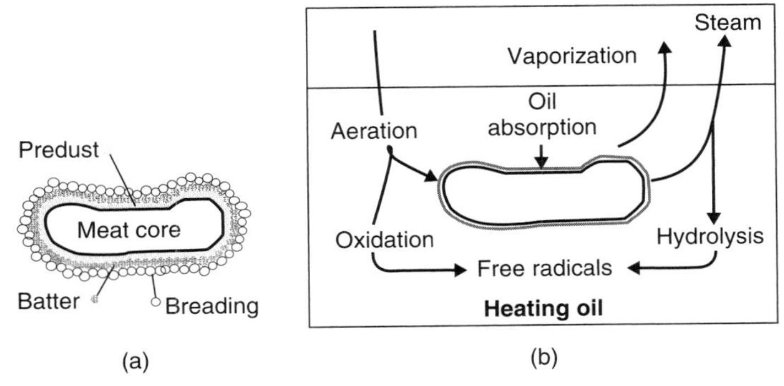

FIGURE 3 Schematic representation of the two most important operations during nugget processing: (a) coating; (b) frying.

In general, battered and breaded products are coated products in which the meat protein component as whole muscle, ground meat, or mechanically deboned meat is the core, surrounded by a cereal base such as wheat flour or cornstarch and coating (Figure 3a). The coating operation can range from a very simple home-style operation to very complex production lines that require expensive equipment. During the process, dry and/or wet ingredients are applied to the moist surface of a regular or marinated (moisture-enhanced) meat. Proper adhesion represents a challenge to the processor, depending on the roughness (skin on, skin off), temperature (partially frozen, thawed), amount of moisture on the surface (semidry or wet after forming), fat on the surface, and other factors. A simple homemade process consists of dipping slices of chicken breast muscle in dry flour, followed by quickly dipping in an egg batter and then breading from both sides while pressing the crumbs into the meat. This is followed by frying in a pan filled with hot oil. In Europe the final product was called *chicken schnitzel*. The schnitzel is served right away and has a very attractive fried smell and crispy texture (Barbut, 2002).

Coating consists of three main operations that give the characteristic texture, color, and flavor to poultry nuggets: predust, battering, and breading, described in the following sections.

Predusting Predusting is the process of coating the nuggets with a fine coating of flour or a dry batter mix, and sometimes seasonings, very fine breadcrumbs, or combinations. The predust is commonly used as the first layer prior to the application of the batter and breading. The predust adheres to the surface by absorbing free water at the surface and is later used to form an intermediate layer between the product and subsequent batter (Owens, 2001; Barbut, 2002).

Battering Batters are a suspension of dry ingredients used to coat the product. The main categories in nugget batter are listed in Table 1. Batters consist of a mixture of various ingredients, which can include flours, starches, eggs, milk,

TABLE 1 Batter Categories Employed in Nugget Elaboration

Batter Type	Description
Adhesion	Designed to adhere to the meat product
Cohesion	Designed to form a shell around the product
Tempura	Used to create a puffed layer around the product and usually not breaded later

spices and seasoning, leavening agents, and stabilizers. In all these batter types it is important that a certain degree of binding be achieved between the outer layer and the product. The rate of surface drying (the rate at which the batter, or later the breading, can absorb moisture) is another crucial factor in maintaining an adequate batter layer (Owens, 2001; Barbut, 2002).

Breading Breading is usually applied on top of the batter and is used to enhance the appearance and texture of the product as well as to increase its volume and weight. The type of breading can range from simple, unbaked flour to very structured baked crumbs. Usually, the breading is a cereal-based product that has been baked and ground into fine, medium, or large crumbs (Barbut, 2002). There is no standard formulation; many different types of breading can be used for nugget coating, varying in size, shape, texture, color, and flavor. Table 2 lists the main breading types employed for coating poultry nuggets.

Frying

Deep-fat frying is one of the most commonly used procedures for the preparation and production of foods. The presence of food moisture and atmospheric oxygen cause various chemical changes at high temperatures, such as polymerization and degradation. These reactions lead to changes in the functional, sensory, and nutritional quality of fried foods (Pokorný and Schmidt, 2001).

In nugget processing, rapid heat transfer quickly sets the coating structure, allowing little time for excess moisture infiltration. Oven heating produces a moderately acceptable product in terms of crispness, color, and flavor. Although the heating rate is slower than that of deep-fat frying, the elevated chamber temperature of the oven causes some evaporative drying of the coating, resulting in a perception of crispness. Microwave heating does not appear to be a suitable method for coated foods, since microwave oscillations cause molecular vibrations and frictional heating within the food so that evaporative drying does not occur. The result is a soggy coating with minimal crispness (Antonova, 2001). The increased frying time results in decreased product lightness (L^*) and increased redness (a^*) and yellowness (b^*). In the same manner, maximum load to puncture increased with increasing frying time. With an increasing degree of oil hydrogenation, the surface color of the fried chicken nugget samples were lighter, the texture increased, the oil and moisture contents decreased, but the oil and moisture contents had a negative correlation (Ngadi et al., 2007). Oil

TABLE 2 Principal Types of Breading Employed to Coat Poultry Nuggets

Breading Type	Description
American breadcrumbs	American breadcrumbs are similar to homemade breadcrumbs, with a distinct crust that provides nice highlighting during the frying operation. There is usually a two-tone appearance to this crumb, representing the inner crumbs and surface crumbs from the loaf of bread; it is a durable breading and is generally midrange in cost. In terms of pickup, medium-sized to large quantities can be used to coat the product.
Japanese breadcrumbs	Japanese breadcrumbs are made from crustless bread and have a sliver shape; the texture of the crumb is fairly open and porous and is produced white or colored, available in a range of fine to coarse granules. The amount of pickup can be controlled from medium to high, and the degree of browning during the frying operation can be controlled from medium light to dark.
Cracker meal	Cracker meal is a fine, flat, dense crumb that is similar to cracker crumbs; white or colored breadcrumbs are generally used with minimal or no crust on the surface, and it can be used in a predust or conventional application where the flakes are fairly dense, and hence the final product will have a crunchy texture.
Flour breaders	Flour provides an economical way to coat a product and often is used for full-fry products, where the resulting fried coating provides relatively low browning of the surface and a very dense coating matrix, and the pickup is fairly low (minimal increase in product weight). Can be used to achieve many different surface textures. When combined with batter, breading balls are formed, which result in the unique homestyle appearance.
Extruded crumbs	Extruded crumbs are generally used as a low-cost alternative where shapes can be achieved by extruding crumbs (American or Japanese breadcrumb shape, or like cracker meal). These crumbs may be used as a predust or breader.

Source: Adapted from Owens (2001); Barbut (2002).

absorption by fried foods may range from 10 to 40%, depending on conditions of frying and the nature and size of the food (deMan, 1999).

Lipid Oxidation During Frying Deep frying is a complex process involving both the oil and the food to be fried. During nuggets frying, steam and moisture escape in the form of steam, removing volatile compounds. The presence of steam results in hydrolysis, with the production of free fatty acids. In the same manner, aeration produces oxidation that together with the liberated steam results in free radicals that can rapidly oxidize lipids in hot frying oil (deMan, 1999). Heat is transferred by frying oil. At high frying temperatures antioxidants may be partially lost by evaporation, especially in combination with water vapor originating from water present in the fried material. Steam enhances losses of nonpolar volatile material. All oxygen originally dissolved in frying oil has already been consumed by oxidation in the time taken to heat frying oil to frying temperatures. Additional oxygen can enter frying oil only by diffusion from air, which is sometimes prevented by special metal cover sheets on the frying bath. Under these conditions autooxidation chains are short, so that antioxidants are used up relatively rapidly (Pokorný and Schmidt, 2001) (Figure 3b).

Synthetic antioxidants, such as butylated hydroxytoluene (BHT) or butylated hydroxyanisole (BHA), are lost relatively easily. BHA does partially dimerize during frying, forming dimers with carbon-to-carbon and ether linkages. The dimers retain certain antioxidant activity. Because of the volatility of common synthetic antioxidants, less volatile synthetic antioxidants suitable as additives to frying oils have been developed. They are more polar, containing hydroxyl groups in side chains, such as a methylol group instead of a methyl group, or are dimers or trimers of common antioxidants. They are not used in industry or are used only to a very limited extent. Losses of natural antioxidants are only small, as their volatility is much lower than that of common synthetic antioxidants. They are also changed by oxidation reactions; for example, carnosol is converted into miltirone (a quinone) and dehydrorosmariquinone (Pokorný and Schmidt, 2001). Rosemary oleoresin has been shown to be as effective as polyphosphate and a combination of BHT, BHT, and citric acid in mechanically deboned poultry meat (Cuppet, 2001).

Packaging and Storage

When the cooking process is completed, the nuggets can be frozen. After freezing, the products are packaged and prepared for distribution. Because these coated products are cooked and frozen before distribution, bacterial growth does not usually limit shelf life. Instead, dehydration and lipid oxidation or rancidity are more important factors. Dehydration can be greatly reduced by moisture-barrier packaging with good integrity and cold tolerance. Rancidity is reduced by using fresh frying oils that contain antioxidants and using modified-atmosphere packaging (Owens, 2001).

MOST COMMON PROBLEMS IN NUGGET PROCESSING

Most common problems during poultry nugget elaborations occur during battering and breading operations. These problems will be reflected in the finished products but may be detected and corrected during the process itself prior to frying, when it is not too late to correct the problem. Some of the common problems and potential improve are described in Table 3 (Barbut, 2002).

MICROBIOLOGY OF POULTRY MEAT AS RAW MATERIAL FOR NUGGETS

Poultry meat and its derivatives are among the food products that cause the most concern to public health authorities, owing to the associated risk of bacterial

TABLE 3 Principal Problems in Nugget Processing

Problem	Potential Improvement
Excess or insufficient pickup	Mismanaged batter with too high or too low viscosity, resulting in a thick or thin coating layer, respectively. Batter viscosity and temperature should be monitored.
Uneven coating	Low-viscosity batter results in an uneven batter deposition, causing no breading sticking or insufficient breading sticking to the area.
Marriages or doubles	Fried products sticking to each other due to inadequate line speed, or product fragments floating in the batter equipment can attach to each other.
Flares and tails	Excess batter staying attached to the fried product due to high batter viscosity can result in tails remaining on the product. Requires changing the type of breading used.
Dark color	High temperature in the fryer can result in rapid deterioration and darkening of the oil, which will then be transferred onto the product. An excessive frying period will result in burning of the surface. The components added to the breading can also be adjusted to control the browning rate.
Ballooning	After the frying operation, separation of the coating from the product can later cause cracking and cause the breading material to fall off. Increasing the breading size can help by creating a more porous surface. Adjusting the batter's ingredients, such as by adding fat or gums, can also modify the porosity of the coating system and allow moisture to come out easily during frying.
Shelling	Usually occurs in tempura-type batters where a hard shell is formed prior to allowing the hot water vapor to escape from the product. A thick batter deposited on the product can form a hard shell around the product during the frying operation. Viscosity should be checked and adjusted as needed, and if the temperature is too high prior to the frying operation, an excessive amount of gas can be released.

Source: Adapted from Barbut (2002).

food poisoning (Beli et al., 2001). Food animals may be infected, contaminated, or be asymptomatic carriers of pathogenic microorganisms, and together with the environment they serve as sources of contamination for carcasses during the slaughtering process and for meat products during processing, storage, and handling (Sofos et al., 2003). The process of mechanical deboning has made removal of all the meat left on the carcass, neck, and back of a chicken feasible commercially, thus providing a new raw material for processed meat products, mechanically deboned chicken meat. This is a common ingredient in poultry nuggets, due principally to its low price.

Salmonella, *Listeria*, and *Staphylococcus* are the microorganisms most frequently associated with chicken. Bonato et al. (2006) studied the effect of 20% washed mechanically deboned chicken meat as a substitute for breast meat in chicken nugget characteristics during frozen storage. Oxidative phenomena in nuggets with 20% mechanically deboned chicken meat is similar to 100% breast muscle formulation. No difference was found in microbiological counts between formulations or packing atmosphere.

Salmonella

Salmonella are gram-negative, facultatively anaerobic, non-spore-forming rods. There are approximately 2600 *Salmonella* serotypes. They can grow at temperatures as low as 5.2°C and as high as 46.2°C, at pH values of 3.8 to 9.5 and at water activity (a_w) levels above 0.93 (Bacon and Sofos, 2003). Greenwood and Swaminathan (1981) studied two methods for determined the presence of *Salmonella* in MDPM: a conventional cultural method and a modified membrane filter disk inmunobilization (MFDI) procedure. The modified MFDI procedure was found to be more sensitive than the conventional cultural method. Of 100 samples analyzed, 22 were positive for *Salmonella* by the conventional cultural method, and 61 were positive by the modified MFDI procedure. *Salmonella* is one of the microorganisms most frequently associated with outbreaks of illness spread by food. De Freitas et al. (2004) reported *Salmonella* strains in four of 30 samples of MDPM (13.33%), with *S. albany* the main serotype isolated. This serotype has been identified in chicken carcass, but this is the first recorded account of its isolation from MDPM. The contamination of two samples of emulsion shows that *Salmonella* was able to survive after the addition of preservatives, a fact that could signify a serious risk in the final product.

Listeria

Listeria are non-spore-forming, aerobic, microaerophilic, or facultatively anaerobic, Gram-positive rods that are motile by means of peritrichous flagella. They are psychrotrophic and can grow at temperatures as low as 4°C and up to 45°C (Bacon and Sofos, 2003). In meat products this bacterium is effectively inactivated by pasteurization. Recontamination may occur during slicing and packaging, and the microorganism can often be isolated in small numbers from vacuum-packed sliced meat products (Schmidt and Kaya, 1990). *Listeria* is highly resistant to various factors and procedures used in the manufacture of meat products: pH values and the influence of NaCl, phosphates, sodium nitrite (Kostenko et al., 1999). Marinsek and Grebenc (2002) reported a 15.78% incidence of *L. monocytogenes* in samples of mechanically deboned chicken meat, concluding that *L. monocytogenes* contamination of some meat products is significant.

Staphylococcus

Staphylococcus aureus is a mesophile; some strains can grow at a temperature as low as 6.7°C. In general, growth occurs over the range 7 to 47°C, and enterotoxins are produced between 10 and 46°C. It grows well in culture media; without NaCl it can grow well in 7 to 10% concentrations, and some strains can grow in a

20% concentration. *S. aureus* can grow over the pH range 4.0 to 9.8, but its optimum is in the range 6 to 7. With respect to a_w, the staphylococci are unique in being able to grow at values lower than for any other nonhalophilic bacteria (0.86 is the generally recognized mìmimum a_w) (Jay, 2000). *S. aureus* is an important source of food poisoning throughout the world. This microorganism was reported as a contaminant of poultry carcasses and mechanically deboned poultry meat (Kumar et al., 1986). Bijker et al. (1987) reported lax standards of hygiene during the production of mechanically deboned meat, resulting in high numbers of *S. aureus* (10^4 to 10^5 CFU/g). These data indicate that measures of hygiene observed during boning of carcasses and during collection, storage, and transport of bones or poultry parts should be tightened markedly. Use of bones of poor sensory quality generally results in mechanically deboned meat of inferior bacteriological quality.

CONCLUSIONS

Poultry nuggets represent a nutritive animal protein food. The incorporation of carbohydrates and lipids during processing enhances their nutritional value. Poultry is chosen as the raw material in nugget processing because it is a cheap source of meat.

REFERENCES

Antonova I. 2001. Determination of crispness in breaded fried chicken nuggets using ultrasonic technique. M.S. in Biological Systems Engineering thesis, Virginia Polytechnic Institute and State University, Blacksburg, VA.

Bacon RT, Sofos JN. 2003. Biological food hazards: characteristics of biological food hazards. In: *Current Issues in Food Safety*. Hoboken, NJ: Wiley, pp. 155–193.

Barbut S. 2002. Battering and breading. In: *Poultry Products Processing*. Boca Raton, FL: CRC Press, Chap. 10.

Beli E, Duraku E, Telo A. 2001. *Salmonella* serotypes isolated from chicken meat in Albania. Int J Food Microbiol 71:263–266.

Bijker PGH, van Logtestijn JG, Mossel DAA. 1987. Bacteriological quality assurance (BQA) of mechanically deboned meat (MDM). Meat Sci 20(4):237–252.

Bonato P, Perlo F, Teira G, Fabre R, Kueider S. 2006. Nuggets formulados con carne de ave mecánicamente recuperada y lavada: estabilidad durante el almacenamiento en congelación. Cienc Tecnol Aliment 5(2):112–117.

Cuppet S. 2001. The use of natural antioxidants in food products of animal origin. In: Pokorny J, Yanishlieva N, Gordon M, eds., *Antioxidants in Food*. Abington, UK: Woodhead Publishing, Chap. 12.

De Freitas LA, Campiteli MF, de Fátima CE, Pasetto FD. 2004. Monitoring of the dissemination of *Salmonella* in the chicken frankfurt–sausage production line of a sausage factory in the state of São Paulo, Brazil. Mem Inst Oswaldo Cruz 99(5):477–480.

deMan JM. 1999. Chemical and physical properties of fatty acids. In: Chow CK, ed., *Fatty Acids in Foods and Their Health Implications*. New York: Marcel Dekker, Chap. 2.

Greenwood DE, Swaminathan B. 1981. Rapid detection of salmonellae in mechanically deboned poultry meat. Poult Sci 60(10):2253–2257.

Jackson V, Schilling MW, Coggins PC, Martin JM. 2006. Utilization of rice starch in the formulation of low-fat, wheat-free chicken nuggets. J Appl Poult Res 15:417–424.

Jay JM. 2000. *Modern Food Microbiology*, 6th ed. Gaithersburg, MD: Aspen Publishers, pp. 441–445.

Kostenko YG, Shagova TS, Yankovsky KS. 1999. Listeriosis: technological factors and safety of meat products during their manufacture. In: *45th ICoMST*, Yokohama, Japan, pp. 556–557.

Kumar S, Wismer-Pedersen J, Caspersen C. 1986. Effect of raw materials, deboning methods and chemical additives on microbial quality of mechanically deboned poultry meat during frozen storage. J Food Sci Technol India 23:217.

Kumar BJ, Kalaikannan A, Radhakrishnan KT. 2007. Studies on processing and shelf life of pork nuggets with liquid whey as a replacer for added water. Am J Food Technol 2:38–43.

Marinsek SJ, Grebenc S. 2002. *Listeria monocytogenes* in minced meat and thermally untreated meat products in Slovenia. Slov Vet Res 39(2):131–136.

McWatters KH. 1977. Performance of defatted peanut, soybean and field pea meals as extenders in ground beef patties. J Food Sci 42:1492–1495.

Ngadi M, Lia Y, Olukaa S. 2007. Quality changes in chicken nuggets fried in oils with different degrees of hydrogenatation. Food Sci Technol 40:1784–1791.

Owens CM. 2001. Coated poultry products. In: Sams AR, ed., *Poultry Meat Processing*. Boca Raton, FL: CRC Press, Chap. 13.

Pokorný J, Schmidt S. 2001. Natural antioxidant functionality during food processing. In: Pokorny J, Yanishlieva N, Gordon M, eds., *Antioxidants in Food*. Abington, UK: Woodhead Publishing, Chap. 14.

Prinyawiwatkul W, McWatters KH, Beuchat LR, Phillips RD. 1997. Optimizing acceptability of chicken nuggets containing fermented cowpea and peanut flours. J Food Sci 62(4):889–905.

Schmidt U, Kaya M. 1990. Verhalten von *Listeria monocytogenes* in vakuumverpakten Bruhwurstaufschnitt. Flesichwirtschaft 70:236–240.

Sofos JN, Skandamis P, Stopforth J, Bacon T. 2003. Current issues related to meat-borne pathogenic bacteria. In: *Proceedings of the 56th American Meat Science Association Reciprocal Meat Conference*, Columbia, MO, June 15–18, pp. 33–37.

16

PASTE PRODUCTS (PÂTÉ)

ALFONSO TOTOSAUS-SÁNCHEZ

Tecnológico de Estudios Superiores de Ecatepec, Laboratorio de Alimentos, Ecatepec de Morelos Estado de México, México

Handbook of Poultry Science and Technology, Volume 2: Secondary Processing, Edited by Isabel Guerrero-Legarreta and Y.H. Hui

INTRODUCTION

Paste products are related to emulsified meat products made primarily with liver, lard, or pork backfat, sometimes meat, and a great variety of species. These types of products originated in the French cuisine, and some examples are *pâté* (cooked in casings), *terrine* (hot-molded in recipients), *mousse* (including eggs to form a foamy texture) and *rillete* (made with meat and liver). The oily and soft textures allow the mixture of many tasty flavors. Liver sausage or pâté is the paste product most commonly consumed around the world. Paste products, like many other traditional meat products, were formulated initially more like a craft than a science. Many of the foods that are familiar to us today are the result of a long and complex history of development. Today, consumers can choose meat products from a wide variety among a few hundred different products (McClements, 2005). This makes it challenging to develop or maintain certain meat paste characteristics when these products are processed on a large scale.

LIVER SAUSAGE OR PÂTÉ

Liver pâté has the particular characteristics that diverse sources of poultry liver (broiler, ostrich, goose, duck) are employed in its formulation and it has a relatively higher fat content (from 30 to 50% depending on particular formulation). Fat is an important part of the formulation, since fat in pâté is responsible for product spreadability. The fat must be dispersed in the batter and entrapped and emulsified by the protein matrix. In liver pâté the amount of lean meat (usually precooked) is lower than it is in a meat sausage formulation. This reduces the availability of functional muscle protein, making the product less thermostable; fat can be released or color and flavor affected by excessive heat. Consequently, liver proteins have become the main emulsifying agent. Nonetheless, other proteins, particularly milk protein concentrates, can be added to the formulation to improve the emulsification (Pérez-Chabela and Totosaus, 2004; Totosaus and Pérez-Chabela, 2005).

Poultry Liver

Poultry liver has great potential to be used as a raw material for processing into many different paste products, and it is an excellent source of protein and other vital nutrients in the human diet (Hazarika et al., 2006). Since slaughter begins by catching the birds and preparing them for transport from farmyards, stress problems can be present at this very first stage of poultry processing, where among other welfare problems is that of underfeeding (Pérez-Chabela and Totosaus, 2006). Stress levels in birds had a marked influence on liver weight. Corporal weight lost and increases in liver weight are characteristic changes associated with stressful situations. The increase in liver weight is due to lipid accumulation, caused by changes in alimentation and protein conversion. When the liver cannot eliminate triglycerides in the form of lipoproteins, hepatic lipogenesis causes steatosis, due to fatty acid accumulation (Terreas et al., 2001). This is the basis for foie gras manufacture.

Liver composition can be altered without stress by modifying the feeding composition to enhance the nutritional characteristics of pâté. As in another species, lipid and triacylglycerol concentrations in livers can be reduced by dietary conjugated linoleic acid (CLA) supplementation. Feeding broilers proportional diets increased saturated fatty acids (SFAs) in liver lipids, whereas monounsaturated fatty acids (MUFAs) decreased. Feeding CLA to broilers results in a substantial reduction in liver fat accumulation and promotes CLA incorporation into hepatic lipid pools, where the changes in fatty acid profiles were due mainly to increases in SFAs and concurrent decreases in MUFAs in the liver (Badinga et al., 2003). The chemical composition of liver lipids can be determined and predicted with near-infrared reflectance spectroscopy (NIRS) (Molette et al., 2001). Organic arsenical salts such as roxarsone (3-nitro-4-hydroxyphenylarsonic acid) can be included to modify fatty acid contents in mule duck livers, with an increase in very low density lipoprotein and decreased high-density lipoprotein (Chen and Chiou, 2001). As a final point, before slaughter, feed withdrawal affects the metabolic and hence the physical properties of poultry meat and liver. The estimation of liver pH seemed to be a more reliable indicator of the length of feed withdrawal. Longer periods of feed withdrawal resulted in higher ultimate liver pH values (6.10 to 6.20 in full-fed broilers, increasing to 6.60 in livers from broilers deprived of food). Feed withdrawal must cause a change in the concentration of stored glycogen in the liver, which is converted anaerobically to lactic acid postmortem, with declining pH when sufficient glycogen is available (Van Der Wal et al., 1999).

Emulsion

The technological basis of pâté and related products is emulsion of the added fat. Meat emulsion is a biphasic system, formed by the solid dispersing in a liquid in which the solid is not miscible. This definition must always be viewed with some reservation, because although, for example, it takes into account the behavior of particles that are insoluble in the aqueous phase, this is far from being the case in terms of fats. This type of biochemical constituent forms a structure that is closer to that of an emulsion than that of a dispersion. Within the area of chopped products grouped together under the name *meat emulsions*, we can define *fine paste* as a mixture composed essentially of lean meat, "fat," and water, whose homogeneity is such that the grain of the constituents added is no longer visible to the naked eye. Mortadella, frankfurters, and liver pâté are some of the products typical of this category (Linden and Lorient, 1999).

Fat

The fat content of meat products can vary considerably, depending on the proportion of lean and fat from the original meat as well as the level of inclusion of other ingredients. Traditional meat products such as sausages, pastry-covered pies, and salami are high in fat (up to 50%), but modern products include ready-to-eat and prepared meats that can be low in fat (5%) (Moloney, 2000). Fats and oils influence the nutritional, organoleptic, and physicochemical properties of food emulsions in a variety of ways. They are a major source of energy and essential nutrients; however, overconsumption of certain types of these constituents leads to obesity and human health concerns. The perceived flavor of a food emulsion is determined by the type and concentration of lipids present. Like an emulsion, the oil phase may also act as a solvent for various other important food components, including oil-soluble vitamins, antioxidants, preservatives, and essential oils. Reducing the fat content of an emulsion can therefore have a profound influence on its flavor profile, stability, and nutritional content (McClements, 2005). Although nutritional concerns relating to the excessive levels of fat being consumed in Western cultures are warranted, it is also important to acknowledge the essential beneficial role that lipids and lipid-related compounds play in disease prevention and growth. Phospholipids, glycolipids, and cholesterol play a structural role in cell membranes, with lipids also having an important function in biochemical regulatory systems. These lipid fractions can serve as precursors for beneficial biologically active compounds such as prostaglandins, steroid hormones, and bile acids. Animal fat contribution to overall calorific intake and the proportion of fat (and type) present within meat products is a further cause of dietary concern for consumers (Kerry and Kerry, 2006). Figure 1 represents the average composition of poultry liver with an emphasis on fat content and fatty acid composition. Unsaturated fats represent 55% of total fat, where monounsaturated fatty acids are the main proportion of the fat content (41%, which corresponds to 27.5% of total composition), with a relatively higher amount of oleic acid.

LIPID OXIDATION

Lipid oxidation in muscle foods is one of the major degradative processes responsible for losses in quality. Autooxidation leads to the formation of short-chain aldehydes, ketones, and fatty acids, as well to development of some polymers, all of which are believed to contribute to the oxidized flavor in meat and poultry products. Unsaturated fatty acids are especially vulnerable to oxidation, being readily cleaved at the double bond to produce various oxidation products (Pearson et al., 1983). Pâté products, with their high fat content, are very susceptible to oxidation. Exposure of pâté to light and air during display, between manufacture and consumption, can result in a series of changes in fat and meat pigments, reducing the product quality.

Autoxidation

Fatty acids are susceptible to autoxidation, where unsaturated bonds present in all fats and oils represent actives sites that can react with oxygen. This reaction leads

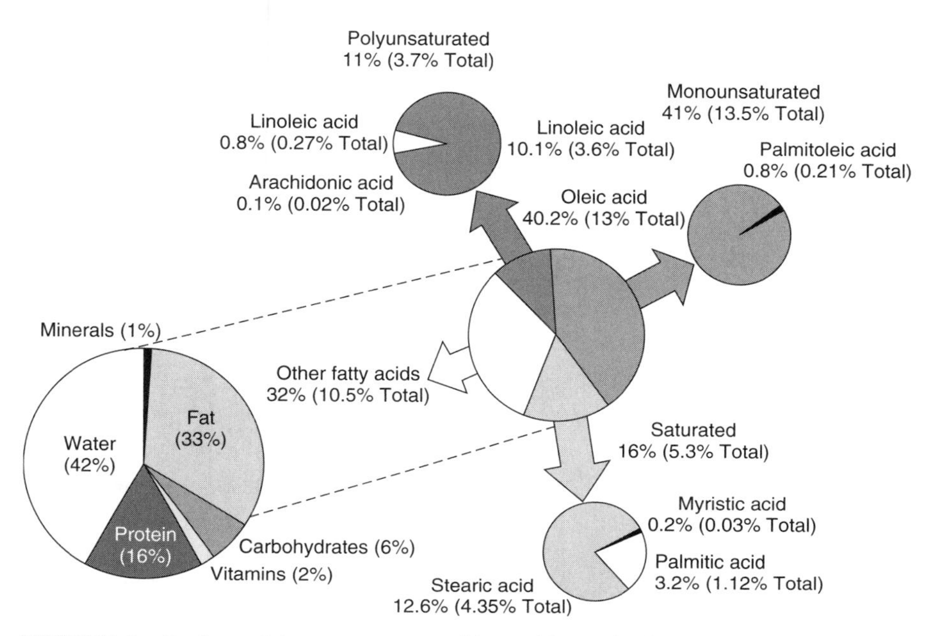

FIGURE 1 Poultry pâté average composition with emphasis on fat composition and content (percent of total composition in parentheses).

to the formation of primary, secondary, and tertiary oxidation products that may make the fat-containing food unsuitable for consumption. Among the factors that affect the rate of autoxidation are the amount of oxygen present, the degree of unsaturation, the presence of antioxidants, the presence of prooxidants (especially copper and some organic compounds, such as heme-containing molecules and lipoxidase), the storage temperature, and exposure to light (deMan, 1992).

The autoxidation reaction can be divided in three stages: initiation, propagation, and termination. The first reaction, *initiation*, takes place when light or metals produce the removal of hydrogen from a carbon atom next to a double bond on the fatty acid chain, yielding a free radical. This free radical will be combined with oxygen to form a peroxy free radical, which in turn will abstract hydrogen from another unsaturated molecule to yield peroxide and a new free radical, starting the *propagation* reaction. The propagation ends when the free radicals react with each other after reacting with all the possible double bonds in fatty acid chains due to environmental conditions, forming nonactive products (Fig. 2) (deMan, 1992).

Photooxidation

Photooxidation is an oxidation process resulting from exposure to light in the presence of oxygen and a sensitizer, such as heme compounds, which accept energy from light in the visible spectrum. Two types of processes can

Initiation

Hydrogen is abstracted from an olefinic compound to yield a free radical.

$$RH \rightarrow R^{\bullet} + H^{\bullet}$$

Propagation

The removal of hydrogen takes place at the carbon atom next to a double bond by the action of light or metals, forming a free radical.

$$R^{\bullet} + O_2 \rightarrow RO_2^{\bullet}O_2$$

$$RO_2^{\bullet} + RH \rightarrow ROOH + R^{\bullet}$$

Termination

After promoting a free-radical-forming chain reaction, free radicals react with themselves to yield nonactive compounds.

$$R^{\bullet} + R^{\bullet} \rightarrow RO_2^{\bullet}$$

$$R^{\bullet} + RO_2 \rightarrow RO_2R$$

$$nRO_2^{\bullet} \rightarrow (RO_2)_n$$

FIGURE 2 Sequential reactions involved in fatty acids oxidation. (Adapted from deMan, 1992.)

be distinguished: sensitized photooxidation and photolytic autooxidation. The latter can be initiated by ultraviolet-catalyzed decomposition of peroxides and hydroperoxides. Photooxidation involves singlet oxygen ($^1O_2$) to form monohydroperoxides from unsaturated fatty acids by a mechanism that does not involve free radicals. Secondary oxidation products can be formed by the free-radical side reactions. This type of reaction can be inhibited by quenchers, which are able to deactive the singlet oxygen (deMan, 1992).

Compounds such as heme and nonheme iron have been reported to act as catalysts of lipid oxidation. Muscle tissues contain a considerable amount of iron bound to proteins, mainly as hemoglobin residues from residual blood. In addition, muscle cells also contain cytochromes. All these proteins contain prosthetic group heme, which has an iron atom at its center. The iron atom itself promotes autooxidation of fats (Pearson et al., 1983).

Thermal Oxidation

Thermal oxidation of fatty acids can be produced during a heating process such as cooking or frying. Great care is taken to minimize thermal oxidation of fats and oils during processing by removing oxygen by vacuum processing, although even the most careful techniques can result in some thermal decomposition (deMan, 1992).

Enzymatic Oxidation

In animal systems enzymatic oxidation involves mainly the oxidative transformation of arachidonic acid to prostaglandins, thromboxanes, and leukotrienes. These compounds are formed in animal tissues and have a board range of biological activities. The first reaction catalyzed by the enzyme cyclooxygenase from arachidonic acid is prostaglandin G_2, which contains hydroxyperoxide and cyclic peroxide functions. The 5-hydroperoxide formed reacts further to form leukotrienes (deMan, 1992).

Antioxidant Supplementation

Due to their fat content, pâtés need the application of antioxidants to extend shelf life, since lipid oxidation is one of the primary mechanisms associated with quality deterioration (Russell et al., 2001; Fernández-López et al., 2004). Lipid oxidation is a process by which molecular oxygen reacts with unsaturated lipids to form lipid peroxides. This free-radical chain reaction process can be overcome by initiators or by initiating variables such as temperature, photosensitizers, radiation, singlet oxygen, oxygen-transition metal complexes, or by lipoxygenase-like enzymatic catalysis (Monahan, 2000). The use of antioxidants in this type of product is important to maintain quality. Supranutritional amounts of α-tocopheryl acetate and dietary oils fed to ducks improve the oxidative and color stability of duck liver pâté (Russell et al., 2001).

OTHER PÂTÉ CHARACTERISTICS

Microbiology

Since liver is part of the digestive tract, and they have to be separated manually, there is a risk of contamination. The importance of an adequate thermal process ensures a microbial count reduction. Pâté boiling greatly reduces the number of bacteria, especially gram-negative bacteria, and the relationship between the initial and final bacterial content of liver pâté (fresh, after boiling, and after 6 days of storage at 0°C) shows clearly the influence on the microbiological quality of marketed products of contamination of the raw materials employed (Gesche and Ordoñez, 1984).

Species

Chicken or broiler liver is used since that is the most common poultry species. But alternatives are available, such as ostrich or more regionally bred poultry, such as duck or goose. The manufacture of pâtés from ostrich liver is a feasible option for an industry that has largely released its products to the fresh meat and meat products market with good general quality and acceptability, based on their chemical composition and sensory scores (Fernández-López et al., 2004). In this respect, and because some of the fraudulent or unintentional mislabeling occurs may go undetected, resulting in lower-quality pâté, and because some population groups, for philosophical or religious reasons, do not wish to eat meat from certain species, randomly amplified polymorphic DNA (RAPD) has been used in species detection to generate fingerprint patterns for pork, chicken, duck, turkey, and duck meats in order to detect adulterations (Calvo et al., 2001).

MOST COMMON DEFECTS

During pâté production many defects can be present, due to a lack of control during the process. For example, fat separation and jelly pockets are due to higher temperature of the cooking water. In the same manner, dark rings in the product are caused by a loss of heat after cooking and before being placed into the smokehouse. A burst casing on cooked pâté is due to overcooking, since liver tends to expand during cooking (Keeton, 2001).

CONCLUSIONS

In recent years, due to an increasing tendency toward removing or reducing the amounts of food constituents that have been associated with human health concerns (such as fat, salt, and cholesterol), it is important to understand the role that each ingredient plays in determining the overall physicochemical and organoleptic properties of foods, so that this role can be carried out by a healthier alternative ingredient (McClements, 2005). In this view, the fat reduction and fat replacer incorporation can be subject to investigation in order to offer healthy paste products, the main concern being related to pâté spreadability and texture. Since the feed chosen has an effect on liver composition, the opportunity to change or modify the fatty acid composition to enhance the nutritive properties of paste products, together with lard (pork backfat) replacement, seems to be a good alternative to improving the functional, technological, and nutritional characteristics of this type of meat product.

REFERENCES

Badinga L, Selberg KT, Dinges AC, Comer CW, Miles RD. 2003. Dietary conjugated linoleic acid alters hepatic lipid content and fatty acid composition in broiler chickens. Poult Sci 82:111–116.

Calvo JH, Zaragoza P, Osta R. 2001. Random amplified polymorphic DNA fingerprints for identification of species in poultry pâté. Poult Sci 80:522–524.

Chen KL, Chiou PWS, 2001. Oral treatment of mule ducks with arsenicals for inducing fatty liver. Poult Sci 80:295–301.

deMan JM. 1992. Chemical and physical properties of fatty acids. In: Chow CK, ed., *Fatty Acids in Foods and Their Health Implications*. New York: Marcel Dekker, pp. 17–45.

Fernández-López J, Sayas-Barberá E, Sendra E, Pérez-Alvarez JA, 2004. Quality characteristics of ostrich liver pâté. J Food Sci 69:SNQ85–SNQ91.

Gesche E, Ordoñez T. 1984. Flora bacterial de paté de hígado en tres etapas de su elaboración. Arch Latinoam Nutr 34:384–390.

Hazarika P, Kondaiah N, Anjaneyulu ASR. 2006. Chicken liver: a versatile but often overlooked product. Poult Int 45(3):18, 20.

Keeton JT. 2001. Formed and emulsion products. In: Sams AR, ed., *Poultry Meat Processing*. Boca Raton, FL: CRC Press, Chap. 12.

Kerry JF, Kerry JP. 2006. Producing low-fat meat products. In: Williams C, ed., *Improving the Fat Content of Foods*. Boca Raton, FL: CRC Press, Chap. 14.

Linden G, Lorient D. 1999. Meat products. In: *New Ingredients in Food Processing Biochemistry and Agriculture*. Boca Raton, FL: CRC Press, Chap. 7.

McClements DJ. 2005. Emulsion ingredients. In: *Food Emulsions: Principles, Practices, and Techniques*, 2nd ed. Boca Raton, FL: CRC Press, Chap. 4.

Molette C, Berzaghi P, Zotte AD, Remingnon H, Babile R. 2001. The use of near-infrared reflectance spectroscopy in the prediction of the chemical composition of goose fatty liver. Poult Sci 80:1625–1629.

Moloney AP. 2000. The fat content of meat and meat products. In: Kerry JF, Kerry JP, Ledward D, eds., *Meat Processing*. Boca Raton, FL: CRC Press, Chap. 7.

Monahan FJ. 2000. Oxidation of lipids in muscle foods: fundamental and applied concerns. In: Decker E, Faustman C, Lopez-Bote CJ, eds., *Antioxidant in Muscle Foods*. New York: Wiley–Interscience, pp. 3–23.

Pearson AM, Gray JI, Wolzak AM, Horenstein NA. 1983. Safety implications of oxidized lipids in muscle foods. Food Technol 37(7):121–129.

Pérez-Chabela ML, Totosaus A. 2004. Poultry pâté. In: Smith JS, Hui YH, eds., *Food Processing: Principles and Applications*. Ames, IA: Blackwell Publishing, pp. 433–438.

Pérez-Chabela ML, Totosaus A. 2006. Poultry carcass evaluation. In: Hui YH, ed., *Handbook of Food Technology and Food Engineering*, vol. 4, *Food Technology and Food Processing*. Boca Raton, FL: CRC Press, pp. 165–1 to 165–6.

Russell EA, Lynch A, Lynch PB, Kerry JP. 2001. Quality and shelf life of duck liver pâté as influenced by dietary supplementation with α-tocopheryl acetate and various fat sources. J Food Sci 68:799–802.

Terraes JC, Sandoval GL, Fernández RJ, Revidatti FA. 2001. Respuesta a una maniobra inductora de estrés y al tratamiento como producto hepatoprotector en pollos de engorde. Vet Mex 32(3):195–200.

Totosaus A, Pérez-Chabela ML. 2005. Poultry products: pâté and nuggets. In: Hui YH, ed., *Handbook of Food Technology and Food Engineering*, vol. 4, *Food Technology and Food Processing*. Boca Raton, FL: CRC Press, pp. 167–1 to 167–8.

Van Der Wal PG, Reimert HGM, Goedhart HA, Engel B, Uijttenboogaart TG. 1999. The effect of feed withdrawal on broiler blood glucose and nonesterified fatty acid levels, postmortem liver pH values, and carcass yield. Poult Sci 78:569–573.

17

POULTRY HAM

VANDANA SOHLIA AND AMARINDER S. BAWA
Defence Food Research Laboratory, Siddartha Nagar, Mysore, India

INTRODUCTION

Poultry meat is one of the most accepted muscle foods in the world. In the past few years, an increase in consumption of poultry meat has been recorded in both developed and developing countries (Anon., 2001; Oliveira et al., 2005). In 1999, global production of broiler chicken first reached 40 billion, and based on predictions, poultry is expected to become the overall meat of choice by 2020 (Bilgili 2002). The higher consumption of poultry (Figure 1) and tremendous growth of the industry over the last 20 years, which is the result of various factors, such as its high-quality protein, relatively low fat content, new products

Handbook of Poultry Science and Technology, Volume 2: Secondary Processing, Edited by Isabel Guerrero-Legarreta and Y.H. Hui

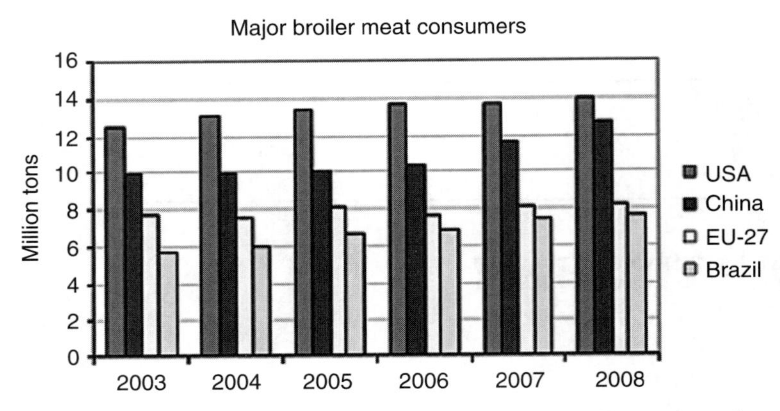

FIGURE 1 Trends in broiler meat consumption. (From http://www.fas.usda.gov/dlp/livestock_poultry.asp.)

and generally low selling price, favorable feed conversion, and absence of cultural and religious taboos, have resulted in increased production and consumption worldwide (Van Horne, 2002).

POULTRY MEAT QUALITY FOR PROCESSING PRODUCTS

Processed poultry meat products are somewhat similar to their meat counterparts in protein content but are higher in moisture content and lower in fat and energy. This favorable nutritional profile has made chicken and turkey alternatives in the manufacture of healthy low-fat meat products. The types of products available range from cut-up portions to re-formed roasts, breasts, rolls, steaks, hams, burgers, frankfurters, bolognas, coarse ground sausages, salamis, and bacons (Maurer, 2003). In the preparation of processed meat products, the quality of meat plays a significant role; moreover, in the preparation of ham and other formed poultry products, skeletal muscle proteins are of vital importance (Damodaran, 1994). The primary poultry skeletal muscle proteins are listed in Table 1 (Lawrie, 1998), and these proteins play a significant role in determining the quality of processed poultry products. The myofibrillar, sarcoplasmic, and stroma proteins present in poultry muscle are essential in imparting desirable functional properties such as protein hydration, water-holding capacity, emulsification, and tenderness to processed products (Jimenez-Colmenero et al., 1994; Lawrie, 1998; Froning and McKee, 2001; Smith, 2001; Ponce-Alquicira, 2002). Although sarcoplasmic proteins play a minor role in meat protein functionality, they have a marked influence on the color of the processed product, due to the presence of myoglobin and other heme proteins. The color of the product can also be influenced by various animal- and processing-related factors (Flotcher, 1999).

TABLE 1 Main Poultry Skeletal Muscle Proteins

Protein Fractions	Content (% total protein)
Myofibrillar (salt-soluble proteins)	
Myosin	29
Actin	13
Tropomyosin	3.2
Troponins C.I.T.	3.2
Actinins	2.6
Desmin	2.1
Conectin	3.7
Sarcoplasmic (water-soluble proteins)	
Myoglobin and other heme proteins	1.1
Glycolytic enzymes	12
Mitochondrial enzymes	5
Lysosomal enzymes	3.3
Stroma (insoluble proteins)	
Collagen	5.2
Elastin	0.3
Reticulin	0.5

INGREDIENTS AND ADDITIVES USED IN POULTRY HAM PROCESSING

In the production of ham a number of additives and ingredients are required to stabilize the mixture, to add specific characteristics, and to improve organoleptic quality as well as microbiological stability (Kemp et al., 1980; Murphy et al., 2003; Hwang and Tamplin, 2007; Marcos et al., 2008). According to Freixanet (2004), in cooked ham a particular distinction lies between the ingredients and the additives. Ingredients are chosen from a variety of compounds naturally present which are consumed as part of a normal diet; additives are selected on the basis of specific technological and functional requirements, they are added intentionally to serve particular purposes, and they may not necessarily be consumed as a regular item of the diet.

Ingredients

Meat Selection of meat for ham preparation is a very crucial step because the final quality of poultry ham will depend on the quality of the raw meat. Meat selection for the production of ham varies widely depending on production goals. Poultry ham can be processed from boneless breast, legs, thighs, desinewed meat, drumsticks, mechanically deboned poultry meat, meat with or without skin, and bone-in or boneless meat. According to the formula designed by the processor and consumer taste preference, the poultry meat may have varying degrees of fat, nerves, and tendons trimmed. For high-quality ham preparation, meat should have a pH of 5.7, be at low temperature, and have higher microbial quality. Pale, soft, and exudative (PSE) and dark, firm, and dry (DFD) hams are not suitable for processing dry-cured hams because of excessive water loss due to their lower water-binding capacity (Arnau et al., 1995) and high pH, respectively. However, DFD hams can be used to process cooked ham since cooking eliminates the risk of microbial spoilage from wet-cured DFD hams; moreover, high pH facilitates higher water retention during heating.

Water The second most important ingredient for cooked ham processing is water. Water is used primarily in the preparation of brine, and manufacturers also add water to the formula in a specific amount to improve the consistency of the mixture and to dissolve solid ingredients. Water should be potable, of high chemical and microbial quality. It should be free of Ca^{2+} and Mg^{2+} salts as well as minerals such as iron, copper, and other metals because a high ion concentration can influence such functional properties as the water-holding capacity of the finished product. Metal may result in toxicological risk as well as possibly interfering with brine constituents such as ascorbates, reducing the antioxidant effect and color stability of the finished product. Hence, water quality is of great importance in getting good-quality poultry ham, and it should be as soft as possible for processing ham.

Salt Table salt, sodium chloride, has been used since ancient times in the preparation of meat products. There are several important roles for salt, such as reduction of water activity (a_w) and inhibition of the growth of spoilage microorganisms, enhancing the flavor and solubilizing myofibrillar protein, thus improving the texture and water content as well as fat binding. Sometimes when for health reasons a low-sodium ham is desired, salt is partially replaced by other substituents. Potassium chloride can be used, but it imparts a bitter metallic flavor that doesn't go well with other flavoring compounds present in the ham.

Sugar Sugars are used in ham primarily to reduce the water activity and enhance the product taste. In the case of fermented ham, sugar also acts as a food source, to enable microbial fermentation. Sugars are commonly used in mixture form, and the composition may vary according to the manufacturer's formulation. Most commonly, mixtures of sucrose, dextrose, lactose, fructose, glucose syrup, and dextrins are used in the preparation of ham.

Proteins Proteins and their hydrolyzed products play two major roles: improving the nutritional quality and improving the functional properties of processed ham. They are used in restricted amounts because of legislation; they may also mask the flavor of product if used in excess. But because of their desirable functional properties, manufacturers use a variety of functional proteins from various sources, such as milk (milk whey, lactalbumin, caseinates), blood (blood plasma), collagen (partially hydrolyzed proteins, dried skin powder), egg (egg albumin), vegetable sources (soy protein isolates and concentrates), and protein hydrolysates (hydrolyzed collagen and protein from mechanically recovered meat, hydrolyzed vegetable protein).

Starches or Binders and Extenders Manufacturers use starches derived from wheat, potato, corn, and manioc to increase overall yield, to improve binding properties and slicing characteristics, and for good texture and flavor. Most important, starches are basically polysacchrides, which form a gel when exposed to heat and hence hold large amounts of water.

Fibers Fibers are basically polysacchrides, such as cellulose, hemicellulose, pectin, and lignin. These fibers can be extracted from the cellular walls of cereals and vegetables, and depending on the source and extraction process, their yields may vary from 55 to 85%. Fibers provide good functional properties, such as an improved water-holding capacity and texture, to processed meats. In addition to these functions, fiber is also used as a substitute for fatty materials and hence to reduce the caloric value.

Flavorings Other important ingredients in poultry ham processing are spices, seasoning, and flavoring. The types of flavoring used vary widely and include liqueurs, wines, fruit juices, hydrolyzed vegetable proteins, Maillard reaction products, volatile oils, resins, oleoresins derived from natural spices, fruits, vegetables, herbs, roots, meats, seafood, and smoke extracts. Common spices and seasonings include allspice, pepper, cardamom, caraway, coriander, cumin, garlic, mustard, nutmeg, paprika, rosemary, sage, thyme, and turmeric.

Additives

Colorings As the name implies, coloring agents from various natural and chemical sources are used to give a good appearance to poultry ham. Cochineal carmine is the most commonly used coloring agent in the manufacture of ham; it gives the ham a natural pink tone. The coloring agent in carmine is carminic acid ($C_{22}H_{20}O_{13}$), which is stable on exposure to light, pH variations, and thermal processing. Other coloring agents used in ham processing are norbixin sodium salt from bixa or annatto, betanin constituting beet red, stabilized hemoglobin (sterilized and dehydrated), and caramel (only for outer covering, to simulate a smoked appearance), but due to their poor stability and other technological problems, their use is limited. Artificial colorings (e.g., Red 2G, Red 40, Ponceau 4R) are also used sometimes, but consumer demand for the use of naturally colored hams has restricted their use.

Nitrites Nitrite is a multifunctional ingredient: It is an important preservative in ham and, in addition, is responsible for the development of cured color, texture, and flavor; it serves as a strong antioxidant to protect flavor; and it acts as an antimicrobial (Cassens, 1995; Ferreira and Silva, 2007) to control *Clostridium botulinum* (Shahidi and Pegg, 1992), *C. perfringens, Staphylococcus aureus*, and other enterobacteria. Nitrite controls and stabilizes the oxidative states of lipids in meat products (Shahidi and Hong, 1991; Cassens, 1995), thus preventing lipid oxidation and subsequent warmed-over flavors (Yun et al., 1987; Vasavada and Cornforth, 2005). Nitrite as such does not act on the meat (Pegg and Shahidi, 2000). It is the nitrous oxide that is responsible for the beneficial effects produced. Nitrite reacts with many of the components in the matrix (Ferreira and Silva, 2007), and most of it is present as nitrous oxide bound with myoglobin (5 to 15%), sulfydryl groups (5 to 15%), lipids (1 to 15%), and proteins (20 to 30%), partially present as nitrate (<10%) and nitrite (10 to 15%).

Nitrates Potassium nitrate was the first nitrifying agent to be used in manufacturing salted meat products; however, these substances were originally present in salts as contaminants, and now thery are added intentionally in the form of saltpeter. Nitrate that is added as a protective agent against botulism does not have a nitrifying action as such on meat, but its effects are seen after slow reduction from nitrate to nitrite through the action of nitrate reductase (Cassens, 1995), an enzyme present in the natural microflora of ham (micrococcaceae, lactobacilli, enterobacteria, and other microorganisms). The use of nitrates by large manufacturers is rare because the process of conversion of nitrate to nitrite in the product is much slower, and the nitrate reductase–forming bacteria are reduced to a very low level after cooking. So nitrites or nitrous oxide are added directly for an enhanced reaction rate. Since the excessive use of nitrites, nitrates, and nitrous oxide can be toxic to humans, the use of these additives is carefully controlled.

Preservatives The use of preservatives has been substantially reduced due to legislation as well as because of advances in thermal processing, refrigeration networks, and manufacturing conditions.

Cure Accelerators Cure accelerators (Table 2) such as ascorbates and erythorbates are used to speed the curing process. They also stabilize the color of the final product.

TABLE 2 Cure Accelerator Ingredients and Their Maximum Amount

Ingredient	Maximum Amount
Ascorbic acid	$\frac{3}{4}$ oz per 100 lb of meat
Erythorbic acid	$\frac{3}{4}$ oz per 100 lb of meat
Sodium erythorbate	$\frac{7}{8}$ oz per 100 lb of meat
Citric acid	May replace up to 50% of the ingredients listed above
Sodium citrate	May replace up to 50% of the ingredients listed above
Sodium acid pyrophosphate	Alone or in combination with others may not exceed 8 oz (0.5%)
Glucono delta lactone	8 oz per 100 lb of meat

Antioxidants Antioxidants are approved for use to retard the oxidative rancidity and subsequently, protect flavor and color. Of all the approved antioxidants, L-ascorbate and its optical isomer sodium erythorbate are in most common use. They are generally added in the form of a salt such as sodium ascorbate. It serves three main functions in ham, based on its behavior as a powerful reducing agent. The first main function is its action as a nitrite reducer, thereby accelerating the formation of pink color by providing an enhanced effect along with a natural reducing agent. The second important function is the stability of color in the finished product, and finally, it contributes to preventing the formation of nitrosamines by blocking the formation of nitrostating agents. Nitrosamines are well documented as a cancer-causing agent. Since sodium ascorbate is insoluble in fatty tissues, it imparts little antioxidizing effect. Commercially available antioxidants such as tocopherols, butylhydroxyanisole (BHA), and butylhydroxytoluene (BHT) are not generally used; however, the antioxidant reinforcing agents such as trisodium citrate and sodium lactate are used in ham preparation. Trisodium citrate contributes buffering and chelating properties, while sodium lactate reduces water activity as well as inhibiting bacterial growth, particularly lactobacilli.

Phosphates Phosphates serve two main functions in cooked ham: increased water-holding capacity and enhanced solubility and extraction of myofibrillar protein, thereby improving the texture in ham. Another effect of phosphates in ham is attributed to their buffering capacity by homogenizing the pH in different muscles, thereby lowering the exudative effects of PSE muscle. Phosphates are generally effective in the form of pyrophosphate, so the mixtures of tripolyphosphates, pyrophosphates, and hexametaphosphates are generally used in cooked ham.

Stabilizers Stabilizers are used to enhance viscosity in brine and retention of water by forming gels, reduce syneresis in processed ham, and improve water-holding capacity. Of these substances, the most commonly used in the manufacture of ham are carrageenans, gums extracted from grains such as caruba gums, guar gum, and alginates, with a small amount of salt, usually potassium chloride.

Flavor Enhancers The most universally used flavor enhancer in meat products as well as in ham processing is monosodium glutamate. It is produced by the fermentation of molasses, and in cooked ham it ranges from 0.2 to 1 g/kg of the finished product.

PROCESSING OF POULTRY HAM

Ham is the thigh or rump of any animal that is slaughtered for meat. Ham processing and cooking is a traditional and well-worn craft. Poultry ham is typically manufactured from poultry thigh meat. The product is usually leaner than the traditional pork ham (Maurer, 2003). Ham can be either dry or wet cured, based on the different processing technologies employed (Ponce-Alquicira, 2005; Toldrá, 2005, 2007). In dry-cured hams, fresh ham is rubbed with a dry-cure mixture of salt and other ingredients, followed by ripening and drying. In wet-cured or cooked ham, fresh ham is either soaked in or injected with brine or curing solution, followed by cooking or heat processing. There are various types of hams throughout the world based on country of origin, duration of the process for preparation, smoking, storage, and cooking; and so on.

Dry-Cured Ham

The processing of dry-cured ham is of ancient origin, when the purpose was to preserve food for periods of scarcity. Now, with advances in preservation methods such as refrigeration as well as in scientific knowledge of the basic biochemical mechanism of curing, the production of dry-cured ham has been improved and simplified (Toldrá, 2002, 2004). The main steps involved in the production of dry-cured ham are shown in Figure 2.

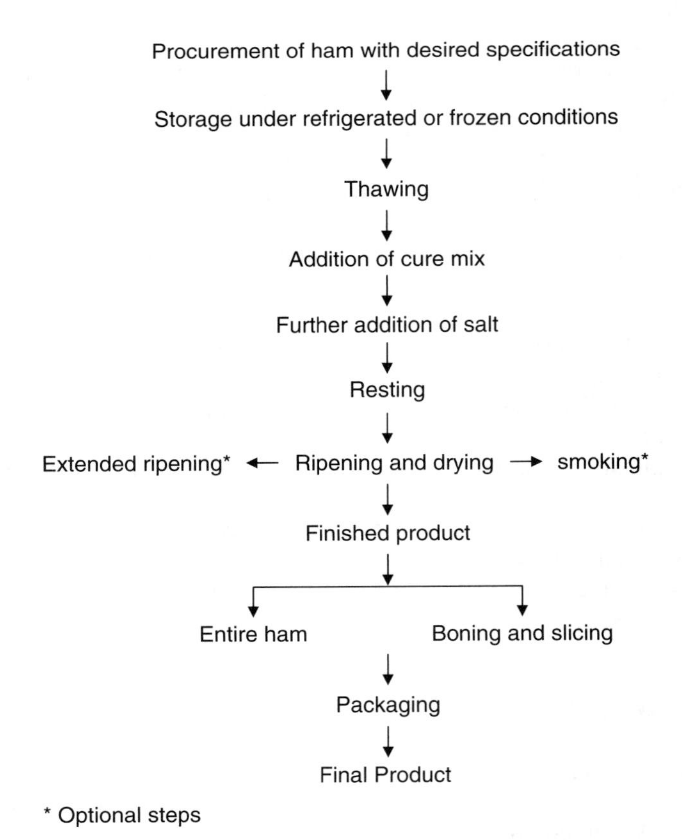

FIGURE 2 Main steps involved in the production of dry-cured ham.

Procurement of Hams Meat of precise specifications is procured and classified, usually on the basis of weight, pH, and fat (Toldrá, 2002). As soon as cuts of the required specifications are procured, they are tested for pH, temperature, microbial load, off-odor and flavor, as well as other quality parameters. The iodine index and acid value are also determined, to measure the degree of unsaturation and freshness, respectively (Toldrá, 2002). Raw hams may be purchased chilled or frozen. The internal temperature of fresh chilled ham should not be above 4.4°C, while that of frozen hams should be below −18°C when received. Before processing, frozen hams are thawed, usually by refrigeration or a cold-water method until a temperature of about 4°C is reached inside the ham. Thawing is carried out in proper packaging to prevent dehydration as well as to reduce the risk of microbial contamination.

Addition of Cure Mix Refrigerated hams and frozen/thawed hams are kept for 1 to 2 days at a temperature of 2 to 4°C to obtain a uniform temperature throughout the ham. Presalting is the step where nitrite or nitrate salts of sodium or potassium (E_{249}, E_{250}, E_{251}, E_{252}) are rubbed on the external surface of the ham. Sometimes, sugars, seasoning, phosphates, and cure accelerators (e.g., sodium ascorbates) are added. Maximum levels of nitrate and nitrate that can be added to meat products, and residual amounts, have been established (EC, 2006). According to European legislation for ham, a maximum of 150 mg/kg of sodium nitrite or potassium nitrate (Ferreira and Silva, 2007) can be added and a 300-ppm combination of sodium and potassium nitrate is allowed. In the United States, 156 ppm of sodium nitrite is permitted. For uniform distribution of dry-cure ingredients, hams are sometimes time-rotated in a rotary drum after rubbing with dry-cure ingredients. These curing ingredients slowly diffuse throughout the entire ham, and reduction of nitrate to nitrite also takes place.

Further Addition of Salt This step is the addition of salt with the objective of penetration of the required amount of salt into the ham surface. This can be achieved either by sprinkling or by hand rubbing the exact amount of salt. Salt addition is determined on the basis of the weight of the ham (Parolari, 1996). Sometimes an excess of salt is added, but the time required for salt penetration into the piece is monitored (Toldrá, 2002). After addition of the required quantity of salt, hams are placed in shelves or bins with holes under refrigerated conditions for variable durations, as required for different types of ham production. Since dry curing draws out moisture, it reduces the ham weight by about 3 to 4%. After the required salting stage, an excess of salt can be removed either by rinsing or brushing.

Resting Resting is carried out to stabilize salt concentration thoughout the ham. Salt and dry-cure ingredients penetrate and diffuse into the inner surfaces of ham. The time required may vary from 40 to 60 days, which depends on the size of the ham, pH, amount of fat, temperature, and environmental conditions of the chamber in which hams are placed. After this stage, weight losses of around 4 to 6% are common.

Smoking Smoking is more common practiced where drying is difficult. It is an optional step usually done to impart characteristic flavor and color to the hams. For example, American country style or German Westphalian ham are processed by the use of smoke. In addition to a particular flavor, the smoke components exhibit a preservation effect, due to their bactericidal properties (Ellis, 2001).

Ripening and Drying Ripening and drying are carried out under highly controlled conditions of airspeed, temperature, and relative humidity to achieve a high-quality ham. The duration of ripening and drying periods are very specific for different types of ham produced, and it is also determined on the basis of criteria of final weight loss of around 32 to 36% of the initial weight of the ham (Martín et al., 1998; Ventanas and Cava, 2001). Other criteria for conclusion of this stage are the monitoring and evaluation of ham quality by experts. The common practice followed by ham processors to monitor ham quality is a sniffing test, consisting of the insertion of a small probe into the ham followed by rapid sniffing by an expert to detect any off-flavor (Parolari, 1996; Spanier et al., 1999). Usually, ripening and drying are carried out for six to nine months, while some types of hams are ripened further for few more months to generate more intense flavor.

Extended Ripening Extended ripening is an optional step, carried out only to intensify flavor generated through chemical and enzymatic reactions. It is done in ripening cellars at mild temperatures for long periods of time, up to 24 to 30 months. Iberian hams are processed with extended ripening to produce an excellent rich flavor.

Final Product Hams may be distributed commercially as an entire piece, as boned pieces, or in the form of slices. Boned ham slicing can be done either by retailers or finally by consumers at home. For better quality, sliced ham may be either vacuum packed or kept under a controlled atmosphere (Blom et al., 1997; Toldrá et al., 2004).

Physical and Biochemical Changes During the Processing of Dry-Cured Ham All the physical and biochemical reactions occurring in dry-cured ham are the consequence of water loss and salt penetration. The diffusion of water from inside the ham to the surface is very important in determining the quality of the ham. The rate of water diffusion also determines the weight loss; slow diffusion rates result in lower weight loss and soft-textured ham due to accumulation of water inside ham, while a high diffusion rate results in excessively dried ham and a tougher texture (Toldrá, 2005). To have a good optimum diffusion rate, optimum relative humidity (RH) and temperature in the drying chamber should be maintained. Computer-controlled drying curves or sorption isotherms should be modeled to predict the time, temperature, and RH required for optimum drying.

Salt diffusion is another important aspect determining the finished product quality. Salt diffusion is a very slow process and is influenced by various internal and external environmental conditions, such as size of ham, its temperature, pH, moisture content, and the presence of intramuscular fat. Salt equalization through the entire ham requires about three to four months, but salt concentration may vary within the various muscles of ham, although to a lesser extent.

pH plays an important role in all the biochemical reactions occurring in a ham. The pH increases from an initial value of around 5.6–5.8 to 6.4 in the finished product. The range of pH is quite narrow (Toldrá, 1998) and most of the biochemical reactions takes place within this range only; however, proteolysis may also take place in low-pH ham, as reported by Buscailhon et al. (1994c). PSE ham shows a similar pattern for salt diffusion, as seen in normal hams (Bellati et al., 1985; Arnau et al., 1995), but they tend to take more salt, which could be attributed to their lower pH and higher moisture content.

Biochemical Changes During the Dry-Curing Process Enzymatic reactions play a significant role in most biochemical changes during dry-cure ham processing (Toldrá, 1992). As a consequence of biochemical changes, adequate texture and desirable flavor are obtained due to generation of various compounds (Parolari, 1996; Toldrá, 1998; Toldrá and Flores, 1998). Two major biochemical changes responsible for flavor generation are proteolysis and lipolysis (Toldrá, 2006a).

Proteolysis Proteolysis is a chain of enzymatic reactions resulting in the generation of peptides and free amino acids through the degradation of major meat proteins (sarcoplasmic and myofibrillar). These chain reactions result in the generation of aroma and taste compounds along with the breaking of muscle network, and the reaction rates are determined by the various inherent factors of the ham (Ramos et al., 2007) as well as processing conditions. Inherent factors may include the activity of endogenous muscle enzymes, which depend on the genetic traits (Armero et al., 1999a,b; Soriano et al., 2005) and age (Toldrá et al., 1996; Rosell and Toldrá, 1998). Main muscle proteases are cathepsins (Rico et al. 1990, 1991; Toldrá et al., 1993), the endoprotease enzyme acting on the proteins resulting in the production of polypeptides and amino acids. Another category of endoprotease enzymes includes calpain (Rosell and Toldrá, 1996), which acts on proteins to generate polypeptides. Tripeptidyl peptidases (Toldrá, 2002) and dipeptidyl peptidases (Sentandreu and Toldrá, 1998, 2000, 2001a,b) act as exoproteases which transform polypeptides into tripeptides and dipeptides, respectively. Another group of proteases of concern in proteolysis are methyonyl (Flores et al., 2000), alanyl (Flores et al. 1996), leucyl (Flores et al., 1997a), proglutamyl (Toldrá et al., 1992), and arginyl (Flores et al., 1993), which convert peptides into amino acids by their amino peptidase action. These enzymes have good stability, ranging from days to years in a long dry-curing process (Toldrá and Etherington, 1988; Toldrá et al., 1993, 2004). Cathepsin (Rico et al., 1990, 1991; Toldrá et al., 1992a,b; Arnau et al., 1998) and peptidase (Sentandreu and Toldrá, 2001c) activities have been reported to be inhibited by salt addition. Varying levels of salts have also been reported to influence the generation of free amino acids (Toldrá et al., 2000), particularly lysine and tyrosine, for improved taste in the Parma ham variety (Careri et al., 1993), and salt content can also influence the sensory quality, mainly in the taste and texture of hams (Parolari et al., 1994; Andres et al., 2004, 2005). Proteolysis reactions are also influenced by processing conditions such as temperature, duration of drying and ripening, and amount of salt added, which consequently determine the final quality of ham.

Lipolysis Lipolysis is an enzymatic chain reaction similar to that of proteolysis, the difference being that the enzymes involved, and their substrate and products formed, are different. Lipolysis results in the generation of free fatty acids through the action of lipases and phospholipases on triacylglycerols and phospholipids, respectively. Immediate degradation products (i.e., fatty acids) result in the generation of flavor components, and further oxidation of fatty acids may also result in the production of aroma components. The release of fatty acid by the action of enzymes is determined by various factors, such as the initial composition of fatty acids, which in turn depend on breed (Armero et al., 2002; Cava et al., 2004), dietary fat (Ruiz and Opez-Bote, 2002; Pastorelli et al., 2003), and age (Toldrá et al., 1996). Lipolytic enzymes are located primarily in muscle (acid lipase, neutral lipase, phospholipase, acid esterase, and neutral esterase) and adipose tissue (hormone-sensitive lipase, monoacyl glycerol lipase, lipoprotein lipase, acid esterase, and neutral esterase), respectively (Motilva et al., 1992, 1993a,b; Toldrá and Novarro, 2002). These enzymes show good stability throughout the process (Motilva, 1992, 1993a,b). Their activity is also influenced by pH, salt concentration, and water activity; also, the conditions prevailing in the ham favor lipolytic enzyme activity (Motilva and Toldra 1993). During the dry-curing process, generation of free fatty acids continues up to 10 months in muscles (Vestergaard et al., 2000; Motilva et al., 2003a) and up to six months in adipose tissues (Buscailhon et al., 1994b; Motilva et al., 2003b), which results in an increase in the concentration of, especially, oleic, linoleic, stearic, and palmitic fatty acids, primarily through phospholipid degradation (Toldrá, 2002).

Oxidation Fatty acids generated with single and double bonds are highly unstable and prone to further oxidative reactions (Motilva et al., 1993a; Buscailhon et al., 1994a), which liberate various volatile compounds that can contribute to desirable characteristic flavor or off-flavor if excessive oxidation takes place (Lillard, 1978; Berdague et al., 1991; Buscailhon et al., 1993). Muscle oxidative enzymes such as peroxidases and cyclooxygenases play an important role in flavor generation (Coutron-Gambotti, 1999) and the formation of secondary oxidation products, which also contribute toward the specific flavor of ham (Flores et al., 1998). For example, French ham (Buscailhon et al., 1994c) and Spanish ham (Flores et al., 1997b; Sabio et al., 1998) are characterized by aldehydes and ketones, while esters are involved in the specific aged Parma ham odor (Careri et al., 1993; Bolzoni et al., 1996). Chinese Jinhua ham flavor is contributed by volatile compounds of alkanes and alkenes as well as branched aldehyde and sulfur compounds (Du and Ahn, 2001).

TABLE 3 Proximate Composition of Turkey Ham per 100-g Edible Portion

Nutrient	Composition (Wet Weight Basis)
Water, g	71.38
Protein, g	18.93
Energy, kJ	538
Total lipids, g	5.08
Saturated, g	1.70
Unsaturated, g	2.67
Ash, g	4.23
Sodium, mg	996

Wet-Cured Ham

Another important type of ham is cooked ham, or canned ham. It is a boneless product made from cured meat pieces bound together into a specific shape in a sealed container and heat processed (Ham Processing, 2004a). Cooked or canned poultry ham is prepared primarily from cured thigh meat of poultry, especially turkey. The term *turkey ham* is always followed by the statement "cured turkey thigh meat." Canned poultry meat can also be prepared from meat cited in the ingredient section. It is a ready-to-eat, shelf-stable, convenience product with a high nutritive value (Keeton, et al., 2001; Maurer, 2003), as shown in Table 3. The principal steps involved in the processing of canned poultry ham are shown in Figure 3. All the operations are carried out below 10°C, to prevent microbial contamination.

Procurement Procurement proceeds as discussed in the section on dry-cured ham.

Curing or Brine Immersion and Injection Hams are generally cured by injection of brine under pressure through a multineedle system which facilitates and accelerates diffusion of curing brine through the entire piece (Marriott et al., 1992; Pearson and Gillett, 1999; Graumann and Holley, 2007). Basic brine ingredients include water, sodium chloride, dextrose, sodium tripolyphosphate, carrageenan, sodium nitrite, and ascorbates. However, the curing formula may vary according to the processor's choice. The basic curing brine formula cited by Ponce Alquicira (2002) has the composition salt (2%), phosphate (0.5%), sucrose (0.5%), carrageenan (0.3%), sodium nitrite (156 ppm), and sodium erythrobate (450 ppm), respectively.

Vacuum Tumbling and Massaging After injection of brine, tumbling or massaging of hams is done to facilitate uniform diffusion and absorption of curing ingredients, to extract salt-soluble proteins, and to prepare a tender and juicy ham (Marriott et al., 1992; Larouse and Brown, 1997; Pearson and Gillett, 1999;

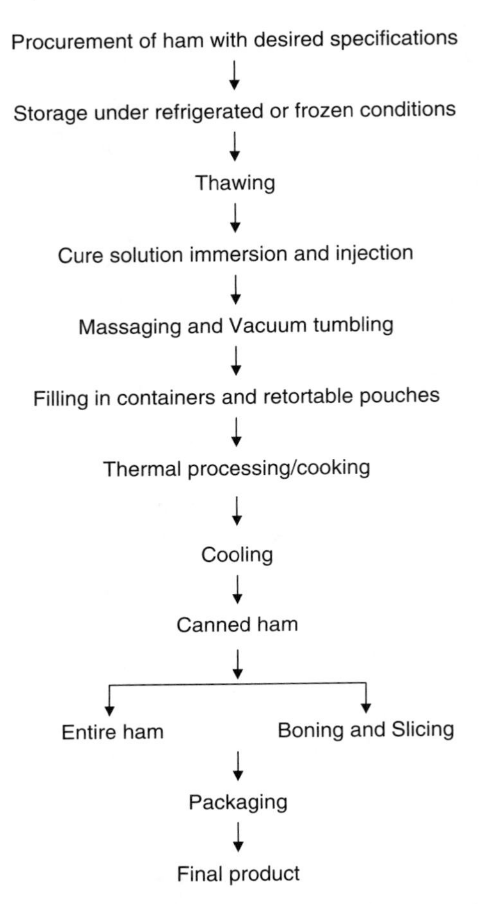

FIGURE 3 Main steps involved in the production of wet-cured and cooked ham.

Keeton, 2001). For this purpose, hams are massaged for a few hours in mixers using paddles at low speed, which results in the slow agitation of ham, preventing disruption or damage of desirable full muscle texture. In tumbling, hams are placed in vacuum tumblers consisting of large rotating tanks with baffles inside and jacketed walls to cool the product while tumbling. During the tumbling process, the refrigeration temperature (4 to 8°C) is maintained and the rotation rate is kept between 3 and 25 rpm, because high speed may result in cell breakdown. Vacuum conditions accelerate the brine uptake by ham and the incorporation of air bubbles into the brine, which may cause undesirable oxidation (Claus et al., 1994; Larrouse and Brown, 1997; Pearson and Gillett, 1999).

Filling After overnight tumbling, hams are placed in shaped molds by hand or by automatic vacuum fillers. The meat is forced in tightly under vacuum to prevent air pockets (Larouse and Brown, 1997; Turner, 1999; Guerrero-Legarreta,

2001). Containers used for poultry ham are available in a variety of forms and shapes, such as flat rectangles or pot-bellied shapes. Containers can be metal cans (aluminum or steel), retortable pouches (preferred), or glass jars (used less often because of fragility).

Cooking/Thermal Processing The final processing stages of cooked ham are considered the most critical of all the processing steps. Rigorous temperature control and timing are required to achieve complete food safety and wholesomeness of the product along with the established technological goals of cooking (Ham Processing, 2004b; Ponce Alquicira, 2005). The cooking process is designed to fulfill various technological requirements (Cabeza et al., 2007): protein coagulation (Pearson, 1994; Smith, 2001), gelation (Claus et al., 1994; Pearson, 1994; Smith, 2001; Smith and Acton, 2001), color stabilization, microbiological stabilization (Claus et al., 1994; Van Laack, 1994; Guerrerro-Legarreta, 2001; Murphy et al., 2003), and enzyme deactivation, to develop sensory characteristics (color, flavor, taste, and texture) (Cambero et al., 1992; Bailey, 1994; Roos, 1997; Mottram, 1998; Adams et al., 2001). Enzymatic, oxidation, Maillard, and other reactions take place in the ham during cooking (Toldrá, 2006b). Excellent information on cooking and cooling ham has been published by Chris Harris (2006), including the views of Josep Lageras from his technical article "Manufacturing Process for Whole Muscle Cooked Meat Products." Hams in their special molds for different ham varieties are cooked in a series of ovens and cookers (Marcos et al., 2008) with either steam or hot-water baths at constant temperature and time monitoring to prevent under- and overcooking of the product. In general, color stability of cooked ham is reached at temperatures between 65 and 70°C, and to reduce the risk of microbiological contamination the product must be maintained at a core temperature of 68 to 70°C for 30 to 60 min. The cooking step is similar to the pasteurization process. Another factor that must be taken into account is the rate at which the temperature increases during cooking. A slow heating rate can result in thermal resistance strains. The addition of salt and nitrite have been reported to reduce the thermal resistance of microorganisms (Claus et al., 1994).

Cooling Cooling is a more critical stage than cooking. It determines the microbiological integrity of a product. Slow cooling rates can result in the growth of pathogenic microorganisms, spore formers, and toxin-producing species. It is well documented that cooking and cooling in a sealed container reduce the microbial risk, and the ham may contain only the bacteria already present in the meat before cooking or that survived the cooking process. As a matter of good practice, the research association recommends that cooling to 50°C should not take more than 1.15 to 3.15 h, that the maximum time for cooling from 50°C to 12°C should be 7.30 h, and that cooling from 12°C to 5°C should range from 1.15 to 1.45 h. In general, the cooling process should not take more than 10 h to reach 5°C at the center of the ham. Cooling can be accomplished by placing cans in refrigerators or in chilled water for small-scale ham production or for hams processed in roasting or boiling bags; however, in large-scale production systems, the use of water sprays or showers or immersion in cold water is practiced to cool the temperature down quickly to between 50 and 60°C. After prechilling, processed ham cans are placed in a chilling room for a minimum of 24 h, and hams are not removed from the mold before 48 h in order to have maximum stability of color and other organoleptic properties.

Final Product Once poultry ham is cooled, it can be sold either as a canned product or taken out of molds or a retort pouch to prepare sliced, roasted, and smoked hams (Marcos et al., 2008). After opening the cans, surface heat treatment of hams is essential to prevent recontamination. Smoking is sometimes done to provide typical color, flavor, and appearance (Ellis, 2001). At the final consumption stage, the cooked ham product can be processed similar to the method used for dry-cured ham.

Physical and Chemical Changes During Cooked Ham Processing Cooked ham also shows some biochemical reactions as a result of enzymatic action, the major ones being proteolysis and lipolysis, but to the lesser degree than for dry-cured hams since most of proteases and lipases are inactivated during the cooking process (Toldrá and others, 1992a,b). So the contribution toward the flavor of ham by the action of lipases and proteases is restricted to the resting time given to hams before cooking. The major chemical reactions that take place in cooked ham are the Maillard reaction, thermal degradation of lipids, and degradation of thiamine, which results in the generation of various flavoring compounds, such as aldehydes, ketones, and sulfur compounds (Bailey, 1994; Roos, 1997; Adams et al., 2001).

CONCLUSIONS

Poultry meat is one of the most widely accepted muscle foods in the world. Poultry ham is a processed, formed poultry meat product similar to pork ham in protein content but with a higher moisture content and lower fat and energy contents. Poultry ham can be of two types based on the processing technology involved, and a variety of ingredients and additives with their specific roles can be used in the preparation of poultry ham. Both types of hams undergo various biochemical reactions that influence their sensory characteristics, as summarized in Table 4 (Toldrá, 2007).

Dry-cured poultry ham basically involves curing with a dry-salt cure mixture, followed by drying and ripening, which results in a shelf-stable product with low water activity (Lücke, 1986). Various physical and biochemical changes take place during the processing of dry-cured ham, the major one being water and pH changes, and protelysis and lipolysis, which are necessary for the desired final quality of dry-cured ham. Two of the most important texture problems in dry-cured ham are excessive softness (Parolari et al., 1994; Virgili et al., 1995) and pastiness (Arnau, 1991; Arnau et al., 1998; Garcí a-Garrido et al., 2000; García-Rey et al., 2004).

TABLE 4 Main Groups of Biochemical Reactions Affecting Sensory Properties of Dry-Cured and Cooked Hams

Groups of Reactions	Dry-Cured Ham	Cooked Ham
Protein degradation	Intense by enzymatic proteolysis	Intense by heat denaturation
Generation of small peptides and free amino acids by proteolysis	Large	Poor
Lipid degradation	Intense by enzymatic lipolysis	Medium by heat damage
Generation of free fatty acids by lipolysis	Large	Poor
Oxidation of free fatty acids	Intense	Medium
Generation of volatile compounds	Intense	Medium
Strecker degradation of amino acids	Intense	Scarce
Maillard reactions	Medium	Intense
Cured color generation	Nitrosomyoglobin	Nitrosohemochrome

Wet-cured or cooked ham is processed through immersion or injection of curing solution and involves a thermal treatment at above 75°C, the temperature for various technological purposes (protein coagulation, gelation, color stabilization with nitrite, and destruction of spoilage bacteria). The quality of cooked ham is influenced by many factors. One of these factors is the technology: the type of meat cut, the composition and quantity of brine injected, the rate and extent of tumbling or massaging, and the cooking time and temperature (Delahunty et al., 1997). The most desirable physical properties of a high-quality cooked ham are cohesiveness, textural firmness, and juiciness (Katsaras and Budras, 1993).

At present, the market for both types of presliced, prepackaged dry-cured and cooked ham is developing at a rapid rate, making their consumption immediately after opening the package virtually risk-free. However, it is also worth noting that many processed poultry hams are handled prior to retail commercialization and sale, and any action, such as cutting, slicing, and repackaging may increase the risk of contamination by a variety of pathogenic microorganisms from the environment, mechanical equipment, tools used, and handlers.

REFERENCES

Adams RL, Mortram DS, Parker KJ, Brown HM. 2001. Flavour-protein binding: disulfide interchange reactions between ovalbunia and volatile disulfides. J Agric Food Chem 49(9):4333–4336.

Andres AI, Cava R, Ventanas J, Thovar V, Ruiz J. 2004. Sensory characteristics of Iberian ham: influence of salt content and processing conditions. Meat Sci 68:45–51.

Andres AI, Ventanas S, Ventanas J, Cava R. 2005. Physicochemical changes throughout the ripening of dry-cured hams with different salt content and processing conditions. Eur Food Res Technol 221:30–35.

Anon. 2001. http:/www.gov.on.ca/OMAF. . .estock/swine/facts/info_qs_species.htm.

Armero E, Barbosa JA, Toldrá F, Baselga M, Pla M. 1999a. Effect of the terminal sire and sex on pork muscle cathepsin (B, B + L and H), cysteine proteinase inhibitors and lipolytic enzyme activities. Meat Sci 51:185–189.

Armero E, Flores M, Toldrá F, Barbosa JA, Oliver J, Pla M, Baselga M. 1999b. Effect of pig sire types and sex on carcass traits, meat quality and sensory quality of dry cured ham. J Sci Food Agric 79:1147–1154.

Armero E, Navarro JL, Madal ML, Baselga M, Toldrá F. 2002. Lipid composition of pork muscle as affected by sire genetic type. J Food Biochem 26:91–102.

Arnau J. 1991. Aportaciones a la calidad tecnológica del jamón curado elaborado por procesos acelerados. Thesis, Universitat Autónoma de Barcelona, Facultat de Veterinaria, Barcelona, Spain.

Arnau J, Guerrero L, Casademont G, Gou P. 1995. Physical and chemical changes in different zones of normal and PSE dry cured ham during processing. Food Chem 52:63–69.

Arnau J, Guerrero L, Sárraga C 1998. The effect of green ham pH and NaCl concentration on cathepsin activities and sensory characteristics of dry-cured ham. J Sci Food Agric 77:387–392.

Bailey ME, 1994. Maillard reactions and meat flavor development. In: Shahidi F, ed., *Flavor of Meat and Meat Products*. New York: Blackie Academic & Professional, Chap. 9.

Bellatti M, Dazzi G, Chizoline R, Palmia F, Parolari G. 1985. Modifications chimiques et physiques des proteins au cours de la maturation du jambon de Parme. Viandes Prod Carnees 6:142–145.

Berdague JL, Denoyer C, le Quere JC, Semon E. 1991. Volatile compounds of dry-cured ham. J Agric Food Chem 39:1257–1261.

Bilgili SF. 2002. Poultry meat processing and marketing: What does the future hold? Poult Int, Sept., pp. 12–22.

Blom H, Nerbrink E, Dainty R, Hagtvedt T, Borch E, Nissen H, Nesbakken T. 1997. Addition of 2.5% lactate and 0.25% acetate controls growth of *Listeria monocytogenes* in vacuum-packed, sensory-acceptable servelant sausage and cooked ham stored at 4°C. Int J Food Microbiol 38:71–76.

Bolzoni L, Barbieri G, Virgili R. 1996. Changes in volatile compounds of Parma ham during maturation. Meat Sci 43:301–310.

Buscailhon S, Berdague JL, Monin G. 1993. Time-related changes in volatile compounds of lean tissue during processing of French dry-cured ham. J Sci Food Agric 63:69–75.

Buscailhon S, Gandemer G, Monin G. 1994a. Time-related changes in intramuscular lipids of French dry-cured ham. Meat Sci 37:245–255.

Buscailhon S, Berdague JL, Bousset J, Cornet M, Gandemer G, Touraille C, Monin G. 1994b. Relations between compositional traits and sensory qualities of French dry-cured ham. Meat Sci 37:229–243.

Buscailhon S, Berdague JL, Gandemer G, Touraile C, Monin G. 1994c. Effects of initial pH on compositional changes and sensory traits of French dry-cured hams. J Muscle Foods 5:257–270.

Cambero ML, Seuss I, Honikel KO. 1992. Flavor compounds of beef broth as affected by cooking temperature. J Food Sci 57:1285–1290.

Careri M, Mangia A, Barbieri G, Bolzoni L, Virgili R, Parolari G. 1993. Sensory property relationships to chemical data of Italian type dry-cured ham. J Food Sci 58:968–972.

Cassens RG. 1995. Use of sodium nitrite in cured meats today. Food Technol 49:72–81.

Cava R, Ferrer JM, Estevez M, Morcuende D, Toldrá F. 2004. Compositon and proteolytic and liplytic enzyme activity in muscle longissimus dorsi from Iberian pigs and industrial genotype pigs. Food Chem 88:25–33.

Claus JR, Colby JW, Flick GJ. 1994. Processed meats/poultry/seafood. In: Krinsman DM, Kotula AW, and Breideastein BC, eds., *Muscle Foods: Meat, Poultry, and Seafood Technology*. New York: Chapman & Hall, Chap. 5.

Concepción CM, Cambero I, de la Hoz L, Ordóñez JA. 2007. Optimization of e-beam irradiation treatment to eliminate *Listeria monocytogenes* from ready-to-eat (RTE) cooked ham. Innov Food Sci Emerging Technol 8:299–305.

Coutron-Gambotti C, Gandemer G. 1999. Lipolysis and oxidation in subcutaneous adipose tissue during dry-cured ham processing. Food Chem 64:95–101.

Damodaran S. 1994. Structure–functional relationship of food proteins. In: Hettiarachchy NS, Ziegler GR, eds., *Protein Functionality in Food Systems*. IFT Basic Symposium Series. New York: Marcel Dekker, Chap. 1.

Delahunty CM, McCord A, O'Neill EE, Morrissey PA. 1997. Sensory characterisation of cooked hams by untrained consumers using free-choice profiling. Food Qual Pref 8(5–6):381–388.

Du M, Ahn DU. 2001. Volatile substances of Chinese traditional Jinhua ham and Cantonese sausage. J Food Sci 66:827–831.

EC (European Commission). 2006. Directive 2006/52/CE of the European Parliament and of the Council.

Ellis DF. 2001. Meat smoking technology: meat fermentation technology. In: Hui YH, Nip WK, Rogers RW, Young OA, eds., *Meat Science and Applications*. New York: Marcel Dekker, pp. 509–519.

Ferreira IMPLVO, Silva S, 2007 Quantification of residual nitrite and nitrate in ham by reverse-phase high performance liquid chromatography/diode array detector. Talanta, doi:10.1016/j.talanta.2007.10.004.

Flores M, Aristoy MC, Toldrá F. 1993. HPLC purification and characterization of porcine muscle aminopeptidase B. Biochimie 75:861–867.

Flores M, Aristoy MC, Toldrá F. 1996. HPLC purification and characterization of soluble alanyl aminopeptidase from porcine skeletal muscle. J Agric Food Chem 44:2578–2583.

Flores M, Aristoy MC, Toldrá F. 1997a. Curing agents affect aminopeptidase activity from porcine skeletal muscle. Z Lebensm-Unters-Forsch 2005:343–346.

Flores M, Grimm CC, Toldrá F, Spanier AM. 1997b. Correlations of sensory and volatile compounds of Spanish Serrano dry-cured ham as a function of two processing times. J Agric Food Chem 45:2178–2186.

Flores M, Spanier AM, Toldrá F. 1998. Flavour analysis of dry-cured ham. In: Shahidi de F, ed., *Flavor of Meat Products and Seafoods*. London: Blackie Academic & Professional, pp. 320–341.

Flotcher DL. 1999. Poultry meat colour. In: Richarson RL, Mead GC, eds., *Poultry Meat Science*. Wallingford, UK: CAB International, pp. 159–173.

Friexanet L. 2004. Additives and ingredients in ham production. Meat Process Global, Nov.–Dec., pp. 16–22.

Froning GW, McKee SR. 2001. Mechanical separation of poultry meat and its use in products. In: Sams AR, ed., *Poultry Meat Processing*. Boca Raton, FL: CRC Press, Chap. 14.

García-Garrido J, Quiles-Zafra R, Tapiador J, Luque de Castro M. 2000. Activity of cathepsin B, D, H and L in Spanish dry-cured ham of normal and defective texture. Meat Sci 56:1–6.

García-Rey R, García-Garrido J, Quiles-Zafra R, Tapiador J, Luque de Castro M. 2004. Relationship between pH before salting and dry-cured ham quality. Meat Sci 67:625–632.

Graumann GH, Holley RA. 2007. Survival of *Escherichia coli* O157:H7 in needle-tenderized dry cured Westphalian ham. Int J Food Microbiol 118:173–179.

Guerrero-Legarreta I. 2001. Meat canning technology. In: Hui YH, Nip WK, Rogers RW, Young OA, eds., *Meat Science and Applications*. New York: Marcel Dekker, pp. 521–535.

Ham Processing. 2004a. Company profile: high quality ham. Meat Process Global, Nov.–Dec., pp. 8–10.

Ham Processing 2004b. Cooking for perfection. Meat Process Global, Nov.–Dec., pp. 12, 14.

Harris C. 2006, Cooking for perfection. Meat Process Global, Mar.–Apr., pp. 8–10.

Hwang CA, Tamplin ML. 2007. Modeling the lag phase and growth rate of *Listeria monocytogenes* in ground ham containing sodium lactate and sodium diacetate at various storage temperatures. J Food Sci 72(7).

Jímenez-Colmenero F, Careche J, Carballo J. 1994. Influence of thermal treatments on gelation of actomyosin from different myosystems. J Food Sci 59(1):211–215, 220.

Katsaras K, Budras KD. 1993. The relationship of the microstructure of cooked ham to its properties and quality. Lebensm-Wiss Technol 26(3):229–234.

Keeton JT. 2001. Formed and emulsion products. In: Sams AR, ed., *Poultry Meat Processing*. Boca Raton, FL: CRC Press, Chap. 12.

Kemp JD, Abidoye DFO, Langlois BE, Dranklin JB, Fox JD. 1980. Effect of curing ingredients, skinning and boning on yield, quality and microflora of country hams. J Food Prot 45:244–248.

Larousse J, Brown BE. 1997. *Food Canning Technology*. New York: Wiley–VCH, pp. 235–264, 297–332, 383–424, 489–530.

Lawrie RA. 1998. *Lawrie's Meat Science*, 6th ed. Cambridge, UK: Woodhead Publishing, pp. 11–22, 58–94, 212–254.

Lillard DA. 1978. Chemical changes involved in the oxidation of lipids in foods. In: de Supran MK, ed., *Lipids as a Source of Flavor*. ACS Symposium Series 75. Washington, DC: American Chemical Society, pp. 68–80.

Lücke F, 1986. Microbiological processes in the manufacture of dry sausage and raw ham. Fleischwirtschaft 66:1505–1509.

Marcos B, Jofré A, Aymerich T, Monfort JM, Garriga M. 2008. Combined effect of natural antimicrobials and high pressure processing to prevent *Listeria monocytogenes* growth after a cold chain break during storage of cooked ham. Food Control 19:76–81.

Marriott NG, Graham PP, Claus JR. 1992. Accelerated dry curing of pork legs (hams): a review. J Muscle Foods 3:159–168.

Martin M. 2001. Meat curing technology. In: Hui YH, Nip WK, Rogers RW, Yong OA, eds., *Meat Science and Applications*. New York: Marcel Dekker, pp. 491–508.

Martín L, Antequera T, Ruiz J, Cava R, Tejeda JF, Córdoba JJ. 1998. Influence of the processing conditions of Iberian ham on proteolysis during ripening. Food Sci Technol Int 4:17–22.

Maurer AJ. 2003. Poultry. In: Caballero B, Trugo LC, Finglas PM, eds., *Encyclopedia of Food Science and Nutrition*, 2nd ed. San Diego, CA: Academic Press, pp. 4680–4694.

Motilva MJ, Toldrá F. 1993. Effect of curing agents and water activity on pork muscle and adipose subcutaneous tissue lipolytic activity. Z Lebensm-Unters-Forsch 196:228–231.

Motilva MJ, Toldrá F, Flores J. 1992. Assay of lipase and esterase activities in fresh pork meat and dry-cured ham. Z Lebensm-Unters-Forsch 195:446–450.

Motilva MJ, Toldrá F, Nieto P, Flores J. 1993a. Muscle lipolysis phenomena in the processing of dry-cured ham. Food Chem 48:121–125.

Motilva MJ, Toldrá F, Aristoy MC, Flores J. 1993b. Subcutaneous adipose tissue lipolysis in the processing of dry-cured ham. J Food Biochem 16:323–335.

Motilva MJ, Toldrá F, Nieto P, Flores J. 2003a. Muscle lipolysis phenomena in the processing of dry-cured ham. Food Chem 48:121–125.

Motilva MJ, Toldrá F, Aristoy MC, Flores J. 2003b. Subcutaneous adipose tissue lipolysis in the processing of dry-cured ham. J Food Biochem 16:323–335.

Mottram DS. 1998. Flavour formation in meat and meat products: a review. Food Chem 62(4):415–424.

Murphy RY, Duncan LK, Beard BL, Driscoll KH. 2003. *D* and *Z* values of *Salmonella, Listeria innocua*, and *Listeria monocytogenes* in fully cooked poultry products. J Food Sci 68:4.

Oliveira KAM, Santos Mendonça, RC, deMiranda Gomide LA, Dantas Vanetti MC. 2005. Aqueous garlic extract and microbiological quality of refrigerated poultry meat. J Food Process Preserv 29:98–108.

Parolari G. 1996. Review: achievements, needs and perspectives in dry-cured ham technology: the example of Parma ham. Food Sci Technol Int 2:69–78.

Parolari G, Virgili R, Schivazzappa C. 1994. Relationship between cathepsin B activity and compositional parameters in dry-cured hams of normal and defective texture. Meat Sci 38:117–122.

Pastorelli G, Magni S, Rossi R, Pagliarini E, Baldini P, Dirinck P, Van Opstaele F, Corino C. 2003. Influence of dietary fat, on fatty acid composition and sensory properties of dry-cured Parma ham. Meat Sci 65:571–580.

Pearson AM, Gillett TA. 1999. *Processed Meats*, 3rd ed. Gaithersburg, MD: Aspen Publishers, pp. 53–78, 372–413.

Pegg RB, Shahidi F. 2000. *Nitrite Curing of Meat*. Trumbull, CT: Food and Nutrition Press, pp. 23–66.

Ponce-Alquicira E. 2002. Canned turkey ham. In: Smith IJS, Christen GL, eds., *Food Chemistry Workbook*. West Sacramento, CA: Science Technology System, pp. 135–143.

Ponce-Alquicira E. 2005. Canned poultry products: turkey ham. In: Hui YH, ed., *Handbook of Food Science, Technology and Engineering*, vol. 4. Boca Raton, FL: CRC Press, pp. 166-1 to 166-12.

Ramos AM, Serenius TV, Stalder KJ, Rothschild MF. 2007. Phenotypic correlations among quality traits of fresh and dry-cured hams. Meat Sci 77:182–189.

Rico E, Toldrá F, Flores J. 1990. Activity of cathepsin D as affected by chemical and physical dry-curing parameters. Z Lebensm-Unters-Forsch 191:20–23.

Rico E, Toldrá F, Flores J. 1991. Effect of dry-curing process parameters on pork muscle cathepsins B, H and L activities. Z Lebensm-Unters-Forsch 191:20–23.

Roos KB. 1997. How lipids influence food flavor. Food Technol 51(1):60–62.

Rosell CM, Toldrá F. 1996. Effect of curing agents on m-calpain activity throughout the curing process. Z Lebensm-Unters-Forsch 203:320–325.

Rosell CM, Toldrá F. 1998. Comparison of muscle proteolytic and lipolytic enzyme levels in raw hams from Iberian and White pigs. J Sci Food Agric 76:117–122.

Ruiz J, Opez-Bote C. 2002. Improvement of dry-cured ham quality by lipid modification through dietary means. In: Toldrá F, ed., *Research Advances in the Quality of Meat and Meat Products*. Trivandrum, India: Research Signpost, pp. 255–271.

Sabio E, Vidal-Aragon MC, Bernalte MJ, Gata JL. 1998. Volatile compounds presents in six types of dry-cured ham from south European countries. Food Chem 61:493–503.

Sentandreu MA, Toldrá F. 1998. Biochemical properties of dipeptidylpeptidase III purified from porcine skeletal muscle. J Agric Food Chem 46:3899–3984.

Sentandreu MA, Toldrá F. 2000. Purification and biochemical properties of dipeptidylpeptidase I from porcine skeletal muscle. J Agric Food Chem 48:5014–5022.

Sentandreu MA, Toldrá F. 2001a. Importance of dipeptidylpeptidase II in postmortem pork muscle. Meat Sci 57:93–103.

Sentandreu MA, Toldrá F. 2001b. Dipeptidylpeptidase IV from porcine skeletal muscle: purification and biochemical properties. Food Chem 75:159–168.

Sentandreu MA, Toldrá F. 2001c. Dipeptidylpeptidase activities along the processing of Serrano dry-cured ham. Eur Food Res Technol 213:83–87.

Shahidi F, Hong C. 1991. Evaluation of malonaldehyde as a marker of oxidative rancidity in meat products. J Food Biochem 15:97–105.

Shahidi F, Pegg RB. 1992. Nitrite-free meat curing systems: update and review. Food Chem 43:185–191.

Smith DM. 2001. Functional properties of muscle proteins in processed poultry products. In: Sams AR, ed., *Poultry Meat Processing*. Boca Raton, FL: CRC Press, Chap. 11.

Smith DP, Acton JC. 2001. Marination, cooking, and curing of poultry products. In: Sams AR, ed., *Poultry Meat Processing*. Boca Raton, FL: CRC Press, Chap. 15.

Soriano A, Quiles R, Mariscal C, Barcia A. 2005. Pig sire type and sex effects on carcass traits, meat quality and physicochemical and sensory characteristics of Serrano dry-cured ham. J Sci Food Agric 85:1914–1924.

Spanier AM, Flores M, Toldrá F. 1999. Flavor differences due to processing in dry-cured and other ham products using conducting polymers (electronic nose). In: Shahihi F, Ho C-T, eds., *Flavour Chemistry of Ethnic Foods*. New York: Kluwer Academic/Plenum Publishers, pp. 169–183.

Toldrá F. 1992. The enzymology of dry-curing of meat products. In: Smulders JM, Toldrá F, Flores J, Prieto M, eds., *New Technologies for Meat and Meat Products*. Nijimegen, The Netherlands: Audet, pp. 209–231.

Toldrá F. 1998. Proteolysis and lipolysis in flavour development of dry-cured meat products. Meat Sci 49:S101–S110.

Toldrá F. 2002. *Dry-Cured Meat Products*. Trumbull, CT: Food and Nutrition Press.

Toldrá F, 2004. Curing: (b) Dry. In: Jensen W, Devine C, Dikemann M, eds., *Encyclopedia of Meat Sciences*. London: Elsevier Science, pp. 360–365.

Toldrá F. 2005. Dry-cured ham. In: Hui YH, ed., *Handbook of Food Science, Technology and Engineering*, vol. 4. Boca Raton, FL: CRC Press, pp. 164–1 to 164–11.

Toldrá F. 2006a. Biochemical proteolysis basis for improved processing of dry-cured meats. In: Nollet LML, Toldrá F, eds., *Advanced Technologies for Meat Processing*. Boca Raton, FL: CRC Press, pp. 329–351.

Toldrá F. 2006b. Biochemistry of processing meat and poultry. In: Hui YH, Nip WK, Nollet LML, Paliyath G, Simposon BK, eds., *Food Biochemistry and Food Processing*. Ames, IA: Blackwell Publishing, pp. 315–335.

Toldrá F. 2007. Ham. In: Hui YH, ed., *Handbook of Food Products Manufacturing*, Hoboken, NJ: Wiley, pp. 233–249.

Toldrá F, Etherington DJ. 1988. Examination of cathepsins B, D, H and L activities in dry cured hams. Meat Sci 23:1–7.

Toldrá F, Flores M. 1998. The role of muscle proteases and lipases in flavor development during the processing of dry-cured ham. CRC Crit Rev Food Sci Nutr 38:331–352.

Toldrá F, Navarro JL. 2002. Action of muscle lipases during the processing of dry-cured ham. In: Toldrá F, ed., *Research Advances in the Quality of Meat and Meat Products*. Trivandrum, India: Research Signpost, pp. 249–254.

Toldrá F, Cervero MC, Part C. 1992a. Porcine aminopeptidase activity as affected by curing agents. J Food Sci 58:724–726, 747.

Toldrá F, Rico E, Flores J. 1992b. Activities of pork muscle proteases in cured meats. Biochimie 74:291–296.

Toldrá F, Rico E, Flores J. 1993. Cathepsin B, D, H and L activity in the processing of dry-cured-ham. J Sci Food Agric 62:157–161.

Toldrá F, Flores M, Aristoy MC, Virgili R, Parolari G. 1996. Pattern of muscle proteolytic and lipolytic enzymes from light and heavy pigs. J Sci Food Agric 71:124–128.

Toldrá F, Aristoy MC, Flores M. 2000. Contribution of muscle aminopeptidases to flavor development in dry cured ham. Food Res Int 33:181–185.

Toldrá F, Gavara R, Lagaron JM. 2004. Packaging and quality control. In: Hui YH, Goddik LM, Josephsen J, Stanfield PS, Hansen AS, Nip WK, Toldrá F, eds., *Handbook of Food and Beverage Fermentation Technology*. New York: Marcel Dekker, pp. 445–458.

Turner TA. 1999. Cans and lids. In: Footitt RJ, Lewis AS, eds., *The Canning of Fish and Meat*. Gaithersburg, MD: Aspen Publishers, Chap. 5.

Van Horne PLM. 2002. Production cost development of broiler meat. Arch Gefluegelkd 66:26–27.

Van Laack RLJM. 1994. Spoilage and preservation of muscle foods. In: Kinsman DM, Kotula AW, Breidenstein BC, eds., *Muscle Foods: Meat, Poultry, and Seafood Technology*. New York: Chapman & Hall, Chap. 14.

Vasavada MN, Cornforth DP. 2005. Evaluation of milk mineral antioxidant activity in meat balls and nitrite-cured sausage. J Food Sci 70:250–253.

Ventanas J, Cava R. 2001. Dinámica y control del proceso de secado del jamón ibérico en secaderos y bodegas naturales y en cámaras climatizadas. In: *Tecnología del Jamón Ibérico*. Madrid: Editorial Mundi Prensa.

Vestergaard CS, Schivazzappa C, Virgili R. 2000. Lipolysis in dry-cured ham maturation. Meat Sci 55:1–5.

Virgili R, Parolari G, Schivazappa C, Bordini CS, Borri M. 1995. Sensory and texture quality of dry-cured ham as affected by endogenous cathepsin B activity and muscle composition. J Food Sci 60(6):1183–1186.

Yun J, Shahidi F, Rubin LJ, Diosady LL. 1987. Oxidative stability and flavor acceptability of nitrite-free meat curing systems. Can Inst Food Sci Technol J 40:246–251.

18

LUNCHEON MEAT INCLUDING BOLOGNA

BACILIZA QUINTERO SALAZAR
Centro de Investigación y Estudios Turísticos, Universidad Autónoma del Estado de México, Ciudad Universitaria, Toluca, México

EDITH PONCE-ALQUICIRA
Departamento de Biotecnología, Universidad Autónoma Metropolitana–Unidad Iztapalapa, México D.F., México

Handbook of Poultry Science and Technology, Volume 2: Secondary Processing, Edited by Isabel Guerrero-Legarreta and Y.H. Hui

INTRODUCTION

In contrast to the past, the demand for convenient meat products such as luncheon meats and bologna, as well as other refrigerated processed meat, is increasing today. The principal reasons for this increment include changes in the methods of food preparation and consumption as a result of accelerated lifestyles, the demographic explosions in big cities, and the growing interest of consumers for healthy and more diversified ready-to-eat meat products.

Poultry luncheon meats and bologna are convenient meat products that offer a great advantage, as poultry is perceived by consumers as healthy, being low in saturated fat and cholesterol. Therefore, poultry luncheon meats are gaining popularity all around the world, and people are spending more money on them. For example, it was reported that sales of refrigerated sliced luncheon meats increased from $3.25 million in 2003 to $3.413 million in 2005 (5.1% more) (Anon., 2005). It has also been reported that households in the United States spent about $3.75 billon for luncheon meats in 2006, double the sales in 2003 (Eyre, 2007). An increase of 14% from 2005 to 2010 has been predicted in the consumption of luncheon meats (Anon., 2005). For these reasons, it is important to discuss aspects such as classification, ingredients, processing stages, microbiology, and new tendencies in the development of luncheon meats.

DEFINITION AND CLASSIFICATION OF LUNCHEON MEATS AND BOLOGNA

At present there are several definitions for luncheon meats. For example, in the United States, according to the *Code of Federal Regulations* § 319.260 (2003b), luncheon meat is defined as a cured, cooked food product made from comminuted meat. A wider definition was given by the CODEX Alimentarius (CAC, 1991): foods prepared from meat or poultry meat or a combination of these, which has been comminuted and cured, and which may have been smoked and that may or may not contain binders. A traditional definition of luncheon meats is given by the Cambridge Dictionary Online (2008): very small pieces of meat pressed into a block which is usually sold in metal containers that can be sliced and eaten cold. However, from a commercial point of view, the term is frequently used to describe processed and prepackaged meats, often molded into a loaf and that may be sold sliced, that generally are eaten cold, and that frequently are used to prepare sandwiches or salads. Luncheon meats are also known as lunch meats, cold cuts, sandwich meats, cooked meats, sliced meats, cold meats, or deli meats. There is a great variety of this sort of product, and frequently, bologna is included.

Poultry Luncheon Meats and Their Classification

The classification of meat products is made according to various criteria, such as the type and meat source, technology of processing, ingredients, shape of

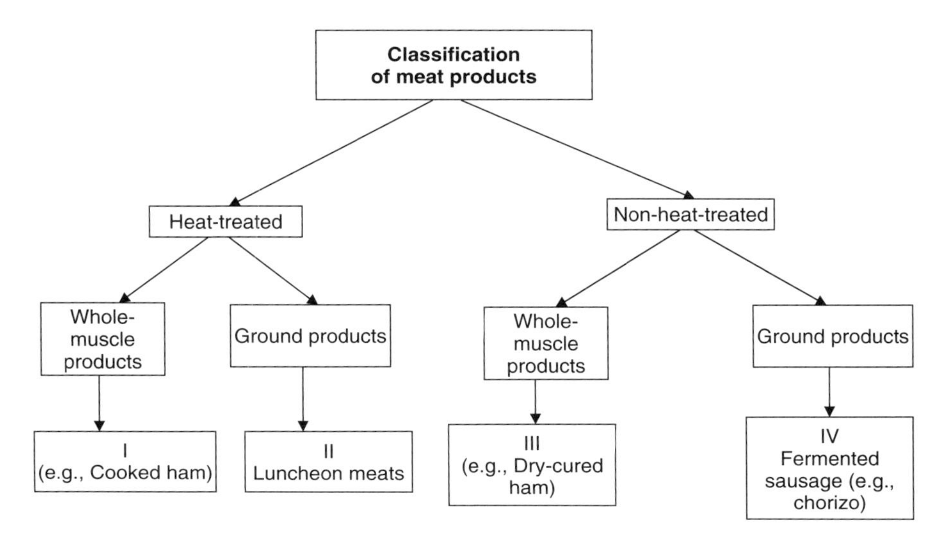

FIGURE 1 Classification of meat products according to Vandendriessche (2008).

the final product, temperatures applied during processing, and time of maturation. For example, Vandendriessche (2008) recently proposed a classification for meat products based on heat treatment (heated or not heated) in combination with the integrity of muscle (whole-muscle products and ground products). As a result, this combination of characteristics gives four categories (I, II, III, and IV) of meat products; luncheon meat and bologna are situated in category II, as heat-treated and ground products (see Figure 1). But, using this classification, reconstructed cooked ham or poultry breast are situated between categories I and II.

However, meat product classifications can also vary within countries. For example, the *Code of Federal Regulations* (2003b), in Title 9, Animal and Animal Products, part 319, Definitions and Standards of Identity or Composition (9CFR319), has classified the meat products in more than 30 sections. In this document, luncheon meats and bologna are classified separately as follows: "Subpart K, Luncheon Meat, Loaves and Jellied Products (§ 319.260, Luncheon meat and meat loaf)" and Subpart G, Cooked sausage (§ 319.180 Frankfurter, wiener, bologna, garlic bologna, knockwurst, and similar products).

An official classification of poultry luncheon meats has not been proposed, even though chicken and turkey are the principal meats used to prepare several products, such as breast, salami, pastrami, ham, loaf, bologna, and other variants derived from them. Nevertheless, these products are generically classified as luncheon or deli meats by the principal poultry processing companies in the United States. In this chapter the term *luncheon meats* includes all these products, in agreement with the meat industry.

Commercial Types of Poultry Luncheon Meats

A great variety of luncheon meats are offered in the market; however, the number of breast poultry–derived products is greater than that of other luncheon meats. Also, there is a strong tendency to offer functional and ecological luncheon products that include fat-free, 0% trans fat, and organic varieties, among others. Some of the most common luncheon meat products and their variants present in the U.S. market are presented in Table 1.

It was mentioned earlier that salami, pastrami, loaf, ham, bologna, as well as poultry breast are generically termed luncheon meats in the market. However, there are two categories of products within this denomination: formed (turkey/chicken ham) and emulsified meat products (bologna). According to Keeton (2001), formed products include formed, restructured, and sectioned products. Keeton also defines *formed meat products* as items produced by sectioned muscle pieces combined with ground or emulsified myofibrillar protein binders and chilled brine. *Restructured meat products* are characterized by their small particle size through grinding, chopping, and emulsifying operations. *Sectioned products* have more whole-muscle texture than restructured products.

Emulsified poultry products are represented by bologna, loaf, and cooked salami. The principal characteristics of these products is that they are typically manufacturing using mechanically deboned poultry meat and may also contain a maximum of 30% fat (Keeton, 2001; CFR, 2003b). A proposed commercial classification for luncheon meat is presented in Figure 2.

Descriptions of Some Poultry Luncheon Meats

Pastrami The word *pastrami* is derived from the Romanian word *pastra*, which means "to preserve." Pastrami is a delicatessen meat product produced by immersion of meat in brine, partially dried, seasoned with several herbs and spices, such as marjoram, basil, allspice, garlic, black pepper, cloves, and smoked flavoring. This product was originally prepared using only beef meat; however, it can also be prepared using turkey breast.

Turkey or Chicken Breast Poultry breast is one of the most popular luncheon meats: recognized as a high-quality product low in fat and cholesterol and in nitrite and salts. It is made from whole boneless breast muscle portions freed of fat and injected or marinated with brine, containing salt, phosphates, and herbs, and may be smoked (see Figure 3).

Turkey or Chicken Ham Turkey ham is a ready-to-eat product made from cured thigh meat of turkey. The size and shape of turkey ham depends on how the meat is processed, and it is generally available in whole or half portions. Turkey ham is approximately 95% fat-free, which makes it a low-fat alternative to pork ham. It can be served hot or cold, sliced thick or thin. The term *turkey ham* is always followed by the statement "cured turkey thigh meat."

TABLE 1 Examples of Poultry Luncheon Meats in the U.S. Market

Luncheon Meat	Examples of Poultry Products: Chicken	Examples of Poultry Products: Turkey
Bologna	Chicken bologna Light With garlic	Turkey bologna Light
Ham	Cooked chicken ham	Turkey hams Hickory-smoked turkey ham
Breast	Oven-roasted chicken breast Buffalo style BBQ chicken breast Oil-browned chicken breast Honey-roasted chicken breast Smoked chicken breast Rotisserie-flavor chicken breast	Smoked turkey breast Mesquite Hickory Peppered Sun-dried tomato Cajun style Lemon pepper flavor Honey Cured turkey breast Smoked turkey breast Buffalo style Roasted turkey breast Dry roasted Oven roasted Delicatessen specialties Functional turkey breast Turkey fat-free 0 trans fats Organic
Pastrami	Chicken pastrami Fat-free	Turkey pastrami Hickory-smoked
Salami	Chicken salami Cotto salami Cooked salami	Hickory-smoked cooked turkey salami
Meat loaf	Olive loaf Pickle loaf	
Canned luncheon meat	Canned chicken luncheon meat Canned turkey luncheon meat (Spam)	Canned turkey luncheon meat (Spam)

Source: Data from the home pages of the following meat processing companies: Honeysuckle-white, Cargill Meat Solutions (http://www.honeysucklewhite.com/productFamilies.do?categoryId=9), Farmland Foods (http://www.farmlandfoods.com/products/deli-lunchmeats.asp), and Tyson Foods (http://www.tyson.com/Recipes/Product/SandwichMeats.aspx).

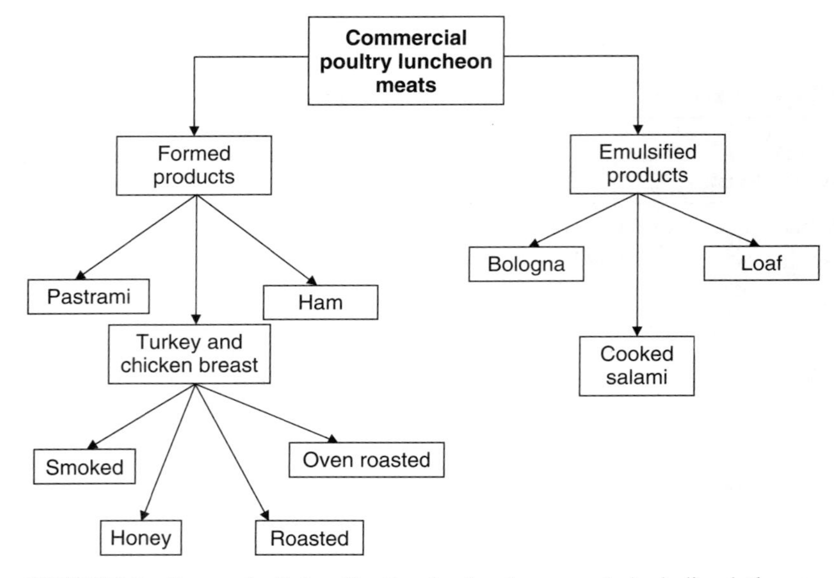

FIGURE 2 Proposal of classification for luncheon meats including bologna.

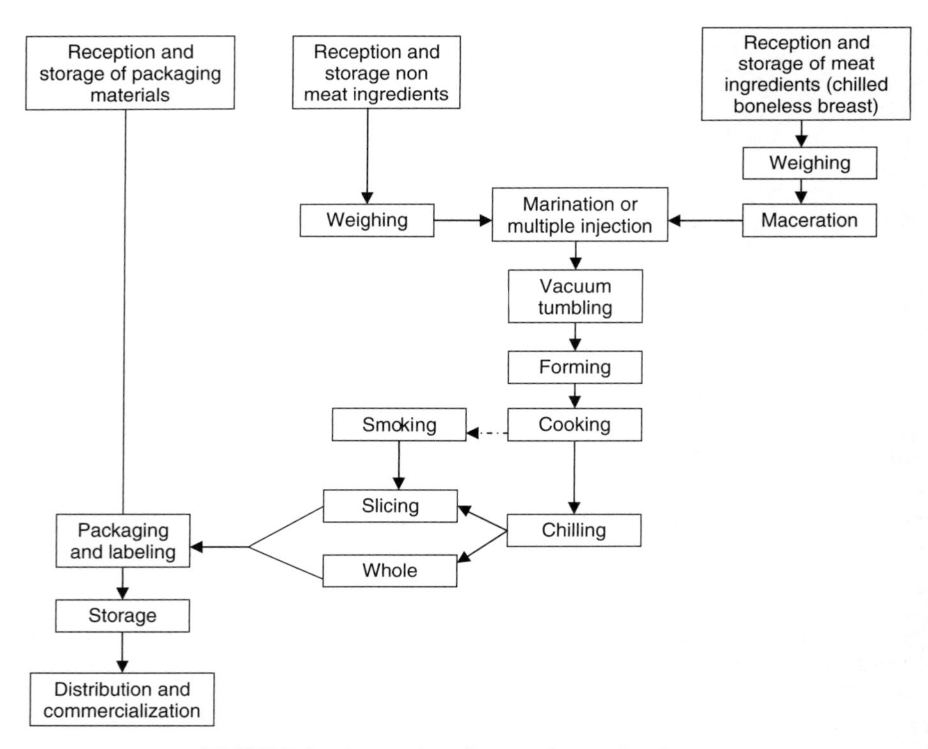

FIGURE 3 Processing diagram for poultry breast.

Bologna Bologna is an American sausage similar to the Italian mortadella. Specifications from the *Code of Federal Regulations (2003b)* in § 319.180 indicate that bologna is a comminuted, semisolid sausage prepared from one or more types of raw skeletal muscle meat or raw skeletal muscle meat and raw or cooked poultry meat, seasoned and cured, using one or more of the approved curing agents and orthophosphates. This regulation establishes that this sausage may contain raw or cooked poultry meat and/or mechanically deboned poultry without skin and without kidneys and sex glands. The final product may not contain more than 30% water and no more than 40% of a combination of fat and added water. Nonmeat binders and extenders or isolate soy protein may be used up to 3.5% or 2%, respectively.

Meat Loaf According to the *Code of Federal Regulations* (2003b), meat loaf (§ 319.261) is a cooked meat food product in loaf form made from comminuted meat and/or mechanically deboned poultry or turkey. The loaf shape may be formed by hand or during cooking into an ordinary saucepan. A great variety of ingredients may be added either to improve cohesiveness (breadcrumbs, milk, egg, cereals, etc.) or for variety (herbs, spices, vegetables, etc). Water or ice may also be added, at a maximum of 3% of the total ingredients used.

INGREDIENTS: FUNCTIONALITY AND QUALITY

According to the CODEX Alimentarius (CAC, 1991), the ingredients used to prepare luncheon meats may be divided into two categories: essential and optional. Essential ingredients comprise meat or poultry meat, water, and curing ingredients. Optional ingredients include edible offal; fat per se; cured and uncured pork rind per se; carbohydrates and protein binders; sweeteners, spices, seasonings, condiments, and hydrolyzed protein.

Main Ingredients

Meat The main meat raw materials used to manufacture luncheon meats and bologna include skeletal muscle from breasts, thighs, and legs. Also, mechanically deboned skeletal muscle from poultry of high quality can be used f or up to 100% of the lean source (Lawrence and Mancini, 2004). However, according to Abdullah (2007), the use of up to 40% of good-quality mechanically deboned chicken meat can give luncheon meats acceptable quality. On the other hand, The CODEX Alimentarius (CAC, 1991) establishes that edible offal, fat per se, as well as cured and uncured pork rind per se, can be added in formulations as optional ingredients. In this standard, poultry skin is included within the edible offal classification. However, the *Code of Federal Regulations* (2003b), § 319.180, indicates that bologna, may contain mechanically separated poultry without skin, kidneys, and sex glands.

Raw materials must be of high quality with no off-odors and no apparent change in color. It should be low in microbial counts, with no sign of evident spoilage (e.g., off-odor, superficial viscosity). To maintain poultry meat quality it is necessary to control the temperature during storage. According to Keeton (2001), cold cuts and raw materials should not be above 4.4°C (40°F), while frozen materials should be below −17°C (0°F) when received. It is important to point out that control of thawing is necessary to prevent microbial growth and excess purge loss. Also, strict control of operations related to handling raw materials most be observed to prevent cross-contamination and the presence of any chemical and physical risk.

In relation to the functionality of raw materials, poultry meat provides myofibrillar proteins responsible for maintaining the structure of meat products, to retain water (the water-holding capacity), for juiciness, and to bind the meat particles (cohesiveness). These functional properties can vary within species, and such factors as stress during slaughter, the postmortem state, pH, and the addition of salts may deteriorate, or improve meat functionality, and therefore need to be considered carefully.

To increase meat functionality and yield during the manufacture of luncheon meat products, it is recommended to use *hot meat* (i.e., meat in prerigor). Under this condition, the myofibrillar proteins actin and myosin are more extractable and have better functionality than that of *postrigor* meat (Troeger, 1992). However, postrigor meat is more common and is easy to use. Finally, the use of meat with a high proportion of connective tissue should be limited because it will have lower functional properties, giving poor-quality products (Lawrence and Mancini, 2004). Fat is another fundamental ingredient in luncheon meat and bologna formulations. Their functionality is related to palatability and acts as vehicle, and it is a source of flavor-related compounds. Their physical and chemical characteristics can affect the processing characteristics of emulsified sausages and product stability. A reduction in fat levels can produce a rubbery and firm texture, and often, less juiciness (Claus et al., 1994). Fat concentrations can vary according to the type of meat products. For example, for bologna, the *Code of Federal Regulations* (2003b), § 319.180, has established that bologna may not contain more than 30% fat.

Water During the preparation of luncheon meat and bologna, water and ice are used. Their principal functions include ingredient distribution, solubilization and extraction of meat proteins, temperature control during chopping and emulsification, machinability, improving various sensory qualities, and cost reduction by compensation of evaporation during cooking; also, water facilitates the manufacture of low-fat and low-calorie products (Claus et al., 1994; Lawrence and Mancini, 2004). Finally, water must be of high quality, as it can be a source for microbial contamination; in addition, dissolved calcium and iron salts most be present at very low levels, as these ions are involved in lipid oxidation reactions (Claus et al., 1994).

Salt Salt (NaCl) or an NaCl·KCl mixture is considered to be the most basic curing ingredient for luncheon meats and bologna. It has three basic functions:

as (1) a flavor enhancer; (2) a texture developer: extracting myofibrillar proteins to increase binding properties; and (3) a shelf-life and safety enhancer: retarding microbial growth by reducing water activity (Claus et al., 1994) and because salt acts in combination with sodium nitrite to prevent the growth of *Clostridium botulinum* (Claus et al., 1994; Keeton 2001; Desmond, 2006). The common salt concentration present in meat product formulations varies from 1.5 up to 3% (Keeton, 2001). On the other hand, there is a tendency to reduce sodium levels in meat products by using salt substitutes such as KCl in combination with flavor enhancers (e,g., yeast extracts, lactates, monosodium glutamate, nucleotides) to reduce the bitter taste of potassium salts (Desmond, 2006). Finally, there is evidence that NaCl levels in phosphates added to cooked bologna sausages may be reduced to 1.4% of added NaCl without loss of a pleasant flavor (Ruusunen and Puolanne, 2005).

Nitrites Addition of sodium and/or potassium nitrites to luncheon meats and bologna as well as other meat products is called *curing*. Nitrites may be purchased as curing salts containing 6.25% sodium nitrite mixed with sodium chloride crystals (Keeton, 2001; Lawrence and Mancini, 2004). Nitrites have several functions: (1) they react with myoglobin, giving the characteristic pink color; (2) contribute to the flavor development associated with curing meats, (3) act as antioxidants by decreasing heme iron oxidation; and (4) prevent botulism by inhibiting the growth of *C. botulinum* as well as the outgrowth of its spores and prevent the development of other microorganisms (Keeton, 2001; Lawrence and Mancini, 2004; Honikel, 2008). The level of nitrite is limited to 156 ppm (Keeton, 2001; Lawrence and Mancini, 2004). The nitrite levels permitted are similar in several countries; for example, the *Code of Federal Regulations* (2003d) establishes in § 424.21: "The use of nitrites, nitrates or combination shall not result in more than 200 ppm of nitrite, calculated as sodium nitrite in finished product except that nitrites may be used in bacon." In the CODEX Alimentarius (CS, 1991) the level of nitrites permitted for standard luncheon meats, calculated on the total of the final product, is 125 mg/kg total nitrite expressed as sodium nitrite. Addition of nitrites is strictly regulated because they can eventually form carcinogenic amines in the acidic ambient of the human stomach. Fortunately, nitrite concentration in meat products decreases during storage. According to Honikel (2008), the biggest decrease in nitrite concentration occurs during manufacture, at the end of the heating process, and this early loss is usually about 65% of the initial concentrations. It is important to point out that according to the *Code of Federal Regulations* (2003d), § 424.21, "Nitrates are not permitted and may not be used in baby, junior, and toddler foods." Finally, because of the negative perception of nitrites by some customers, there are now several luncheon meats, such as ham, frankfurters, bologna, salami, and Polish sausage, in "uncured" natural and organic versions. However, according to Sebranek and Bacus (2007), more studies are necessary in relation to the nitrite and nitrosamines formation in processed meats manufactured with natural nitrate sources, to assure that nitrite levels cannot affect consumers.

Optional Ingredients

Several ingredients can be added optionally during manufacture. These ingredients give value-added poultry formulations, and their principal functions are (1) to reduce the product cost; (2) to extend product storage life; (3) to increase meat particle binding, giving a more musclelike texture; (4) to increase the meat water-holding capacity; (5) to improve juiciness and tenderness; (6) to modify textural characteristics, and (7) to act as fat replacers in the product (Keeton, 2001).

Cure Accelerators Cure accelerators are substances that reduce the curing time, as they promote formation of nitric oxide hemochrome. The principal cure accelerators used are sodium ascorbate, sodium erythorbate, and sodium citrate. They accelerate the reaction of sodium nitrite reduction into nitric oxide (NO). When NO concentration increases, a greater proportion of myoglobin is transformed into nitrosomyoglobin, responsible for the characteristic color of raw cured meats (Wirth, 1992; Lawrence and Mancini, 2004). This reaction is promoted in acid media. On the other hand, sodium ascorbate and sodium citrate have antioxidant activity (Claus et al., 1994). Sodium ascorbate reduces the production of nitrosamines. The level of cure accelerators is limited to 550 ppm (Lawrence and Mancini, 2004).

Seasonings Seasonings are spices, vegetables, and herbs used to prepare meat products. They impart the specific and unique flavor that differentiates each meat product.

Sweeteners Sweeteners such as sucrose, glucose, corn syrup, and sorbitol are frequently used in poultry luncheon meats and bologna to enhance flavor, to compensate for the harshness of salt, to increase browning, and to reduce costs. In fermented sausages, glucose and sucrose act as a source or carbon for lactic acid bacteria (LAB). Glucose (dextrose) is preferred because this is more easily metabolized by LAB (Keeton, 2001). On the other hand, honey is frequently used to prepare roasted bread and ham with honey. The proportion of sweeteners in the meat formulations vary within 0.5 to 2% (Lawrence and Mancini, 2004).

Spices and Flavorings Spices improve flavor; there is an innumerable supply of flavorings that could be added to poultry luncheon meats to impart or modify flavor, such as herbs, chili peppers, oils, and synthetic flavors. Several spices, such as thyme and rosemary, also possess antimicrobial and antioxidant activity and may improve product shelf life. Spices can be sold unground or finely ground; however, they are commonly contaminated, and it is recommended that they be irradiated before being included in a formulation, to prevent food spoilage. Therefore, spice extracts, essential oils, oleoresins, or microencapsulated flavorings are preferred in industry, as they are free of microorganisms, uniform, and easy to handle (Lawrence and Mancini, 2004; Ponce-Alquicira, 2007). Some of the spices commonly used in meat products are black and white pepper, onion, pimiento, garlic, cloves, and nutmeg, as shown in Table 2.

TABLE 2 Spices Used in Some Ethnic and Nontraditional Formulations of Luncheon Meats

Luncheon Meat	Spices
Len's bologna[a]	Pepper, onion, garlic, mustard, coriander, white pepper, nutmeg
Krakauerwurst: a German version of the Polish krakauer[a]	Garlic, coriander, mustard, caraway
Mortadella[a]	Coriander, white pepper, garlic, black pepper, paprika, mace
Mortadella di Bologna[a]	Coriander, black pepper, garlic, white pepper, anise, mase, nutmeg, caraway, cinnamon, clove, pistachio (optional)
Mortadella di Prato: a Tuscan-style mortadella[a]	Garlic, black pepper, coriander, cinnamon, clove, alkermes
Turkey pastrami roll[a]	Black pepper, cloves, garlic, allspice, coriander, bay leaf, cinnamon
Cooked salami	White pepper, spicy Spanish pimento, chili, garlic
Turkey pastrami	Garlic, black pepper, marjoram, basil, allspice, cloves

[a]From Sonoma Montain Sausages (http://home.pacbell.net/lpoli/index.htm).

Smoke The smoke in a the product is obtained by the partial burning of wood, predominantly hardwood such as beech, hickory, and oak (Stolyhwo and Sikorski, 2005). It contains carbonyls, organic acids, alcohols, and phenols which have specific flavor properties. Acids contribute to the product skin formation, to flavor, and accelerate the curing reaction. Alcohols act as a carrier of flavors and as a bacteriostatic. Fenols are the principal contributors to smoke taste and are responsible for antioxidant, color, and antimicrobial activities. Carbonyls are the principal compounds related to color formation and also serve to promote protein cross-linking (Keeton, 2001). Industrially, smoke can be applied to meat products in special smoke chambers or by direct addition of smoke flavoring into a formulation. These flavorings are generally smoke extracts, filtered and separated from the resinous material that contains most of the polycyclic aromatic hydrocarbons (PAHs), which are associated with carcinogenic activity (Stolyhwo and Sikorski, 2005). Smoke from mesquite and hickory are frequently used in luncheon meat products in the U.S. market.

Ingredients Increasing Yield and Functionality

Phosphates Phosphates are used principally to improve water-holding capacity. This property is obtained as a result of a synergist action between phosphates and salt to extract myofibrillar proteins. Phosphates are also implicated in the dissociation of the actomyosin complex. In addition, they reduce loss by cooking or purge. Their antioxidant action is derived from their ability to sequester metals such as copper that can promote lipid oxidation (Claus et al., 1994). In addition, phosphates can be used to preserve the color of cured products and to enhance flavor. The majority of formulations of meat products include between 0.3 and 0.4% phosphates (Keeton, 2002). Meat products with levels of phosphates greater of 0.4% may present a soapy flavor. U.S. regulations permit the addition of 5000 ppp of meat block weight (Lawrence and Mancini, 2004). Pyrophosphates, diphosphates, tetrasodium phosphates, and sodium acid pyrophosphate at pH 7 are recommended for sausage emulsions, while sodium tripolyphosphate and sodium hexametapolyphosphate are recommended for curing brine (Keeton, 2001).

Proteins and Carbohydrates Proteins and carbohydrates are used as extenders and binders in luncheon meats to reduce the cost of the final products. They improve water-holding capacity and yield, also help in sliced operations, improve texture, and modify color and flavor (Lawrence and Mancini, 2004). The most common proteins used in the meat industry to enhance functionality are soy proteins (concentrates and isolates), milk proteins (whey and casein), hydrolyzed plant and animal proteins, and gelatin. The carbohydrates used most include starches and hydrocolloids (e.g., carrageenan, alginates).

Others Other ingredients that may be used to increase functionality for luncheon meats are transglutaminase and Fibrimex. Transglutaminase is a dependent calcium enzyme capable of cross-linking muscle proteins. Fibrimex is a bovine blood plasma protein used to produce meat products with a structure similar to that of whole-muscle cuts that can be sliced raw and prepared as desired (Keeton, 2001).

PROCESSING OF LUNCHEON MEAT

Commercial luncheon meats are ready-to-eat meat products and include poultry bologna, boiled and baked ham, pastrami, turkey and chicken salami, turkey and chicken breast, and poultry pastrami. They are typically sliced, either in an official establishment or after distribution from establishment. There is not a unique process that involves all the variety of luncheon poultry products, due to the differences in nature and formulation. For this reason the manufacturing processes were separated into two groups: emulsified and chopped products (bologna and loaf), and formed products (see Figure 4).

Emulsified Products

For emulsified products such as bologna and loaf, the process includes the selection and conditioning of raw materials, grinding, mixing, blending, stuffing, linking, cooking, cooling, packaging, and storage (Figure 4).

Formulation and Selection of Raw Materials These products may be prepared using fresh breasts, legs, thigh muscle, and drumstick meat, as well as mechanically deboned poultry, or in combination with other species. These materials

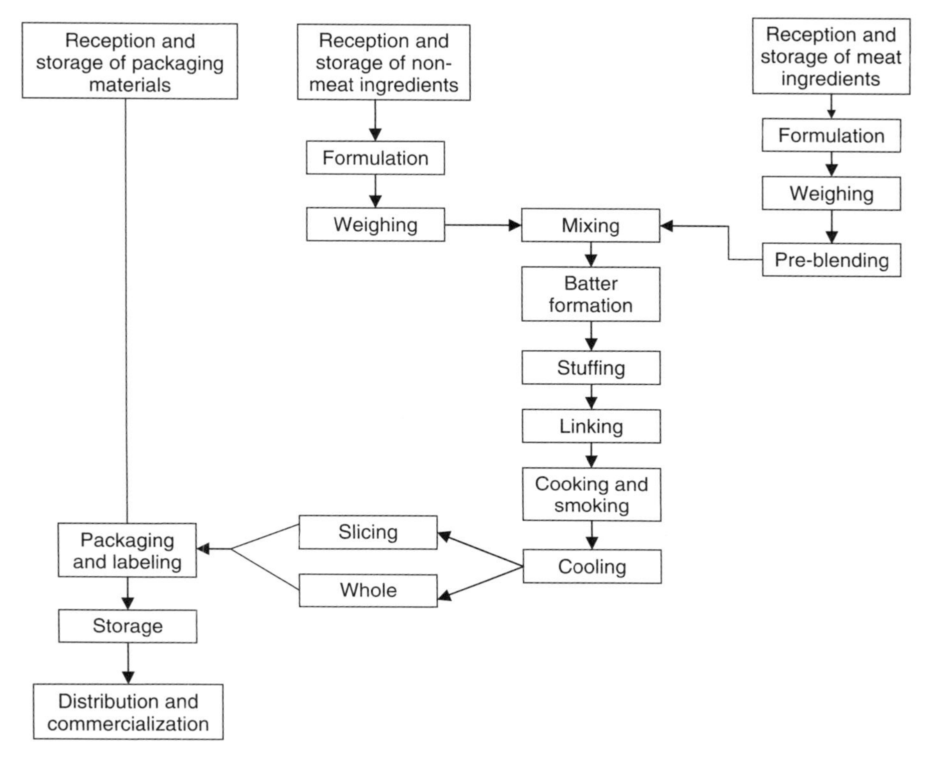

FIGURE 4 Processing diagram for bologna and other cooked sausages.

present excellent microbiological (low microbial count) and functional qualities. The temperature of raw materials has to be in the range of -18 to 0°C and the pH below 6.0. Fat, moisture, protein content, and low microbial counts are parameters that can affect the quality of the final product and have to be analyzed prior to the elaboration of meat products. In industries that produce a high volume of emulsified products, a system called *least cost formulation* is used. This is a computerized program that makes it possible to determine how to allocate the ingredients required for a given product at minimum cost (USDA–FSIS, 2007).

Weighting or Measuring of Ingredients The proportion of meat, fat, species, and curing agents is next weighted, under hygienic conditions, according to a formula established previously.

Comminuting This processing consists of a reduction in meat particle size. This process assists in the homogeneous distribution of ingredients and increases meat tenderness. A grinder, chopper, flaker, or a combination can be used for comminutation (Judle et al., 1989; USDA–FSIS, 2007). During this stage it is important to control temperature to avoid fat disaggregation and microbial growth.

Preblending A preblend is formed by mixing lean ground meats with salt, sodium nitrites, and water. It is maintained in storage for 6 to 24 h to promote the extraction of miofibrillar proteins. During this stage, protein, moisture, and fat content may be analyzed to ensure the lean/fat radio. Preblends of different compositions may be combined by using composition data (Judle et al., 1989; Lawrence and Mancini, 2004).

Blending Blending consists of combining preblends or lean meats with salt, curing salts, phosphates, and water in the form of ice in a vacuum chopper. This additional mixing helps to extract the meat proteins responsible for cohesiveness.

Mixing and Batter Formation In this stage, product blending is subjected to high-speed cutting and shearing action. Fat, muscle, and connective tissue are reduced, and myofibrillar proteins are solubilized in the continuous phase to form a fat-in-water emulsion in which myofibrillar proteins such as actin, myosin, and actomyosin act as emulsified agents. Formation of a typical batter includes two transformations: shelling proteins with the formation of a viscose matrix, and emulsification of solubilized proteins, fat globules, and water (Judle et al., 1989). It is important to monitor the batter temperature below 10 to 12°C to avoid fat globule melting. Overchopping makes the protein fibers too short. Factors that affect batter formation and stability include temperature during matrix formation, fat particle size, pH, concentration and type of soluble protein, and batter viscosity (Judle et al., 1989).

Stuffing Stuffing casings may be manufactured from synthetic materials; however, natural casings are commonly used in cooked sausages. It is recommended that vacuum systems be employed to reduce air voids and maintain the consistency within the product. It is also important to consider the permeability according to the product. After stuffing, the product is linked by pinching and twisting the casing to form separate units of sausages.

Smoking and Cooking Most products are both smoked and cooked (e.g., bologna). Smoking and cooking involves a short drying period, followed by smoking and cooking (Lawrence and Mancini, 2004). The objective of smoking is to improve aroma, flavor, and color characteristics. Smoking promotes coagulation of meat proteins on the product surface. Smoke can be applied by the incomplete combustion of specific hardwoods or by using smoke concentrates (liquid smoke), which can be applied before cooking by atomized spraying (Lawrence and Mancini, 2004).

Cooking Typical heat processing of cooked sausages is carried out between 65 and 77°C in the center of the product. The objectives of cooking are (1) to fix the emulsion and form a rigid structure, (2) to reduce spoilage and pathogenic flora by pasteurization of the product, (3) to separate casings, and (4) to produce typical coloration of cooked sausages by transformation of nitrosomyoglobin to nitrosylhemochrome.

Chilling A chilling or cooling process is known as *stabilization*. Bologna is normally chilled with a brine spay or shower (Lawrence and Mancini, 2004). Chilling takes place using air, cold water, or brine.

Slicing and Packaging After cooling, casings have to be stripped away. The products are then sliced and/or packaged in retail consumer-sized portions. Most are vacuum packaged to improve the appearance of the product, minimize weight loss, and maximize shelf life (Lawrence and Mancini, 2004). The slicing operation must be performed under hygienic conditions to avoid the entrance of pathogenic microorganisms during the postprocessing stage. Luncheon meats are sold using several packaging presentations (e.g., variety packs).

Storage Luncheon meats and bologna are stored under refrigeration (2 to 4°C) to prevent microbial growth.

Formed Products

Formed products are prepared using breasts, legs, and thigh muscles as well as mechanically deboned poultry or in combination with other meat species. Raw meats are either marinated previously or injected with curing solution to extract myofibrillar proteins. Nonmeat binders (protein isolate, starch, carrageenan) may be incorporated into a brine or marinade for injection-formed products. Vacuum tumbling is used to extract a major proportion of myofibrillar proteins. Afterward, the mixture is molded into a specific shape and cooked fully stuffed at 71.1 to 73.8°C (Keeton, 2001). Formed products are often sliced and vacuum packaged (Figure 4).

MICROBIOLOGY OF LUNCHEON MEAT

Spoilage Microorganisms

Luncheon meats are commonly packaged under vacuum and stored at chill temperatures. This microaerophilic condition, as well as the presence of curing salts and reduced water activity (a_w), promotes the development of lactic acid bacteria (Samelis et al., 2000a). According to Bell and Gill (1982), *Bacillus* and *Enterococcus* species are the leading spoilage microflora that develop on pasteurized luncheon meats. However, Samelis et al. (2000a) reported that the type of product and the packaging conditions determinate the growth of the various species of lactic acid bacteria on cooked meat stored at 4°C. It has also been reported that *Leuconostoc mesenteroides* subsp. *mesenteroides* is the predominate microorganism on vacuum-packed and sliced turkey breast fillets. In other studies, Samelis et al. (2000b) reported that *Lactobacillus sakei* subsp. *carnosus* was predominant in the spoilage flora of sliced, vacuum-packed, smoked, oven-cooked turkey breast fillets. In contrast, *Leuconostoc mesenteroides* subsp. *mesenteroides* predominated in the spoilage flora of sliced, vacuum-packed, unsmoked, boiled turkey breast fillets also stored at 4°C. Additionally, they observed that the spoilage flora of unsmoked breasts grew faster than those of smoked breasts and was more diverse.

Although lactic acid bacteria are not pathogenic, their metabolic activity in cooked meat products can produce sour off-flavors and off-odors, milky exudates, slime, discoloration, swelling of the pack, and/or greening (Borch et al., 1996; Samelis et al., 2000a). For this reason, control of the bacterial load is very important in this type of product. Bell and De Lacy (1987) reported that nisin (500 IU/g), sorbic acid (0.125%), lauricidin (0.5%), and lauric acid (0.25%) were effective in controlling the growth of the natural *Enterococcus* spoilage microflora in luncheon meats.

Pathogens

According to Mead (2000), foodborne diseases cause approximately 76 million illnesses, 325,000 hospitalizations, and 5000 deaths in the United States each year. The principal pathogenic flora associated with ready-to-eat meat products are *Salmonella, Escherichia coli* O157:H7, *Staphylococcus aureus, Clostridium, Campylobacter, and Listeria monocytogenes*. Ready-to-eat (RTE) foods, including turkey deli meats, have been implicated in several listeriosis outbreaks in the United States (Yang et al., 2006). Contamination with *L. monocytogenes* in RTE meat occurs primarily during slicing and packaging after cooking (Zhu et al., 2005). Outbreaks of listeriosis have frequently been associated with ready-to-eat meat products. For this reason, the Food Safety and Inspection Service of the U.S. Department of Agriculture (USDA-FSIS) and U.S. Food and Drug Administration (FDA) enforce a zero-tolerance policy for *L. monocytogenes* in RTE foods (Voetsch et al., 2007). Additionally, the FDA and USDA-FSIS have developed action plans to prevent contamination with this microorganism in RTE foods. These plans include strategies for guidance, training, research, education, surveillance, and enforcement. Fortunately, there is a marked decrease in the incidence of listeriosis, which may be related to the decrease in the prevalence of *L. monocytogenes* contamination of RTE foods since 1996 (Voetsch et al., 2007).

An alterative to reducing or eliminating pathogens such as *L. monytogenes* may include the use of pressurized steam treatment immediately after vacuum sealing of retail packages (Murphy et al., 2005) or of radiation or hydrostatic pressure. Finally, formulating meat products with antimicrobial additives is another common approach to controlling *L. monocytogenes* in RTE meat (Zhu et al., 2005).

Currently, there is a trend toward improving the safety of cooked meat products, including luncheon meats by using antimicrobial packaging, which is made by inclusion of antimicrobial agents from natural origins such as bacteriocins (Quintero-Salazar et al, 2005), spices and essential oils, and enzymes on the surface of packaging materials (Quintero-Salazar and Ponce-Alquicira, 2007; Coma, 2008), as an alternative to inhibiting outbreaks of the principal microorganisms associated with the consumption of RTE products, with a special emphasis on meat products.

NUTRITIONAL FACTS

It is important to understand that the poultry meat used to prepare luncheon meats has great advantages over meat from other sources. For example, Vandendriessche (2008) reports that poultry fat contains lower saturated fatty acid levels and a higher content of polyunsaturated acids (27 and 23%, respectively) than those of white pork (37.1 and 11.3%) and beef (44.6 and 4.8%). However, as do other processed meat products, luncheon meats present positive and negative nutritional effects on human health, as they are rich in high-quality proteins, low in sugars, and are a good source of vitamin B_6/B_{12} (riboflavin, thiamine, cobalamin), zinc, and iron. Some negative aspects include the presence of salts and PACs in smoked products, nitrosamines, as well as being low in fiber and calcium (Vandendriessche, 2008). Due to the increasing number of health problems arising from food consumption, such as cardiovascular diseases, diabetes, cancer, hypertension, obesity, and osteoporosis, there is a trend toward decreasing the levels of ingredients associated with these health problems in luncheon meats. For this reason, the meat industry has focused its attention on the development of healthier meat products that are low in carbohydrates and saturated fatty acids, as well as low in calories and sodium levels, with added probiotics and prebiotics, and "healthy" substances such as phytosterols, vitamin A, provitamin A, lycopene, special oils (olive, amaranth), wellness herbs, ginseng, fibers, and calcium; and with no addition of nitrates (Vandendriessche, 2008).

According to Jiménez-Colmenero et al. (2001), strategies to produce healthier meat and meat products include the modification of carcass composition, the manipulation of raw materials, and the reformulation of meat products. However, the development of healthy poultry meat products is not an easy job because of the reduction of ingredients such as salt and fat and the addition of new ingredients that can affect the sensorial characteristics (e.g., flavor, texture, color) of the final products and subsequently, acceptance by consumers. Some strategies to produce healthy meat products are presented in Table 3.

NEW TRENDS

Before describing the new tendencies in poultry luncheon meats, it is important to note that the meat industry had experienced important changes during recent years. Vandendriessche (2008) has divided the meat processing technology of the last 25 years into three periods:

1. Period of quality
2. Period of quality and food safety
3. Period of quality, food safety, and nutrition/health

The *period of quality* began in the 1980s–1990s with the implementation of the Japanese-based ISO9000 quality system standard. The *period of quality and food*

TABLE 3 Strategies Used to Produce Healthier Luncheon Meats

Characteristic	Effect on Sensorial and Microbiological Quality	Alternatives	Effect
Low salt levels	Decreased myofibrilar protein extraction and fat functionality	Addition of magnesium and potassium salts.	
		Use of sodium, magnesium, and potassium salts in combination	
		Lactates, monosodium glutamate, and nucleotides	Inhibition of microbial growth and flavor enhancement
		Addition of alginates and transglutaminase	Textural properties improved
Reduction of nitrite levels	Flavor, color, and antioxidant activity	Ascorbate or erythorbate	Inhibitors of *N*-nitrosamine production
		Erythrosine	Generation of color by formation of mononitrosyl ferrohemochrome
		Chemical antioxidants and chelating agents	Generation of cured flavor
		Sorbic acid and potassium sorbate, sodium hypophosphite, fumaric acid, even lactic acid bacteria	Microbial action of nitrites
		Reduction of nitrites concentration	Bacteriocin nisin
Low fat levels	Affects palatability		
Reduction of calories	Affects palatability	Reduction of fat concentration	
Low levels of cholesterol	Affects palatability	By replacing animal fat by vegetable oils from peanuts, canola, sunflower, olive, etc.	Reduction of cholesterol levels
Incorporation of functional ingredients	In some cases, could affect texture and flavor	Phytosterols, vitamin A, lycopene, special oils (olive, amaranth), wellness herbs, ginseng, fiber, and calcium	
		Pre- and probiotics (for fermented meat products)	

[a] *Source:* Jiménez-Colmenero et al. (2001); Desmond (2006); Vandendriessche (2008).

safety was characterized by the introduction of hazard analysis of critical control points (HACCPs). The introduction of HACCPs as a method universally accepted for food safety assurance has been used to reduce the frequency of foodborne diseases caused by the presence of such food pathogens as *L. monocytogenes, E. coli* O157:H7, *Salmonella*, and *Campylobacter* in RTE and convenient meat products, including luncheon meats, whose demand has increased constantly due to changes in eating habits and methods of food preparation as a result of an acceleration in the rhythm of life.

According to Vandendriessche (2008), due to the increasing number of health problems associated with food consumption (e.g., cardiovascular diseases, diabetes, cancer, hypertension, obesity, osteoporosis), the *period of quality, food safety, and nutrition/health* in the meat industry has begun. However, there also exists great interest in ethnic flavors as well as organic meats and animal welfare. Natural and organic products are perceived as being more healthful because natural products contain no preservatives or artificial ingredients, and organic meats are derived from cattle raised on organic grains under more healthful conditions (Anon., 2005).

CONCLUSIONS

Poultry luncheon meats and bologna are very popular precooked cured meat loafs or sausages that are generally eaten cold, that may be sold sliced, and are frequently used to prepare sandwiches or salads. These sorts of poultry products represent a new opportunity to the meat industry to offer healthy, convenient, and functional foods specially formulated to attend to consumer demands. However, formulation and processes involved in the manufacture of poultry luncheons meats must be in agreement with nutritional and legislation aspects.

REFERENCES

Abdullah BM. 2007. Properties of five canned luncheon meat formulations as affected by quality of raw materials. Int J Food Sci Technol, 42:30–35.

Abdullah B, Al-Najdawi R. 2005. Functional and sensory properties of chicken meat from spent-hen carcasses deboned manually or mechanically in Jordan. Int J Food Sci Technol 40:537–543.

Anon. 2005. Breakfast and sandwich meats, U.S., October 2005: executive summary. In: National Provisioner Online. http://provisioneronline.com/FILES/Executive_Summary.doc. Accessed Jan. 18, 2008.

Baggio SR, Bragagnolo N. 2006. Cholesterol oxide, cholesterol, total lipid and fatty acid contents in processed meat products during storage. Lebesm-Wiss Technol 39:513–520.

Bell RG, De Lacy KM. 1987. The efficacy of nisin, sorbic acid and monolaurin as preservatives in pasteurized cured meat products. Food Microbiol 4:277–283.

Bell RG, Gill CO. 1982 Microbial spoilage of luncheon meat prepared in an impermeable plastic casing. J Appl Bacteriol 53:97–102.

Borch E, Kant-Muermans, Blixt Y. 1996. Bacteria spoilage of meat and meat products. Int J Food Microbiol 33:103–120.

CAC (CODEX Alimentarius Commission). 1991. Standard for Luncheon Meat. Codex Standard 89–1981 (Rev. 1–1991). http://www.codexalimentarius.net/download/standards/158/CXS089e.pdf. Accessed Jan. 7, 2008.

Cambridge Dictionary Online. 2008. http://dictionary.cambridge.org/. Accessed Jan. 16, 2008.

CFR (*Code of Federal Regulations*). 2003a. Title 9, Animal and Animal Products, Chap. III, Pt. 317, Labeling, Marking Devices, and Containers. http://www.access.gpo.gov/nara/cfr/waisidx_03/9cfr317_03.html. Accessed Jan. 22, 2008.

CFR. 2003b. Title 9, Animal and Animal Products, Chap. III, Pt. 319, Definitions and Standards of Identity or Composition. http://www.access.gpo.gov/nara/cfr/waisidx_03/9cfr319_03.html. Accessed Jan. 22, 2008.

CFR. 2003c. Title 9, Animal and Animal Products, Chap. III, Pt. 380, Poultry Products Inspection Regulations. http://www.access.gpo.gov/nara/cfr/waisidx_03/9cfr381_03.html. Accessed Jan. 22, 2008.

CFR. 2003d. Title 9, Animal and Animal Products, Chap. III, Pt. 424, Preparation and Processing Operations. http://www.access.gpo.gov/nara/cfr/waisidx_03/9cfr424_03.html. Accessed Jan. 22, 2008.

Claus JR, Colby JW, Flick GL. 1994. Processed meats/poultry/seafood. In: Kinsman DM, Kottula AW, Breidenstein BC, eds., *Muscle Foods: Meat, Poultry, and Seafood Technology*. New York: Chapman & Hall, pp. 106–162.

Coma V. 2008. Bioactive packaging technologies for extended shelf life of meat-based products. Meat Sci 78:90–103.

Desmond E. 2006. Reducing salt: challenge for the meat industry. Meat Sci 74:188–196.

Eyre C. 2007. Packaging wields the power in luncheon meat market. http://www.foodproductiondaily.com/news/ng.asp?n=79851-nielson-luncheon-meat-pouches. Accessed Jan. 31, 2008.

Honikel K-O. 2008. The use and control of nitrate and nitrite for processing of meat products. Meat Sci, 78:68–76.

Jiménez-Colmenero E. 1996. Technologies for developing low-fat products. Trends Food Sci Technol 7:41–47.

Jiménez-Colmenero F, Carballo J, Cofrades S. 2001. Healthier meat and meat products: their role as functional foods. Meat Sci 59:5–13.

Judle MD, Aberle ED, Forrest JC, Hedrick HB, Merkel RA. 1989. Principles of meat science. In: *Principles of Meat Science*, 2nd ed. Dubuque, IA: Kendall/Hunt, pp. 135–174.

Keeton JT. 2001. Formed and emulsion products. In: Sams AR, ed., *Poultry Meat Processing*. Boca Raton, FL: CRC Press, pp. 195–226.

Lawrence T, Mancini R. 2004. Meat: hot dogs and bologna. In: Smith JS, Hui HY, eds., *Food Processing: Principles and Applications*. Ames, IA: Blackwell Publishing, pp. 339–398.

Mead PS, Slutsker L, Dietz V, McCaig LF, Bresee JS, Shapiro C, Griffin PM, Tauxe RV. 2000. Food related illness and death in the United States: CDC synopses. Emerg Infect Dis 5(5). http://www.cdc.gov/ncidod/eid/vol5no5/mead.htm. Accessed Jan. 30, 2008.

Murphy RY, Hanson RE, Duncan LK, Feze N, Lyon BG. 2005. Considerations for post-lethality treatments to reduce *Listeria monocytogenes* from fully cooked bologna using ambient and pressurized steam. Food Microbiol 22:359–365.

Ponce-Alquicira E. 2007. Poultry marination. In: Hui YH, ed., *Handbook of Food Products Manufacturing*. Hoboken, NJ: Wiley, pp. 749–770.

Quintero-Salazar B, Ponce-Alquicira E. 2007. Edible packaging for poultry and poultry products. In: Hui YH, ed., *Handbook of Food Products Manufacturing*. Hoboken, NJ: Wiley, pp. 797–815.

Quintero-Salazar B, Vernon Carter EJ, Guerrero-Legarreta I, Ponce-Alquicira I. 2005. Incorporation of the antilisterial bacteriocin-like inhibitory substance from *Pediococcus parvulus* MXVK133 into film forming protein matrices with different hydrophobicity. J Food Sci 70(9):M398–M403.

Ricke SC, Kundinger MM, Miller DR, Keeton JT. 2005. Alternatives to antibiotics: chemical and physical antimicrobial inventions and foodborne pathogenic response. Poult Sci 84(4):667–675.

Ruusunen M, Puolanne E. 2005. Reducing sodium intake from meat products. *Meat Sci* 70:531–554.

Samelis J, Kakouri A, Rementzis J. 2000a. Selective effect of the product type and the packaging conditions on the species of lactic acid bacteria dominating the spoilage microbial association of cooked meats at 4°C. Food Microbiol 17:329–340.

Samelis J, Kakouri A, Rementzis J. 2000b. The spoilage microflora of cured, cooked turkey breasts prepared commercially with or without smoking. Int J Food Microbiol 56:133–143.

Sebranek JG, Bacus JN. 2007. Cured meat products without direct addition of nitrate or nitrite: What are the issues? Meat Sci 77:136–147.

Sinha R, Peters U, Cross AJ, Kulldorff M, Weissfeld LJ, Pinsky PF, Rothman N, Hayes RB, Lung P. 2005. Meat, meat cooking methods and preservation, and risk for colorectal adenoma. Cancer Res 65(17):8034–8041.

Stolyhwo A, Sikorski ZE. 2005. Polycyclic aromatic hydrocarbons in smoked fish: a critical review. Food Chem 91:303–331.

Troeger K. 1992. Tratamiento y almacenamiento de la material prima. In: Wirth F, ed., *Tecnología de los Embutidos Escaldados*. Zaragoza, Spain: Acribia, pp. 21–39.

USDA–FSIS (U.S. Department of Agriculture–Food Safety and Inspection Service). 2007. The Regulated Industries: Characteristics and Manufacturing Processes. http://www.fsis.usda.gov/PDF/PHVt-Regulated_Industries.pdf. Accessed Jan. 27, 2008.

Vandendriessche F. 2008. Meat products in the past, today and in the future. Meat Sci 78:104–113.

Voetsch AC, Angulo FJ, Jones TF, Moore MF, Nadon C, McCarthy P, Shiferaw B, Megginson MB, Hurd S, Anderson BJ, et al. 2007. Reduction in the incidence of invasive listeriosis in foodborne diseases active surveillance network sites, 1996–2003. Clin Infect Dis, 44:513–520.

Wirth F. 1992. Curado-formación y conservación del color. In: Wirth F, ed., *Tecnología de los Embutidos Escaldados*. Zaragoza, Spain: Acribia, pp. 127–147.

Yang H, Mokhtari A, Jaykus L, Morales RA, Cates SC, Cowen P. 2006. Consumer phase risk assessment for *Listeria monocytogenes* in deli meats. Risk Anal 26(1): 89–103.

Zhu M, Du M, Cordray J, Dong, Ahn DU. 2005. Control of *Listeria monocytogenes* contamination in ready-to-eat meat products. Compr Rev Food Sci Food Saf 4:34–42.

19

PROCESSED EGG PRODUCTS: PERSPECTIVE ON NUTRITIONAL VALUES

MAHENDRA P. KAPOOR, MOLAY K. ROY, AND LEKH R. JUNEJA
Taiyo Kagaku Co. Ltd., Yokkaichi, Japan

INTRODUCTION

In modern times, the egg industry has become an important part of the food industry and has been held up as a powerhouse of nutrition. Eggs are recognized as an important commodity in human diets because of their exceptional nutrition

Handbook of Poultry Science and Technology, Volume 2: Secondary Processing, Edited by Isabel Guerrero-Legarreta and Y.H. Hui

profile as a nutrient-rich food containing high-quality protein and a substantial quantity of several essential minerals, lipids, and vitamins. In this chapter we discuss the brief history of the egg industry and current practices and approaches used by the industry, with an emphasis on the use of egg and egg components in food processing, especially in operations related directly to processed poultry products. Specific aspects of egg-processing technologies on a commercial level have been discussed elsewhere; this chapter is a comprehensive summary on the diverse steps taken in systematic secondary egg processing. This includes the functional properties of eggs, where applicable, in breading, gelation, emulsification, frying, and so on. Most of these approaches are in industrial processing, so much of this chapter is focused on potential opportunities. Discussion is confined to nutritional and functional roles of eggs, along with emerging issues of educating consumers on changing views in the area of food habits and health benefits from egg consumption.

BACKGROUND AND PARADIGMS

History indicates that jungle fowl were domesticated in Asia 3000 years ago. Since then, the egg industry has undergone a major technological and sociological revolution. The poultry industry is vertically integrated, which means that the industry has a tremendous amount of control over their products. A relatively few specialized poultry breeders supply hatching eggs for commercial laying flocks, and so far nearly 200 breeds and varieties of chickens have been established around the world. The industry is distinctly different from many other animal industries in that egg producers own and manage nearly every aspect of their businesses: rearing of birds, feeding, housing, husbandry, and marketing of their product, and are capable of meticulous monitoring of the entire process.

Poultry producers usually do not own the primary breeding stock; their birds are purchased from primary breeders. It is generally accepted that worldwide egg production has expanded by about 47% over the past 30 years. Also, to maximize the return on investment, the processors, handlers, and distributors of eggs and egg products require a thorough understanding of egg production practices to cope more intelligently with problems of product quality and characteristics. The application of basic genetic principles to breeding programs is extremely important in gaining dividends. The areas to which breeders have been directing increased attention include the quality of freshly laid eggs, the disease-resistant general viability of the progenies, and shell quality, which is a major economic problem because of losses from cracked and broken eggs. Other areas include the infrastructure, wherein temperature control is a most important environmental factor in the early life of chicks. Floor brooding and rearing becomes more common with good feeding and management programs. This prevents starvation and dehydration and results in birds of uniform size and body growth.

For larger commercial operations, automation is also being considered, since supplying feed and water to the birds and gathering the eggs require considerable labor if performed manually. Mechanical feeders in which the feed is conveyed and evenly distributed in a trough are becoming increasingly popular for all type of brooding operations. General conditions such as day length and lighting for layers also influence the rate of egg production. Also, birds kept in cages or at high density on the floor may develop cannibalistic habits, which can lead to increased mortality and loss in performance. Waste management and poultry health are the other important issues being considered. The preservation of shell egg quality is also of significant importance, followed by frequent gathering, proper cooling, and handling under controlled humidity. Management differences in the rearing of layers may be accounted for by economics, producer preference, and/or geography.

MICROBIOLOGICAL ASPECTS AND RELATED CHEMISTRIES

An understanding of the microbiology of eggs is also necessary in evaluating the antimicrobial defense system of an egg. Shell eggs consist of about 9.5% shell, 63% albumen, and 27.5% yolk [1]. The total solids of albumen, yolk, and whole egg are around 12, 52, and 24%, respectively. Knowledge of the chemistry of eggs and egg products provides useful information on chemical composition and physicochemical properties (e.g., viscosity, surface activity, pH) of native albumen and yolk. This information is vital for interpreting the changes that occur during shell egg storage and during pasteurizing, drying, and freezing. Several reviews are available in the literature on the chemistry and components of avian eggs [2–5]. Further, alternation in egg components may be reflected in a loss of functionality of albumen (foaming ability) and yolk (emulsifying characteristics). Many publications have reviewed the diverse biological properties, including antimicrobial activity, vitamin-binding properties, protease inhibitory action, and antigenic or immunogenic characteristics of egg components. The textural compositions of egg components are presented in Table 1.

Recent research not only returns eggs to their golden past, but also acknowledges it as a functional food and supports its values as a rich array of nutrients. From the beginning, eggs have been promoted for their high-quality proteins and high nutritional density in proportion to their energy content. Eggs remain an important food source as well as components of many other foods consumed throughout the world. As a high-protein food, eggs contribute about 1.3% of the total calories in the human diet. Eggs are also an important source of essential

TABLE 1 Textural Composition (%) of Yolk, Albumen, and Whole Egg

Egg				Elements							
Component	Lipids	Proteins	Carbohydrates	K	S	Na	Fe	P	Mg	Ca	Ash
Yolk	31.8–35.6	15.6–16.7	0.2–1.1	0.236	0.016	0.081	0.008	0.761	0.080	0.191	1.1
Albumen	0.03–0.04	9.8–10.7	0.4–1.0	0.155	0.195	0.165	0.0009	0.018	0.009	0.015	0.55
Whole egg	10.6–12.0	12.7–13.3	0.3–0.9	—	—	—	—	—	—	—	0.9

TABLE 2 Contribution of Egg Components to the Human Diet

Egg Component	Contribution to Diet (%)
Protein	3.9
Fats (saturated)	1.7
Fat (total)	2.0
Cholesterol	33.2
Folate	5.1
Vitamin A	4.3
Vitamin E	4.3
Riboflavin	2.1
Vitamin B_6	2.1
Vitamin B_{12}	3.7
Calcium	1.7
Iron	2.4
Zinc	2.8
Calories (kcal)	1.3

unsaturated fatty acids, iron, phosphate, minerals, fat-soluble vitamins A, D, E, and K, and several water-soluble B vitamins. The substantial amount of high-quality protein, folate, and riboflavin and a number of other nutrients are in excess of its caloric contribution. Table 2 lists the contribution of eggs to the human diet.

In addition, researchers have shown that eggs supply considerable amounts of bioavailable forms of carotenoids, lutein, and zeaxanthin, which may play a critical role in disease prevention. The aforementioned antioxidant compounds have been reported to help in the prevention of macular degeneration, which is a leading cause of blindness in elderly people, and have been associated with a limited risk of cataract development. Choline, another nutrient found in eggs, is useful for brain and memory development. Egg choline is also recommended as a very important nutrient for pregnant women, to ensure normal development of the fetal brain.

Nutrition compositions of eggs are usually varied upon processing raw eggs. Cotterill and Glauert [6] showed that the nutritional composition of a large whole raw egg and a boiled egg is almost the same, of course, but that the composition of a scrambled egg varies slightly due to the milk and fat added during preparation. Some useful information is also released by the U.S. Department of Agriculture (USDA) on the cooked content of eggs. However, an understanding of external factors influencing the nutritional content of eggs is of utmost importance. Stadelman and Pratt [7] have reviewed many factors that directly influence the concentration of nutrients in eggs. The list of such factors include age, breed, and strain of hens, differences in eggs produced by individual hens, rations, and environmental conditions for hens. Additionally, the egg-storage conditions and duration of storage also influence the nutrient values in eggs. The processing, preparation, and cooking methods are secondary contribution factors for alteration of the nutritional content of eggs.

Since the purpose of this chapter is to sketch the background of egg product practices, and particularly to relate such practices to the overall nutritional values of processed products, to ensure the motifs we have updated the content to match the rapidly changing industry. Also, food scientists involved in the creation of new products, quality control, or compositional studies of egg products can benefit from a knowledge of how processed egg products are being produced with retained nutritional values. Some of the selected references at the end of the chapter will also be helpful in this respect.

SECONDARY PROCESSING FOR FUNCTIONALIZED EGG PRODUCTS

The egg products industry makes a major contribution to the world economy. Initially, a dried egg industry was developed, and later, a frozen egg industry came into existence. Currently, the egg products are available in many liquid, frozen, and dried forms. From early in the twentieth century until 1940, China and the United States were the largest egg product processing countries. China was first to process dried eggs and export them to the rest of the world. Actually, the production of dried eggs expanded dramatically during World War II. Other major development was also achieved in the early 1920s, when it was found that the addition of sugar (up to 10%) to yolk prevents gelation. Thus, freezing of egg products was possible, since freezing usually causes irreversible gelation of an egg yolk. Therefore, dehydration and freezing processes made possible preserved eggs of high nutritional value. Later, de-sugaring of egg products before drying was achieved and considered a major technical breakthrough, because the shelf life of dried products could be enhanced significantly. Additive inclusion was suggested to alter the textural and functional properties of egg products. For example, the functional properties of egg products could be modified via adding esters, gums, and surfactants. Many breakthroughs in the process of controlled viscosity and functional performance were possible through better utilization of available information about egg product developments for nutritional purposes. The details on the aforementioned egg product development processes are illustrated in subsequent sections.

Dehydration Processing

Dehydration processing can be applied successfully to preserving eggs. Research has demonstrated that removal of water from eggs stops the growth of microorganisms and slows chemical reactions. This evidently solves the problems, which involve the functional, chemical, and microbiological properties of end products, including moisture, ash, fat, protein, whipping tendency, glucose, viscosity, total plate counts of coli forms, molds, and yeast. There are certain advantages of dried egg products, including low-cost storage, transportation cost, easy handling, better uniformity, and the ability to be used in various convenience foods:

for example, mayonnaise, salad dressing, pastas, ice cream, confections, bakery mixes, and bakery foods.

Depending on the functional properties required, various drying process are being adapted for dried egg white products. The most common methods to dry egg white are spray drying–whipping and spray drying–nonwhipping, wherein the whipping aid is sodium lauryl sulfate, used primarily in angel food cake (and layer cake). Three major forms (granules, powder, or flakes) of pan-drying egg white are in common use. However, because of a solubility problem using spray-dried egg white in specific applications, due to the particles tendency to clump together upon the addition of water, instant egg whites are used, which are an agglomerated product with the ability to disperse rapidly and dissolve readily when added to water [8, 9].

Other dried egg products are the plain whole egg and yolk, which is spray-dried without any other ingredients to prevent the original functional properties. However, to stabilize or extend shelf life, the glucose was eliminated or converted to acid prior to drying. In a blend of whole egg and yolk with carbohydrates, the products possess certain functional properties, including foaming for bakery foods. Another specially dried egg product is scrambled egg mix, which has exceptional storage stability, which is low in cholesterol (one-fifth of the initial amount), contains 51% whole egg, 30% skim milk, 15% vegetable oil (with antioxidants), 1.5% salt, and 2.5% moisture. Solvent extraction and supercritical CO_2 extraction (controlled coagulation) procedures are also proposed for cholesterol removal. Alternatively, starch derivative β-cyclodextrin can be employed, which has the ability to absorb cholesterol selectively from the egg yolk [10]. Also, dried egg products co-dried with several other food ingredients, such as protein, soy, shortenings, and emulsifiers, are being developed and could be used in puffed snacks, for example [11].

In fact, dried egg products are uncooked, and every effort is made to preserve this state during processing. Also, native properties, which include stable foaming when whipped, the ability to coagulate with heat, emulsifying power, flavor, and color can be maintained. However, the rate of heat transfer to egg and the rate of water elimination from egg during drying are certainly governed by the specific heat, density, viscosity, and surface tension of the egg materials. Dried whole egg and yolk products are more susceptible than dried egg white to heat damage, especially the whipping properties. However, the emulsifying ability of egg yolk, which is attributed to its lecithoproteins, is well recognized, whereas the emulsifying properties of whole egg can be enhanced by drying [12].

Coagulation is another important functional property of egg products because egg proteins denature and coagulate over a wide temperature range and are able to retain this ability. Good examples of the use of eggs to thicken foods are cakes, omelets, puddings, and custards. The flavor of eggs can be changed by adverse drying conditions and also by storage, due to certain oxidation changes [13].

However, the flavor stability of whole eggs and yolk products can be controlled by the addition of carbohydrates (sucrose or corn syrups) as additives. The equilibrium moisture content of whole egg as related to temperature and humidity is also important, considering the negligible effect on the moisture–pressure relationship of protein denaturation in egg whites and yolks. Nutritional properties are also sensitive to drying conditions and may be lost on improper drying. Vitamins such as thiamine, riboflavin, pantothenic acid, and nicotinic acid remain essentially intact in dried whole egg products [14]. In another study, nutritional components and protein values of spray-dried whole egg, egg yolk, and egg white were found to be quite comparable to those of fresh eggs. Other physical properties, such as color, flow and water sorption ability, viscosity, and density, are also being thoroughly considered. Subsequent changes in chemical and microbiological properties occur upon drying, and the role of chemical additives is being verified.

Freeze Processing

Frozen food products abound in the market. According to a survey, over one-fifth of total food products are sold in a frozen state. Among these, frozen egg products, including yolk, egg white, whole egg, and a variety of blends of yolk and white, are being marketed as the base ingredient for use in other food products. In this section we feature the changes that occur in egg products as a result of thawing and freezing, which may also cause a large reduction in bacterial growth along with textural changes.

Owing to the thermal properties of egg products, the moisture in the solid product is also an important factor; here latent and specific heat are the governing factors. For example, compared to raw egg yolk, freezing causes minor changes in raw egg white. Upon freezing raw egg white, members of the sulfhydryl group are said to decrease, whereas when freezing raw egg yolks ($<-6^{\circ}$C), the viscosity increases, resulting in gelation, and all fluidity is lost, which cause a major problem in applications. Sodium chloride and sugars (sucrose syrups, glycerin, phosphates, etc.) are the additives most commonly used in frozen egg products (up to 10%) to control gelation in products consisting of egg yolk. However, disadvantages associated with the use of additives is that it may restrict ingredients use to specific food applications. In addition to the aforementioned additives, proteolytic enzymes such as papin, rhizome, or trypsin are also reported to inhibit gelation [15]. However, these enzymes were found to break down the egg component responsible for gelation except papin, which has a mild effect on organoleptic properties [16].

The egg yolk gelation mechanism is complicated and still not well understood. The freezing temperature and rate of thawing also affect egg-yolk gelation. Apart from this, the storage time and rate of temperature fluctuation also greatly alter the gelation properties. However, the functional properties of plain egg yolk are not much affected upon freezing. Water plays an important role in gelation; the ice crystal formation could dehydrate the protein, resulting in an increase in salt concentration and helping promote the rearrangement and aggregation of yolk lipoproteins. The addition of fructose reduces the gelation tendency as well as preventing the aggregation of lipoproteins [17]. Later, Wakamatu et al. wrote a detailed review of the characteristic of gelation processes [18].

Raw whole eggs are also responsive to gelation when subjected to freezing and thawing; however, the characteristics are less drastic than those of the yolk. Most of the functional properties of raw whole eggs are not affected severely by freezing; however, few exceptions have been reported on the considerable changes [19]. The cooked yolk is little affected by freezing, whereas cooked egg white and cooked whole egg become tough, due to the separation of water from the clumps and layers. Frozen scrambled egg mixes are the most common egg product produced by the industry. However, a number of low-cholesterol egg-substitute products are also available in the retail market. In these the egg yolk is replaced by vegetable oil, soy protein, gums, vitamins, minerals, and fat-free dry milk, among others.

The microbial changes that occur due to freezing are also amazing. Freezing reduces the number of bacteria in egg products. Shafi et al. [20] have estimated the bacteria (e.g., thermophiles, anaerobes, and phychrophiles) in different pasteurized frozen commercial egg products. It was also postulated that pasteurized unfrozen whole eggs had a better shelf life than that of a similar frozen product [21].

Pasteurization

Pasteurization is an important technique used in the egg industry. In the United States, pasteurization has been virtually mandatory since 1966. The bacterium *Salmonella enteritidis*, known as a major food-poisoning substance, is found in nearly one in every 20,000 eggs. According to the egg product inspection act of 1970, all egg products are to be rendered free of *S. enteritidis*. Consumers are also advised not to eat raw eggs. However, currently, pasteurized shell eggs, which defy the current conventional wisdom of the danger of eating raw eggs, are available.

Whole liquid eggs and liquid yolks can be pasteurized by heating them in warm water. The temperature of the yolk must be controlled between 128 and 138.5°F and held for no less than 3.5 min. At lower temperatures, the egg is not pasteurized, and at higher temperature, the albumen loses its functionality. The important advantages of the combination of time–temperature conditions are important in regulating the enzymatic test used to determine the adequacy of the pasteurization. The pasteurization of egg white has posed bigger problems, due to substantial instability to heat in the aforementioned range of effective pasteurization. In a very recent process, each egg is weighed and directed to a series of warm-water baths. The combination of time and water temperature heats the eggs enough to kill *S. enteritidis* without cooking the eggs. The entire process takes about 1 h.

Many other pasteurization methods designed separately for the pasteurization of egg white, yolk, and whole egg products have been adopted. For egg whites the heat treatment method described above is recommended at a normal pH of about 9.0. However, from the standpoint of mechanical problems in equipment design and operating constraints, change in the behavioral viscosity may be another factor for logical consideration of pasteurization controls. In contrast, the lactic acid–aluminum sulfate process at pH 7 is considered more suitable for the stabilization of egg whites, where an aluminum sulfate solution is added prior to heat treatment. Another option is the heat plus hydrogen peroxide process, in which the bacterial agent hydrogen peroxide is allowed to react for 2 min at an elevated temperature that results in elimination of the problem with excessive foam formation. The residual peroxides can be removed via cooling, and successful applications have already been adopted in commercial egg processing plants. In addition to these, the heat plus vacuum process, which permits air to be removed from albumen at lower temperatures, with similar microbiological results. It also eliminates the problem of coagulation of eggs on the plates during the heating process. Conversely, the pasteurization method designed for yolk products required more effort, because *S. enteritidis* is heat resistant in yolk compared to egg whites, believed to be due to the lower pH and higher solid content of egg yolks. Therefore, the conventional plate-type pasteurization units cannot be employed.

The heat effects on egg products are also a crucial problem that influences egg product performance. Changes in product physical properties upon heat treatment involve primarily the viscosity and microbiological properties. Holding times and flow characteristics can be influenced by viscosity. A heavy viscous product may exhibit laminar flow. Considering these factors, it is wise to determine the reasonable holding times in the individual plant pasteurization equipment.

It is also well known that denaturation and performance impairment are functions of time and temperature. The functional properties, such as denaturization of proteins, are little affected and perform satisfactorily. The electrophoretic and chromatographic changes in egg yolk proteins due to heat were studied by Dixon and Cotterill [22]. Matsuda et al. [23] reported a similar fractional or stepwise study on the aggregation of egg white proteins. They observed that ovoinhibitor (ovotransferrin) was much more unstable than ovomucoid under heat treatment, and the time dependency of heat-induced aggregation of flavoprotein was greater than those of other proteins in egg white. The gelatin properties of egg proteins were also studied [24], with the finding that lysozyme produced the strongest gel followed by globulins, ovalbumin, and conalbumin.

There has been considerable debate as to the destruction of bacteria in egg products, since the compositional differences in egg products account for the wide range of pasteurization conditions recommended. The heat resistance of bacteria also depends on the pH of the product. For example, salmonellae have greater heat resistance in yolk (pH 6.0) than in egg whites (pH 9.1). Also, the type of acid (acetic, lactic, or hydrochloric) or additive is another important factor in heat resistance. Thus, egg yolks, egg whites, and whole eggs have different effects on the heat resistance of salmonellae bacteria, apparently due to the differences in pH and in composition. For example, heat destruction of salmonellae in egg yolk (pH 6.2) is half that of whole egg. This also indicates that less severe pasteurization conditions are required for egg white than for whole egg. Also, yolk must be pasteurized at higher temperature than either egg white or whole

TABLE 3 Thermal Resistance (*D*-Values) for *Salmonella* in Plain Yolk

Storage (25°C) Time (days)	Pasteurization Temperature (°C)	*D*-Value (Thermal Resistance)
0	57	3.50
4	57	2.50
0	60	0.40
0	61.1	0.57
0	63.3	0.20

TABLE 4 Thermal Resistance (*D*-Values) for *Salmonella* and *Oranienburg* in Egg Yolk After Addition of Salt or Sugar

Microorganism	Storage (25°C) Time (days)	Pasteurization Temperature (°C)	*D*-Value (Minimum) (Thermal Resistance)
Salmonella	0	57	7.00
	4 (at 6°C)	57	4.50
	0	61.1	0.74
	0	63.3	0.72
	0	64.4	0.20
Oranienburg	0	61	6.00
	4	61	11.50
	0	63.3	11.50
	0	64.4	6.44
	0	65.5	3.85
	0	66.7	2.07

egg. Cotterill and Glauert, Garibaldi et al., and Palumbo et al. [25–27] studied the thermal resistance of microorganisms and calculated *D*-values at different pasteurization temperatures using the equation

$$D = \frac{t}{\log N_0 - \log N} \tag{1}$$

Table 3 lists the *D*-values of *Salmonella* for plain yolk under identical storage conditions.

Usually, sugar or salt additives enhance the thermal resistance of microorganisms (*Salmonella*) in egg products and influence the flow characteristics of egg yolk products [25]. Acetic acid is recommended for reducing thermal resistance in yolk products. However, studies on whole egg products confirmed that heat treatment with or without homogenization had no significant effect on the viscosity of the whole egg. Table 4 presents the *D*-values for *Salmonella* or *Oranienburg* in egg yolks under various pasteurization conditions.

Among the variety of types of pasteurization equipment used in the egg-processing industry, batch pasteurizers are commonly used, even without USDA approval. A conventional batch pasteurizer system contains a processing vessel, a stirrer, a heating source, a temperature sensor, and a supply of refrigerated cooling water. Also, high-temperature short-time pasteurization equipment is in general use. The use of a tubular laboratory-scale pasteurizer with controlled performance has also been reported.

Apart from these, several nonthermal technologies have been considered. The main emphasis is given to electroheating [28], wherein high-frequency radio waves (200 kHz) in combination with heat (60°C for 3.5 min) were used to manufacture a whole egg product with an extended shelf life. The USDA approved this process for commercial use. In another process the use of high hydrostatic pressure (450 MPa at 20°C for 15 min) was reported to obtain a 5-log reduction of *Listeria innocua*, wherein microbial inactivation was found increased at advancing pressure [29]. However, in high-pressure processes, the temperature seems to be the least important factor. Irradiation is another choice for the production of liquid egg products. Wong et al. [30] employed an electron-beam linear accelerator to pasteurize egg white at 2.5 to 3.3 kGy and noticed no adverse affect on angel cake volume. Huang et al. [31] used electron-beam linear irradiation at 2.5-kGy dosages to egg yolk, followed by storage for up to 90 days at 4°C. Identical inferences were reported. Irradiated egg yolk was found with a higher emulsifying capacity. Characteristics that were much superior to those of unpasteurized eggs were summarized with enhanced plant safety.

The effect on whole egg pasteurization of high-intensity pulsed electric fields was studied by Qin et al. [32]. Liquid whole egg was mixed with 0.15% citric acid prior to subjection to high-intensity pulses. Further, no adverse effects on physical and chemical properties were found, and an enhanced shelf life of nearly 4 weeks was reported. Later, Ma et al. [33] summarized the effects of high-intensity pulsed electric fields as a technological advance in the nonthermal pasteurization of eggs.

ALTERNATIVE PROCESSING TECHNIQUES TO INDUCE FUNCTIONALITY

Several alternative methods are being proposed to improve the functionality of egg components. Kato et al. [34] presented a physical approach to improving the gelling and surface functional properties of egg white proteins. No chemicals or preservatives were used during functionalization. In their first attempt they used controlled dry heating of spray-dried egg white proteins, by which partial denaturation was considered to take place without any loss of solubility. Also, the conjugation of proteins with polysaccharides in the dry state was proposed, resulting in covalent cross-linking between amino groups in proteins and reduction of the end carbonyl moieties in polysaccharides. These conjugates showed better emulsifying properties than those of commercial emulsifiers. Further, the major function that regulates the functional properties of eggs is *coagulation*, a term used to describe changes from the fluid to the solid or semisolid state. The adequacy of performing coagulation is necessary in determining the value of eggs in food products, which is dependent on protein coagulation. The eggs are also capable of foaming with greater capacities. This can be described as the unfolding of the protein molecules, wherein the polypeptides chain exists with the long axes parallel to the surface [35]. Such changes in the molecular configuration results in the loss of solubility or precipitation of some of the albumen, which collects at the air–liquid interface. Therefore, the adsorption film is essential to the stability of the foam. Also essential for the formation of foam is a liquid with a high tensile strength or elasticity [36]. Several factors that influence the functional properties of egg foams are the method of beating, the blending of components, homogenizing, as well as the temperature of foaming. In addition, the influence on the egg foams of added ingredients such as acid or base, water, salt, sugar and egg yolk, oil, surfactant, esters, chemical modifiers, emulsifiers, and stabilizers on has been studied extensively. Germs' group [37, 38] has developed a method to improve the solubility and foaming properties of egg yolk protein and its separation from yolk lipid by lipid–lipid extraction. They modified the egg yolk protein by partial proteolysis using the enzyme specified, endopeptidase, of serine type. Finally, the protease-treated egg yolk protein was isolated by delipidation. Further, the emulsion properties that particularly influence the functional characteristics of egg components are found to have a major impact on functional food products such as cakes, sponge, candy, custards, mayonnaise, and salad dressings.

Recently, researchers in the egg-processing industry have placed considerable emphasis on altering the composition of egg components through new processing techniques. Due to primary concern about adverse effects on serum cholesterol level, eggs are often not considered a safe functional food. This negative image of eggs has resulted in a significant decline in per capita egg consumption over a number of years. One distinctive characteristic of the functional food drift is a paradigm shift away from elimination of negative ingredients (fat, salt, and cholesterol) to enhancement with positive ingredients (calcium, vitamins, and antioxidants). Although this has long been recognized, recent extensive studies have investigated the influence of hens' diet on the composition of the eggs. Hargis and Van Elswyk [39] have reviewed much of the past work on the development of functional eggs accomplished by the nutritional manipulation of the diets of laying hens. Van Elswyk [40] has reported that feeding hens with flaxseed and fish oil promotes the incorporation of omega-3 fatty acid into the egg yolk, which is very important in cardiovascular disease (CVD) prevention. Currently, several companies are marketing eggs fortified with omega-3 and omega-6 fatty acids [41] and vitamin E, since it is also well accepted as nature's most effective lipid-soluble chain-breaking antioxidant [42] and has been shown to inhibit low-density lipoprotein (LDL) cholesterol oxidation in vitro [43].

Several other goals may be of particular interest, including altering the lipid composition, separating certain desirable egg fractions, such as avidin and lysozyme, and concentrating the solids through ultrafiltration or reverse osmosis techniques. Removal of cholesterol and lipids through organic solvents

[44] was also suggested; however, the process has several disadvantages for use in the food industry because they may denature protein and adversely affect the functional properties of the resulting products. Further, the use of supercritical carbon dioxide extraction [45] or the use of absorbents to remove cholesterol was also recommended [46]. β-Cyclodextrin was employed to remove cholesterol, although no fat could be removed through this process. On the other hand, supercritical carbon dioxide extraction could remove both cholesterol and other lipids as well [47]. Also, there has been some interest in degrading or modifying cholesterol through utilization of certain microorganism or enzymes [48]. Therefore, functional egg consumption is a leading trend in the food industry today, and this trend will undoubtedly continue as consumers seek new opportunities to enhance their health through diet.

PROCESSED EGGS FOR CUSTOM-DESIGNED PRODUCTS

End users were first to recognize the advantages of processed eggs and moved functional egg products into the marketplace as value-added convenience products. Institutional users were quick to use liquid, dehydrated, and frozen eggs. Recently, ready-to-cook and ready-to-eat egg preparations have become very common in the marketplace. This shows readily how attitudes toward egg products are changing. In 1960, pioneering studies of a stagnant egg market by Cornell University researchers indicated the need for new and convenient egg products that should have reduced preparation time, be available at a competitive price, and be easy to mass-produce and handle for distribution. Some of the initial egg products marketed were frozen omelettes, apple juice drink, instant French toast, liquid eggs, egg pies, hard-cooked egg roll, egg crust pizza, and Easter eggs. Later, the quality and processing of traditional egg products came into existence, with famous Chinese pidan eggs (also known as alkaline-gelled eggs), salted eggs, flavored eggs, sour eggs, and egg tofu among the list. Further, new developments in egg processing, with an emphasis on nutritional values, are recommended not only to solve the problem of overproduction of raw eggs but also to increase agricultural profit. For example, in traditional pidan, the albumen is a translucent gel of brownish color. However, for the curiosity and multiple tastes of consumers, the transparent pidan was developed [49]. In other series of processed functional eggs, an egg white drink [50], egg white jelly [51], and egg white snacks, which are an egg white powder, potato starch, sugar, and leavening agent blend in water cooked in an electric or microwave oven [52], were commercialized. Texturized egg products using extrusion technology are also recommended [53]. Seasoned shell eggs and alkaline fermented egg production technology are also available. Apart from these, egg products from processed liquid eggs currently available in the market are available in the confectionery and bakery industry.

Recently, a new type of processed functional eggs has been in marketed in Asian Pacific countries, wherein eggs fortified with microcomponents such as minerals and vitamins are top consumer products. These nutritional eggs contain at least 20% more vitamins D, E, and B_{12} than are in ordinary eggs. However, these products have a poor shelf life (three days). Vita-iodine eggs containing nearly 7 mg % vitamin E and 1.3 mg % iodine are a comparatively stable product with a somewhat longer shelf life. The other main category of functional eggs is omega eggs, fortified with omega-3 fatty acids and claimed to be a healthy egg product. Omega eggs can serve as a functional food alternative to fish and supplements for the consumption of docosahexaenoic acid (DHA). This can easily be inserted into egg yolk through manipulation of the laying hens' diet. Further, Sim and Qi [54] have supported the designer egg concept, where the author focuses on perfecting the egg through a diet enriched with omega-3 long-chain polyunsaturated fatty acid (PUFA) and cholesterol stability and its nutritional significance. To avoid the misconception that all omega-3–enriched functional eggs have similar properties, Farrell [55] has described further insights on enrichment and revealed that all omega-3–enriched eggs are not the same and that there are considerable differences in the balance of omega-3 in enriched eggs produced by various methods. The author reviewed methods of enrichment of hen eggs. Flaxseed [about 42% oil, nearly 53% of which is α-linolenic acid (LNA)] is commonly used as laying hen diets for enriched eggs [56]. Brettschneider et al. [57] reported the use of rapeseed or canola seed, which contains about 42% oil, nearly 11% of which is LNA. Collins et al. [58] have reported on the enrichment of eggs from laying hens on 68% pearl millet grains, which contain 5% oil, of which 4% is LNA. Fish oils are the other alternative to enriched eggs from laying hens [59, 60]. Enrichment with marine algae [61], vegetable oils [62], and oil mixtures have also been reported [41]. The omega-3–enriched functional eggs are currently being produced by several dozen companies around the world and a vast literature exists on the health benefits of these processed functional eggs.

Egg flakes and dried egg sheets are also becoming popular because the egg flakes are being used mainly in instant noodles, wherein the principal component is processed egg yolk. Egg flakes can be produced by two drying procedures: hot-air-blast drying at 55 to 80°C or extrusion technology. Taiyo Kagaku Co. in Japan is the largest producer of egg flakes, for multiple commercial uses. The microwave drying process is being used to produce egg flakes. Figure 1 shows an automated microwave drying processing line. Egg flakes can also be used in frozen fish paste. The second product is dried and strip-type egg sheets for use in cooked noodles and several other dishes. This product is quite similar to egg flakes, but the production procedure is somewhat different, as it is manufactured using a combination of a blender and a spray dryer (Figure 2). For the commercial manufacture of egg sheets, extrusion technology can be employed, or a stuffer with a specially designed nozzle can be used.

Several other functional egg products are currently in use in diverse industries, ranging from cosmetics to biotechnology to the pharmaceutical industry. In the cosmetic field they can be used in egg shampoo, lipstick, moisturizing creams, and makeup lotions. In biotechnology and pharmaology, egg albumen is used as

FIGURE 1 Automated microwave drying processing line to produce nutritional egg flakes. (Courtesy of Taiyo Kagaku.)

(a) (b)

FIGURE 2 Dried and strip-type nutritional egg sheets are being manufactured using a combination of (a) a blender and (b) a spray dryer for cooked noodles. (Courtesy of Taiyo Kagaku.)

a carrier material for immobilized bacteria and yeast in the production of new drugs. Also, the isolation of bioactive components of eggs, such as lysozymes, avidin, and lecithin, has already been achieved on a large industrial scale [63, 64]. Separation of sialic acid from chalaza and yolk is also achieved commercially, and is used to regulate a variety of important biological phenomena [65]. The use of egg yolk antibodies [66] and egg yolk immunoglobulin (IgY) for the prevention of disease is also well documented [67, 68].

CONCLUSIONS

Many manufacturing processes are based on the huge amount of scientific data available; however, developing new commercial functional egg products has never been easy, and the competitive environment of the future will make it even more challenging. Modification of egg components for use as functional foods and in value-added health-oriented dietary products is a relatively new way to encourage people to consume eggs. Marketing experts in the field predict that future consumers will demand that more attention be directed to nutrition, especially for nonadditive egg products, and convincing consumers continues to be a complex task. Nevertheless, steady progress is being made and the industry is focusing on new initiatives, such as the development of novel egg products enriched with nutrients such as omega-3 fatty acids, egg yolk antibodies, and egg yolk immunoglobulin for the prevention of disease. Through emerging research on secondary processing of eggs and egg components, it is imperative to educate consumers on changing views in the area of diet and disease prevention. More research is still needed regarding findings on the absence of health risks from dietary cholesterol and health benefits from egg-derived functional foods manufactured via secondary processing.

REFERENCES

1. Cotterill OJ, Geiger GS. 1977. Egg product yield trend from shell eggs. Poult Sci 56:1027–1031.
2. Brooks J, Taylor DJ. 1955. *Eggs and Egg Products*. Special Report 60. London: Great Britain Department of Science and Industrial Research, Food Investigations Board.
3. Feeney RE. 1964. Egg proteins. In: Schultz HW, Anglemeir AF, eds., *Symposium on Foods: Proteins and Their Reactions*. Westport, CT: AVI Publishing.
4. Vedehra DV, Nath KR. 1973. Eggs as a source of protein. In: Furia TE, ed., *Critical Reviews in Food Technology*, vol. 4. Boca Raton, FL: CRC Press.
5. Powrie WD, Nakai S. 1985. Characteristics of edible fluids of animal origin: eggs. In: Fennema O, ed., *Food Chemistry*, 2nd ed. New York: Marcel Dekker.
6. Cotterill OJ, Glauert JL. 1979. Nutrient values for shell, liquid/frozen and dehydrated eggs derived by linear regression analysis and conversion factors. Poult Sci 58:131–134.
7. Stadelman WJ, Pratt DE. 1989. Factors influencing composition of the hen's egg. World's Poult Sci J 45:247–266.

8. Bergquist DH. 1978. Instant dissolving egg white. U.S. patent 4,115,592.
9. Bergquist DH, Lorimor GD, Wildy TE. 1992. Tower spray drier with hot and cool air supply. U.S. patent 5,096,537.
10. Warren MW, Ball HR, Jr, Froning GW, Davis DR. 1991. Lipid consumption of hexane and supercritical carbon dioxide reduced cholesterol dried-egg yolk. Poult Sci 70:1991–1997.
11. Froning GW, Clegg J, Long C. 1981. Factors affecting puffing and sensory characteristics of extruded egg products. Poult Sci 60:2091–2097.
12. Chapin RB. 1951. Some factors affecting the emulsifying properties of hen's egg. Ph.D. dissertation, Iowa State College, Ames, IA.
13. Bate-Smith EC, Hawthorne JR. 1945. Dried-egg: the nature of the reaction leading to loss of solubility of dried-egg protein. J Soc Chem Ind 64:297–302.
14. Everson GJ, Souder HJ. 1957. Composition and nutritive importance of eggs. J Am Diet Assoc 33:1244–1254.
15. Lopez A, Fellers CR, Powrie WD. 1954. Enzyme inhibition of gelation on frozen egg yolk. J Milk Food Technol 18(3):77–80.
16. Feeney RE, MacDonnell JR, Fraenkel-Conrat H. 1954. Effects of crotoxin on egg yolk and yolk constituents. Arch Biochem Biophys 48:130–140.
17. Mayer DD, Woodburn M. 1965. Gelation of frozen-defrosted egg yolk as affected by selected additives: viscosity and electrophoretic findings. Poult Sci 44: 437–446.
18. Wakamatu T, Sato Y, Saito Y. 1983. On sodium chloride action in the gelation process of low density lipoprotein (LDL) from hen egg yolk. J Food Sci 48:507–516.
19. Miller C, Winter AR. 1950. The functional properties and bacterial content of pasteurized and frozen whole egg. Poult Sci 29:88–97.
20. Shafi R, Cotterill OJ, Nichols ML. 1970. Microbial flora of commercially pasteurized egg products. Poult Sci 49:578–585.
21. Vadehara DV, Steele FR Jr, Baker RC. 1969. Shelf life and culinary properties of thawed frozen pasteurized whole egg. J Milk Food Technol 32:362–364.
22. Dixon DK, Cotterill OJ. 1981. Electrophoretic and chromatographic changes in egg-yolk proteins due to heat. J Food Sci 46:981.
23. Matsuda T, Watanabe K, Sato Y. 1981. Heat induced aggregation of egg white proteins as studied by vertical flat-sheet polyacrylamide gel electrophoresis. J Food Sci 46:1829–1834.
24. Johnson TM, Zabik ME. 1981. Gelation properties of albumen proteins, singly and in combination. Poult Sci 60:2071–2083.
25. Cotterill OJ, Glauert J. 1971. Thermal resistance of salmonellae in egg yolk containing 10% sugar or salt after storage at various temperatures. Poult Sci 50:109–115.
26. Garibaldi JA, Straka RP, Ijichi K. 1969. Heat resistance of *Salmonella* in various egg products. Appl Microbiol 17:491–496.
27. Palumbo MS, Beers SM, Bhaudri S, Palumbo SA. 1995. Thermal resistance of *Salmonella* spp. and *Listeria monocytogenes* in liquid egg yolk and egg yolk products. J Food Prot 58:960–966.
28. Reznik D, Knipper A. 1994. Method of electroheating liquid egg and product thereof. U.S. patent 5,290,538.
29. Ponce E, Pla R, Mor-Mur M, Gervilla R, Guamis B. 1998. Inactivation of *Listeria innocua* inoculated in whole egg by high hydrostatic pressure. J Food Prot 61:119–122.

30. Wong YC, Herald TJ, Hachimeister KA. 1996. Comparison between irradiated and thermally pasteurized liquid egg white on functional, physical and microbiological properties. Poult Sci 75:803–806.
31. Huang S, Herald TJ, Mueller DD. 1997. Effect of electron beam irradiation on physical, physicochemical and functional properties of liquid egg yolk during frozen storage. Poult Sci 76:1607–1615.
32. Qin B, Pothakamury UR, Vega H, Martin O, Barbosa-Cánovas GV, Swanson BG. 1995. Food pasteurization using high intensity pulsed electric fields. Food Technol 49(12):55–60.
33. Ma L, Chang FJ, Gongora-Nieto MM, Barbosa-Cánovas GV, Swanson BG. 2000. Food pasteurization using high intensity pulsed electric fields: promising new technology for non-thermal pasteurization for eggs. In: Sim JS, Nakai S, Guenter W, eds., *Egg Nutrition and Biotechnology*. London: CABI Publishing.
34. Kato A, Ibrahim HR, Watanabe H, Honma K, Kobayashi K. 1989. New approach to improve the gelling and surface functional properties of dried egg white by heating in dry state. J Agric Food Chem 37:433–437.
35. Cunningham FE. 1963. Factors affecting the insolublization of egg white proteins. Ph.D. dissertation, University of Missouri, Columbia, MO.
36. Poole S. 1989. Review: the foam enhancing properties of basic biopolymers. Int J Food Sci Technol 24:121–137.
37. Germs AC. 1994. Improvement of solubility and foaming properties of egg yolk protein and its separation from yolk lipid by liquid–liquid extraction. In: Sim JS, Nakai S, eds., *Egg Uses and Processing Technologies*. London: CABI Publishing, pp. 329–339.
38. Germs AC, 1991. Isolation of proteins from egg yolk. In: Oosterwoud A, de Vries AW, eds., *Quality of Poultry Products*, vol. II, *Eggs and Egg Products*. Proceedings of the 4th European Symposium on Quality of Eggs, Beekbergen, The Netherlands, pp. 243–248.
39. Hargis PS, Van Elswyk ME. 1991. Modifying yolk fatty acid composition to improve health quality of shell eggs. In: Haberstroh C, Morris CE, eds., *Fat and Cholesterol Reduced Foods Technologies and Strategies*. Advances in Applied Biotechnology, vol. 12. Woodland, TX: Portfolio Publishing, pp. 249–260.
40. Van Elswyk ME. 1997. Comparison of *n*-3 fatty acid sources in laying hen rations for improvement of whole egg nutritional quality: a review. Br J Nutr 78(1): s61–s69.
41. Farrell DJ. 1998. Enrichment of hen eggs with *n*-3 long chain fatty acids and evalution of enriched eggs in human. Am J Clin Nutr 68:538–544.
42. Packer L. 1991. Protective role of vitamin E in biological system. Am J Clin Nutr 53:1050–1055.
43. Esterbauer H, Striegl G, Puhl H, Oberreither S, Rothender M, El-Saadini M, Jurgens G. 1989. The role of vitamin E and carrotenoids in preventing oxidation of low-density lipoproteins. Ann NY Acad Sci 570:254–267.
44. Larsen JE, Froning GW. 1981. Extraction and processing of lipid components from egg yolk solids. Poult Sci 60:1124–1125.
45. Froning GW, Wehling RL, Cuppett SL, Pierce MM, Niemann L, Siekman SK. 1990. Extraction of cholesterol and other lipids from dried egg yolk using supercritical carbon dioxide. J Food Sci 55:95–98.
46. Haggin J. 1992. Cyclodextrin research focuses on variety of applications. Chem Eng News, May 18, pp. 25–26.

47. Oakenfull DG, Pearce RJ, Sidhu GS. 1991. Low cholesterol dairy products. Aust J Dairy Technol 11:110–112.
48. Dehal SS, Frier TA, Young JW, Hartman PA, Beitz DC. 1991. A novel method to decrease the cholesterol of foods. In: Haberstroh C, Morris CE, eds., *Fat and Cholesterol Reduced Foods Technologies and Strategies*. Advances in Applied Biotechnology, vol. 12. Woodland, TX: Portfolio Publishing, pp. 203–220.
49. Su HP, Lin CW. 1993. A new process for preparing transparent alkalized duck egg and its quality. J Sci Food Agric 61:117–120.
50. Chang HS, Lin SM. 1986. Studies on the manufacturing of flavored chicken eggs. J Chin Soc Anim Sci 15:71–82.
51. Kitabatake N, Shimizu A, Doi E. 1988. Preparation of transparent egg white gel with salt by two-step heating method. J Food Sci 53:735–738.
52. Huang YC, Yang SC. 1993. Studies on the effect of processing conditions on the qualities of egg white snack. Food Sci (Taiwan) 20:291–300.
53. Hsu SY. 1993. Processing effect on the qualities of texturized egg products made of fresh egg white. Food Sci (Taiwan) 20:161–167.
54. Sim JS, Qi GH. 1995. Designing poultry products using flaxseed. In: Thompson LU, Cunnane S, eds., *Flaxseed in Human Nutrition*. Champaign, IL: American Oil Chemists' Society Press, pp. 315–333.
55. Farrell DJ. 2000. Not all ω-3-enriched eggs are the same. In: Sim JS, Nakai S, Guenter W, eds., *Egg Nutrition and Biotechnology*. London: CABI Publishing, pp. 151–161.
56. Van Elswyk ME 1997. Nutritional and physiological effects of flax seed in diets for laying fowl. World's Poult Sci J 53:253–264.
57. Brettschneider S, Danicke S, Jeroch H. 1995. The influence of graded levels of rapeseed in lying hen diets on egg quality and special consideration of hydrothermal treatment of rapeseed. In: Briz RC, ed., *Egg and Egg Products Quality*. Proceedings of the 6th European Symposium on the Quality of the Egg and Egg Products, Zaragoza, Spain, pp. 227–232.
58. Collins VP, Cantor AH, Pescatore AJ, Straw ML, Ford MJ. 1997. Pearl millet in layer diets enhances egg yolk *n*-3 fatty acids. Poult Sci 66:326–330.
59. Hargis PS, Van Elswyk ME, Hargis BM. 1991. Diet modification of yolk lipid with menhaden oil. Poult Sci 70:874–883.
60. Oh S, Ryue J, Hsieh CH, Bell DE. 1991. Egg enriched in ω-3 fatty acid and alteration in lipid concentrations in plasma and lipoproteins and in blood pressure. Am J Clin Nutr 79:407–412.
61. Yongmanitchai W, Ward OP. 1991. Screening of algae for potential alternative sources of eicosapentaenoic acid. Phytochemistry 30:2963–2967.
62. Farrell DJ. 1994. The fortification of hen's egg with ω-3 long chain fatty acids and their effects in human. In: Sim JS, Nakai S, eds., *Egg Uses and Processing Technologies: New Developments*. Wallingford, UK: CABI Publishing, pp. 386–401.
63. Durance TD, Nakai S. 1988. Simultaneous isolation of avidin and lysozyme from egg albumen. J Food Sci 53:1096–1102.
64. Juneja LR, Sugino H, Fujiki M, Kim K, Yamamoto T. 1994. Preparation of pure phospholipids from egg yolk. In: Sim JS, Nakai S, eds., *Egg Uses and Processing Technologies: New Developments*. Wallingford, UK: CABI Publishing, pp. 139–149.
65. Juneja LR, Kaketsu M, Kim K, Yamamoto T, Itoh T. 1991. Large-scale preparation of sialic acid from chalaza and egg yolk membrane. Carbohydr Res 214: 179–189.

66. Gassmann M, Thommes P, Weiser T, Hubscher U. 1990. Efficient production of chicken egg yolk antibodies against a conserved mammalian protein. FASEB J 4:2528–2532.
67. Akita EM, Nakai S. 1992. Immunoglobulins from egg yolk: isolation and purification. J Food Sci 57:629–634.
68. Hatta H, Kim M, Yamamoto T. 1990. A novel isolation method for hen egg yolk antibody IgY. Agric Biol Chem 54:2531–2535.

20

DIETARY PRODUCTS FOR SPECIAL POPULATIONS

JORGE SORIANO-SANTOS

Departamento de Biotecnología, Universidad Autónoma Metropolitana–Unidad Iztapalapa, México D.F., México

INTRODUCTION

Consumers are becoming more aware of the nutritional value of the foods they eat. Poultry products are natural choices to meet the emerging desire to keep in shape because of their high nutrient content and relatively low caloric value. Most nutrition programs can be improved by the inclusion of poultry meat, since

Handbook of Poultry Science and Technology, Volume 2: Secondary Processing, Edited by Isabel Guerrero-Legarreta and Y.H. Hui

poultry products are of the highest nutritional quality and are marketed at prices that even low-income people can afford.

The production and consumption of chicken meat has increased significantly throughout the world in recent years. The largest producers are the United States, China, and Brazil. An increased percentage of chicken meat produced has also been used as raw material for the processing of chicken-based meat products. These products are becoming more popular, due to their sensory characteristics, ease of preparation, and improved shelf life compared to fresh meat (Silva, 1995). Science and technology have made possible a great expansion of the poultry industry. Broiler production satisfies a large portion of the population in various countries with enriched sources of nutrients and is recognized as one of the best ways of supplying good-quality animal protein for consumption, thereby fulfiling people's nutritional requirements.

Processing of poultry meat can be divided into primary processing (stunning, scalding, plucking, chilling, postmortem aging, and cold storage) and further processing (heating, storage, freeze-drying, irradiation, and creation of restructured or ready-to-eat products). The term *further processed* is used in the poultry industry in a manner similar to the way the term *processed meats* is used in the red meat industry. Examples of methods used in preparing further-processed poultry products are size reduction, deboning, restructuring, emulsifying, battering and breading, heating, and freezing. Many products are "ready to eat" when they leave the processing plant, in contrast to the "ready to cook" status of non-further-processed whole birds. Further processing reduces the preparation efforts of consumers—hence the term *convenience foods* frequently used for such products. Researchers have considered whether the further processing involved in preparing poultry meat products, in such additional steps as cooked emulsions and comminuted and frozen products, reduces their nutritional value. However, studies indicate that nutrient loss during primary or further processing of poultry is minimal.

In this chapter we focus on the nutritional value of poultry products designed to meet the dietary needs of various special populations. For poultry meat, the current focus is on reducing the fat content of the final product to get a "leaner" product, such as light, low-fat, reduced-cholesterol, enriched omega-3 fatty acid products for infants, children, the elderly, and healthy or diseased populations. Some aspects of processing that may further enhance the nutritional value of special dietary poultry products are the use of blood, giblets, and bone residue protein; hot-deboning; removal of the abdominal fat pad in ready-to-cook carcasses; and reduction of fat and sodium in further-processed products.

NUTRIENT REQUIREMENTS AND DIET CHARACTERISTICS FOR INFANTS AND YOUNG CHILDREN

For infants and small children the amount of nutrients required from complementary foods depends on the quantity provided by human milk and varies markedly by nutrient, ranging from nearly 100% iron to 0% for vitamin C. Calculation of the nutrients required from complementary foods is based on the recommended intake for each nutrient minus the amount of the nutrient consumed daily from human milk (WHO, 1998). Although conceptually straightforward, this calculation is complicated by the fact that no single set of nutritional requirements for infants and young children has been agreed upon. Three are currently in use: the recommended nutrient intakes in the 1998 WHO report (WHO, 1998); the dietary reference intakes (DRIs) published by the US Institute of Medicine (1997, 1998, 2000), and recommended nutrient intake requirements in the WHO/FAO preliminary report (Joint FAO/WHO Expert Consultation, 2002). WHO identified two groups of micronutrients: those that do not vary with maternal nutritional status and those that do (WHO, 1998). The B vitamins (except folate) and vitamin A, iodine, and selenium are in the latter category.

For the young infant in particular, the monotonous nature of the diet results in a somewhat greater risk for nutrient imbalance. For example, in the exclusively breast-fed infant, the balance of nutrients in human milk is ideally suited to meet the infant's nutritional needs. If, however, the intake of any of the nutrients is significantly altered, there may be effects on other nutrients that cannot be compensated for by other dietary constituents. The young formula-fed infant is also dependent on a single "food," resulting in a critical dependence on the provision of adequate concentrations of all essential nutrients in forms that are optimally bioavailable. The introduction of complementary foods for the older infant begins the process of dietary diversification, but infants and young children commonly continue consuming diets more limited in variety than those of adults. During adolescence, diets often become less than ideal from a micronutrient standpoint, whereas micronutrient needs are increasing in conjunction with increased growth rates, bone deposition, menarche, and so on. Teenagers may choose fad diets or diets with limited variability and may rely on high-energy, low-nutrient-dense food choices. Such dietary practices obviously have implications for micronutrient bioavailability. The nutrition status will influence nutrient bioavailability in children as it does in adults.

The recommended duration of exclusive breast feeding, defined as human milk being the only source of infant food and liquid, is 6 months (WHO, 2001a). After this age, complementary foods should be introduced and breast feeding continued up to the child's second birthday or beyond. The period of complementary feeding is defined as the period when foods other than human milk are provided to infants and young children who are still breast feeding (WHO, 1998). Complementary foods are defined as those non-human-milk food-based sources of nutrients that are offered during this period. Fortified complementary foods are defined as centrally produced and specially formulated complementary foods to which one or more micronutrients have been added to increase their nutrient content.

In the young infant, iron and zinc are highly bioavailable from human milk. By about 6 months of age, other dietary sources are needed to maintain continued normal status. For the older infant and toddler, iron and zinc are also important for normal growth and development. During adolescence, adequate calcium intake is critical to normal bone mineralization. In girls, peak calcium absorption and calcium deposition in bones occur at or near menarche (American Academy of Pediatrics, 1998).

Historically, both the WHO and the United Nations Children's Fund (UNICEF) have emphasized the use of local foods formulated at home rather than centrally produced fortified foods for complementary feeding (WHO, 1979). However, in recognition of recent scientific information showing the potential limitations of purely home-based and local approaches to satisfy the requirements for some nutrients, the WHO/UNICEF Global Strategy for Infant and Young Child Feeding (WHO, 2002) states that "industrially processed complementary foods also provide an option for some mothers who have the means to buy them and the knowledge and facilities to prepare and feed them safely" and "food fortification and universal or targeted nutrient supplementation may also be required to ensure that older infants and young children receive adequate amounts of micronutrients."

Bioavailability of Dietary Nutrients

Bioavailability can be defined as including the absorption and utilization of a nutrient (Fairweather-Tait, 1997). The concept of bioavailability as applied to nutrients is critically important for understanding nutrient metabolism, homeostasis, and ultimately, nutrient requirements. Numerous factors affect bioavailability, such as the chemical form of the nutrient, the food or supplement matrix in which the nutrient is consumed, and other foods in the diet. Several host factors also influence bioavailability, including age, gender, physiologic state, and coexisting pathologic conditions.

Several factors that may affect bioavailability apply primarily to infants, children, and adolescents: increased nutrient requirements to support growth and development, maturation of the gastrointestinal tract and the digestive and absorptive process, a monotonous diet, age, and rates of growth during certain critical periods. It should be noted that other recognized factors that affect bioavailability of nutrients, including such inhibitors of nutrient absorption as phytic and oxalic acids, affect bioavailability in adults but not in children. Considerations for nutrient bioavailability for pediatric populations include maturation of the gastrointestinal tract, growth, character of the diet, and nutritional status.

NUTRIENT REQUIREMENTS FOR THE MATURE ADULT

Governments have a real interest in keeping their citizens healthier, and many countries have developed dietary guidelines. For example, the purpose of the *Dietary Guidelines for Americans* is to provide science-based advice to promote health and to reduce the risk for chronic diseases through diet and physical activity, as well as the Food Guide Pyramid (MyPyramide), which represents the variety, moderation, and proportionality needed for a healthful diet. The system provides many options to help Americans make healthy food choices and be active everyday. MyPyramide is based on both the *Dietary Guidelines for Americans* and the dietary reference intakes (reference values for nutrient intakes to be used in assessing and planning diets for healthy people). The *Dietary Guidelines for Americans* are important for two major reasons: they serve as a tool for educators, and they serve as a guide in setting food and nutrition policies. The inclusion of food safety in the 2000 edition is an important step toward ensuring their continued relevance for health promotion and disease prevention (USDA–USDHHS, 2000). It also better reflects current knowledge about diet and long-term health. Guidance of consumers toward nutritionally adequate diets must include research-based knowledge on foods for the table, to assure retention of both nutritional and eating qualities and to avoid foodborne illness (U.S. Congress, 1977).

The number of people aged 60 years old or older is escalating rapidly worldwide. The United Nations Population Division estimated that this age group represented about 10% of the world's population, or about 600 million people, in 1999. They project that by the year 2050, this proportion will increase to 20% and will include more than 2 billion people. These changes will be most dramatic in the less developed countries, where the population age structure will change rapidly from one that is predominantly young, with few elderly, to one with more balanced numbers across age groups (WHO, 1999).

Nutrition status has a major impact on disease and disability and offers great promise for minimizing this oncoming burden. However, the current trend in developing countries is toward higher-fat, more refined diets that contribute to increased risk of chronic disease, and the prevalence of chronic disease is already increasing rapidly. At the same time, social and demographic changes are placing the elderly at even greater risk of food insecurity and malnutrition. Total energy intake declines with age, but requirements for many nutrients go up to maintain organ systems with declining functionality (Martin et al., 1996). It is therefore more difficult for the elderly to meet their nutrient requirements than for younger adults, and the selection of nutrient-dense foods becomes of even greater importance.

The leading cause of death among older people worldwide is vascular disease and associated chronic conditions (WHO, 2001b). There is great potential for prevention of these diseases through healthy lifestyles that include physical activity, nutritious diets, and avoidance of smoking or substance abuse. Unfortunately, along with a dramatic change in age structure, there is evidence of a characteristic sequence of changes in dietary behavior and physical activity patterns that lead to increased risk of chronic disease. Called the *nutrition transition* by Popkjn (1994), this process appears to be occurring rapidly and predictably in countries throughout the world. Basically, with a change from traditional, rural communities to more population-dense, urban environments, there is a change in diet from one high in fiber and low in fat to one rich in animal fats, sugars, and refined products that are low in fiber. Although overall nutrient intake adequacy improves with an increasing variety of foods, the movement toward more fats, sugars, and refined foods quickly moves beyond this more optimal state to one in which diets contribute to rapidly escalating rates of obesity and chronic disease (Tucker and Buranapin, 2001).

COMMON DISEASES RELATED TO DIET

Overweight and obesity are common in older adults and are increasing in prevalence (Mokdad et al., 2001). Adults aged 65 years who are overweight have a body mass index (BMI; in kg/m^2) between 25 and 29.9, and the obese have a BMI above 30. Both stages are associated with cardiovascular disease risk factors or co-morbidities. Excess body weight is an independent risk factor for cardiovascular diseases, which are the leading cause of death and disability in the elderly, as well as causing other risk factors, such as hypertension, dyslipidemia, and type 2 diabetes (Sowers, 2003).

Overweight and obesity are associated with increased risk for cancers at numerous sites, including breast, colon, endometrium, esophagus, gallbladder, liver, prostate, ovarian, pancreas, and kidney (Calle et al., 1999). Although it is not clear whether losing weight reduces the risk of cancer, there are physiological mechanisms which suggest that weight loss may be beneficial, since overweight or obese persons who lose weight intentionally have reduced levels of circulating glucose, insulin, bioavailable estrogens, and androgens (Calle et al., 2003). Despite some uncertainty about weight loss and cancer risk, it is nonetheless clear that people who are overweight or obese should be strongly encouraged and supported in their efforts to reduce their weight. Hence, poultry products would meet this emerging demand because of their relatively low caloric value. The current focus is on reducing the fat content of the final product to get a leaner product that may be used in preparing a weight loss diet. Cardiovascular disease, cancer, diabetes, and hypertension undermine health, shorten life expectancy, and cause enormous suffering, disability, and economic cost.

Gout, a disease found more frequently in older males, may be defined as a common type of arthritis caused by an increased concentration of uric acid in biological fluids. Uric acid is created from the breakdown of purine, a molecule found in DNA and RNA. In gout, uric acid crystals are deposited in joints, tendons, kidneys, and other tissues, where they cause considerable inflammation and damage. Gout may lead to debilitation of the joints and tendons by the uric acid deposits around them and by deposits in the kidney. The dietary treatment of gout involves a low-purine diet, and poultry is a good option.

CONSIDERATIONS FOR PRODUCING DIETARY POULTRY PRODUCTS

Older broilers are being marketed today because of industry's demand for deboned meat, which can be produced more efficiently from a larger bird. Consequently, three general sizes of broilers are marketed: small birds (including

Cornish hens), which are used for some institutional markets (e.g., precooked); midsized birds for whole and cut-up portions for consumers, fast-food chains, and deli operations; and larger birds for deboning and as roasters (Hamm and Ang, 1984). They are good sources of high-quality, easily digested proteins and also good sources of complex B vitamins and minerals. Singh and Essary (1971) reported differences in the amounts of niacin, riboflavin, and thiamine found in breast and thigh meat from 8- and 10-week-old broilers due to age, sex, and the part being examined. The older birds differ from the younger in taste and texture, and their nutrient composition is definitely lower (Hamm and Ang, 1984).

Przybyla (1985) lists a number of processed poultry products considered "ready to eat": poultry hotdogs, luncheons meats, chicken sausages, patties, chicken rolls (similar to franks), nuggets, chicken fingers, chicken planks, fried strips of chicken breasts, chicken pastrami, chicken bologna, chicken shavani kabab, chicken samosa, chicken lollypops, tandori, turkey ham, turkey bacon, turkey luncheon meat, turkey ham, and cured turkey thigh meat. All of them can be manufactured for a special diet to fulfill nutrient requirements for a specific population. Although nutrient loss during primary or further processing of poultry is minimal, the addition of blood, giblets, and bone residue protein, omega-3 fatty acids, or reduction of sodium or removal of the abdominal fat pad in ready-to-cook carcasses may further enhance the nutritional value of poultry products.

Another aspect to consider in special dietary further-processed poultry products is the bioactive amine content, which is formed by bacterial enzymatic decarboxylation of free amino acids. Amines have been considered to be useful indexes of the quality of fresh and processed meat, reflecting the quality of the raw material used and of hygienic conditions prevalent during its processing. Amines can impart putrid odors and off-flavors that can affect food acceptance (Geornaras et al., 1995). Biogenic amines constitute a potential health risk, especially when coupled with additional factors, such as monoamine oxidase inhibitor drugs, alcohol, and gastrointestinal diseases (Stratton et al., 1991). Chicken-based meat products such as mortadela, frankfurters, sausage, meatballs, hamburger, and nuggets contain bioactive amines. Nuggets contain amine profiles similar to those of fresh chicken meat. However, there is a prevalence of spermidine over spermine in most of the products, suggesting the incorporation of significant amounts of vegetable protein in the formulations (Silva and Glória, 2002). Overall, chicken-based meat products had lower biogenic amine levels than those of beef and pork. The histamine and tyramine levels in these products are low and probably unable to elicit direct adverse effects.

Reduced-Fat, Low-Fat, and Omega-3 Fatty Acid–Enriched Products

Consumer demand for food products of superior health quality has renewed interest in modifying the lipid composition of poultry meat. While work involving reduction of the cholesterol content of poultry products has met with less success, dietary fatty acid modification has proved to be a feasible way to add value poultry products for the health-conscious consumer. In particular, the composition of lipids in the diets of infants and young children has been studied extensively and has shown the key role in neurological development of the essential fatty acids, especially the omega-6 and omega-3 long-chain polyunsaturated fatty acids (PUFAs) (Uauy and Hoffman, 1991). These PUFAs are likely to be low in the milk of women consuming little food from animal sources as well as in cereal-based complementary food diets.

As a dietary staple, chicken muscle should ideally provide the essential fatty acids (omega-3 and omega-6; *n*-3 and *n*-6) that humans cannot synthesize. The parent fatty acids, linoleic acid (LA; C18, *n*-6) and α-linolenic acid (ALA; C 18:3, *n*-3), must be provided in the diet (Trautwein, 2001). Insufficient *n*-3 fatty acid intake influences health negatively. For a healthy diet, there should be a consumption of marine fish products, such as certain oily fish, which have high concentrations of *n*-3 fatty acids, to contribute to the prevention of cardiovascular disease and to assist with the growth and functional development of the central nervous system in the newborn. Eicosapentaenoic acid (EPA, C 20:5, *n*-3) and docosahexaenoic acid (DHA, C 22:6, *n*-3), have roles in the prevention of atherosclerosis (Moreno and Mitjavila, 2003), coronary heart disease (Yaqoob, 2004), hypertension, inflammatory (Calder, 2001) and autoimmune disorders (Zamaria, 2004), cancers (Terry et al., 2004), and diabetes (Nettleton and Katz, 2005). However, many consumers prefer not to eat such fish, but most will accept chicken.

Because of the association with a decreased risk of coronary heart disease, recent dietary fat studies have centered on the manipulation of the specific fatty acids EPA and DHA found in marine sources. Enrichment of poultry meat with these fatty acids might provide an excellent alternative source. The *n*-3 fatty acid content of meat can be increased readily by the inclusion of marine oils in the diet. However, off-flavors associated with carcass enriched in this way have prompted the use of terrestrial sources of *n*-3 fatty acids. While effective in enriching meat products with ALA, plant sources result in only minor changes in the content of 20-carbon *n*-3 fatty acids (Hargis and Van Elswyk, 1993).

Consumers who have special dietary regimes and do not like to buy chicken containing abdominal fat pad can remove it themselves before preparing a chicken. The average abdominal fat pad weighs about 40 g, which constitutes 2.5% of the total weight of the carcass and 10% of the total body fat. Several large poultry companies are currently removing this fat at the processing plant in an effort to sell a product that is lower in total fat than their competitor's chicken.

Among special dietary poultry products that can be fabricated are sausages and frankfurters. Sausages are emulsions of the oil-in-water type; the continuous phase is made up of water and soluble compounds, the dispersant is oil, and the emulsifier is protein. Technically, meat emulsions can be characterized as food obtained from a homogeneous finely triturated mixture of muscular tissue, blood, viscera, and other animal products or subproducts authorized for human consumption, such as fat, water (Hansen, 1960; Mucciolo and Gomes 1981), and nonflesh components (Mucciolo et al., 1980). Adding fat to an emulsion aims to produce, for example, the texture and softness preferences of the population, such as a fat average of 25% in sausages (Takino et al., 1974). Chicken and/or turkey frankfurters traditionally contain 18 to 22% fat, compared to the usual 25 to 30% in pork and beef franks. Some producers of poultry franks have lowered the fat content of their product to 13 to 16% by using mechanically deboned meat from portions of the poultry that contain less fat than backs or legs, such as the front quarter, breast cage, or skinless necks. Poultry frankfurters ranged in caloric content from 180 to 300 kcal/100 g. From a sensory standpoint, fat is an important component in increasing the palatability in a food such as frankfurters. If the fat content is too low, the resulting product tends to be rubbery and tough. Therefore, although consumers may think they want a much leaner frankfurter, they might not find such a product acceptable.

The breed of chicken and the type of feed have influences on fatty acid composition. In comparison to other meats, chicken has been reported to be relatively abundant in PUFAs, as diets of fast-growing broilers are generally rich in PUFAs (Asghar et al., 1990).

Battered and breaded deep-fried poultry products have been a mainstay of the further-processed and fast-food industry for many years. The current emphasis is toward boneless products such as nuggets and patties. Nuggets are a breaded coarsely comminuted product usually prepared with manually deboned poultry meat. Recently, it has been made possible to manufacture low-fat chicken nuggets by using skinless chicken breast meat and various fat replacers. The meat content ranged from 70 to 84.5%. Fat replacers have been used, such as tapioca starch, wheat flour, and carrageenan. The fat content of the products ranged from 0.26% texture profile analyses indicated no significant differences in the cohesiveness and springiness compared to those of commercial products (Chuah et al., 1998). Chicken skin connective tissue, a by-product of fabrication operations, could be a potential water binder and texture-modifying agent for use in reduced-fat comminuted meat products as a less expensive water binder or texture-modifying agent (Osburn and Mandigo, 1998). A sodium bicarbonate washing process has been used to remove fat from chicken skin to concentrate the protein content prior to incorporation in bologna (Bonifer et al., 1996). Preheating chicken skin connective tissue to a gelatinous state converts collagen to gelatin, a strong water binder, forming a gel which if incorporated into reduced-fat meat products, may improve product yield, texture, and palatability (Osburn and Mandigo, 1998).

Baker et al. (1986) evaluated four cooking methods for battered and breaded broiler parts. The three most commonly used methods for commercial preparation of retail frozen, fully cooked and browned, battered and breaded chicken are (1) water cooking, which is a thorough cooking in hot water followed by 54 s of frying; (2) fully frying in 177°C oil; and (3) oven cooking, which is frying for 2.5 min, followed by thorough heating in a 218°C oven. The fat content is higher in breast cooked by fully frying than in breasts cooked by water cooking and oven cooking.

Stadelman (1985) demonstrated that breaded chicken products can be produced with reduced caloric content using hot-air cooking instead of deep-fat frying, which resulted in a 23 to 31% decrease in the fat content of parts and a 13 to 15% decrease in calories, and by removing the skin before breading and hot-air cooking, which resulted in a 42 to 65% decrease in calories. Cooking systems such as the aforementioned or the broiling method will become more commonplace in the future as the demand increases for poultry products with less fat and fewer calories.

Frying with olive oil increases the percentage of oleic acid in fried chicken meat products. Labeling of chicken-based products should include the fat content and fatty acid composition. The final content in fat and fatty acids in chicken and fried chicken–based products depends on the fat content of the raw product, the presence of an edible coating which limits the penetration of fat during frying, and the composition of the oil used (García et al., 2003).

To preserve the above-mentioned poultry products it is necessary that antioxidants delay lipid oxidation by reducing free-radical activities in meat, and adding natural antioxidants from foods is an efficient method of increasing oxidative stability. There are two categories of endogenous antioxidants. The antioxidant enzymes catalase, glutathione peroxidase, glutathione reductase, and superoxide dismutase can be considered as preventive antioxidants. Other antioxidants include α-tocopherol, glutathione, ascorbic acid, and β-carotene (Halliwell and Gutteridge, 1989). The effects of α-tocopherol on oxidative propagation in chicken meat have been studied, but the influence of antioxidant enzymes has received little attention (O'Neil et al., 1999).

There is a chicken breakfast sausage manufactured by combining mechanically deboned chicken meat with hand-deboned skinless chicken thigh meat, fat replacer ingredients (blend or modified food starch and oat flour), and rosemary concentrate as antioxidant. The use of mechanically deboned chicken meat in frankfurters, various loaf products, fermented sausages, and restructured chicken products poses a major problem because of the rapid onset of oxidative rancidity, which results in off-flavors and odors, but the use of antioxidants effectively controls oxidative rancidity (Lee et al., 1997).

The American Heart Assotiation has recommended the ingestion of 300 mg of cholesterol daily for men (Krzynowek, 1985). Consumers are aware that the consumption of animal food that contains saturated fat and cholesterol raises blood cholesterol (Flynn et al., 1985). Under metabolic conditions, carbohydrate substitution for lauric, myristic, and palmitic acids raises the low-density lipoprotein (LDL cholesterol) as well as the high-density lipoprotein (HDL cholesterol), while stearic acid has a small effect. The oleic and linoleic acids raise HDL and lower LDL (Katan et al., 1994).

PROTEIN ENRICHMENT OF SPECIAL POULTRY MEAT PRODUCTS

A number of studies have demonstrated that protein modification can improve the functional properties of various tissues: beef (DuBois et al., 1972), fish (Spinelli et al., 1972), beef heart (Smith and Brekke, 1984), and mechanically deboned fowl (Smith and Brekke, 1985a). Modification refers to the intentional alteration of the physicochemical properties of proteins by chemical, enzymatic, or physical agents to improve functional properties. According to Brekke and Eisele (1981), acylation reactions involving the direct addition of chemical groups to functional groups of amino acid side chains via substitution have the most potential for modifying food proteins chemically. The anhydrides of acetic and succinic acids are usually the acylating agents, since they are easy to use, safe, and inexpensive and produce acylated derivatives that are functionally important. When a protein reacts with acetic anhydride, the acylation reaction is termed *acetylation*; when succinic anhydride is used, the reaction is referred to as *succinylation*. Succinylation affects the physical character of proteins by increasing the net negative charge, changing the conformation, and increasing the propensity of proteins to dissociate into subunits, breaking up protein aggregates and increasing protein solubility (Tarek, 2000; Pato et al., 2005).

For acylated proteins to be incorporated into foods, they will need to be safe, digestible, and probably approved by the U.S. Food and Drug Administration and U.S. Department of Agriculture as food ingredients, since the protein has been modified. Similar techniques may also be useful in improving the functional properties of poultry giblets, thereby making these products, with good nutritional properties, more usable by the poultry further-processing industry. Furthermore, animal blood is a potential source of high-quality protein. The composition of dried poultry blood is 80% protein, 8% moisture, 1% fat, and 11% fiber or ash. Broiler chickens contain about 7.5% of their body weight in blood, 45% of which is collectible during slaughtering operations (Kotula and Helbacka, 1966). Although blood protein has a strong taste and odor and the distinct red color of hemoglobin, all of which may be disagreeable to consumers, it may be used to enrich foods with proteins (Calvi et al., 1984). The off-flavor, which is probably due to lipid breakdown, can be minimized with newer, low-temperature drying methods. Blood from larger animals is collected routinely, decolorized when desired, and used in foods such as blood sausage in some countries (Stevenson and Lloyd, 1979).

Poultry products can be enriched with proteins by using poultry giblets (heart, gizzard, and liver), which are a good source of protein, iron, and niacin. In addition, liver is high in vitamins A and C. However, the undesirable texture of gizzard and heart tissue has been a factor in the underuse of these foods. The meat is removed mechanically and the giblets are used in canned goods in general, including sausages (Pereira et al., 2000).

Bone residue is the material remaining when mechanically deboned poultry is prepared. Bone residue has characteristics that make it a valuable potential source of protein for human food. It contains 20% protein, 7.7% fat, 11.7% ash, and 60% moisture (Opiacha et al., 1991). Bones from slaughtered animals, especially larger animals such as beef and swine, are generally used for animal feed, gelatin, and glue. However, they could be used as ingredients in certain processed products; they are high in protein and provide a dietary source of minerals such as calcium.

Bone products are used as food ingredients in some European countries. Some countries consider bone-derived protein added to a meat product to be meat.

Freeze-dried protein isolates from bone residue using sodium chloride contains 60 to 65% lipid, 5 to 10% ash, and 4 to 6% moisture. The freeze-dried protein extract obtained by using alkali contains 45% protein, 47% fat, and 14% ash (Young, 1976). Lawrence and Jelen (1982) state that severe alkali treatments of protein may cause racemization or destruction of certain amino acids; in addition, unusual new amino acids may be produced, such as lysinoalanine, lanthionine, and ornithinoalanine. However, this method conducted with bone residue should not produce amino acids that pose health hazards for consumers. Protein extracts from bone residue have relatively good functional properties (water-holding capacity, emulsifying capacity, solubility) and could serve as ingredients in other poultry proteins. The poultry industry should be encouraged to explore the economic feasibility of using this protein source, which is currently underutilized or discarded.

Technological processes used in food manufacture affect the functional, nutritional, and biological properties of food proteins. On the other hand, proteins may be added as functional ingredients to foods to emulsify, bind water or fat, form foams or gels, and alter flavor, appearance, and texture (Anantharaman and Finot, 1993). In recent years, the role of proteins in the diet as physiologically active components has been acknowledged increasingly. Such proteins or their precursors may occur naturally in raw food materials, exerting their physiological action directly or upon enzymatic hydrolysis in vitro and in vivo.

Much scientific interest has focused on physiologically active peptides derived from food proteins. These peptides are inactive within the sequence of precursor protein and can be released by enzymatic proteolysis. Bioactive peptides are usually short-chain peptides that can be absorbed by intestinal enterocytes and transported from the luminal to the basolateral side. After this discharge into the intestinal interstitial space, a transfer across the capillary wall to blood can occur, leading to biological activity that may be beneficial (Gardner, 1988). Bioactive peptides with different activities have been described, including immunomodulatory, antimicrobial, antithrombotic, and antihypertensive activities (Smacchi and Gobbertti, 2000). In this sense the nutritional and functional value of a protein is determined not only by its total amino acid composition and digestibility but also by the presence of bioactive peptides in its sequence. Hence, poultry products could be manufactured that would affect a specific physiological activity in the human body.

REDUCED SODIUM CONTENT IN POULTRY MEAT PRODUCTS

Because high blood pressure or hypertension is an established risk factor for cardiovascular disease and is highly prevalent in the elderly population, reductions in sodium intake are an essential component of countries' public health policies. Differences in sodium intake of 2300 mg have been associated with 5 to

10 mmHg lower systolic and 2 to 5 mmHg lower diastolic blood pressure, with the largest differences occurring at older ages (Law et al., 1991). The estimated effect on stroke and cardiovascular disease risk from the resulting downward shifts in the distribution of blood pressure would be substantial. Results from clinical trials to lower blood pressure suggest that a decrease of 11 to 12 mmHg systolic and 5 to 6 mmHg diastolic blood pressure would yield 38 and 16% reductions in stroke and coronary heart disease, respectively (Cutler et al., 1995).

The degree of concern within the scientific community (Putnam and Reidy, 1981) and by many consumers warrants the production of food products containing less sodium. Poultry meat itself is not high in sodium content; cooked breast meat contains 63 mg of sodium per 100 g of meat, and cooked thigh meat contains 75 mg/100 g. However, during further processing of poultry meat into products, the sodium content may increase dramatically as sodium chloride and various sodium phosphates are added to the product. Processed meat products comprise a major source of sodium in the form of sodium chloride (salt). Salt has an essential function in meat products in terms of flavor, texture, and shelf life. Sodium chloride is generally used in further-processed products such as frankfurters at levels of 1.5 to 2.5%. In conclusion: Salt influences the flavor, may affect the shelf life, and affects the functional properties of myofibrillar proteins.

Apart from lowering the level of salt added to products, there are a number of approaches to reduce the sodium content in processed foods, including the use of salt substitutes, such as calcium chloride, magnesium chloride, and in particular, potassium chloride(Hand et al., 1982; Maurer, 1983), all of them in combination with masking agents and flavor enhancers (Desmond, 2006). Smith and Brekke (1985b) varied the sodium chloride content of frankfurters prepared from enzyme-modified, mechanically deboned fowl. They found that 0.5% salt was the smallest amount that could be added to produce a satisfactory frankfurter from which the casing could easily be removed. Barbut et al. (1986) reported that turkey frankfurters with 1.5% salt combined with phosphate were as acceptable as reference frankfurters that contained 2.5% salt. Replacing 100% of the sodium chloride with magnesium chloride or potassium chloride was detrimental to the flavor of frankfurters prepared from mechanically deboned turkey. Only 35% of the sodium chloride could be replaced successfully by potassium chloride; magnesium chloride caused off-flavors, even at the same level. The sodium chloride in poultry frankfurters could be reduced to at least 1.5% (590 mg of sodium/100 g of meat) without detracting from the flavor and to as low as 0.5% (197 mg of sodium/100 g of meat) if additional spices can be found to improve flavor.

REFERENCES

American Academy of Pediatrics, Committee on Nutrition. 1998. In: Kleinman RE, ed., *Pediatric Nutrition Handbook*, 4th ed. Elk Grove Village, IL: AAP.

Anantharaman K, Finot PA. 1993. Nutritional aspects of food protein in relation to technology. Food Rev Int 9:629–655.

Asghar A, Lin CF, Buckley DJ, Booren AM, Flegal CJ. 1990. Effects of dietary oils and α-tocopherol supplementation on membranal lipid oxidation in broiler meat. J Food Sci 55(1):46–50.

Baker RCD, Scott-Kline J, Jutchinson, Goodman A, Charvat J. 1986. A pilot plant study of the effect of four cooking methods on acceptability and yields of prebrowned battered and breaded broiler parts. Poult Sci 65:1322.

Barbut SA, Maurer J, Lindsay RC. 1986. Effects of reduced sodium chloride and added phosphates on sensory and physical properties of turkey frankfurters. Poult Sci 65(Suppl 1):10–15.

Bonifer LJ, Froning GW, Mandigo RW, Cuppett SL, Meagher MM. 1996. Textural, color and sensory properties of bologna containing various levels of washed chicken skin. Poult Sci 75:1047–1055.

Brekke CJ, Eisele TA. 1981. The role of modified proteins in the processing of muscle foods. Food Technol 35(5):231–234.

Calder PC. 2001. Polyunsaturated fatty acids, inflammation, and immunity. Lipids 36(9):1007–1024.

Calle EE, Thun MJ, Petrelli JM. 1999. Body mass index and mortality in a prospective cohort of US adults. N Engl J Med 341(15):1097–1105.

Calle EE, Rodriguez C, Walker-Thurmond K. 2003. Overweight, obesity, and mortality from cancer in a prospectively studied cohort of US adults. N Engl J Med 348(17):1625–1638.

Calvi BG, Kasaoka AJ, Kueseter G. 1984. *Animal Blood Protein as a Food Ingredient*. Memorandum of Screening and Surveillance. Washington, DC: U.S. Department of Agriculture.

Chuah EC, Normah O, Mohd YJ. 1998. Development of low-fat chicken nuggets. J Trop Agric Food Sci 26:93–98.

Cutler JA, Psaty BM, MacMahon S, Furberg CD. 1995. Public health issues in hypertension control: what has been learned from clinical trials. In: Laragh JH, Brenner BM, eds., *Hypertension: Pathophysiology, Diagnosis, and Management*. New York: Raven Press, pp. 253–270.

Desmond E. 2006. Reducing salt: a challenge for the meat industry. Meat Sci 74(1):188–196.

DuBois MW, Anglemier AF, Davidson WD. 1972. Effect of proteolysis on the emulsification characteristics of bovine skeletal muscle. J Food Sci 37(1):27–28.

Fairweather-Tait SJ. 1997. From absorption and excretion of minerals to the importance of bioavailability and adaptation. Br J Nutr 78(Suppl 2):S95–S100.

Flynn MA, Naumann HD, Nolph GB, Krause G, Ellersiek M. 1985. The effect of meat consumption on serum lipids. Food Technol 39(2):58–64.

García AMT, García LMC, Capita R, Garcia FMC, Sánchez MFJ. 2003. Deep-frying of chicken meat and chicken-based-products: changes in the proximate and fatty acid compositions. Ital J Food Sci 15(2):225–239.

Gardner MLG. 1988. Gastrointestinal absorption of intact proteins. Am Rev Nutr 8:329–350.

Geornaras I, Dykes GA, Holy AV. 1995. Biogenic amine formation by chicken-associated spoilage and pathogenic bacteria. Lett Appl Microbiol 21:164–166.

Halliwell B, Gutteridge JMC. 1989. *Free Radicals in Biology and Medicine*, 2nd ed. New York: Oxford University Press.

Hamm D, Ang CYW. 1984. Effect of sex and age on proximate analysis cholesterol and selected vitamins in broiler breast meat. J Food Sci 49:286–287.

Hand LW, Terrell RN, Smith GC. 1982. Effects of chloride salts on physical, chemical and sensory properties of frankfurters. J Food Sci 47(6):1800–1802.

Hansen LJ. 1960. Emulsion formation in finely comminuted sausage. Food Technol 14(3):565–569.

Hargis PS, Van Elswyk ME. 1993. Manipulating the fatty acid composition of poultry meat and egg for the health conscious consumer. World's Poult Sci J 49(3):251–264.

Institute of Medicine. 1997. *Dietary Reference Intakes for Calcium, Phosphorus, Magnesium, Vitamin D and Fluoride*. Washington, DC: National Academies Press.

Institute of Medicine. 1998. *Dietary Reference Intakes for Vitamin A, Vitamin K, Arsenic, Boron, Chromium, Copper, Iodine, Iron, Manganese, Molybdenum, Nickel, Silicon, Vanadium, and Zinc*. Washington, DC: National Academies Press.

Institute of Medicine. 2000. *Dietary Reference Intakes for Vitamin C, Vitamin E, Selenium, and Carotenoids*. Washington, DC: National Academies Press.

Joint FAO/WHO Expert Consultation. 2002. *Vitamin and Mineral Requirements in Human Nutrition*. Geneva, Switzerland: World Health Organization.

Katan MB, Zock PL, Mensink RP. 1994. Effects of fats and fatty acids on blood lipids in humans: an overview. Am J Clin Nutr 60(Suppl 6):1017S–1022S.

Kotula AW, Helbacka NV. 1966. Blood volume of live chickens and influence of slaughter technique on blood loss. Poult Sci 45:684–688.

Krzynowek J. 1985. Sterol and fatty acids in seafood. Food Technol 39(2):61–68.

Law MR, Frost CD, Wald NJ. 1991. By how much does dietary salt reduction lower blood pressure? I. Analysis of observational data among populations. Br Med J 302(6780):811–815.

Lawrence RA, Jelen P. 1982. Formation of lysino-alanine in alkaline extracts of chicken protein. J Food Prot 45:923–924.

Lee TG, Williams SK, Sloan D, Littell R. 1997. Development and evaluation of a chicken breakfast sausage manufactured with mechanically deboned chicken meat. Poult Sci 76(2):415–421.

Martin C, Barker DJ, Osmond C. 1996. Mothers' pelvic size, fetal growth, and death from stroke and coronary heart disease in men in the UK. Lancet 348(9037):1264–1268.

Maurer AJ. 1983. Can sodium be reduced in poultry products? Turkey World, July–Aug., pp. 34–37.

Mokdad AH, Browman BA, Ford ES, Vinicor F, Marks JS, Koplan JP. 2001. The continuing epidemics of obesity and diabetes in the United States. JAMA 286:1195–1200.

Moreno JJ, Mitjavila MT. 2003. The degree of unsaturation of dietary fatty acids and the development of atherosclerosis (review). J Nutr Biochem 14(4):182–195.

Mucciolo P, Gomes MCG. 1981. A relação umidade:protein (u:p) na repressão de fraudes de salsichas enlatadas. Bol SBCTA 15:379–393.

Mucciolo P, Meira DR, Gomes MCG. 1980. A relação umidade:proteina de salsichas enlatadas eseu comportamento em função do tempo de processamento. Rev Inst Adolfo Lutz 40:129–134.

Nettleton JA, Katz R. 2005. *n*-3 long-chain polyunsaturated fatty acids in type 2 diabetes: a review. J Am Diet Assoc 105(3):428–440.

O'Neill, LM, Galvin K, Morrissey PA, Buckley DJ. 1999. Effect of carnosine, salt and dietary vitamin E on the oxidative stability of chicken meat. Meat Sci 52(1):89–94.

Opiacha JO, Mast MG, MacNeil JH. 1991. In Vitro protein digestibility of dehydrated protein extract from poultry bone residue. J Food Sci 56(6):1751–1752.

Osburn WN, Mandigo RW. 1998. Reduced-fat bologna manufactured with poultry skin connective tissue gel. Poult Sci 77:1574–1584.

Pato C, Vinh T, Marion D, Douliez JP. 2005. Effects of acylation on the structure, lipid binding, and transfer activity of wheat lipid transfer protein. J Protein Chem 21(3):195–201.

Pereira NR, Tarley CRT, Matsushita M, deSouza NE. 2000. Proximate composition and fatty acid profile in Brazilian poultry sausages. J Food Comp Anal 13(6):915–920.

Popkjn BM. 1994. The nutrition transition in low-income countries an emerging crisis. Nutr Rev 52(9):285–298.

Przybyla A. 1985. Prepared chicken items offer versatility, low cost. Prepared Foods 154(8):159–164.

Putnam JJ, Reidy K. 1981. Sodium: Why the concern? Natl Food Rev 15:27–29.

Silva CMG, Glória MBA. 2002. Bioactive amines in chicken breast and thigh after slaughter and during storage at $4 \pm 1^{\circ}C$ and in chicken-based meat products. Food Chem 78(2):241–248.

Silva JCT. 1995. Por que a avicultura se expande. Rev Nac Carne 222:50–57.

Singh SP, Essary EO. 1971. Vitamin content of broiler as affected by age, sex, thawing and cooking. Poult Sci 50:1150–1155.

Smacchi E, Gobbertti M. 2000. Bioactive peptides in dairy products: synthesis and interaction with proteolytic enzymes. Food Microbiol 17(2):129–141.

Smith DM, Brekke, CJ. 1984. Functional properties of enzymatically modified beef heart protein. J Food Sci 49(6):1525–1528.

Smith DM, Brekke CJ. 1985a. Enzymatic modification of the structure and functional properties of mechanically deboned fowl proteins. J Agric Food Chem 33(4):631–637.

Smith DM, Brekke CJ. 1985b. Characteristics of low-salt frankfurters produced with enzyme-modified mechanically deboned fowl. J Food Sci 50(2):308–312.

Sowers JR. 2003. Obesity as a cardiovascular risk factor. Am J Med 115(Suppl 8A):37S–41S.

Spinelli JB, Koury B, Miller R. 1972. Approaches to the utilization of fish for the preparation of protein isolates: enzymatic modifications of myofibrillar fish proteins. J Food Sci 37(4):604–608.

Stadelman W. 1985. This chicken product breaks "grease barrier." Broiler Ind 48:46.

Stevenson TR, Lloyd GT. 1979. Better uses for abattoir blood. Agric Gaz NS Wales 90:42–45.

Stratton JE, Hutkins RW, Taylor SL. 1991. Biogenic amines in cheese and other fermented foods: a review. J Food Prot 54:460–470.

Takino M, Komatsu I, Galli S. 1974. Relação umidade: proteina de salsichas e mortadelas consumidas em São Paulo. Atual Vet 19:4–10.

Tarek A. 2000. Functional properties and nutritional quality of acetylated and succynilated mung bean protenin isolate. Food Chem 70(1):83–91.

Terry PD, Terry JB, Rohan TE. 2004. Long-chain (*n*-3) fatty acid intake and risk of cancers of the breast and the prostate: recent epidemiological studies, biological mechanisms, and directions for future research. J Nutr 134(12):3412S–3420S.

Trautwein EA. 2001. *n*-3 fatty accids: physiological and technical aspects for their use in food. Eur J Lipid Sci Technol 103(1):45–55.

Tucker KL, Buranapin S. 2001. Nutrition and aging in developing countries. J Nutr 131(9):2417S–2423S.

Uauy R, Hoffman DR. 1991. Essential fatty acid requirements for normal yet and brain development. Semin Perinatol 15:449–455.

U.S. Congress, Senate Select Committee on Nutrition and Human Needs. 1977. *Dietary Goals for the United States*. Washington, DC: U.S. Goverment Printing Office.

USDA–USDHHS (U.S. Department of Agriculture–U.S. Department of Health and Human Services). 2000. *Nutrition and Your Health: Dietary Guidelines for Americans*, 5th ed. Washington, DC: USDA–USDHHS, pp. 24–26.

WHO (World Health Organization). 1979. *Meeting on Infant and Young Child Feeding: Statement and Recommendations, 1979*. Geneva, Switzerland: WHO.

WHO. 1998. *Complementary Feeding of Young Children in Developing Countries: A Review of Current Scientific Knowledge*. WHO/NUT/98.1. Geneva, Switzerland: WHO.

WHO. 1999. *Health and Development in the 20th Century*. World Health Report. Geneva, Switzerland: WHO.

WHO. 2001a. *Infant and Young Child Feeding*. World Health Assembly 2001. Document 54.2. Geneva, Switzerland: WHO.

WHO. 2001b. *Ageing and Nutrition: A Growing Global Challenge*. Geneva, Switzerland: WHO.

WHO. 2002. *Global Strategy for Infant and Young Child Feeding*. World Health Assembly 2002. Document 55. Geneva, Switzerland: WHO.

Yaqoob P. 2004. Fatty acids and the immune system: from basic science to clinical applications. Proc Nutr Soc 63(1):89–104.

Young LL. 1976. Composition and properties of animal protein isolate prepared from bone residue. J Food Sci 41(3):606–608.

Zamaria N. 2004. Alteration of polyunsaturated fatty acid status and metabolism in health and disease. Reprod Nutr Dev 44(3):273–282.

PART IV

PRODUCT QUALITY AND SENSORY ATTRIBUTES

21

SENSORY ANALYSIS

Maria Dolors Guàrdia, Carmen Sárraga, and
Luis Guerrero
IRTA, Finca Camps i Armet s/n, Monells, Girona, Spain

INTRODUCTION

Sensory analysis is a scientific discipline that allows us to measure the characteristics of a product in an objective and reproducible way through the human senses. The instruments for measurement are human beings, and for this reason it is very important to describe in an exhaustive and detailed manner the methodology used to reduce the intrinsic error common to this type of measuring. Aspects such as the uniformity in the temperature of the samples during

Handbook of Poultry Science and Technology, Volume 2: Secondary Processing, Edited by
Isabel Guerrero-Legarreta and Y.H. Hui

tasting or the order of presentation during evaluation can increase extensively the variability between tasters and/or replications hindering the detection of differences. In general, obtaining a quality sensory measurement depends on two basic elements: the tasters and the characteristics used in the execution of a test. We discuss the most important aspects to consider for each of these elements and propose a detailed methodological guide for the sensory evaluation of poultry and poultry-meat products.

SELECTION OF JUDGES

As in the selection of any other equipment for laboratory analysis, the selection of future judges is decisive and must be carried out very carefully. In general, this is done in two stages: a preliminary selection and a specific selection.

Preliminary Selection

The selection of candidates to participate in a sensory analysis panel is probably one of the most important parts of the methodological process in this analytical procedure. However, no training program is useful if the characteristics of the people selected do not surpass the necessary minimums. The criteria of selection can be assigned to one of two groups, depending on the psychological or physiological aptitudes of the candidates. Normally, the first, which are more difficult to quantify, are not evaluated in depth. However, during selection it is more important to be sure that the panel has good psychological aptitude rather than good physiological capacity (Civille and Szczesniack, 1973). In-depth knowledge of the personality characteristics of the persons selected allows us to detect, for example, subjects with either excessively timid or hostile dominant features, who should be avoided for the benefit of good management of the group.

Generally, the criteria used for a preliminary selection do not include sensory aptitudes. The most important aspects to be evaluated in this first selection should be motivation and interest, repulsion or attitude to specific foods, and the person's health, availability, and features of personality (intellectual capacity, power of concentration, ease of communication, verbal fluency, creativity). During this period, other general data or personal aspects, such as age, gender, name, nationality, profession, religion, and experience in sensory analysis, can be obtained generally through a questionnaire and/or oral interview. Meilgaard et al. (1987) proposed some models of questionnaires for the gathering of this information.

Test results, used as a basis for selecting panelists, are intended primarily to detect any sensory incapabilities, to determine sensory sensibility, and to evaluate the potential of candidates in describing and communicating their perceptions. After this process, only those who are clearly inadequate (e.g., with important health problems, inability to attend the majority of sessions, or who refuse to taste one of the products to be evaluated throughout the evaluation of samples) should be eliminated, the others should be maintained as candidates.

Specific Selection

It is important to highlight that the selection of subjects according to their sensory aptitudes is not advisable until they have all received a certain degree of training. Every selection process should be preceded by preliminary formation with practical and theoretical sessions. This will provide candidates with sufficient knowledge of sensory evaluation and how they should use their senses, especially in a first contact with the products to be used in the selection process. In addition, this prior training allows new judges, who have no experience in this type of testing, to be in the same condition as those who have already taken part in some type of sensory analysis. Furthermore, the confidence of the most insecure candidates will increase and a more comfortable atmosphere during the trials will be established.

After the preliminary formation, the selection is carried out in two ways:

1. Trials to determine any incapability of perception, such as the Ishihara color test (Ishihara, 1971) or the detection of gustatory or olfactory insensibility.
2. Trials of sensory aptitudes, which can be split up into two groups:
   a. General tests such as the capability to understand and answer a questionnaire, or the capability to express themselves (their descriptive ability).
   b. Specific tests such as the capability to discriminate between stimuli, the capacity to learn these stimuli and memorize them, and the ability to differentiate intensity.

Even though a large selection program might seem excessive, it must be acknowledged that all work carried out continues to train the group. According to Amerine et al. (1965), fast methods of selection based on a few tests are not usually satisfactory. For this reason, efforts to design a good program of selection should not be short-changed.

Generic and Specific Training

The main purpose of training is to familiarize tasters with the various sensory techniques, with the most frequent attributes, as well as with the scale of measurement for each. During this process, tasters must develop their sensory memory, which will allow them to evaluate the samples in a reliable way. Training is usually carried out in two stages: general or generic training and applied or specific training. In both it is advisable to do practical and theoretical sessions together, because sometimes it is not easy to group the tasters together simultaneously at the start of a session. In consequence, once the group is formed, it is productive to prolong the session as much as possible without provoking the fatigue of the persons concerned. In this sense, the addition of discussions and short theoretical explanations allows for an increase in the number of products to be sensory-evaluated and for the recuperation of the tasters. Also, comprehensive knowledge of sensory techniques, types of tests, most common errors, and so on, notably improves the group's preparation and often helps to reduce sources of variation that usually occur due to the ignorance of some of the tasters.

The basis of sensory evaluation must be reached during the generic training. So, it is important that tasters become accustomed to the methodological rigor that this type of testing requires. They must evaluate, with practical examples, different types of sensory attributes: visual, not oral, texture, olfactory response, flavor (a complex set of olfactory and gustatory properties that are perceived during the tasting process and that can be influenced by tactile properties, thermal, painful, and even kinesthetic effects), and oral texture. At this stage the use of scales of reference is important, since the multiproduct character of a sample is very useful as a didactic tool. Meilgaard et al. (1978) described various scales of reference; ISO 1036 (1994b) also provides some examples. However, as pointed out by Issanchou et al. (1997), each product needs its own adequate references, which can be completely different from one type of food product to another, thus making it impossible to generalize a set of standards for every food.

At the end of the generic training the tasters should know the principles of sensory evaluation, and their sensory memory should have improved with an increased number of attributes (definition, normal procedures for evaluation, products of reference for specific characteristics, and scoring of the samples in an adequate way). The specific training should be oriented on fixing and consolidating only those aspects of the sensory analysis applicable to each particular case.

In meat products a satisfactory way to start the sensory characterization is to taste meat from different species in the same session and then discuss them in an open session with all the members of the panel. The aim of these initial sessions is to obtain the main sensory descriptors of the meat and their intensity. Afterward, it is important to focus on the species under study. In the particular case of poultry meat and in order to obtain a large number of sensory descriptors to discuss, it is useful to taste meat from animals of different genetic strains, genders, ages, animals fed on different diets, different muscles or anatomical areas (legs, breast), and meat with different storage times, because in poultry meat, lipid oxidation has been associated with adverse changes in appearance, flavor, and texture (Jensen et al., 1998; Lyon, 1987). The shared discussion of the descriptors makes it possible to understand the meaning of each descriptor and to learn the scoring intensity for each. The use of reference scales is generally very useful in this phase, although if possible, they must be prepared based on meat that has a different intensity for each specific descriptor (see the example in Table 1). At the end of this period, the key descriptors will be selected and included in the final descriptive profile. After this selection, the training will focus on these attributes. The procedure for selection of the main descriptors to be included in a sensory profile is available in ISO 11035 (ISO, 1994a).

The training period ends when the tasters know exactly every descriptor to evaluate, the procedure to be used, and how to score each descriptor accurately. To verify the quality of the selection and of the training, it is advisable to carry out a study of both individual and collective repeatability of the group and also the

TABLE 1 Example Reference Scale for Rancidity in Poultry Meat

Score	Sample Used as Reference[a]
0	Breast chicken (with no storage)
4	Breast chicken stored 1 week
7	Breast chicken stored 2 weeks

[a]Samples are vacuum-packed breast of chicken cooked for 30 min in a 80°C water bath. Samples were stored at 4°C. Before tasting, samples were reheated for 25 min at 65°C in an electric oven.

reliability of the panel. In the scientific literature there are different methodologies for checking the quality of sensory panels. Guerrero and Guàrdia (1998) proposed a simple method based on using the analysis of variance (ANOVA) to estimate the reliability of a panel. These authors propose applying an individual ANOVA for each judge and sensory descriptor, including the product and the session as main effects in the model, and a global ANOVA, including the product, the taster, the session nested to the taster, and the interaction product × taster as main effects. By means of this information, it is possible to study both the person and the overall repeatability. Furthermore, this method provides a good tool to use to study the degree of accordance between tasters for each descriptor of the profile. It also allows us to decide which tasters need additional training, and for which sensory descriptors.

SENSORY METHODOLOGY USED TO EVALUATE POULTRY SAMPLES

In this section we describe a useful sensory methodology for preparing poultry meat and products to be sensory-evaluated. The methodology includes preparation of the samples, the sensory profile to be used, and the conditions for execution of the panel test. Due to the considerable variability that exists in animal foods, it is very important to standardize the tasting conditions, which results in a reduction in experimental error and an increase in the ability for detection of differences due to the various treatments tested.

Sample Preparation of Raw and Cooked Poultry Meat

A brief overview of the literature of sensory analysis of poultry products demonstrates a high diversity of sensory procedures for sample preparation. Such diverse approaches make it difficult to draw comparisons among studies. In this part we propose a sample methodology based on our own experience. The sensory evaluation proposed is to be carried out in both breast and leg samples of poultry.

Gastrocnemius lateralis and gastrocnemius medialis are the poultry leg muscles selected as objects of the sensory analysis. In broilers, quails, and other small

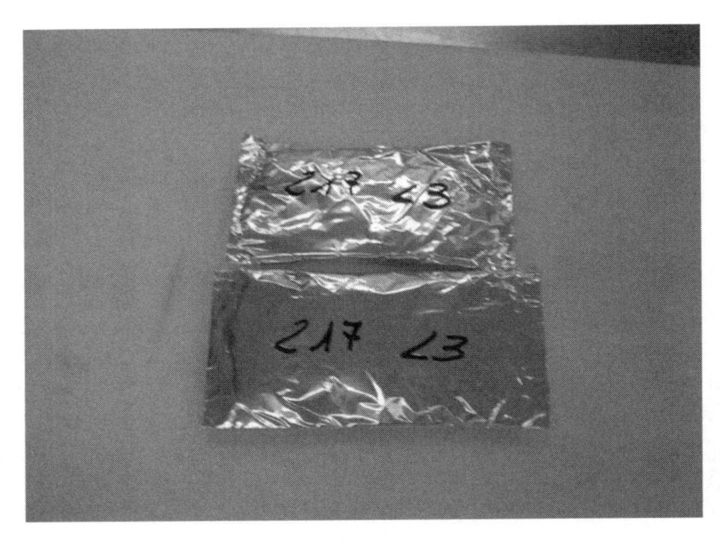

FIGURE 1 Sample preparation before cooking in an electric oven.

birds, we suggest using different muscles because of their size and to extend the same criterion to the other poultry species. So, flavor descriptors could be evaluated on the gastrocnemius lateralis muscle and texture descriptors on the gastrocnemius medialis muscle. Sensory analysis of breast will be carried out on the pectoralis major muscle.

Samples of poultry meat do not really need aging. The storage of raw samples of poultry meat must be very short, and if the sensory evaluation is not done soon after, it is preferable to freeze them to prevent alteration. The process of freezing will be carried out at −20°C for a maximum of two months (Sebranek et al., 1979; Akamittath et al., 1990). The samples will be thawed at 4°C for 24 h.

Samples will be cooked in a convection oven preheated to 180–200°C until they reach a temperature of 80°C at the core of the sample. The cooking process can also be carried out by means of a water bath, even though in this case the flavor characteristics will be very different from those perceived by consumers at home. In the former case, samples are cooked wrapped in aluminum foil to prevent excessive loss of juiciness (Figure 1). To block possible effects due to the cooking batch, it is very important to cook one sample of each treatment in each oven batch. For the breast samples, the external part of the M. pectoralis major is rejected, and then the central part is cut into homogeneous cubes according to the number of tasters and descriptors (Figure 2). Afterward, each cube is wrapped in aluminum foil and codified with a random three-digit number. The use of electric heaters is recommended to keep the samples warm until their sensory evaluation (Figure 3). Depending on the number of descriptors, panelists will use one cube for the flavor and one cube for the texture attributes.

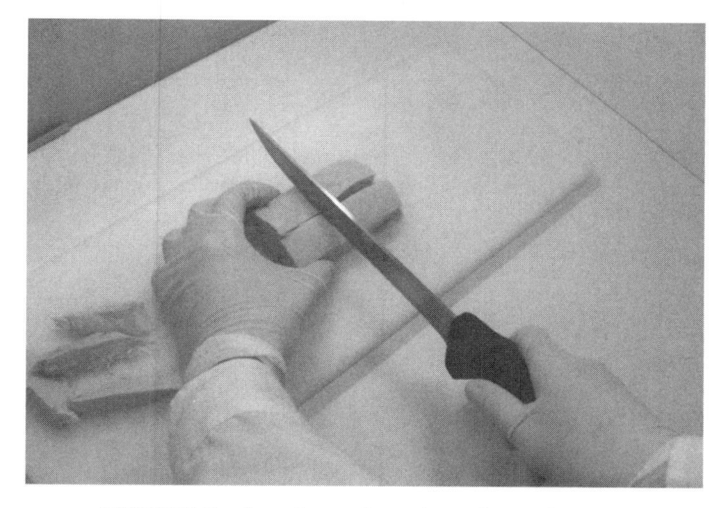

FIGURE 2 Sample cut into pieces for each taster.

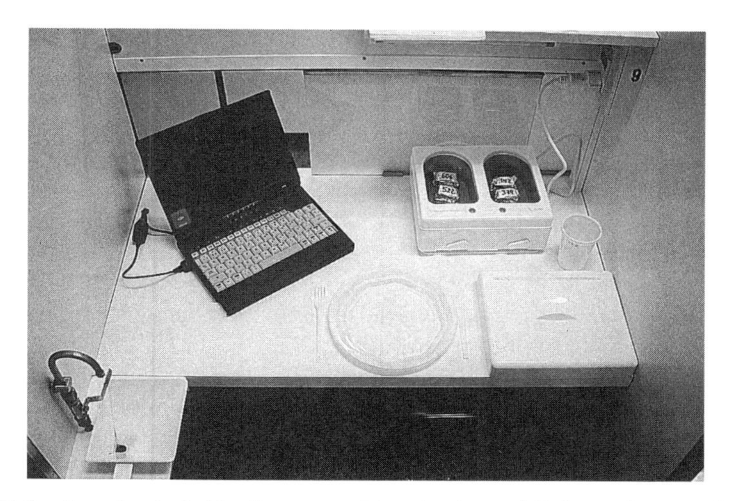

FIGURE 3 Samples individually wrapped in aluminum foil for each taster in a standardized booth.

Similarly, from the cooked-leg samples, the muscles gastrocnemius lateralis and gastrocnemius medialis are dissected from each individual sample and then prepared as the breast samples. Analyses made in small poultry birds means that each assessor needs one sample (i.e., it is only possible to analyze one leg per assessor). This fact is very important in designing the trial (primarily to calculate the number of samples needed). Tables 2 and 3 show the sensory descriptors that can be used in the sensory characterization of poultry meat, with their definition and analysis procedure. The sensory profile can be extended to

TABLE 2 Sensory Attributes Used in the Characterization of Raw Poultry Meat Products

Aspect: evaluated using standard artificial daylight

White/yellow skin color: yellow color intensity in the skin (white = 0, yellow = 10)

White/pink skin color: pink color intensity in the skin (white = 0, pink = 10)

Lightness/darkness of the meat: quantification of the darkness of the meat after raising the skin

Color uniformity of the meat: quantification of the color variations on the sample surface

Internal fat color: yellow color intensity in the internal fat of the sample (white = 0, yellow = 10)

Conformation: quantification of the muscle building and the shape of the piece (leg or breast) being evaluated

Amount of feathers: quantification of the feathers that remain adhering to the skin

Odor: evaluated by direct nasal evaluation of the surface of the product

Rancid: intensity of rancid odor perception, similar to old olive oil odor

Raw meat/poultry meat: intensity of raw meat and poultry meat

Blood/liver/metallic: quantification of metallic odor, similar to that of metallic surfaces

Nonoral texture: evaluated using the sense of touch

Hardness of the internal fat: quantification of the firmness of the internal fat

include other descriptors according to each case, but it is important to avoid the use of hedonic and other nonpertinent attributes. According to Byrne et al. (1999), the phenomenon of warmed-over flavor (WOF) is complex and its mechanism is not fully understood. Probably, more specific terminology is required to describe the flavor notes that arise as WOF develops. However, in some cases it can be useful to employ the descriptor WOF as a single attribute, since a good level of agreement is normally reached when scoring different samples during the training period and also for their commercial importance. Panelists evaluate the flavor and texture of the samples, scoring each attribute using a numerical intensity scale ranging from 0 (none) to 15 (extreme). Table 4 shows the main attributes and methods of sample preparation for poultry meat available in the literature.

Sample Preparation of Elaborated Poultry Products

Consumption of elaborated poultry products has increased due to the fact that they are included in most diets considered as healthy. The high yield of turkey carcass makes it suitable for the elaboration of different products, some of the best known being cured turkey ham and cooked brine-injected breast. The preparation of samples for sensory analyses of these products is similar to that for cooked meat regarding general parameters (e.g., temperature, order of presentation), but in this case samples are presented to tasters in slices 2 mm thick. Again, it is advisable to provide each assessor with two slices, one for evaluation of the flavor attributes and the other for texture evaluation. Based on our own experience, the

TABLE 3 Sensory Attributes Generally Used in the Characterization of Cooked Poultry Meat Products[a]

Odor: evaluated immediately after opening a wrapped sample
- *Rancid:* intensity of rancid odor perception, similar to old olive oil odor
- *Toasted:* typical odor of toasted meat
- *Blood/liver/metallic:* quantification of metallic odor, similar to that of metallic surfaces
- *Other attributes:* hazelnut, ...

Flavor: evaluated through mastication of a sample
- *WOF* (warmed-over flavor): intensity of the oxidized flavor of reheated meat
- *Rancid:* intensity of rancid flavor perception, similar to old olive oil flavor
- *Blood/liver/metallic:* quantification of metallic flavor, similar to that produced by a solution of iron sulfate(II)
- *Poultry flavor:* quantification of the intensity of poultry flavor
- *Sweetness:* basic taste produced by a water solution with sugar
- *Other attributes:* acid taste, bitter taste, ...

Oral texture: evaluated through mastication of a sample between the molar teeth
- *Hardness:* force required to bite through the sample
- *Initial juiciness:* moisture released by the product in the mouth as a result of initial chewing
- *Pastiness:* pasty feeling inside the mouth
- *Stringiness:* the perception of long and parallel coarse particles in the meat during chewing
- *Final juiciness:* moisture released by the product in the mouth as a result of final chewing
- *Tooth adhesion:* perception of mouth residuum that remains stuck to the teeth once chewing is finished
- *Other attributes:* gumminess, adhesiveness, tenderness, ...

[a] Sensory assessments should be made under controlled lighting conditions to mask differences in product appearance.

main descriptors to be included in this sensory profile are the basic tastes, poultry flavor and off-flavors, hardness, juiciness, crumbliness, stringiness, and pastyness.

CARRYING OUT SENSORY ANALYSIS

A panel of 8 to 10 trained tasters should be used to sensory-analyze a maximum of 8 to 10 samples per session according to the tasters' experience and the number of attributes to be evaluated. At least, four replications of each treatment will be carried out because of the high variability existing among animals even if they belong to the same treatment. For this reason it is advisable to do a high number of replications. In addition, and to balance the order of sample presentation, the design of McFie et al. (1989) is recommended. Sensory evaluation will be conducted in a sensory testing room equipped with positive air pressure, individual booths, and red lights to mask obvious color differences (ISO, 1988) except for visual evaluation. It is important to rinse the

TABLE 4 Sensory Descriptors of Poultry After Various Preparation Methods

References	Sample Preparation	Sensory Descriptors
Bou et al. (2001)	Deboned legs with skin, and vacuum-packed, were cooked at 80°C for 35 min in a pressure cooker. Other samples were also cooked in an oven (90°C and 90% RH) until the center of the leg reached 80°C. Samples stored at −20°C until sensory evaluation. Samples thawed by reheating 10 min at 85°C in a water bath.	Rancid aroma, rancid flavor, acceptability
Byrne et al. (1999)	Patties of chicken breast muscle (pectoralis major), frozen and vacuum-packed, were placed in a 25°C water bath until the sample core temperature reached 15 to 18°C. Patties were than removed from their vacuum bags and heated in convection ovens at 160, 170, 180, and 190°C. Final internal temperature ranged between 78 and 82°C. Storage time after cooking was 0, 2, and 4 days. Patties reheated before sensory evaluation.	*Odor:* cardboard-like, linseed oil–like, sulfur-like, chicken meat–like, roasted-like *Flavor:* chicken meat–like, rancid-like, vegetable oil–like, bread-like, toasted-like, nut-like *Taste:* umami, metallic, bitter, sweet, salt, sour *Aftertaste:* astringent
Carreras et al. (2004)	Vacuum-packed breast cooked for 30 min in an 80°C water bath and stored 6 days at 4°C before sensory evaluation. On day 6, samples were reheated for 25 min at 65°C in an electric oven and served at 70 ± 5°C.	*Flavor:* rancidity, WOF *Texture:* initial juiciness, hardness, pastiness, stringiness, tooth adhesion
Chartrin et al. (2006)	Breast samples of duck at 4°C overnight. Samples were grilled (2 to 4 min on each side at 170°C) cut into 12 pieces.	*Raw meat:* Color intensity (darkness, intermediate color, lightness) *Cooked meat:* juiciness, tenderness, stringiness, duck flavor, overall flavor

Kennedy et al. (2005)	Chicken wrapped in aluminum foil and placed breast side up in Pyrex roasting dishes for 2 h at 200°C using a fan-assisted oven until reaching 80°C internal temperature. Sensory evaluation of the breasts.	Sensory evaluation using a consumer panel; products evaluated in terms of flavor, juiciness, texture, appearance, freshness, and overall liking
Larmond et al. (1983), Salmon et al. (1988)	Oven roasting at 162°C until an internal turkey breast temperature of 85°C was attained.	Flavor and off-flavor
Liu et al. (2004a, 2004b), Lyon et al. (2004, 2005)	Right breast muscles. Individually frozen and bagged samples were cooked by immersing the bags in 85°C water for about 25 min to achieve a maximum breast internal temperature of 80°C. Breast cut in 1.9 × 1.9 cm samples for sensory evaluation. Samples served monadically to panelist at 55°C.	*Sensory flavor:* brothy, chicken-meaty, cardboardy, wet feathers, bloody-serumy, sweet, salty, sour *Sensory texture:* springiness, cohesiveness, hardness, moisture release, particle size, bolus size, chewiness, toothpack *Afterfeel–aftertaste:* metallic, oily-greasy
Lyon (1988)	Chicken patties made of 50% white meat and 50% dark meat frozen at −34°C. Thawed raw samples were cooked in a 177°C oven to an internal temperature of 75 to 80°C and then sensory-evaluated. The same precooked samples were reheated in a 163°C conventional oven for 23 to 25 min and then sensory-evaluated.	*Aromatic/taste:* chickeny, meaty, brothy, liver/organy, browned, burned, cardboard/musty, warmed-over, rancid/painty *Primary taste:* sweet, bitter *Feeling factor on tongue:* metallic
Perlo et al. (2006)	Nuggets of chicken meat evaluated after one month in frozen storage at −22 ± 2°C. Samples thawed at 4 ± 1°C and heated in a warm oven at 100°C.	Internal color, hardness, chewiness, cohesiveness, and appearance

(*continued overleaf*)

TABLE 4 (*Continued*)

References	Sample Preparation	Sensory Descriptors
Rababah et al. (2005)	Chicken breasts were cooked in an electric oven preheated to 190°C until an internal temperature of 74°C. Samples were cooled to 43°C at room temperature and sectioned into cubes.	Raw chicken *Aroma:* fresh chickeny, sweet aromatic, oxidized *Appearance:* color, moistness Cooked chicken *Appearance:* color *Taste:* sweet, salty, sour, bitter *Aroma:* chickeny, brothy, sweet aromatic, cardboard *Texture/mouth feel tenderness*
Ruiz et al. (2001)	Legs cooked hanging in an electric oven preheated to 170°C until and internal temperature of 80°C was reached.	*Raw meat:* Visual (white/yellow skin color, white/pink skin color, lightness/darkness of meat, color uniformity of meat, internal fat color, leg conformation (quality of feathers); odor (rancid, raw meat/poultry meat, blood/liver/metallic); texture (hardness of internal fat) *Cooked meat:* odor (rancid, cooked skin, peanut/hazelnut, toasted); flavor (rancid, blood/liver/metallic, peanut/hazelnut, poultry flavor); texture (initial juiciness, tenderness, pastiness, stringiness, final juiciness, teeth adhesion)

Sárraga et al. (2006)	Breast and legs of turkey vacuum-packed and cooked for 80 and 120 min, respectively, in an 80°C water bath. M. gastrocnemius lateralis and M. gastrocnemius medialis of each leg were dissected and used for flavor and texture evaluation. Samples stored for 1 day at 4°C. Before sensory analysis, samples were reheated for 25 min at 65°C in an electric oven.	*Flavor:* rancidity and WOF *Texture:* initial juiciness, hardness, pastiness, stringiness, and tooth adhesion
Vara-Ubol and Boweres (2002)	Ground turkey meat with additives. Evaluation of the patties. Patties cooked in a rotary hearth oven at 163°C to an internal temperature of 80°C. Samples stored at 4°C for 2 and 4 days. Reheated in a microwave oven at a power level of 7 for 2 min.	*Aroma:* turkey, meaty, stale *Flavor:* turkey, meaty, stale, bloody-serumy, salty Juiciness Slick mouthfeel *Aftertaste:* metallic and soapy, bitter
Vermerein et al. (2006)	Cooked chicken fillet and cooked turkey fillet. Comparison between fillets (inoculated and noninoculated products) at different storage times.	Attributes of odor and taste (fresh to spoiled) and fitness for human consumption
Williams and Damron (1998)	Breast meat of broilers 48 h at 7°C. Breast wrapped in aluminum foil and cooked at 176.7°C in a conventional preheated gas oven to 80°C internal muscle temperature. Cooked chicken cooled at room temperature for 10 min. The skin was removed and the breast was separated into pectoralis major and pectoralis minor muscles. Samples of pectoralis major were cut in 1.25 × 1.2 cm cubes. Samples were kept warm until served.	Flavor, juiciness, tenderness, and off-flavor

mouth at the beginning of the sensory evaluation and between samples. This should be done with mineral water, unsalted bread toasted, or apples.

STATISTICAL ANALYSIS

The statistical analysis of the data varies according to the part of the animal on which we focus. As stated before, the procedures for sensory evaluation of legs or breasts are notably different. Something similar happens with other poultry products, such as sausages and ground meat, since each of them may have a specific method of data analysis, depending on the experimental design used. In any case, the first step is the definition of the experimental unit, the basis on which the study or experiment is carried out. In the case of poultry legs or breasts, the experimental unit is normally each animal. However, when dealing with poultry products made up of mixtures of minced parts commonly from different animals, the experimental unit is usually each piece (e.g., each sausage).

Once the experimental unit is defined, it is important to point out that the final number of observations used for drawing conclusions should be equal to or less than the number of experimental units. For example, if we have eight animals that are all evaluated by eight assessors, we will obtain a data matrix of 64 observations. However, as the experimental unit is the animal, conclusions should be drawn from a statistical analysis that considers only eight observations, not 64. The simplest way to do this is by using the average of the eight assessors for each experimental unit or animal instead of the original data. Generally speaking, the most convenient, powerful, and commonly used technique for analyzing sensory data is the analysis of variance.

In general, when the sensory properties of poultry legs are analyzed, each assessor evaluates one leg. In this situation the ANOVA model should include the treatment, the assessors, the tasting session, and the interaction treatment $\times$ assessor as fixed effects (if there are replicates). The principal inconvenience of this design is the confusion of the animal effect and assessor effect. This inconvenience may be eliminated by using two legs from the same animal, one for each of two different assessors. In this case, the simplest way to prepare data for statistical analysis is to calculate the mean value of two legs provided by two different assessors, this mean value representing the value of the experimental unit. The ANOVA model is simpler, as it includes only the treatment and taste session as fixed factors.

When analyzing breast samples, the same sample is normally used for several tasters. Again, in this case the most convenient solution is to compute the mean value of all the assessors for each breast or animal. For poultry products the idea is the same: The mean value of all the assessors when they evaluate the same experimental unit is the most frequent situation.

The statistical models described here are the simplest way to analyze sensory data. However, more sophisticated models can also be used. A more complex model can include the animal as a random effect or can include other interactions between the main factors.

REFERENCES

Amerine MA, Pangborn RM, Roessler EB. 1965. *Principles of Sensory Evaluation of Foods*. New York and London: Academic Press.

Bou R, Guardiola F, Grau A, Grimpa S, Manich A, Barroeta A, Codony RF. 2001. Influence of dietary fat source, alpha-tocopherol, and ascorbic acid supplementation on sensory quality of dark chicken meat. Poult Sci 80:800–807.

Byrne DV, Wender LP, Bredie WLP, Martens M. 1999. Development of a sensory vocabulary for warmed-over flavor: II. In chicken meat. J Sensory Stud 14:67–78.

Carreras I, Guerrero L, Guàrdia MD, Esteve-Garcia E, García Regueiro JA, Sárraga C. 2004. Vitamin E levels, thiobarbituric acid test and sensory evaluation of breast muscles from broilers fed α-tocopheryl acetate- and β-carotene-supplemented diets. J Sci Food Agric 84:313–317.

Chartrin P, Méteau K, Juin H, Bernadet MD, Guy G, Larzul C, Rémignon H, Mourot J, Duclos MJ, Baéza E. 2006. Effects of intramuscular fat levels on sensory characteristics of duck breast meat. Poult Sci 85:914–922.

Civille GV, Lyon BG. 1996. *Aroma and Flavor Lexicon for Sensory Evaluation: Terms, Definitions, References, and Examples*. Philadelphia: American Society for Testing and Materials, pp. 46–49.

Civille GV, Szczesniack AS. 1973. Guidelines to training a texture profile panel. Texture Stud 4:204–223.

Guerrero L, Guàrdia MD. 1998. Evaluación de la fiabilidad de un panel de cata. III Jornadas de Análisis Sensorial, Valdediós, Villaviciosa, Asturias, España.

Ishihara S. 1971. *Tests for Colour-Blindness*. Tokyo: Kanehara Shuppan.

ISO (International Organization for Standardization). 1988. *Sensory Analysis: General Guidance for the Design of Test Rooms*. ISO 8589. Geneva, Switzerland: ISO.

ISO. 1994a. *Sensory Analysis: Identification and Selection of Descriptors for Establishing a Sensory Profile by a Multidimensional Approach*. ISO 11035. Geneva, Switzerland: ISO.

ISO. 1994b. *Sensory Analysis: Methodology. Texture Profile*. ISO 1036. Geneva, Switzerland: ISO.

Issanchou S, Schlich P, Lesschaeve I. 1997. Sensory analysis: methodological aspects relevant to the study of cheese. Lait 77:5–12.

Jensen C, Lauridsen C, Bertelsen G. 1998. Dietary vitamin E: quality and storage a stability of pork and poultry. Trends Food Sci Technol 9:62–72.

Kennedy OB, Steward-Know BJ, Mitchell PC, Thurnham DI. 2005. Vitamin E supplementation, cereal feed type and consumer sensory perception of poultry meat quality. Br J Nutr 93:333–338.

Larmond E, Salmon RE, Klein KK. 1983. Effect of canola meal on the sensory quality of turkey meat. Poult Sci 62:397.

Liu Y, Lyon BG, Windham WR, Lyon CE, Savage EM. 2004a. Prediction of physical, color, and sensory characteristics of broiler breasts by visible/near infrared reflectance spectroscopy. Poult Sci 83:1467–1474.

Liu Y, Lyon BG, Windham WR, Lyon CE, Savage EM. 2004b. Principal component analysis of physical, color, and sensory characteristics of chicken breasts deboned at two, four, six, and twenty-four hours postmortem. Poult Sci 83:101–108.

Lyon BG. 1988. Descriptive profile analysis of cooked, stored, and reheated chicken patties. J. Food Sci 53(4):1086–1090.

Lyon BG, Smith CE, Savage EM. 2004. Effects of diet and feed withdrawal on the sensory descriptive and instrumental profiles of broiler breast fillets. Poult Sci 83:275–281.

Lyon BG, Smith CE, Savage EM. 2005. Descriptive sensory analysis of broiler breast fillets marinated in phosphate, salt, and acid solutions. Poult Sci 84:345–349.

MacFie HJ, Bratchell N, Greenhoff K, Vallis L. 1989. Designs to balance the effect of order of presentation and first-order carry-over effects in hall tests. J Sensory Stud 4:129–148.

Meilgaard M, Civille GV, Carr BT. 1987. *Sensory Evaluation Techniques*, vols. 1 and 2, Boca Raton, FL: CRC Press.

Perlo F, Bonato P, Teira G, Fabre R, Kueider S. 2006. Physicochemical and sensory properties of chicken nuggets with washed mechanically deboned chicken meat. Meat Sci 72:785–788.

Poste LM. 1990. A sensory perspective of effect of feeds on flavor in meats: poultry meats. J Anim Sci 68:414–420.

Rababah T, Hettiarachchy NS, Eswaranandam S, Meullenet JF, Davis B. 2005. Sensory evaluation of irradiated and nonirradiated poultry breast meat infused with plant extracts. J Food Sci 70(3):228–235.

Ruiz JA, Guerrero L, Arnau J, Guàrdia MD, Esteve-Garcia E. 2001. Descriptive sensory analysis of meat from broilers fed diets containing vitamin E or beta-carotene as antioxidants and different supplemental fats. Poult Sci 52:213–219.

Salmon RE, Stevens VI, Poste LM, Agar V, Butler G. 1988. Effect of roasting breast up or breast down and dietary canola meal on the sensory quality of turkeys. Poult Sci 67:680.

Sárraga C, Carreras I, García Regueiro JA, Guàrdia MD, Guerrero L. 2006. Effects of α-tocopheryl acetate and β-carotene dietary supplementation on the antioxidant enzymes, TBARS and sensory attributes of turkey meat. Br Poult Sci 47:700–707.

Sárraga C, Guàrdia MD, Díaz I, Guerrero L, García-Reguiero JA, Arnau J. 2007a. Nutritional and sensory quality of porcine raw meat, cooked ham and dry-cured shoulder as affected by dietary enrichment with docosahexaenoic acid (DHA) and α-tocopheryl acetate. Meat Sci 76(2):377–384.

Sárraga C, Guerrero L, Díaz I, Guàrdia MD. 2007b. Implicaciones nutricionales del enriquecimiento con DHA y vitamina E de las dietas de cerdo y pavo: evaluación de la calidad del jamón cocido, paletilla curada y embutido de pavo. Eurocarne 76:55–63.

Sebranek JG, Sang PN, Topel DG, Rust E. 1979. Effects of freezing methods and frozen storage on chemical characteristics of groung beef patties. J Anim Sci 4(5):1101–1108.

Vara-Ubol S, Boweres JA. 2002. Inhibition of oxidative flavor changes in meat by α-tocopherol in combination with sodium tripolyphosphate. J Food Sci 67(4):1300–1307.

Vermein L, Devlieghere F, Vandekinderen I, Rajtak U, Debevere J. 2006. The sensory acceptability of cooked meat products treated with a protective culture depends on glucose content and buffering capacity: a case study with *Lactobacillus sakei* 10A. Meat Sci 74:532–545.

Williams SK, Damron BL. 1998. Sensory and objective characteristics of broiler meat from commercial broilers fed rendered whole-hen meal. Poult Sci 77:329–333.

22

TEXTURE AND TENDERNESS IN POULTRY PRODUCTS

LISA H. MCKEE

Department of Family and Consumer Sciences, New Mexico State University, Las Cruces, New Mexico

INTRODUCTION

Defined as the rheological and structural attributes of a food as perceived by mechanical, tactile, visual, and/or auditory means (ISO, 1981), *textural properties* of food are typically used by the consumer to determine the quality of a product. Texture is one of the most complex organoleptic properties of food and can be difficult to measure and assess. Lawless and Heymann (1999) separated texture into auditory, visual, and tactile properties. Auditory texture is illustrated by the

Handbook of Poultry Science and Technology, Volume 2: Secondary Processing, Edited by Isabel Guerrero-Legarreta and Y.H. Hui

crispiness of a potato chip or the crunch of a fresh apple. The thickness of a milkshake can be assessed visually when stirred with a straw. Tactile texture can be subdivided into oral components, such as mouthfeel, size and shape, and handfeel. Although a number of reviews on textural properties of foods have been published (Bourne, 1982; Moskowitz, 1987), a great deal about human perception of texture remains unknown.

Texture can be evaluated by both instrumental and sensory methods. Most instrumental methods are rheological in nature, where the resistance of a sample to some type of deformation stress is measured. The texturometer, an instrument designed by scientists at General Foods which measured the force required for penetration of plungers into the food (Friedman et al., 1963), was the forerunner of a wide range of instruments, including the Brookfield viscometer, the Instron Universal Testing machine, and the TA.XT2 texture analyzer. Although the objective of most instrumental texture analysis is to provide a mechanical method that will correlate to sensory evaluations, correlation coefficients between the two types of texture measurement often span a very wide range. Sensory texture measurements can be made using standard sensory techniques such as ranking, descriptive analysis, scoring, or discrimination tests. Terminology and panelists' understanding of that terminology is extremely important to appropriate sensory evaluation, and methods such as texture profile analysis have been developed to train and standardize sensory panels to provide more accurate analysis of textural properties.

The textural properties of poultry products are affected by a number of factors. The popularity of portion-controlled boneless fresh poultry products, and single-type packaging such as all legs or breast pieces, have resulted in the need for value-added products that utilize less popular components such as backs and necks as well as carcasses remaining after processing. Removal of the meat from these less popular components significantly affects the textural properties. The addition of marinating solutions to provide tenderization and flavor also influences textural properties. Since little poultry is consumed raw, the effect of heat on the textural properties of poultry is a primary concern in both fresh and value-added poultry products.

EFFECT OF COOKING ON TEXTURE AND TENDERNESS

The addition of heat can cause a variety of reactions that will affect the textural properties of poultry meat. Denaturation and coagulation can cause shortening and hardening of muscle fibers. Changes in the muscle fibers influence the water-holding capacity of the meat, which, along with melting fat, affects juiciness and tenderness. The type of muscle being heated, the length and temperature of heating, and the presence of connective tissues also influence the textural properties of cooked poultry meat.

Murphy and Marks (2000) cooked preformed ground chicken breast patties to 40, 50, 60, 70, and 80°C to determine the effects of heat on the soluble protein content, collagen solubility, Warner–Bratzler shear values, and cooking loss.

Soluble protein content decreased, while soluble collagen increased as temperature increased. Peak shear force increased as temperature increased between 40 and 60°C, then decreased at 70 and 80°C. Cooking loss increased as temperature increased, due to shrinkage of myofibrillar proteins and loss of water-holding capacity. When data were analyzed using a developed linear model, temperature was correlated significantly with soluble protein content, collagen solubility, toughness, and cooking loss.

The quality characteristics of chicken breasts processed by either cook–chill (no vacuum sealing) or *sous vide* (vacuum sealed) methods were evaluated by Church and Parsons (2000). In the initial part of the study, samples were heated at 70°C for 2 min (70/2), 80°C for 10 min (80/10), and 80°C for 30 min (80/30) and then evaluated for organoleptic properties by a 25-member trained panel. In a subsequent study, samples heated at 80°C for 10 min were blast chilled to 3 ± 1°C and then stored at <5°C for 2, 5, and 7 days. Samples in the second study were evaluated by a 10-member trained panel. Samples freshly processed at 80/10 were judged to be more juicy and tender than product processed at either 70/2 or 80/30, regardless of packaging method. Chicken breasts processed by *sous vide* tended to be more juicy and tender than conventional cook–chill breasts, although the significance was variable. Neither chilled storage time nor processing method affected the tenderness of chicken breasts in the second study, although juiciness was judged to be greater in *sous vide* products stored 2 and 5 days. Overall, *sous vide* processing was found to have fewer benefits in terms of textural properties of chicken breasts than had been reported previously.

Zamri et al. (2006) evaluated the effects of combinations of heat and high-pressure treatments on the texture of breast meat. Heat and pressure were found to have a synergistic effect at temperatures through 50°C, with muscle hardness increasing as temperature and pressure increased. Muscle hardness decreased at 60 and 70°C when the pressure exceeded 200 MPa. Differential scanning calorimetry indicated that both myosin and actin completely denatured at 20°C when simultaneous pressure treatments above 200 MPa were applied. The more extreme high-pressure/temperature treatments were also found to induce disulfide bonding between myosin chains.

The effects of brine composition (sodium chloride or sodium chloride plus sodium lactate), internal temperature (77 or 94°C), heating rate (slow or fast), and storage periods (0, 14, and 28 days) on the textural properties of *sous vide* processed chicken breasts were investigated by Turner and Larick (1996). No differences in shear peak force were detected due to brine composition, heating rate, final internal temperature, or days of storage. Sensory panelists, however, found that chicken breasts processed to 77°C were more juicy and more tender than those processed to 94°C.

Barbanti and Pasquini (2005) investigated the effect of cooking on cooking loss and tenderness of raw and marinated chicken breast meat. Treatments included cooking with hot air or hot air plus steam at 130, 150, and 170°C for 4, 8, and 12 min. Marination followed by air–steam cooking produced the most tender chicken slices as determined by a TA.XT2 texture analyzer using a Warner–Bratzler blade. Unlike other reports in the literature, marination was associated with the greatest cooking losses. The authors speculated that this result was probably due to the lack of polyphosphates in the marinade. Cooking loss was correlated significantly with cooking time, and shear force with cooking temperature. Short cooking times (4 min) and low temperatures (130 to 150°C) were associated with lower cooking losses and greater tenderness.

Chicken breast meat infused with grape seed extracts, green tea extracts, or tertiary butylhydroquinone was cooked by either microwave or conventional electric oven and evaluated for pH, color, oxidation, and textural properties by Rababah et al. (2006). Breasts cooked by microwave were found to have greater values of maximum shear force, work of shearing, hardness, cohesiveness, springiness, and chewiness than breasts cooked in a conventional electric oven. Addition of any of the plant extracts did not affect the textural properties. The increased textural values associated with microwave cooking were attributed to less denaturation of the muscle proteins as well as compression and compaction of the chicken breast fibers due to high-speed molecular motion during microwave cooking.

Jiménez-Colmenero et al. (1998) utilized high-pressure processing and different cooking temperatures to prepare low-fat chicken batters. Samples were treated at either 200 or 400 MPa using water at 60, 70, or 80°C as the pressurizing medium. The texture profile, including hardness, peak force required for first compression, cohesiveness, springiness, and chewiness, was analyzed using an Instron Universal Testing machine. Pressure treating was associated with improved water-binding properties at both 70 and 80°C. In nonpressurized samples, hardness and chewiness increased with increasing internal temperature, but cohesiveness decreased when the temperature increased from 60°C to 70°C, particularly in higher-salt batters. Hardness, cohesiveness, springiness, and chewiness increased with increasing cooking temperatures in pressure-treated samples but were lower in pressurized chicken batters compared to the nonpressurized samples at all cooking temperatures. Changes in textural properties due to pressurization were attributed to limited formation of gel structures.

The effects of high-temperature processing and moisture level on texture and collagen solubilization of gels prepared from spent breast meat were investigated by Voller-Reasonover et al. (1997). Gels in the study contained 0, 10, or 20% added water and were first water bath–processed to an internal temperature of 82°C. In experiment 1, gels were either untreated after water cooking or retorted to 121.1°C. In experiment 2, samples were retorted to 115.6, 121.1, and 126.7°C. Processing times at each temperature were calculated to equal an F_0 value of 6.0; samples cooked to 115.6°C were processed for 26 min, those cooked at 121.1°C were processed for 12 min, and gels cooked to 126.7°C were processed for 6 min. Gel hardness was lower in retorted samples than in water-cooked samples and decreased with added water content for both water-cooked and retorted samples in experiment 1. Water-cooked samples had lower fracturability than retorted gels, but greater stress, strain, elasticity, cohesiveness, chewiness, and gumminess. These characteristics were attributed to weakening of the gel structure as a result of the additional thermal processing in retort samples. Gels processed at

115.6°C in experiment 2 had increased collagen solubilization and lower hardness compared to those processed at 126.7°C. Crude protein content was positively correlated with gumminess, chewiness, and elasticity, while higher fat content was associated with lower elasticity and increased cohesiveness. These attributes were thought to be related to differences in moisture content which subsequently affected fat and protein contents.

TEXTURAL PROPERTIES OF SPENT HEN MEAT

Meat from spent hens (layers past egg production age) is typically tough and has poor water-holding capacity, emulsification ability, and other functional properties, due to high collagen content and increased protein cross-links. Since tenderization of such meat using phosphates (Baker and Darfler, 1968), enzymes (Devitre and Cunningham, 1985), and electrolytes (Lyon and Hamm, 1986) has been only partially successful, Naveena and Mendiratta (2001) evaluated ginger extract as a tenderizing agent for spent hen meat. Raw meat chunks treated with 1, 3, and 5% fresh ginger extract were analyzed for a variety of properties, including moisture content, water-holding capacity, collagen solubility, and muscle fiber diameter, while samples cooked to 70°C were evaluated for moisture content, shear force values, and sensory characteristics. Increasing concentration of ginger extract was associated with increased collagen solubility, greater protein proteolysis, decreased shear force values, and improved juiciness and tenderness. An optimal level of 3% ginger extract was then applied to spent breast meat either before or after chilling. Water-holding capacity was greater in post-chill-treated samples, but few other differences in physicochemical properties were detected, due to time of treatment. Juiciness and tenderness scores tended to be higher in post-chill-treated samples. Overall, ginger extract was found to be an acceptable tenderizing agent for use with spent hen meat.

Nurmahmudi and Sams (1997a,b) and Nurmahmudi et al. (1997) conducted a series of studies to determine the effects of calcium chloride on the tenderization of spent chicken meat. In the initial study (Nurmahmudi and Sams, 1997a), 0.2 M and 0.3 M calcium chloride solutions were added to spent breast fillets by tumbling, injection, or soaking methods. Injection of the 0.3 M calcium chloride solution followed by tumbling was the only treatment to reduce mean shear values to acceptable levels. Nurmahmudi et al. (1997) then compared the effects of injecting 0.3 M calcium chloride and 0.6 M sodium chloride into hot-boned spent breast fillets 30 min and 24 h postmortem. Shear values decreased with the addition of either sodium chloride (13.9 kg/g) or calcium chloride (13.0 kg/g) immediately after deboning compared to the hot-boned control (21.4 kg/g) but were not significantly different from the cold-boned control (12.4 kg/g). Replenishment of 0.3 M calcium chloride to 10% at 24 h postmortem did not improve tenderness, but delaying initial application of the chemicals to 24 h postmortem decreased shear values to 9.8 kg/g for calcium chloride and 10.0 kg/g for sodium chloride. Similarities between sodium chloride and calcium chloride suggested that ionic strength may play a role in the tenderization of spent hen meat. Further research by Nurmahmudi and Sams (1997b) confirmed the influence of ionic strength on tenderization. Lack of differences in collagen content between treatments also suggested that tenderization was not due to degradative changes in collagen.

Woods et al. (1997) continued the work of Nurmahmudi and Sams (1997a, 1997b) by investigating the effects of injecting $CaCl_2$ and NaCl and additional aging on the tenderness of spent chicken meat. In the first experiment, spent breast fillets were injected with either 0.3 M $CaCl_2$ or deionized water, vacuum tumbled, and then either cooked immediately or aged for 23 h at 1°C prior to cooking. In the second experiment, spent breasts were injected with 0.3 M $CaCl_2$, 0.6 M NaCl, 0.15 M $CaCl_2$ + 0.3 M NaCl, or water, vacuum tumbled, and aged 23 h at 1°C prior to cooking. As with other studies, injection of $CaCl_2$ was correlated with lower shear values. Additional tenderization was noted when breasts were aged prior to cooking. Similar shear values were noted for the three salt treatments, but the $CaCl_2$ and combination treatments were associated with shorter sarcomeres than was the NaCl treatment. Although the combination treatment resulted in a more tender product with greater acceptability to sensory panelists, shear values remained above the threshold level reported by Lyon and Lyon (1990) to be considered tender by consumers.

The quality characteristics of raw emulsions prepared from spent hens to be marketed as semiconvenient, ready-to-prepare products were investigated by Kala et al. (2007). Prime (67% lean chicken meat), choice (57% lean chicken meat plus by-products), and economy (47% lean chicken meat plus by-products plus extenders) emulsions were stored at 4 ± 1°C and evaluated at days 0, 3, 6, 9, and 12 for emulsion stability, pH, proximate composition, and changes in color and odor. Patties prepared from the emulsions on each of the analysis days were baked at 160 ± 5°C to an internal temperature of 85°C and evaluated for cooking yield, shrink, hardness, adhesiveness, springiness, gumminess, chewiness, shear force, and sensory properties. Patties prepared from the prime emulsion had the lowest shrink percentage and significantly higher values for shear force, hardness, springiness, gumminess, and chewiness than those of the choice and economy emulsions. The addition of egg and cooked potato to the economy emulsion probably contributed to the greater adhesiveness detected in patties prepared from that emulsion. Sensory scores for juiciness were not different between treatments, but texture and binding scores were lower for the economy emulsion patties. Although the economy emulsion was found to have less desirable properties, all three emulsions were considered suitable for marketing as semiconvenient, ready-to-prepare products.

TEXTURAL PROPERTIES OF MECHANICALLY DEBONED POULTRY MEAT

Mechanical separation of poultry meat from bone has been used for many years as a means of obtaining the meat remaining on whole or partial poultry carcasses.

Such processing has significant effects on the meat texture and generally results in a product that contains skin and trimmings as well as meat. The meat emulsion from mechanical deboning has been used as the basis for poultry products such as frankfurters, lunch meat, and other value-added products.

Abdullah and Al-Najdawi (2005) compared the functional and sensory properties of mechanically or manually deboned meat from spent hens. Meat from manually and mechanically deboned whole carcasses and manually and mechanically deboned skinned carcasses was evaluated for pH, emulsifying capacity, water-holding capacity, pigment concentration, and sensory properties. Emulsifying capacity was greatest in meat from mechanically deboned skinned carcasses, while no difference in water-holding capacity was detected between treatments. Emulsification capacity tended to increase during the first two months of frozen storage in manually deboned meat and to decrease in mechanically deboned meat, while water-holding capacity decreased in skinned carcasses throughout frozen storage. No significant differences in texture were detected by the trained sensory panel during the first six weeks of frozen storage, but mechanically deboned poultry meat was rated significantly lower in texture desirability after 12 weeks of frozen storage. The authors concluded that manually deboned poultry meat had keeping qualities superior to those of mechanically deboned product.

Barbut and Somboonpanyakul (2007) evaluated the effect of crude malva nut gum (CMG) added at 0.0, 0.2, and 0.6% levels and sodium tripolyphosphate (TPP) added at 0.0 and 0.5% levels on the yield, textural characteristics, color, and microstructure of mechanically deboned chicken meat batters. Batters containing 0.2 and 0.6% CMG and 0.5% TPP as well as batter containing 0.5% TPP alone had increased moisture retention compared to the control. Hardness was lower in batter containing 0.6% CMG than in the control, possibly due to the excess CMG hindering binding of the meat proteins. Addition of 0.5% TPP plus either 0.2% or 0.6% CMG resulted in harder products compared to those containing 0.6% CMG alone. The presence of 0.5% TPP either alone or in combination with 0.2% or 0.6% CMG was also associated with greater springiness than were products containing only CMG. The authors concluded that the addition of TPP enhanced the ability of CMG to improve the qualities of poultry meat batters.

One frequent outlet for MDPM is in the production of frankfurters. González-Viñas et al. (2004) determined the quality characteristics and consumer preferences of commercially available frankfurters in Spain. Ten frankfurter samples containing various combinations of chicken, turkey, beef, and pork were analyzed for physicochemical composition, including water activity, Warner–Bratzler shear values, and sensory properties using free-choice profiling. Water activity differed little between the samples, ranging from 0.954 to 0.972. Lower breaking-force values were associated with samples containing higher fat contents, while increased dry matter and decreased salt concentration were associated with greater cutting resistance. Generalized Procrustes analysis of sensory data indicated samples 3, 4, and 9, which contained pork, beef, turkey, and chicken, as well as sample 2, which contained pork and turkey, had greater juiciness. Samples 6 and 8, containing pork and turkey, were found to have inferior firmness, hardness, compactness, and smoothness but better juiciness. The most expensive samples [2 (pork and turkey) and 10 (turkey, pork, and beef)] were found to have superior firmness, compactness, and hardness, as well as greater cutting resistance.

Babji et al. (1998) evaluated the effect of chicken skin on the quality characteristics of frankfurters prepared with MDPM. Frankfurters prepared with 80 : 0, 70 : 10, 60 : 20, and 50 : 30 ratios of MDPM to cooked chicken skin were evaluated for a variety of characteristics, including shear force as measured by a Warner–Bratzler shear and gel strength as measured by a folding test. A 30-member untrained sensory panel also evaluated the products for quality characteristics, including hardness and juiciness. Cooking loss was found to be inversely related to the ratio of MDPM to chicken skin, with the formulation containing 30% skin having the lowest cooking loss. Objective measurements indicated that the presence of chicken skin was associated with increased hardness. Frankfurters containing 0 and 10% chicken skin at the zero storage time were the only products scored as “good grade” on the folding test. Gelation quality (folding test score) decreased and shear force increased as the frozen storage time increased. Sensory panelists detected no differences in hardness or juiciness between formulations during initial evaluations, but found that the formulation prepared without chicken skin was more acceptable for these characteristics after frozen storage.

TEXTURE OF PALE, SOFT, EXUDATIVE POULTRY MEAT

Pale, soft, exudative (PSE) poultry meat occurs when acidic conditions in the muscle and high temperatures are present during rigor mortis. Antemortem stress and/or genetics are among the conditions associated with the development of PSE in poultry. PSE meat has been reported to exhibit both lower water-holding capacity and lower gel-forming abilities (Barbut 1997; McKee and Sams, 1997, 1998).

Daigle et al. (2005) investigated the use of turkey collagen, soy protein, and carrageenan to improve the quality characteristics of chunked and formed deli turkey breast prepared from PSE-like meat. Water-holding capacity was determined by weighing samples before and after heat processing. An Instron Universal Testing machine was used to evaluate expressible moisture and protein bind, while a consumer sensory panel was utilized to determine consumer acceptability. Data indicated that water-holding capacity improved with the addition of either 1.5% soy protein concentrate or 1.5% turkey collagen, while 1.5% turkey collagen was associated with decreased expressible moisture. Addition of any of the additives increased protein bind, with the soy protein and turkey collagen improving the bind in PSE-like meat so that it was no longer different from normal meat. Although less effective, carrageenan did improve protein binding in PSE-like turkey meat and the authors speculated that a higher level of carrageenan may have produced a more pronounced effect. No differences in consumer acceptability were detected, indicating that PSE-like meat could be used in the production of acceptable deli turkey roll products.

The effect of regular and modified potato and tapioca starches on the texture and microstructure of batters prepared from normal; pale, soft, exudative (PSE); and dry, firm, dark (DFD) breast meat was investigated by Zhang and Barbut (2005). Cooked batters were evaluated for cooking loss, force required to fracture, and microstructure. A G' value, corresponding to the rigidity of the elastic response of the gelling material, was also determined. The PSE treatment had the lowest fracture values and a more open microstructure than those of the normal and DFD treatments. The G' value increased for all meat batters at temperatures greater than 55°C, indicating an increase in elasticity and gel formation. Both modified and regular potato starch produced higher G' values than for the two types of tapioca starch, but modified tapioca starch was associated with the highest cook yield. Although the addition of any of the starches improved the functionality of the PSE meat, the authors suggested that the choice of starch should be dictated by the specific functional properties required.

EFFECT OF MARINATION ON TEXTURE AND TENDERNESS

Marinade solutions are frequently used in processed poultry products to improve flavor and textural properties. Tumbling products in a marinade or injecting the solution into the poultry are two of the most common methods of marination. Lemos et al. (1999), however, investigated a simpler still-marinating process. Leg/thigh portions and skinless, boneless breast meat were still-marinated for varying lengths of time in solutions containing varying concentrations of NaCl and polyphosphates. Still-marinating was found to improve the tenderness of the boneless breast meat as indicated by the smaller shear values compared to nonmarinated controls. Although more space and time are required for still-marinating, the lower investment cost and simplicity of the process were said to be advantages to using still-marinating.

Marinating solutions may be composed of a variety of ingredients, but often contain salt and phosphates which function as water binders, flavor enhancers, and protein modifiers. Toughness reported in poultry that has been marinated at the processing plant level prompted Young and Lyon (1997b) to study the effect of sodium tripolyphosphate and postchill aging on the texture properties of chicken breast meat. Chicken forequarters were aged for 0, 120, 180, or 240 min postchill, and then one forequarter from each of the 16 birds was marinated in NaCl while the opposite forequarter was marinated in NaCl plus sodium tripolyphosphate. Cooked product texture was evaluated by Warner–Bratzler shear analysis. An interaction between postchill time and marinating was detected. Although shear values decreased as aging time increased regardless of treatment, the average shear value for marinated samples at 0 min (9.14 kg) was 60% higher than the value for nonmarinated controls (5.69 kg). No significant differences in shear values were noted for poultry aged 120 min or more regardless of treatment.

Based on previous research, the authors reported the average shear value of 3.22 kg for meat aged $\geq$ 120 min would correspond to a "very tender" rating if subjected to sensory analysis. Aging poultry parts a minimum of 120 min prior to polyphosphate treatment was recommended based on the results of the study.

The textural and microstructural properties of chicken breasts treated with 10% NaCl, trisodium phosphate (TSP), sodium tripolyphosphate (STPP), or tetrapotassium pyrophosphate (TKPP) during 10 months of frozen storage were investigated by Yoon (2002). All treatments were associated with improved water-binding ability as evaluated by drip loss compared to a water-treated control. After 10 months of storage at -20°C, breasts treated with 10% NaCl were significantly more tender than those treated with any of the phosphates or water. However, no significant toughening was noted for any of the treatments after frozen storage.

Young and Lyon (1997a) also studied the textural properties of peririgor chicken breast meat marinated in 0, 50, 100, 150, and 200 mM $CaCl_2$. Although neither marinated nor cooked pH were affected by calcium treatment, increasing calcium concentration was associated with both increased moisture absorption and increased cooking loss. All levels of calcium treatment improved breast tenderness. The authors concluded that although marination of chicken breasts with calcium may provide a useful means of reducing or eliminating the need for conditioning, methods to improve the water-binding properties damaged by the calcium would need to be developed.

Seabra et al. (2001) also reported increased cooking losses in chicken breasts marinated in calcium chloride. Marination in this study was applied in combination with either carcass aging or hot-boning treatments as well as muscle tensioning. Carcass aging was associated with greater tenderness than was hot-boning. Marination and muscle tensioning, however, did not improve chicken breast meat tenderness as evaluated by either mechanical shear values or by sensory responses.

Allen et al. (1998) determined the relationship between color and functional properties of broiler breast meat tumbled in a commercial marinade. Breast fillets visibly classified as light- or dark-colored were assigned to either marinade or control groups and tumbled for 20 min at 4°C and 80 kPa. Marination pickup and bound moisture were significantly higher, and drip and cooking losses were significantly lower in dark-colored fillets. No differences in shear values were associated with fillet color or marination treatment, but positive correlations between Allo-Kramer shear measurements with initial and tumbled L^* (lightness) values as well as cooking loss were noted.

Although many marinade-type solutions are used to enhance physical and organoleptic properties, other solutions are applied to poultry products to improve the microbiological qualities. Dickens et al. (1994) evaluated the appearance and microbiological quality of raw broiler carcasses as well as the texture and flavor properties of the cooked breast meat after treatment of carcasses with a 0.6% acetic acid solution. Shear values of breast meat cooked in an agitated water bath at 85°C for 30 min were determined using a Warner–Bratzler shear blade attached to an Instron Universal Testing machine. Treated and untreated breast muscles were also cooked by either water cooking in bags or roasting and evaluated by a 10-member trained sensory panel using a triangle test. No differences in Warner–Bratzler shear values were detected between acetic acid–treated carcasses and untreated carcasses, with all samples classified as "very tender." Sensory panelists also detected no differences between treated and untreated meat and no differences between cooking methods used to prepare the chicken breast meat.

ADDITIVES AND TEXTURAL PROPERTIES OF VALUE-ADDED POULTRY PRODUCTS

Formulation of poultry products to meet low-fat, low-carbohydrate, and other consumer dietary demands has led to the need for additives that can provide binding, texturizing, mouthfeel, and moisture retention. Barbut and Choy (2007) studied the effects of whole milk powder, skim milk powder, sodium caseinate, and regular and modified whey protein concentrate added at a 2% level on the yield, texture characteristics, and microstructure of a chicken breast meat system containing 51% water. All dairy products reduced cooking loss and increased values for hardness, fracturability, springiness, cohesiveness, and chewiness compared to the control. On an equivalent protein basis, whole milk powder and modified whey protein concentrate provided the highest moisture retention. Chicken meat batters containing modified whey had significantly higher values for hardness, fracturability, and chewiness compared to the other dairy additives evaluated when compared on an equivalent protein basis. Enhancement of textural properties by the dairy ingredients was thought to be due to development of fine-protein-matrix structures created by gelation of the dairy products in or around the chopped muscle fibers.

Hachmeister and Herald (1998) compared the effects of different starches (acid-thinned dent corn, cross-linked waxy maize, cross-linked dent corn, modified potato, acid-thinned dent corn plus xanthan gum and modified tapioca) on rheological and textural properties of reduced-fat turkey meat batters. The cross-linked starches as well as potato and tapioca starches reduced cooking and reheating losses and resulted in similar hardness properties. Few differences in cohesiveness, chewiness, gumminess, and springiness were noted between the treatments, although samples prepared with acid-thinned dent corn plus xanthan gum were found to be the least firm and most springy. Cross-linked cornstarches were able to provide texture properties similar to modified potato and tapioca starches and can provide useful alternatives in preparation of reduced-fat turkey batter products.

The use of modified cornstarch and water in reduced-fat turkey frankfurter formulations was investigated by Beggs et al. (1997). A Box–Wilson response surface design was used to determine the 20 combinations of modified cornstarch (2.379 to 6.621%) and water (20.93 to 35.07%) to be used in the formulations. Processed frankfurters were stored for 7 days at 5°C prior to analysis of compression characteristics, pH, color, and organoleptic properties. Higher levels of starch and lower levels of water were associated with increased resistance to compression, while springiness decreased with increasing water levels. Moisture release was greater at higher levels of water addition. Cohesiveness of frankfurters was greater at higher levels of starch addition. Increased levels of starch and water were speculated to decrease the protein–protein interactions, resulting in a softer, juicier frankfurter. Optimal levels of 2.3% starch and 33.6% water were determined from the sensory and physical data using regression equations.

Biopolymers produced from transglutaminase modification of soybean protein, casein, whey protein isolate, soybean protein plus casein, and soybean protein plus whey protein isolate were used in the formation of chicken sausages by Muguruma et al. (2003). Treatment with transglutaminase increased the temperature at which protein aggregation occurred and improved the emulsifying properties of the proteins. With the exception of casein, breaking strength increased in chicken sausages formulated with the biopolymers and 0.05% sodium tripolyphosphate compared to sausages prepared with either 0.05% or 0.2% sodium tripolyphosphate alone. The increased firmness of the biopolymer-containing sausages was attributed to increased gel structures composed of thick strands.

The effects of sodium lactate, calcium lactate, and a combination of the two additives on physicochemical properties of microwave-cooked chicken patties were studied by Naveena et al. (2006). Treatments included 3.3% sodium lactate, 0.25% calcium lactate, and 1.65% sodium lactate plus 0.125% calcium lactate as well as a 0% control. Patties were cooked for 4 min in a convection-type microwave oven and analyzed for shear force values, hardness, water-holding capacity, total expressible fluid, and other physicochemical properties at 0, 7, 14, 21, and 28 days of storage at $4 \pm 1^{\circ}C$. Addition of calcium lactate was associated with a lower water-holding capacity, greater expressible moisture, lower shear values, and higher penetrometer readings, indicating a softer texture. Few differences in physicochemical properties were detected between control, sodium lactate, and sodium lactate/calcium lactate patties, and the combination of sodium lactate and calcium lactate was recommended for production of chicken patties.

REFERENCES

Abdullah B, Al-Najdawi R. 2005. Functional and sensory properties of chicken meat from spent-hen carcasses deboned manually or mechanically in Jordan. Int J Food Sci Technol 40:537–543.

Allen CD, Fletcher DL, Northcutt JK, Russell SM. 1998. The relationship of broiler breast color to meat quality and shelf-life. Poult Sci 77:361–366.

Babji AS, Chin SY, Seri Chempaka MY, Alina AR. 1998. Quality of mechanically deboned chicken meat frankfurter incorporated with chicken skin. Int J Food Sci Nutr 49:319–326.

Baker RC, Darfler JM. 1968. A comparison of Leghorn fowl and fryers for pre-cooked battered fried chicken. Poult Sci 47:1550–1559.

Barbanti D, Pasquini M. 2005. Influence of cooking conditions on cooking loss and tenderness of raw and marinated chicken breast meat. Lebensm-Wiss Technol 38(8):895–901.

Barbut S. 1997. Occurrence of pale soft exudative meat in mature turkey hens. Br Poult Sci 38(1):74–77.

Barbut S, Choy V. 2007. Use of dairy proteins in lean poultry meat batters: a comparative study. Int J Food Sci Technol 42:453–458.

Barbut S, Somboonpanyakul P. 2007. Effect of crude malva nut gum and phosphate on yield, texture, color, and microstructure of emulsified chicken meat batter. Poult Sci 86:1440–1444.

Beggs KLH, Bowers JA, Brown D. 1997. Sensory and physical characteristics of reduced-fat turkey frankfurters with modified corn starch and water. J Food Sci 62(6): 1240–1244.

Bourne MC. 1982. *Food Texture and Viscosity: Concept and Measurement*. New York: Academic Press.

Church IJ, Parsons AL. 2000. The sensory quality of chicken and potato products prepared using cook–chill and *sous vide* methods. Int J Food Sci Technol 35:155–162.

Daigle SP, Schilling MW, Marriott NG, Wang H, Barbeau WE, Williams RC. 2005. PSE-like turkey breast enhancement through adjunct incorporation in a chunked and formed deli roll. Meat Sci 69:319–324.

Devitre HA, Cunningham FE. 1985. Tenderization of spent hen muscle using papain, bromelin or ficin alone and in combination with salts. Poult Sci 64:1476–1483.

Dickens JA, Lyon BG, Whittemore AD, Lyon CE. 1994. The effect of acetic acid dip on carcass appearance, microbiological quality and cooked breast meat texture and flavor. Poult Sci 73:576–581.

Friedman HH, Whitney JE, Szczesniak AS. 1963. The texturometer: a new instrument for objective texture measurement. J Food Sci 28:390–396.

González-Viñas MA, Caballero AB, Gallego I, García Ruiz A. 2004. Evaluation of the physico-chemical, rheological and sensory characteristics of commercially available frankfurters in Spain and consumer preferences. Meat Sci 67:633–641.

Hachmeister KA, Herald TJ. 1998. Thermal and rheological properties and textural attributes of reduced-fat turkey batters. Poult Sci 77:632–638.

ISO (International Organization for Standardization). 1981. *Sensory Analysis Vocabulary*, part 4. Geneva, Switzerland: ISO.

Jiménez-Colmenero F, Fernández P, Carballo J, Fernández-Martín F. 1998. High-pressure-cooked low-fat pork and chicken batters as affected by salt levels and cooking temperature. J Food Sci 63(4):656–659.

Kala RK, Kondaiah N, Anjaneyulu ASR, Thomas R. 2007. Evaluation of quality of chicken emulsions stored refrigerated ($4 \pm 1°C$) for chicken patties. Int J Food Sci Technol 42:842–851.

Lawless HT, Heymann H. 1999. Texture evaluation. In: *Sensory Evaluation of Food: Principles and Practices*. Gaithersburg, MD: Aspen Publishers.

Lemos ALSC, Nunes DRM, Viana AG. 1999. Optimization of the still-marinating process of chicken parts. Meat Sci 52:227–234.

Lyon BH, Hamm D. 1986. Effect of mechanical tenderization with sodium chloride and polyphosphates on sensory attributes and shear values of hot stripped broiler breast meat. Poult Sci 65:1702–1707.

Lyon CE, Lyon BG. 1990. The relationship of objective shear values and sensory tests to changes in tenderness of broiler breast meat. Poult Sci 69:1420–1427.

McKee SR, Sams AR. 1997. The effect of seasonal heat stress on rigor development and the incidence of pale, exudative turkey meat. Poult Sci 76(11):1616–1620.

McKee SR, Sams AR. 1998. Rigor mortis development at elevated temperatures induces pale exudative turkey meat characteristics. Poult Sci 77(1):169–174.

Moskowitz H. 1987. *Food Texture: Instrumental and Sensory Measurement*. New York: Marcel Dekker.

Muguruma M, Tsuruoka K, Katayama K, Erwanto Y, Kawahara S, Yamauchi K, Sathe SK, Soeda T. 2003. Soybean and milk proteins modified by transglutaminase improves chicken sausage texture even at reduced levels of phosphate. Meat Sci 63:191–197.

Murphy RY, Marks BP. 2000. Effect of meat temperature on proteins, texture and cook loss for ground chicken breast patties. Poult Sci 79:99–104.

Naveena BM, Mendiratta SK. 2001. Tenderisation of spent heat meat using ginger extract. Br Poult Sci 42:344–349.

Naveena BM, Sen AR, Muthukumar M, Vaithiyanathan S, Babji Y. 2006. The effect of lactates on the quality of microwave-cooked chicken patties during storage. J Food Sci 71(9):S603–S608.

Nurmahmudi, Sams AR. 1997a. Tenderizing spent fowl meat with calcium chloride: 1. Effects of delivery method and tumbling. Poult Sci 76:534–537.

Nurmahmudi, Sams AR. 1997b. Tenderizing spent fowl meat with calcium chloride: 3. Biochemical characteristics of tenderized breast meat. Poult Sci 76:543–547.

Nurmahmudi, Veeramuthu GJ, Sams AR. 1997. Tenderizing spent fowl meat with calcium chloride: 2. The role of delayed application and ionic strength. Poult Sci 76:538–542.

Rababah TM, Ereifej KI, Al-Mahasneh MA, Al-Rababah MA. 2006. Effect of plant extracts on physicochemical properties of chicken breast meat cooked using conventional electric oven or microwave. Poult Sci 85:148–154.

Seabra LM, Zapata JF, Fuentes MF, Aguiar CM, Freitas ER, Rodrigues MC. 2001. Effect of deboning time, muscle tensioning and calcium chloride marination on texture characteristics of chicken breast meat. Poult Sci 80:109–112.

Turner BE, Larick DK. 1996. Palatability of sous vide processed chicken breast. Poult Sci 75:1056–1063.

Voller-Reasonover L, Han IY, Acton JC, Titus TC, Bridges WC, Dawson PL. 1997. High temperature processing effects on the properties of fowl meat gels. Poult Sci 76: 774–779.

Woods KL, Rhee KS, Sams AR. 1997. Tenderizing spent fowl meat with calcium chloride. 4. Improved oxidative stability and the effects of additional aging. Poult Sci 76:548–551.

Yoon KS. 2002. Texture and microstructure properties of frozen chicken breasts pretreated with salt and phosphate solutions. Poult Sci 81:1910–1915.

Young LL, Lyon CE. 1997a. Effect of calcium marination on biochemical and textural properties of peri-rigor chicken breast meat. Poult Sci 76:197–201.

Young LL, Lyon CE. 1997b. Effect of postchill aging and sodium tripolyphosphate on moisture binding properties, color, and Warner–Bratzler shear values of chicken breast meat. Poult Sci 76:1587–1590.

Zamri AI, Ledward DA, Frazier RA. 2006. Effect of combined heat and high-pressure treatments on the texture of chicken breast muscle (pectoralis fundus). J Agric Food Chem 54:2992–2996.

Zhang L, Barbut S. 2005. Effects of regular and modified starches on cooked pale, soft, and exudative; normal; and dry, firm, and dark breast meat batters. Poult Sci 84:789–796.

23

PROTEIN AND POULTRY MEAT QUALITY

MASSAMI SHIMOKOMAKI, ADRIANA LOURENÇO SOARES, AND ELZA IOUKO IDA
Food Science and Technology Department, Londrina State University, Londrina, Paraná, Brazil

INTRODUCTION

Poultry muscle proteins are normally classified into three broad groups based on their solubility in water and salt solutions: sarcoplasmic proteins, myofibril

Handbook of Poultry Science and Technology, Volume 2: Secondary Processing, Edited by Isabel Guerrero-Legarreta and Y.H. Hui

proteins, and stroma proteins. The nature of their amino acid composition dictates their chemical properties. These properties, observed in fresh meat and during meat processing, influence the quality of meat and meat products. Thus, this chapter deals with proteins involved in meat quality, particularly as relative to PSE (pale, soft, exudative) meat and meat texture.

SARCOPLASMIC PROTEINS

The increased consumption of fresh poultry meat and its products in recent years has been accompanied by a rise in raw material quality problems associated particularly with color abnormalities. Consumers may find color abnormalities in broiler breast meat, ranging from yellowish (PSE) to excessive red color [dark, firm, dry (DFD)–like] unacceptable, and it may be inappropriate as meat-processing raw material. According to our rough calculation based on 2003 production, meat quality losses of Brazilian meat companies amount to around $10 million per year (Oda et al., 2003). The intense genetic selection for bird rapid weight gain may result in abnormal muscle biochemical behavior (Anthony, 1998). There is mounting evidence that the thermal stress conditions to which birds are submitted just before sacrifice is one of the causes of PSE meat (Barbut, 1998).

The major cause of PSE pork is already known (Fujii et al., 1991). Malignant hyperthermia associated with skeletal muscle disturbance can lead to death when animals are exposed to heat-stress conditions. Stress conditions are also promoted when pigs are administered anesthetics such as halothane, giving rise to pork stress syndrome (PSS). The biological linkage of the halothane gene *(hal)* or the ryanodine gene (*ryr1*) made it possible to establish a commercially available genetic selection methodology for pigs, thus avoiding the development of PSS by recessive animals (nn) and hybrid dominants (Nn) (Fujii et al., 1991). A gene point mutation capable of coding *ryr1* was found by Fujii et al. (1991). This protein controls Ca^{2+} for excitation–contraction coupling in sarcoplasmic reticulum and thus is responsible for the acceleration of muscle metabolism and the formation of pork PSE (Mickelson and Louis, 1996). The *hal* gene is related to pig sensibility to halothane; normal pigs are dominant (NN), sensitive ones are recessive (nn), and hybrid ones are intermediate (Nn) (Fujii et al., 1991). Recent reports have shown that poultry PSE meat is related to animal nutrition (Olivo et al., 2001), meat phospholipase A_2 activity (Soares et al., 2003), and preslaughter management (Guarnieri et al., 2004).

Halothane Tests

As observed in pigs, broilers are also sensitive in a halothane test, as shown by leg rigidity in Figure 1. The occurrence of positive, intermediate, and negative halothane sensitivity in a commercial broiler lineage is illustrated in Figure 2. From 342 birds tested, 9.1%, classified as *hal*+, were sensitive to halothane,

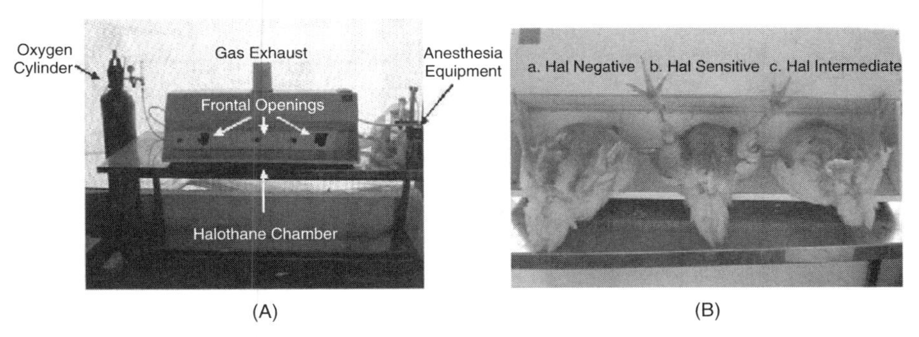

FIGURE 1 (A) Broiler halothane test chamber with anesthesia equipment. The halothane gas concentration was set at 3.0% in pure oxygen. Three birds were placed in the halothane chamber per experiment run. (B) Broiler leg rigidity showing sensitivity to halothane gas: (a) *hal*−, nonresponder, (b) *ha*l+, responder, (c) *ha*l+, −, intermediate (one-leg responder). (From Marchi et al., 2008.)

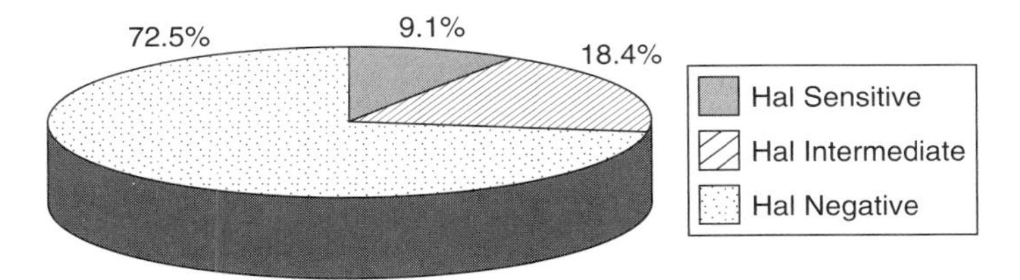

FIGURE 2 Occurrence of broiler positive *hal*+, intermediate *hal*+/−, and negative *hal*—response in a halothane test. (From Marchi et al., 2008.)

18.4%, *hal* +, −, presented partial response; and 72.5%, *hal*−, were nonsensitive to halothane.

Other authors reported somewhat similar results. Cavitt et al. (2004) found from 13.2 to 22.9% *hal*+ in commercial broiler lineages; Owens et al. (2000) reported a variation from 3.5 to 10.0% *hal*+ turkey, and Wheeler et al. (1999) obtained 5.0% *hal*+, 10.0% *hal* +, −, and 85.0% *hal*−, also in turkey. In any case, to prove that the poultry *hal* gene in fact has any role in *hal* response, it is necessary to determine the presence and the nature of the ryanodine genotype as done routinely in PSS pig (Fujii et al., 1991). The results shown herein clearly demonstrate that broiler PSE meat has a genetic origin.

MYOFIBRIL PROTEINS

Myofibril components play a crucial role in meat texture, particularly as related to cold shortening. In poultry, cold shortening does not normally cause toughening in normal industrial meat processing, provided that the breast meat is not excised prematurely or otherwise altered (Papinaho and Fletcher, 1996). Papinaho and

Fletcher (1996) described temperature-induced shortening in intact and excised broiler breast muscle samples. Postmortem tenderization by proteolysis has been studied extensively, and the calpain system is known to initiate the enzymatic digestion of the myofibrillar framework (Goll et al., 1997). Koohmarie (1996) pointed out that the breakdown of the key proteins that maintain the sarcomere structure causes its weakening. Hence, alterations in meat tenderness and Z-line region degradation occur mainly during postmortem tenderization (Taylor et al., 1995). Previous work showed that early breast deboning can toughen the meat (Sams and Janky, 1986; Smith et al., 1992) and that tenderness can be improved by prolonging the time between slaughter and deboning (Northcutt et al., 2001). Sams and co-workers showed the relationship between proteolysis activity and shear force values under cold storage (Sams, 1999; McKee et al., 1997).

Myofibrillar Fragmentation Index

An experiment was designed to observe the proteolitic activity over myofibril proteins during meat maturation under refrigeration by Kriese et al. (2007). Figure 3 shows the myofibrillar fragmentation index (MFI) for intact and excised broiler breast meat. As can be seen, MFI increased from the start up to 24 h under refrigeration and became virtually constant for both samples after 72-h storage. These results indicate that the endogenous muscle protease system reached its maximum activity at $2 \pm 2°C$ after approximately 24 h of refrigeration. However, the entire breast presented higher MFI values in relation to deboned samples throughout the treatment. This difference was 13.2% from 24 h onward.

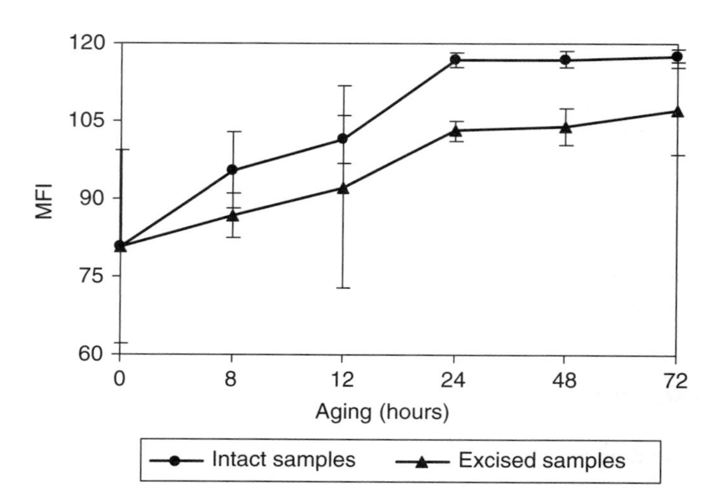

FIGURE 3 Myofibrillar fragmentation index of intact and excised broiler breast samples stored at $2 \pm 2°C$ for 0, 8, 12, 24, 48, and 72 h. (From Kriese et al., 2007.)

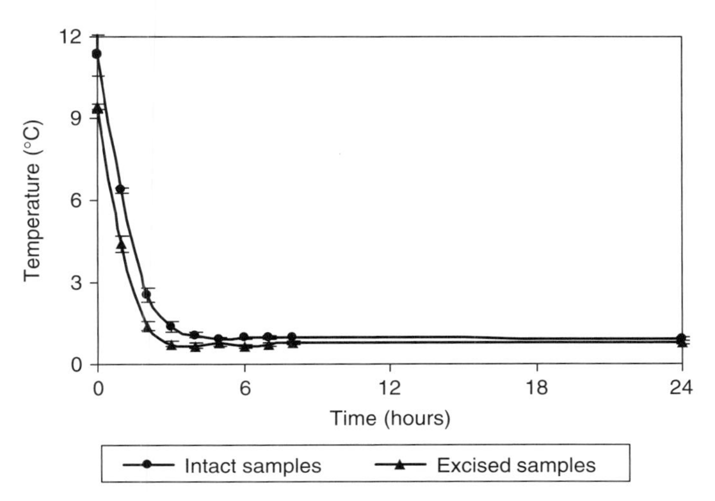

FIGURE 4 Temperature decline rate of intact and excised samples of broiler breast stored at $2 \pm 2^{\circ}$C measured at 0, 1, 2, 3, 4, 5, 6, 7, 8, and 24 h. The temperature values of the intact samples are relatively higher than those of excised samples. (From Kriese and others, 2007.)

Sample Temperature Behavior

Figure 4 shows the temperature decline rate for both intact and excised samples. It can be seen that the initial experimental temperature values were $11.3 \pm 0.8^{\circ}$C for intact and $9.4 \pm 0.1^{\circ}$C for excised samples and that they gradually fell to 0.9 ± 0.1 and $0.8 \pm 0.0^{\circ}$C, respectively ($p \leq 0.05$), after approximately 5 h of storage, and remained constant through the end of the experiment.

Measurement of Breast Meat Tenderness

Figure 5 shows that the decrease in the shear force value of both intact and excised samples is more noticeable at 24 h, becoming virtually constant through the end of the experiment. Again, tenderness was higher in intact samples compared to excised samples ($p \leq 0.05$). A significant negative correlation ($r = -0.93$, $p \leq 0.05$) was observed between MFI and shear force values, suggesting that the higher the protease activity, the more tender the breast meat becomes. One of these enzyme entities, the calpain enzyme system, digests the sarcomere structure, particularly the Z-line region (Taylor et al., 1995; Guarnieri et al., 2004), thus improving meat tenderness. If intact samples were protected against muscle cold shortening and combined synergistically with proteolysis, tenderization might be enhanced. However, as yet, our experiments have not revealed any potential factors in bones involved in the biochemical mechanism of tenderness in relation to the relative increase in enzymatic proteolytic activities. The higher-temperature values of intact samples compared to those of the excised samples in Figure 4 offer another possible explanation. Although the difference began at the beginning

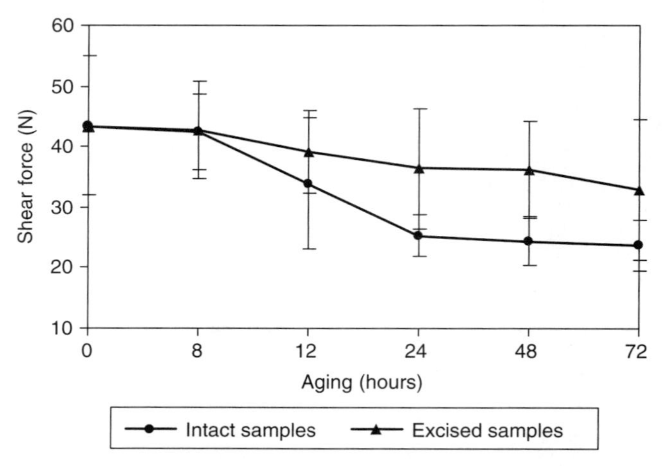

FIGURE 5 Shear force values of intact and excised broiler breast samples stored at $2 \pm 2°C$ for 0, 8, 12, 24, 48, and 72 h. (From Kriese et al., 2007.)

of storage at 1.9°C (Figure 4), these values seem to be critical for protease activity, since the pH values measured (not shown) in both samples were not different ($p \leq 0.05$), leading to more tender meat in excised samples.

COLLAGEN

As of 1999, over 19 types of collagens in poultry tissues were known and, in muscle, there are at least six types of collagens: I, III, IV, V, XI, and XII (McCormick, 1999). Normally, at the age of commercial slaughtering, poultry meat does not present a tenderness problem because of the small quantities of collagen in tissues. However, as the animal ages, the meat gets tough because of the stabilization of collagen molecules. Therefore, it is the quality of collagen that makes meat tough. It has been established that collagen covalent cross-linking mediated by lysyl oxidase changes as an animal ages (Bailey and Shimokomaki, 1971; Robins et al., 1973). The concentration of immature collagen cross-link, hydroxylysinoketonorleucine (HLKNL), decreases with animal age and results in meat toughening (Shimokomaki et al., 1972). The main mature cross-link derived from two reducible HLKNLs is pyridinoline (PYR), which stabilizes the collagen molecule (Fujimoto, 1977; Eyre and Oguchi, 1980; McCormick, 1999). PYR has been shown to be related to meat texture (Nakano et al., 1991; Bosselmann et al., 1995) and to bridge different types of collagens, thus further stabilizing the extracellular macromolecular organization (Shimokomaki et al., 1990). The stabilization of collagen molecules is measured by its increasing insolubility (Robins et al., 1973; Young et al., 1994). Besides commercial broilers being slaughtered at an age when collagen normally does not contribute to texture problems, knowledge of collagen cross-links in poultry intramuscular connective

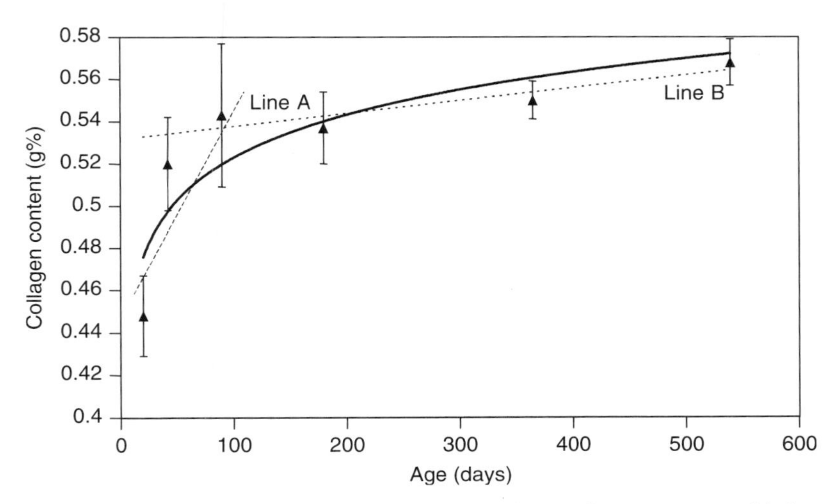

FIGURE 6 Variation in breast meat (pectoralis major) collagen content with hen age from 20 to 100 days (line A) and from 101 to 540 days (line B). These data are averages, with standard deviation bars of four separate experiments in triplicate for each experiment. (From Coro et al., 2002.)

tissue is scarce. Intermediate collagen cross-links in poultry have been shown to change with aging (Nakamura et al., 1974), and although broiler muscle generally presents little connective tissue, it changes as the animals become older.

Collagen Content

Collagen content was measured throughout the hen life span by Coro et al. (2002), and two distinct phases can be observed in Figure 6. First, there is a rapid increase in the concentration of collagen from 20 to approximately 100 days of age, ranging from 0.448 to 0.540 g% (line A). It remained fairly constant up to the age of 540 days after hatching, ranging from 0.540 to 0.568 g% (line B).

Cross-Link Concentration

The concentration of PYR increased from 20 to 40 days of age after hatching, ranging from 0.0091 to 0.0409 mol/mol of collagen, respectively, as shown in Figure 7. Subsequently, a steady increase was observed from 200 to 540 days of age, with PYR values ranging from 0.044 to 0.101 mol/mol of collagen, respectively.

Collagen Solubility

As the birds aged, meat collagen solubility measured in Ringer solution decreased inversely in relation to the concentration of PYR (Figure 7). There was a sharp

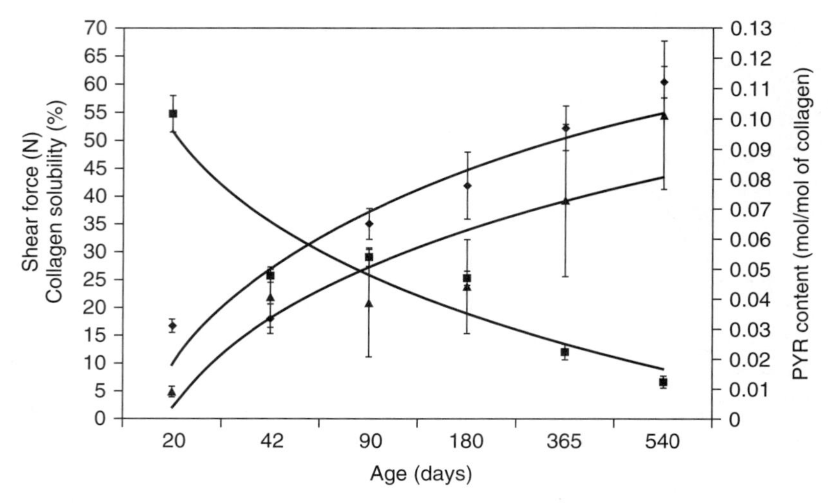

FIGURE 7 Variation in breast meat (pectoralis major) shear value (diamonds), pyridinoline (PYR) content (triangles), and collagen solubility (squares) with hen age. These data are averages, with standard deviation bars of four separate experiments in triplicate for each experiment. (From Coro et al., 2002.)

decrease in solubility from 20 to 40 days after hatching, ranging from 54.73 to 25.70%, respectively. A gradual decrease was observed as the birds grew older from 29.0% down to 6.60% at 90 and 540 days, respectively. These results suggest a straight relationship between the concentration of PYR and collagen solubility (Robins et al., 1973).

Determination of Shear Values

Figure 7 shows a sharp increase in shear values from 16.7 to 35.0 N at ages 20 and 90 days after hatching, respectively. Subsequently, there was a gradual increase as the birds aged, ranging from 41.9 to 60.3 N at 180 and 540 days after hatching, respectively. In contrast to results observed in ovine animals (Young et al., 1994), it seems that shear values in poultry increase concomitantly with the increase in concentration of PYR, as also observed recently in bovine animals (Bosselmann et al., 1995). To be considered tough, the meat shear value must be at least 44 N (4.5 kg), as reported for beef (Johnson et al., 1988; Knapp, 1989). In our experiment, this value was reached six months after hatching (Figure 7). Figure 6 shows two phases for collagen concentration. First, new collagen was laid down sharply up to 42 days of age (line A); thereafter, the relative increase in collagen synthesis slowed down through the end of the experiment (line B). These results were somewhat similar to those obtained for poultry by Nakamura et al. (1974) and pig by Nakano et al. (1991). One explanation for this collagen synthesis behavior is related to the formation of mature collagen cross-links PYR, which start to be laid down biosynthetically from 40 to 50 days after hatching.

The presence of PYR in the collagen fibers would slow down the synthesis of de novo collagen.

Our results also demonstrate that the shear value is not related to the amount of collagen, but rather, to the proportional increase in the concentration of PYR. The collagen solubility and amount of PYR were inversely correlated. These results support the previous assumption that collagen cross-linking is an important factor in regulation of the in vivo catabolism rate (Krane, 1987). Up to 100 days after hatching, the collagen synthesis increased noticeably. As PYR was synthesized in collagen fibrils, it imparted resistance to degradation by collagenase, as measured by the decrease in collagen solubility rate. Whereas the rate of collagen synthesis became proportionally constant as poultry aged, PYR continued being formed (Figure 7), bringing about a slowdown in collagen turnover. The shear value of hen breast meat was measured consistently only after approximately 180 days after hatching. Since there was no relative increase in collagen content after this age, the increase in texture seemed to be related to collagen cross-linking (Figure 7). This was corroborated by the decrease in collagen solubility (Figure 7). Furthermore, it seems that the amount of collagen cross-links is related to muscle growth rate, nutrition status, and poultry genetic lineage specificity (McCormick, 1994). Comparing our results with those reported by Velleman et al. (1996), there was at least five times as much collagen in breasts of White Leghorn hen than of Ross broilers, and PYR was again 10 times as concentrated in young White Leghorn as in aged samples. In fact, it is known that White Leghorn breast presents over 1.0 mol of PYR per mole of collagen by the age of one year, the highest value reported so far in muscle (Velleman et al., 1996). This condition can be explained by the fact that White Leghorn is designed as an egg producer and is not fed a high-energy diet, showing relatively slow growth. In contrast, Ross, a broiler, is submitted to a high-level energy diet, and consequently, it is a fast-growing bird with large protein accretion. In fact, Schreurs et al. (1995) demonstrated higher proteolytic activity for calpain and cathepsins in White Leghorn than in Ross.

The reduced catabolism in fast-growing birds would dictate the lower formation of collagen cross-links, and thus the lower shear values of breast meat. Therefore, the relationship between labile and insoluble collagen in fast-growing hens is favored; however, there was not necessarily a decrease in the amount of insoluble collagen (Etherington, 1987). Our results for poultry are in agreement with those in reports for other animals (McCormick, 1989). The rapid growth and increased collagen synthesis resulted in tenderer meat because of the lower collagen cross-linking.

REFERENCES

Anthony NB. 1998. A review of genetic practices in poultry: efforts to improve meat quality. J Muscle Foods 9:25–33.

Bailey AJ, Shimokomaki M. 1971. Age related changes in the reducible cross-links of collagen. FEBS Lett 16:86–88.

Barbut S. 1998. Estimating the magnitude of the PSE problem in poultry. J Muscle Foods 9:35–49.

Bosselmann A, Möller C, Steinhardt H, Kirchgessner M, Schwarz FJ. 1995. Pyridinoline cross-links in bovine muscle collagen. J Food Sci 60:953–958.

Cavitt LC, Hargis BM, Owens CM. 2004. Gluconeogenesis from lactate in liver of stress-susceptible and stress-resistant pigs. Poult Sci 83:1440–1444.

Coro FAG, Youssef EY, Shimokomaki M. 2002. Age related changes in poultry breast meat collagen pyridinoline and texture. J Food Biochem 26:533–541.

Etherington DJ. 1987. Collagen and meat quality: effects of conditioning and growth rate. Adv Meat Res 4:351–360.

Eyre D, Oguchi H. 1980. The hydroxypyridinium crosslinks of skeletal collagens: their measurement, properties and a proposed pathway of formation. Biochem Biophys Res Commun 92:403–410.

Fujii J, Otsu K, Zorzato F. 1991. Identication of a mutation in porcine ryanodine receptor associated whith malignant hyperthermia. Science 253:448–451.

Fujimoto D. 1977. Isolation and characterization of a fluorescent material in bovine achilles tendon collagen. Biochem Biophys Res Commun 76:1124–1129.

Goll DE, Boehm ML, Geesink GH, Thompson VF. 1997. What causes postmortem tenderization? In: *Proceedings of the 50th Reciprocal Meat Conference*. Ames; IA: American Meat Science Association, p. 60.

Guarnieri PD, Soares AL, Olivo R, Schneider J, Macedo RMG, Ida EI, Shimokomaki M. 2004. Preslaughter handling with water shower spray inhibits PSE (pale, soft, exudative) broiler breast meat in commercial plant: biochemical and ultrastructural observations. J Food Biochem 24:269–277.

Johnson DD, Lunt DK, Savell JW, Smith GC. 1988. Factors affecting carcass characteristics and palatability of young bulls. J Anim Sci 66:2568–2577.

Knapp RH, Savell JW, Cross HR, Mies WL, Edwards JW. 1989. Characterization of cattle types to meet specific beef targets. J Anim Sci 67:2294–2308.

Koohmaraie M. 1996. Biochemical factors regulating the toughening and tenderization processes of meat. Meat Sci 43:193–201.

Krane SM. 1987. Turnover of collagen. Adv Meat Res 4:325–331.

Kriese PR, Soares AL, Guarnieri PD, Prudencio SH, Ida EI, Shimokomaki M. 2007. Biochemical and sensorial evaluation of intact and boned broiler breast meat tenderness during ageing. Food Chem 104:1618–1621.

Marchi DF, Oba A, Soares AL, Ida EI, Shimokomaki M. 2008. Genetic origin for broiler PSE meat. In: *World Poultry Congress*, Brisbane, Australia.

McCormick RJ. 1989. The influence of nutrition on collagen metabolism and stability. Proc Annu Reciprocal Meat Conf 42:137–148.

McCormick RJ. 1994. The flexibility of the collagen compartment of muscle. Meat Sci 36:79–91.

McCormick RJ. 1999. Extracellular modifications to muscle collagen: implications for meat quality. Poult Sci 78:785–791.

McKee SR, Hirschler EM, Sams AR. 1997. Physical and biochemical effects associated with tenderization of broiler breast fillets during aging after pre-rigor deboning. J Food Sci 62:959–962.

Mickelson JR, Louis CF. 1996. Malignant hyperthermia: excitation–contraction coupling: Ca^{2+} regulation defects. Physiol Rev 76:537–592.

Nakamura R, Sekogushi S, Sato Y. 1974. The contribution of intramuscular collagen to the tenderness of meat from chickens with different ages. Poult Sci 54:1604–1612.

Nakano T, Thompson JR, Aherne FX. 1991. Concentration of the crosslink pyridinoline in porcine skeletal muscle epimysium. Can J Inst Food Sci Technol 18:100–102.

Notthcutt JK, Buhr RJ, Young LL, Lyon CE, Ware GO. 2001. Influence of age and postchill carcass aging duration on chicken breast meat. Poult Sci 80:808–812.

Oda SHI, Schneider J, Soares AL, Barbosa DML, Ida EI, Olivo R, Shimokomaki M. 2003 Detecção de cor em filés de peito de frango. Rev Nac Carne 27:30–34.

Olivo R, Soares AL, Ida EI, Shimokomaki M. 2001. Dietary vitamin E inhibits poultry PSE and improves meat functional properties. J Food Biochem 25:271–283.

Owens CM, Matthews NS, Sams AR. 2000. The use of halothane gas to identify turkeys prone to pale, exudative meat when transported before slaughter. Poult Sci 79:789–795.

Papinaho PA, Fletcher DL. 1996. The influence of temperature on broiler breast muscle shortening and extensibility. Poult Sci 75:797–802.

Robins SP, Shimokomaki M, Bailey AJ. 1973. The chemistry of the collagen cross-links: age related changes in the reducible components of intact collagen fibres. Biochem J 131:771–780.

Sams AR. 1999. Meat quality during processing. Poult Sci 78:798–803.

Sams AR, Janky DM. 1986. The influence of brine chilling on tenderness of hot boned chill-boned and age-boned broiler fillets. Poult Sci 65:1316–1321.

Schreurs FJG, Van Der Heide D, Leenstra FR, de Wit W. 1995. Endogenous proteolytic enzymes in chicken muscle: differences among strains with different growth rates and protein efficiencies. Poult Sci 74:523–537.

Shimokomaki M, Elsden DF, Bailey AJ. 1972. Meat tenderness: age related changes in bovine intramuscular collagen. J Food Sci 37:892–896.

Shimokomaki M, Wright DW, Irwin MH, Van Der Rest M, Mayne R. 1990. The structure and macromolecular organization of type IX collagen in cartilage. Ann NY Acad Sci 580:1–7.

Smith DP, Fletcher DL, Papa CM. 1992. Duckling and chicken processing yields and breast meat tenderness. Poult Sci 71:197–202.

Soares AL, Ida EI, Miyamoto S, Blazquez FJH, Pinheiro JW, Shimokomaki M. 2003. Phospholipase A2 activity in poultry PSE (pale, soft, exudative) meat. J Food Biochem 27:309–319.

Taylor RG, Gessink GH, Thompson VF, Koohmaraie M, Goll DE. 1995. Is Z-disk degradation responsible for post-mortem tenderization? J Anim Sci 73:1351–1367.

Velleman SG, Yeager JD, Krider H, Carrino DA, Zimmerman SD, McCormick RJ. 1996. The low avian score normal muscle weakness alters decorin expression and collagen crosslinking. Connect Tissue Res 34:33–39.

Wheeler BR, McKee SR, Matthews NS, Miller RK, Sams AR. 1999. A halothane test to detect turkeys prone to developing pale, soft, and exudative meat. Poult Sci 78:1634–1638.

Young OA, Braggins TJ, Barker GJ. 1994. Pyridinoline in ovine intramuscular collagen. Meat Sci 37:297–303.

24

POULTRY FLAVOR: GENERAL ASPECTS AND APPLICATIONS

José Angel Pérez-Alvarez, Esther Sendra-Nadal, Elena José Sánchez-Zapata, and Manuel Viuda-Martos
Grupo Industrialización de Productos de Origen Animal(IPOA Research Group), Departamento de Tecnología Agroalimentaria, Escuela Politécnica Superior de Orihuela, Universidad Miguel Hernández, Orihuela, Alicante, Spain

GENERAL ASPECTS OF POULTRY FLAVOR

Consumers demand high quality, convenient, innovative, regular, and safe meat products with natural flavor and taste and an extended shelf life (Aymerich

Handbook of Poultry Science and Technology, Volume 2: Secondary Processing, Edited by Isabel Guerrero-Legarreta and Y.H. Hui

et al., 2008). In meat quality, several factor affect consumer acceptability. In some meats, such as free-range poultry meat and poultry skin, color is the main attribute for consumers (Pérez-Alvarez and Fernández-Lopez, 2000, 2006; Pérez-Alvarez, 2006a), and tenderness plays a large role. In addition to meat tenderness, juiciness and flavor are two other organoleptic qualities that consumers demand when eating meat, especially red meats such as ostrich (Sayas-Barberá and Fernández-López, 2000). Flavor is also considered an important part of the eating quality (Ouali et al., 2006). Thus, flavor was found to be the most important factor affecting consumers' meat-buying habits and preferences when tenderness was held constant (Sitz et al., 2005). The flavor quality needs to be addressed in all poultry meats and their meat products.

For flavor concerns, factors that make some poultry meat desirable and others undesirable are influenced by many conditions, including preharvest animal environment and diet, postharvest handling, and consumers' individual preferences. From a structural and ultrastructural point of view, perception of flavor characteristics depends greatly on the nature of the food matrix. The food matrix plays an important role in controlling flavor release at each step of food product preparation and consumption (Mattes, 1984, 1997).

Flavor development in poultry meat and meat products is very complex, and it has to be taken into account that flavor is an attribute of meat palatability (Calkins and Hodgen, 2007). Thus, the technological process, its composition (meat, nonmeat ingredients, and additives), nutritional value, vitamins, and other minority compounds are responsible for flavor retention by the products. As an example, in dry-cured poultry meat products, fermentation and curing are necessary for the development of the characteristic flavors of these products (Pérez-Alvarez, 2006b; Sayas-Barberá et al., 2008). Also, the storage of poultry meat at low temperature is a prerequisite for flavor development, increasing eating qualities (Ouali et al., 2006). Thus, it is important to gain a better understanding of factors that influence poultry meat flavor in order to produce the most flavorful and consistent product. The poultry meat industry must produce the highest-quality product for the targeted market.

Chemical Aspects

From a chemical point of view, meat flavor can be formed by a lot of compounds. Flavor meat compounds include a broad array of compounds, including hydrocarbons, aldehydes, ketones, alcohols, furans, thriphenes, pyrrols, pyridines, pyrazines, oxazols, thiazols, sulfurous compounds, and many others. Sulfurous and carbonyl compounds seem to be the predominant contributors to meat flavor (Shahidi, 1989; Mottram and Madruga, 1994). Bis (2-methyl-3-furyl) has an odor threshold of 0.02 ng/g in water, and methylating 2-methyl-3-furanthiol increases the odor threshold. This suggests that 2-methyl-3-(methylthio) furan is only a minor contributor to the flavor of meat (MacLeod, 1986).

Poultry Meat Processing Flavor

Many compounds that contribute to meat smell and flavor are lipid breakdown products. Thus, oxidative environments in meat postslaughter are known to increase lipid oxidation, resulting in an unfavorable consumer reaction to the flavor and smell (Warner et al., 2005). The onset of oxidation in meat postmortem is well known to produce off-odors, discoloration, and unacceptable flavors associated with rancidity (Warner et al., 2005). Several authors (Gorbatov and Lyaskovskaya,1980; Sentandreu et al., 2003; Campo et al., 2006) have reported that major contributors to meat flavor are lipid peroxidation together with amino acids and peptides generated by proteolysis. But the technologist must take into account that in poultry meat and meat products, lipids play a dominant and multifunctional role in the perception of flavor. According to De Roos (1997) and Mottram (1998), the physical states of lipids (e.g., liquid vs. solid, emulsified vs. free fat, temperature of fat) are among the factors that influence the lipid–flavor interaction of fat products (Tan et al., 2006). Species-specific notes are generally lipid-derived. Differences in fatty acid profile and the resulting carbonyls between species may be the factors responsible for the lipid influence. Specifically, 2-alkenals, including hexenal, heptenal, octenal, nonenal, undecenal, and dodecenal, as well as aldehydes such as octanal, nonanal, and decanal, along with decadienal and *c*-dodecalactone, are related to both chicken-specific aroma and flavor (Ramarathnam et al., 1993).

Sanudo et al. (2000) reported that acceptability and volatile flavor components of cooking poultry meat are affected by cooking. Many flavor compounds isolated recently are altered throughout storage time and cooking.

Lipid Oxidation in Flavor Development

Lipid oxidation in muscle systems is initiated at the membrane level in phospholipid fractions as a free-radical autocatalytic chain mechanism (Labuza, 1971) in which prooxidants interact with unsaturated fatty acids, resulting in the generation of free radicals and propagation of the oxidative chain (Ashgar et al., 1988). Fatty acids such as linoleic and arachidonic acid begin to autooxidize to 9- and 11-hydroperoxide, respectively, which can form 2,4-decadienal, 2-nonenal, 1-octen-3-one, 2,4-nonadienal, and 2-octenal through β-scission, with 2-nonenal and 2,4-decadienal having values as high as those of the sulfur compounds, contributing to a meaty flavor (Calkins and Hodgen, 2007).

Rhee et al. (1996) and Tang et al. (2001) reported that high unsaturated fatty acid concentration favors lipid oxidation, which can affect flavor quality. St. Angelo et al. (1987) reported that omega-3 polyunsaturated fatty acids are positively correlated with aroma, active secondary lipid degradation products, saturated and unsaturated aldehydes (Byrne et al., 2002), alcohols, and ketones.

Beta-carotene oxidation generates a very intense aromatic β-ionone (Sanderson et al., 1971; Gasser and Grosch, 1988). Radical reactions are known to constitute a central process in the oxidation of these compounds even if the nature and origin of the radicals remain unclear (Renerre, 1999). Thus, as apoptosis will start within a few minutes after death, the first oxygen radicals will be generated by mitochondria, and this will initiate the autocatalytic process, which will go on over the entire storage period, even at low temperature, including freezing. This essential initiating phase must be taken into account in studies dealing with the identification of the major determinants of meat flavor (Ouali et al., 2006).

Jahan et al. (2005) found relationships among the antioxidant α-tocopherol, total polyunsaturated fatty acids (PUFAs), and omega-3 fatty acids and flavor components from cooked chicken. These authors reported that two antioxidant enzymes, glutathione peroxidase and glutathione reductase, had relationships with chicken flavor components. Also, Sarraga et al. (2006) reported that dietary antioxidant supplementation demonstrates a possible synergism between antioxidants and cysteine proteinases in the perception of poultry meat quality.

WARMED-OVER FLAVOR

One of the main defects occurring during refrigerated storage of poultry is lipid oxidation, primarily the oxidation of unsaturated fatty acids, and the main volatile originated by oxidation is hexanal. Ertas (1998) reported that lipid oxidation is related to the development of warmed-over flavor (WOF). During storage, WOF can be produced in several types of poultry meat products.

During cooking and storage of meats and meat products, significant quantities of "free" iron is released (Fernández-López et al., 2003a). According to Graf and Panter (1991), in precooked poultry, iron bound to negatively charged phospholipids can suffer site-specific oxidation that generates WOF within 24 h of refrigerated storage. Nitrite completely blocked iron release from hemoproteins, which explained its inhibitory effect on WOF development. Iron sequestration by chelation with phytic acid greatly reduced the formation of WOF. Also, phytic acid substantially inhibited O_2 uptake and malonaldehyde formation (Empson et al., 1991). Similarly, competitive displacement of iron from phospholipids with polyvalent cations substantially lowered the rate of WOF formation. Warmed-over flavor can be quantified by the measurement of TBA substances (TBARS), by sensory analysis (Mielche 1995), and by hexanal content assessment (Shu-Mei-Lai et al., 1995).

TBARS and WOF development in minced thigh meat patties can be reduced by supplemental α-tocopherol during frozen storage, or by cooking and refrigerated storage (O'Neill et al., 1998). Also, lipid oxidation after marinating, cooking, and chill storage is related strongly to WOF using oxidation variables (e.g., volatiles, TBARS) (Mielnik et al., 2008). Sarraga et al. (2006) determined WOF using quantitative descriptive sensory analysis. WOF development was related to rancid and sulfur/rubber sensory notes associated with a concomitant reduction in chicken "meaty" flavor. Increasing cooking temperature was related to increased roasted, toasted, and bitter sensory notes. Roasted flavor was associated with WOF. Lipid oxidation developed rapidly during chilled storage (reflected in volatile compounds and sensory analyses), possibly contributing to the rancid notes observed in WOF. Changes in sulfide compounds were related to progression of oxidation in both lipids and proteins. Increasing cooking temperature increased the formation of Maillard reaction products (MRPs) (Tai and Ho, 1997) without significantly inhibiting the incidence of WOF (Byrne et al., 2002).

Clove and MRPs exhibited very good antioxidative effects in chicken compared to salt and other spices. Studies in a model system, comprising methyl linoleate, indicated that clove and MRPs were very effective in arresting the buildup of secondary oxidation products, formed primarily during refrigerated storage of cooked meat. They were also found to affect the extent of release of nonheme iron during cooking of meat, which is believed to be the primary catalyst accelerating lipid oxidation (Jayathilakan et al., 1997). According to Mielnik et al. (2008), antioxidants could be exploited in marinades to prevent rancidity in stored, heat-treated turkey meat products.

Sarraga et al. (2006) did not find a relationship between dietary antioxidant supplementation and endogenous antioxidant enzyme activities in turkey meat; these authors reported that sensory evaluation showed that longer supplementation might be necessary to prevent rancidity and WOF in turkey meat through feed supplementation. In a study by Higgins et al. (1999) on antioxidant supplementation to livestock, a taste panel showed that poultry leg and breast patties formed from the meat of turkeys given α-tocopheryl acetate–enriched diets developed less WOF than did patties formed from control turkey meat. All samples were tasted after 2 days at 4°C refrigerated display. These authors reported that patties containing 1% salt generally exhibited a greater degree of WOF than did patties without salt. When tocopherol is combined with L-ascorbyl palmitate in cooked, minced turkey meat products, oxidative protection is optimized as a result of indirect synergism (Bruun-Jensen et al., 1996).

Regarding the heat treatment of poultry, Mielche (1995) reported that TBARS increased with increasing heating temperature and storage time. Thus, maximum levels of TBARS were found in poultry meat samples in the order turkey thigh > chicken thigh > turkey breast > chicken breast. The temperature effect was particularly obvious for the chicken samples. Chicken meat heated at 60°C showed little change in TBARS during storage time.

Warmed-over flavor of reheated dark chicken meat patties can be delayed by the chlorination of the chiller water used during processing (Erickson, 1999). Juncher et al. (1998) stated that the development of WOF during retail storage could be avoided when the product was vacuum- or gas-packed. Also, Emrick et al. (2005) reported that increasing the pepper content on precooked chicken patties could also aid decreasing production of malonaldehyde in a precooked meat product, thereby reducing the intensity of WOF perceived by consumers.

Experiments regarding the main changes in odorant compounds of boiled chicken during refrigerated storage and reheating have determined that refrigerated storage and reheating of boiled chicken showing WOF, leading to a loss of meaty, chickenlike, and sweet odor notes and to the formation of green, cardboard-like, metallic off-odors. These changes were caused primarily by a sevenfold increase in hexanal and a sixfold decrease in both (*E,E*)-2,4-decadienal and 2-furfurylthiol (Kerler and Grosch, 1997).

Several compounds have been investigated for the reduction of WOF in poultry as sodium tripolyphosphate, sodium ascorbate monophosphate, sodium lactate, calcium chloride, calcium acetate, and calcium gluconate, which reduced the WOF problem but did not fully prevent other off-flavor development (McKee, 2007). Incorporation of sodium lactate into the brine in a *sous vide* chicken breast did not influence oxidative stability or warmed-over flavor, but did result in enhanced fresh roasted or meaty flavors. A rapid heating rate decreased the levels of sulfur-containing compounds detected but did not influence the levels of other volatile compounds. When chicken breast was processed to 94°C, it contained higher quantities of alcohols and hydrocarbons than that processed at low temperatures (77°C). Storage resulted in a decline in the fresh roasted or meaty flavor attribute and an increase in the WOF note and quantities of alcohols, aldehydes and ketone, hydrocarbons, and total headspace volatile compounds (Turner and Larick, 1996).

DRY-CURED POULTRY MEAT PRODUCTS

The effects of nitrite in cooked and dry-cured meat products have been studied for many years and can be summarized as: formation of a characteristic red color, growth inhibition of spoilage and pathogenic bacteria such as *Clostridium botulinum*, contribution to the development of typical cured meat flavor, and delay in the oxidative rancidity process (Pegg and Shahidi, 2000; Sayas-Barberá et al., 2008). It is easy to distinguish cooked, dry-cured poultry meat products from fresh roast poultry on the basis of flavor, but the chemical identity of distinguishing flavor components in cured meat has been eluded. Some of the flavor differences may be due to the suppression of lipid oxidation by nitrite, but other antioxidants do not produce cured meat flavor. If nitrite does, in fact, form some volatile flavors, these would represent yet another reaction product of nitrite in cured poultry meat (cooked and dry-cured) (Sebranek and Bacus, 2007).

The nitric oxide molecule (NO) itself can easily be oxidized in the presence of oxygen. This means that there is oxygen sequestering and thus an antioxidative action of nitrite in meat batters or dry-cured products. Due to the lack of oxygen, the development of rancidity or WOF is retarded. This reaction is rather important in meat batters, as nitrite acts in this way as an antioxidant (Honikel, 2008).

Flavor and aroma are key attributes that affect the overall acceptance of dry-cured poultry meat products and are affected markedly, as are other dry-cured meat products (e.g., dry-cured loin, dry-cured ham) by raw material, processing techniques, and aging time (Pérez-Alvarez, 2006b). Today, in several Mediterranean countries, poultry meat is used as raw material for dry-cured sausages such as turkey and/or chicken salchichon and chorizo. Also, duck breast is used to obtain dry-cured duck ham. Processing and technology applied is similar to pork and beef dry-cured meat products. In this scenario, ostrich dry-cured meat products have good consumer acceptability.

From a flavor point of view, dry-cured meat products flavor compounds are formed from chemical, biochemical, and enzymatic reactions and microbiological action during the process of fermentation and dry curing. Lipolysis, protelysis, organic acid generation by microbial activity, amino acid decarboxylation, and generation of amines contribute to the flavor of these products.

During the ripening of dry-cured fermented sausages, proteins and lipids experience great changes. Proteolysis influences both texture and flavor development, due to the formation of several low-molecular-mass compounds, including peptides, amino acids, aldehydes, organic acids, and amines, which are important flavor compounds or precursors of flavor compounds (Sayas-Barberá et al., 2008). Lipolysis plays an essential role in the development of dry-cured sausage flavor. Microbial enzymes such as esterase from staphylococci act to esterify alcohols and acids that are present in the microorganism environment (Talon et al., 1998). Esters are very fragrant compounds with very low odor detection thresholds; they give fruity notes to the dry fermented sausage flavor (Montel et al., 1996).

Larrouture et al. (2000) focused on an understanding of the role of staphylococci in relation to amino acid degradation and flavor generation. Degradation of the branched-chain amino acids leucine, isoleucine, and valine into flavor-intensive methyl-branched aldehydes, acids, and the less flavorous alcohols by staphylococci during processing of fermented dry-cured sausages has been linked to the development of fermented sausage flavor.

The flavor and aroma of poultry dry-cured meat products can be determined by sensory descriptive analysis and the composition of aroma impact compounds. As contrasted with pork and beef dry-cured meat products, no studies have been conducted to identify and quantify the volatile compounds in different types of dry-cured poultry meat products. A study on the volatile composition of poultry dry-cured meat products is important, because it may help to relate flavor compounds to sensory descriptors and consumer preferences as well as contributing to the development of systems to monitor flavor quality. To fully comprehend the nature of poultry dry-cured meat product flavor, a sensory language (lexicon) must be established through sensory descriptive analysis in order to differentiate and describe dry-cured products based on their flavor and aroma.

RATITES MEAT FLAVOR

Ratities meat and meat products in general, and ostriches in particular, are comparable to beef in all quality parameters (i.e., color, texture, flavor, etc.) (Walter et al., 2000; Fernández-López et al., 2006a). Thus, Marks et al. (1998) reported, through a sensory panel, that one week of aging of ostrich meat gave higher acceptance scores for flavor than those of lesser aged ostrich meat or a beef control. Meat aging is a well-known process to improve sensorial meat characteristics. The low fat and healthy attributes of ratites meat are emphasized, as they contribute to flavor properties (Darrington, 1996). In ratites meat (ostrich, rhea, and emu), flavor differences are associated primarily with species. Liver and gamy flavors provide the best representation of flavor differences (Rodbotten et al., 2004). Another factor that must be taken into account is the low fat content of ratite meat, which therefore does not have fat-associated flavors (Taylor et al., 1997). Pollok et al. (1997), Girolami et al. (2003), and Hoffman et al. (2008) determined that ostrich muscles (gastrocnemius and iliofibularis) were not different for aroma and flavor between several breed contrasts (Black × Black, Blue × Black, Blue × Blue breeds).

Ostrich farming, especially feeding, can affect sensorial properties. In a study by Hoffman et al. (2005), when ostriches were feed with fish oil enriched with omega-3 fatty acids, ostrich meat aroma and flavor were not influenced significantly by the amount of fish oil in the diet, but increases in fish oil supplementation led to significant increases in the fishy aroma and flavor of the abdominal fat pads.

When an ostrich carcass is deboned, Hoffman et al. (2006) reported that ostrich meat flavor is not affected by hot or cold deboning. Also, technological processes can affect ratites meat and offal flavor. Frozen storage enhances ground emu liverlike giblet flavor. Grilled freshly ground ostrich meat has reduced mealy, beefy, lardy, and liverlike flavors (Andrews et al., 2000). In a comparative flavor characteristics study between ostrich and pork ham, the results showed that ostrich ham tended to be rated lower in salt intensity but had higher flavor intensity ratings than those of pork ham. Similar to Polish sausage products, overall acceptability, flavor acceptability, flavor intensity, salt intensity, texture, and tenderness did not differ significantly between all-pork and ostrich/pork blends (McKenna et al., 2003).

In ostrich dry-cured sausages, flavor characteristics varied, depending on their formulation (ostrich meat or ostrich meat plus pork) (Soriano et al., 2007). Flavor in dry-cured sausages is formed during the ripening stage, when proteins and lipids experience great changes. Proteolysis influences both texture and flavor development, due to the formation of several low-molecular-mass compounds, including peptides, amino acids, aldehydes, organic acids, and amines, which are important flavor compounds or precursors of flavor compounds (Demeyer, 1995; Naes et al., 1995; Fadda et al., 2001). Lipolysis plays an essential role in the development of dry sausage flavor. Lipids are hydrolyzed by enzymes, generating free fatty acids, which are substrates for the oxidative changes that also generate flavor compounds (Samelis et al., 1993; Stahnke, 1995).

FUNCTIONAL POULTRY MEAT PRODUCTS

There is a changing and interesting scenario for the development of new meat products: internationalization of ethnic flavors; trend-setting ingredients that will affect food trends; flavors in the wellness sector, including key functional ingredients; vegetables added to poultry meat products and other natural healthy ingredients; spicy flavors or aromatizated influences; and green poultry products, including natural and organic ingredients (Pérez-Alvarez et al., 2003; Pérez-Alvarez, 2008). In the processing area, one of the main co-products of the chicken meat industry at present is mechanically deboned poultry meat (MDPM). This ingredient has several problems related to its high pH and fatty acid content, the latter being a good substrate for lipid oxidation. Several attempts are currently being made to avoid off-flavors in MDPM-based products (e.g., frankfurters, bologna) and to increase their potentially healthy benefits. To achieve this, several plant extracts, essential oils, and polyphenols are being used. Püssa et al. (2008) added an antioxidant-rich extract from sea buckthorn berries (*Hippophae rhamnoides*) to MDPM products and found that a lower polyphenol (mainly flavonol) content is not sufficient to guarantee complete inhibition of fatty acid oxidation and still leave a part of the added antioxidant polyphenol in the composition. A higher content of sea buckthorn berries extract may reduce the organoleptic properties of patties made from MDPM. This extract produced less lipid oxidation in mechanically deboned turkey meat than in chicken meat during their storage time.

Low-Salt Poultry Meat Products

Sodium chloride is one of the ingredients used most frequently in meat processing. In the modern meat industry, salt is used as a flavoring or flavor enhancer (Ruusunen and Puolane, 2005; Desmond, 2006). Salt imparts a number of functional properties to meat products: It activates proteins to increase hydration and water-binding capacity; it increases the binding properties of proteins to improve texture; it increases the viscosity of meat batters, facilitating the incorporation of fat to form stable batters; and it is essential for flavor (Terrell, 1983). Sodium chloride affects flavor, texture, and the shelf life of meat products. Besides the perceived saltiness, the NaCl brings out the characteristic taste of a meat product, enhancing the flavor (Gillette, 1985). Salt has a flavor-enhancing effect on meat products, with the perceived saltiness due primarily to the sodium ion, with the chloride anion modifying the perception (Miller and Barthoshuk, 1991; Ruusunen and Puolanne, 2005). Fat and salt contribute jointly to most of the sensory properties of processed meats. Matulis et al. (1995) have shown that as salt levels rise, the increase in saltiness is more noticeable in fattier products than in leaner ones. Ruusunen et al. (2001a,b) have shown that the fat content of cooked sausages affects the perceived saltiness, depending on the formulation.

If meat processors decide to reduce salt for health reasons, Price (1997) reported that 25 to 40% salt replacement appears to be the range at which the flavor impact is not as noticeable. As the intensity of some flavors, such as salty, acidic, and spicy, increases, a higher proportion of KCl may be acceptable. This behavior may be a problem in low-salt meat products. The consumer not only perceives reduced saltiness, but the intensity of the characteristic flavor decreases (Ruusunen and Puolanne, 2005) and the product must be reformulated, mainly using spices (Pérez-Alvarez et al., 2000). For low-salt meat products, monosodium glutamate can be used as a flavor enhancer.

Sodium chloride is also a flavor enhancer, increasing the characteristic flavor of meat products (Matulis et al., 1994; Ruusunen et al., 1999, 2001b). Both the perceived saltiness and the flavor intensity depend on the salt content in meat products (Matulis, et al., 1995; Crehan et al., 2000). It must be stressed that a certain amount of salt has to be added to food products before it can even be discerned. A small amount of sodium chloride might taste sweet, which is not appropriate in meat products.

A number of flavor-enhancing and flavor-masking agents are available commercially, and the number of products coming on the market is increasing. These include yeast extracts, lactates, monosodium glutamate, and nucleotides, among others. Taste enhancers work by activating receptors in the mouth and throat, which helps to compensate for the salt reduction (Brandsma, 2006). Other combinations, such as lysine and succinic acid, have been used as salt substitutes (Turk, 1993). This mix has a salty flavor and some antimicrobial and antioxidative properties and from a flavor perspective, may be used to replace up to 75% of the NaCl. However, other water binders, such as phosphates, starches, and gums, may have to be used to maintain the water-binding function lost due to the salt reduction. The use of sodium or potassium lactate with a corresponding reduction in NaCl tends to maintain a certain level of saltiness while reducing to some degree the sodium content in products (Price, 1997).

DEVELOPMENT OF NEW POULTRY MEAT PRODUCTS

Poultry meat product development work is directed to commercializing new products, such as dry-cured poultry sausages (e.g., chicken and turkey salami, salchichon, and chorizo), while managing the inclusion of healthy ingredients (e.g., dietary fiber, omega-3 fatty acids, natural antioxidants). Technologist work with conventional and new ingredients (functional foods), and the fact is that most of them can modify sensory properties, primarily poultry flavor and color. Hard work will be needed to determine the optimal concentration of the new ingredient. Thus, flavor suppliers may also widen their offerings of products and applications. The role of masking flavors, and different types of flavors for different poultry meat product applications, are needed. Consumer tastes are broadening, as exemplified by some up-and-coming flavors and flavor combinations: fried bacon and date palm, and dehydrated fig, almond, cinnamon, and Asian-inspired flavors.

Flavor can play an important role in attracting consumers from the "wellness" sector. Several new ingredients (e.g., functional ingredients, vegetables and other natural health ingredients) are being incorporated into food products and affect food marketing, allowing claims of wellness. Consumers have become more conscious of the possibility to obtain a great variety of novel poultry meat products, with the result that flavor preferences have assumed increasing importance. Across many parts of Europe, consumers are seeking out more unusual varieties of poultry products, such as savory ready-to-eat food (e.g., Cajun hot poultry wings, Thai or tandori chicken and turkey breasts) or poultry and ostrich meat products, such as burgers, dry-cured sausages (e.g., Longaniza de Pascua, salchichon, chorizo) and pâté (Fernández-López et al., 2003b, 2004, 2006b).

New product innovation is focusing primarily on five main aspects of poultry flavor: (1) an understanding of how important flavor considerations are to consumers; (2) identification of national taste differences, with an analysis of their impact on patterns of consumption; (3) flavor preferences and their role; (4) focus on recent flavor innovations; and (5) identification of poultry flavor opportunities. When "designing" a poultry-flavored meat product it is very important for food technologists to take into account that the flavor needs to be accepted by the consumers in a specific country or area, and flavors exogenous to its gastronomic culture should be avoided.

In poultry flavor science, chefs help flavorists to determine future trends and product development. New flavors and combinations for poultry meat products often begin in fine-dining venues and make their way through casual dining to fast food, and eventually to packaged foods, in a process that takes three to five years. Chefs and food technologists help to develop compounds that enhance sweetness or savoriness in ethnic foods and compounds that block or desensitize bitter taste for use in functional poultry meat products.

ETHNIC FOODS

Marketing provides an opportunity to be creative in introducing customers and prospective customers to ethnic foods via trade ads, product data sheets, concept proposals, and prototypes (Rejcek, 2008). Asian and Hispanic flavors are expected to move mainstream, while emerging ethnic cuisines forecasted as gaining a lot of attention include Mediterranean, particularly Greek, and regional Asian influences, such as Thai. All these ethnic foods use poultry in many dishes. Several Mexican and Indian dishes use large amounts of spices (chili and curry, respectively). Mexican food is also expected to move more upscale, and Indian food will also be featured strongly. African flavors will begin to emerge; Scandinavian and German–Slavic tastes are farther out on the horizon.

American consumers are better educated about food and demand sophisticated products and flavors that satisfy their palates as well as their need for health, convenience, and/or adventure. Consumers move toward a wider span of ethnic flavors and speciality products as well as closer scrutiny, even abandonment, of "demonized" ingredients such as salt. It is convenient to have academics working together with the food industry to advance flavor science, which will help to deliver healthier and tastier poultry meat products to consumers.

FLAVOR MEASUREMENT

The relationship between flavor and rancidity must be assessed chemically and by a taste panel (Campo et al., 2006). As rancid flavors develop, there is a loss of desirable flavor notes. However, most studies of rancidity rely on chemical assay methods that determine fatty acid breakdown products, mainly because they are objective, low in cost, and rapid compared with reliable organoleptic assessment (Campo et al., 2006). The disadvantage of such methods is that they measure one among many contributors to rancid flavor, such as hexanal, or secondary breakdown products that do not contribute to flavor, such as the widely used determination of malondialdehyde by the thiobarbituric acid reaction (TBARS) (Tarladgis et al., 1960).

Soriano et al. (2007) ran a study on the sensory assessment of poultry quality according to the following procedure. The assessors were given a brief outline of the free-choice profiling concept and procedure in an initial session. They were asked to smell and taste the samples and to describe, using their own terms, the appearance, color, odor, flavor, texture, and mouthfeel. Individual score cards were prepared and, during additional training sessions, the assessors evaluated the intensity of each sensory attribute.

It is difficult to value the limiting point at which poultry or poultry meat products can be rejected due to lipid oxidation, based on sensory perceptions. Perceptions will depend on personal thresholds, which can vary due to experience, among other factors. But thresholds indicate the point from which stimuli can be perceived, not necessarily from which the stimuli may produce rejection of the product. Clearly, oxidation provokes deterioration of poultry flavor throughout display in atmospheres enriched in oxygen, and this deterioration can be closely related to TBARS. From that point onward, poultry can expect to be rejected, due to a strong sensory perception of lipid oxidation.

Gas chromatography–olfactometry (GC–O) is a very useful technique in studying aroma compounds in poultry meat and meat products. This technique relies on the human nose to elucidate the odor of the GC peak of interest. GC–O techniques have been used to study the aroma profile of chicken meat (Ang et al., 1994). The dynamic headspace technique provides a solvent-free aroma isolate, and it is simple and reproducible. Large amounts of material can be effectively concentrated and injected into the GC column. Other techniques, such as chemometrics, have been reviewed by Martens and Martens (2001), who define chemometrics as the use of mathematics, statistics, and formal logic to provide maximum relevant information from an analysis of chemical data, and who state that valuable food quality studies consist of influences from multiple sources and possibly, nonlinearities. Regarding poultry flavor, further understanding of the relationships among lipid components, antioxidants, and chicken flavor is still necessary. Hoffman et al. (2005, 2008) describe a sensory analysis of ostrich meat (ostrich were fed fish oil–rich omega-3 and the gastrocnemius and iliofibularis muscles were evaluated) in which aroma and flavor were evaluated using a nonuniform scale from 0 to 100. The descriptors used by these authors were (1) aroma (take a few short sniffs as soon as you remove the foil; 100 = extremely intense, 0 = extremely bland) and (2) flavor (This is a combination of taste and flavor experienced prior to swallowing; 100 = extremely intense, 0 = extremely bland). Fernández-López et al. (2008)

evaluated ostrich meat under several packaging conditions (vacuum, air, and modified atmospheres) for off-odors. These authors scored "off-odor" in terms of its intensity associated with meat deterioration (1 = none, 2 = slight, 3 = small, 4 = moderate, and 5 = extreme).

REFERENCES

Andrews L, Gillespie J, Schupp A. 2000. Ratite meat sensory scores compared with beef. J Food Qual 23(3):351–359.

Ang CYW, Liu F, Sun T. 1994. Development of a dynamic headspace GC method for assessing the influence of heating end-point temperature on volatiles of chicken breast meat. J Agric Food Chem 42:2493–2498.

Ashgar A, Gray JI, Buckley DJ, Pearson AM, Booren AM. 1988. Perspectives on warmed-over flavor. Food Technol 42:102–108.

Aymerich T, Picouet PA, Monfort JM. 2008. Decontamination technologies for meat products. Meat Sci 78:114–129.

Azcona JO, Garcia PT, Cossu ME, Iglesias BF, Picallo A, Perez C, Gallinger CI, Schang MJ, Canet ZE. 2008. Meat quality of Argentinean "Camperos" chicken enhanced in omega-3 and omega-9 fatty acids. Meat Sci, doi:10.1016/j.meatsci.2007.12.005.

Brandsma I. 2006. Reducing sodium: a European perspective. Food Technol 60(3):25–29.

Bruun-Jensen L, Skovgaard IM, Madsen EA, Skibsted LH, Bertelsen G. 1996. The combined effect of tocopherols, L-ascorbyl palmitate and L-ascorbic acid on the development of warmed-over flavor in cooked, minced turkey. Food Chem 55(1):41–47.

Byrne DV, Bredie WLP, Mottram DS, Martens M. 2002. Sensory and chemical investigations on the effect of oven cooking on warmed-over flavor development in chicken meat. Meat Sci 61(2):127–139.

Calkins CR, Hodgen JM. 2007. A fresh look at meat flavor. Meat Sci 77:63–80.

Campo MM, Nute GR, Hughes SI, Enser M, Wood JD, Richardson RI. 2006. Flavor perception of oxidation in beef. Meat Sci 72:303–311.

Crehan CM, Troy DJ, Buckley DJ. 2000. Effects of salt level and high hydrostatic pressure processing on frankfurters formulated with 1.5 and 2.5% salt. Meat Sci 55:123–130.

Darrington H. 1996. Meats of the future? Int Food Manuf 13(4):24–25.

Demeyer D. 1995. *Quality and Safety of Fermented Meat Products*. F-FE 169/95. Flair Flow Europe.

De Roos KB. 1997. How lipids influence food flavor? Food Technol 51(1):60–62.

Desmond E. 2006. Reducing salt: a challenge for the meat industry. Meat Sci 74:188–196.

Empson KL, Labuza TP, Graf E. 1991. Phytic acid as a food antioxidant. J Food Sci 56(2):560–563.

Emrick ME, Penfield MP, Bacon CD, van Laack RVL, Brekke CJ. 2005. Heat intensity and warmed-over flavor in precooked chicken patties formulated at 3 fat levels and 3 pepper levels. J Food Sci 70(9):S600–S604.

Erickson MC. 1999. Flavor quality implications in chlorination of poultry chiller water. Food Res Int 32(9):635–641.

Ertas AH. 1998. Oxidation of meat lipids. Gida 23(1):11–17.

Fadda S, Vignolo G, Aristoy MC, Oliver G, Toldrá F. 2001. Effect of curing conditions and *Lactobacillus casei* CRL705 on the hydrolysis of meat proteins. J Appl Microbiol 91(3):478–487.

Fernández-López J, Sevilla L, Sayas-Barberá ME, Navarro C, Marín F, Pérez-Alvarez JA. 2003a. Evaluation of antioxidant potential of hyssop (*Hyssopus officinalis* L.) and rosemary (*Rosmarinus officinalis* L.) extract in cooked pork meat. J Food Sci 68:660–664.

Fernández-López J, Sayas-Barberá ME, Navarro C, Sendra E, Pérez-Alvarez JA. 2003b. Physical, chemical and sensory properties of bologna sausage made with ostrich meat. J Food Sci 68:1511–1515.

Fernández-López J, Sendra E, Sayas-Barberá ME, Pérez-Alvarez JA. 2004. Quality characteristics of ostrich liver pate. J Food Sci 69(2):SNQ85–SNQ90.

Fernández-López J, Jiménez S, Sayas-Barberá E, Sendra E, Pérez-Alvarez JA. 2006a. Quality characteristics of ostrich (*Struthio camelus*) burgers. Meat Sci 73:295–303.

Fernández-López J, Yelo A, Sayas-Barberá E, Sendra E, Navarro C, Pérez-Alvarez JA. 2006b. Shelf life of ostrich (*Struthio camelus*) liver stored under different packaging conditions. J Food Prot 69(8):1920–1927.

Fernández-López J, Sayas-Barberá E, Mu noz T, Sendra E, Navarro C, Pérez-Alvarez JA. 2008. Effect of packaging conditions on shelf-life of ostrich steaks. Meat Sci 78:143–152.

Gasser U, Grosch W. 1988. Identification of volatile flavor compounds with high aroma values from cooked beef. Z Lebensm-Unters-Forsch 186(6):489–494.

Gillette M. 1985. Flavor effects of sodium chloride. Food Technol 39:47–52, 56.

Girolami A, Marsico I, D'Andrea G, Braghieri A, Napolitano F, Cifuni GF. 2003. Fatty acid profile, cholesterol content and tenderness of ostrich meat as influenced by age at slaughter and muscle type. Meat Sci 64:309–315.

Gorbatov VM, Lyaskovskaya YN. 1980. Review of the flavor contributing volatiles and water-soluble non-volatiles in pork meat and derived products. Meat Sci 4:209–225.

Graf E, Panter SS. 1991. Inhibition of warmed-over flavor development by polyvalent cations. J Food Sci 56(4):1055–1058.

Higgins FM, Kerry JP, Buckley DJ, Morrissey PA. 1999. Effects of alpha-tocopheryl acetate supplementation and salt addition on the oxidative stability (TBARS) and warmed-over flavor (WOF) of cooked turkey meat. Br Poult Sci 40(1):59–64.

Hoffman LC, Joubert M, Brand TS, Manley M. 2005. The effect of dietary fish oil rich in *n*-3 fatty acids on the organoleptic, fatty acids and physicochemical characteristics of ostrich meat. Meat Sci 70(1):45–53.

Hoffman LC, Botha Suné St C, Britz T. 2006. Sensory properties of hot-deboned ostrich (*Struthio camelus* var. *domesticus*) muscularis gastrocnemius, pars interna. Meat Sci 72:734–740.

Hoffman LC, Muller M, Cloete SWP, Brand M. 2008. Physical and sensory meat quality of South African Black ostriches (*Struthio camelus* var. *domesticus*), Zimbabwean blue ostriches (*Struthio camelus australis*) and their hybrid. Meat Sci 79:365–374.

Honikel KO. 2008. The use and control of nitrate and nitrite for the processing of meat products. Meat Sci 78:68–76.

Jahan K, Paterson A, Piggott J, Spickett C. 2005. Chemometric modeling to relate antioxidants, neutral lipid fatty acids, and flavor components in chicken breasts. Poult Sci 84:158–166.

Jayathilakan K, Vasundhara TS, Kumudavally KV. 1997. Effect of spices and Maillard reaction products on rancidity development in precooked refrigerated meat. J Food Sci Technol India 34(2):128–131.

Juncher D, Hansen TB, Eriksen H, Skovgaard IM, Knochel S, Bertelsen G. 1998. Oxidative and sensory changes during bulk and retail storage of hot-filled turkey casserole. Food Res Technol 206(6):378–381.

Kerler J, Grosch W. 1997. Character impact odorants of boiled chicken: changes during refrigerated storage and reheating. Food Res Technol 205(3):232–238.

Labuza TP. 1971. Kinetics of lipid oxidation in foods. CRC Crit Review Food Technol 2:355–404.

Larrouture C, Ardaillon V, Pepin M, Montel MC. 2000. Ability of meat starter cultures to catabolize leucine and evaluation of the degradation products by using an HPLC method. Food Microbiol 17:563–570.

MacLeod G. 1986. The scientific and technological basis of meat flavors. In: Birch GG, Lindley MG, eds., *Developments in Food Flavors*. New York: Elsevier Applied Science, pp. 191–223.

Marks J, Stadelman W, Linton R, Schmieder H, Adams R. 1998. Tenderness analysis and consumer sensory evaluation of ostrich meat from different muscles and different aging times. J Food Qual 21(5):369–381.

Martens H, Martens M. 2001. Analysis of two data tables X and Y: partial least squares regression (PLSR). In: *Multivariate Analysis of Quality: An Introduction*. Chichester, UK: Wiley, pp. 275–276.

Mattes RD. 1984. Salt taste and hypertension: a critical review of the literature. J Chron Dis 37(3):195–208.

Mattes RD. 1997. The taste for salt in humans. Am J Cli Nutr 65(Suppl):692S–697S.

Matulis RD, McKeith FK, Brewer MS. 1994. Physical and sensory characteristics of commercially available frankfurters. J Food Qual 17:263–271.

Matulis RD, McKeith FK, Sutherland JW, Brewer MS. 1995. Sensory characteristics of frankfurters as affected by fat, salt and pH. J Food Sci 60:42–47.

McKee L. 2007. General attributes of fresh and frozen poultry meat. In: Nollet LML, ed., *Handbook of Meat, Poultry and Seafood Quality*. Oxford: Blackwell Publishing, pp. 429–437.

McKenna DR, Morris CA, Keeton JT, Miller RK, Hale DS, Harris SD, Savell JW. 2003. Consumer acceptability of processed ostrich meat products. J Muscle Foods 14(2):173–179.

Mielche MM. 1995. Development of warmed-over flavor in ground turkey, chicken and pork meat during chill storage: a model of the effects of heating temperature and storage time. Z Lebensm-Unters -Forsch 200(3):186–189.

Mielnik MB, Sem S, Egelandsdal B, Skrede G. 2008. By-products from herbs essential oil production as ingredient in marinade for turkey thighs. Food Sci Technol 41(1):93–100.

Miller IJ, Barthoshuk LM. 1991. Taste perception, taste bud distribution and spatial relationship. In: Getchell TV, Doty RL, Bartoshuk LM, Snow JB, eds., *Smell and Taste in Health and Disease*. New York: Raven Press, pp. 205–233.

Montel MC, Reitz J, Talon R, Berdagué JL, Rousset AS. 1996. Biochemical activities of micrococcaceae and their effects on the aromatic profiles and odours of a dry sausage model. Food Microbiol 13:489–499.

Mottram DS. 1992. Meat flavor. Meat Focus Int, June, pp. 81–92.

Mottram DS. 1998. Flavor formation in meat and meat products: a review. Food Chem 62:415–424.

Mottram DS, Madruga MS. 1994. Important sulfur containing aroma volatiles in meat. In: Mussinan CJ, Keelan ME, eds., *Sulfur Compounds in Foods*. ACS Symposium Series, vol. 564. Washington, DC: American Chemical Society, pp. 180–187.

Naes H, Holck AL, Axelsson L, Andersen HJ, Blom H. 1995. Accelerated ripening of dry fermented sausage by addition of a *Lactobacillus* proteinase. Int J Food Sci Technol 29(6):651–659.

O'Neill LM, Galvin K, Morrissey PA, Buckley DJ. 1998. Comparison of effects of dietary olive oil, tallow and vitamin E on the quality of broiler meat and meat products. Br Poult Sci 39(3):365–371.

Ouali A, Herrera-Mendez CH, Coulis G, Becila S, Boudjellal A, Aubry L, Sentandreu MA. 2006. Revisiting the conversion of muscle into meat and the underlying mechanisms. Meat Sci 74:44–58.

Pegg RB, Shahidi F. 2000. *Nitrite Curing of Meat: The N-Nitrosamine Problem and Nitrite Alternatives*. Trumbull, CT: Food and Nutrition Press.

Pérez-Alvarez JA. 2006a. Color de la carne y productos cárnicos. In: Hui YH, Guerrero I, Rosmini MR, eds., *Ciencia y Tecnología de Carnes*. Mexico City, Mexico: Limusa, pp. 161–198.

Pérez-Alvarez JA. 2006b. Aspectos tecnológicos de los productos crudo-curados. In: Hui YH, Guerrero I, Rosmini MR, eds., *Ciencia y Tecnología de Carnes*. Mexico City, Mexico: Limusa, pp. 463–492.

Pérez-Alvarez JA. 2008. Overview of meat products as functional foods. In: Fernández-López J, Pérez-Alvarez JA, eds., *Technological Strategies for Functional Meat Products Development*. Kerala, India: Transworld Research Network, pp. 2–17.

Pérez-Alvarez JA, Fernández-López J. 2000. *Aspectos Físicos, Psicológicos, Químicos e Instrumentales para la Determinación del Color en Alimentos*. Elche, Spain: Universidad Miguel Hernández, CD format, pp. 113–147.

Pérez-Alvarez JA, Fernández-López J. 2006. Chemistry and biochemistry of color in muscle foods. In: Hui YH, Wai-Kit Nip, Leo ML, Nollet LML, Paliyath G, Simpson BK, eds., *Food Biochemistry and Food Processing*. Ames, IA: Blackwell Publishing, pp. 337–350.

Pérez-Alvarez JA, Navarro C, Fernández-López J, Sayas-Barberá E. 2000. Aspectos tecnológicos y sensoriales de productos bajos en calorías. In: Rosmini MR, Pérez-Alvarez JA, Fernández-López J, eds., *Nuevas Tendencias en la Tecnología e Higiene de la Industria Cárnica*. Elche, Spain: Universidad Miguel Hernández, pp. 201–217.

Pérez-Alvarez JA, Sayas-Barberá ME, Fernández-López J. 2003. La alimentación en las sociedades occidentales. In: Pérez-Alvarez JA, Fernández-López J, Sayas-Barberá ME, eds., *Alimentos Funcionales y Dieta Mediterránea*. Elche, Spain: Universidad Miguel Hernández, pp. 27–38.

Pollok KD, Miller RK, Hale DS, Angel R, Blue-McLendon A, Baltmanis B. 1997. Quality of ostrich steaks as affected by vacuum-package storage, retail display and differences in animal feeding regime. Ostrich 4:46–52.

Price JF. 1997. Low-fat/salt cured meat products. In: Pearson AM, Dutson TR, eds., *Advances in Meat Research: Production and Processing of Healthy Meat, Poultry and Fish Products*, vol. 11., London: Blackie Academic & Professional, pp. 242–256.

Püssa T, Pällin R, Raudsepp P, Soidla R, Rei M. 2008. Inhibition of lipid oxidation and dynamics of polyphenol content in mechanically deboned meat supplemented with sea buckthorn (*Hippophae rhamnoides*) berry residues. Food Chem 107(2):714–721.

Ramarathnam N, Rubin LJ, Diosady LL. 1993. Studies on meat flavor: 4. Fractionation, characterization, and quantitation of volatiles from uncured and cured beef and chicken. J Agric Food Chem 41(6):939–945.

Renerre M. 1999. Biochemical basis of fresh meat colour. In: *Proceedings of the 45th ICoMST*, Yokohama, Japan, vol. 2, pp. 344–353.

Rhee KS, Anderson LM, Sams AR. 1996. Lipid oxidation potential of beef, chicken, and pork. J Food Sci 61:8–12.

Rhee KS, Anderson LM, Sams AR. 2005. Comparison of flavor changes in cooked–refrigerated beef, pork and chicken meat patties. Meat Sci 71:392–396.

Rejcek P. 2008. Keeping up with the flavor market. Funct Ing, May, p. 24.

Rodbotten M, Kubberod E, Lea P, Ueland O. 2004. A sensory map of the meat universe: sensory profile of meat from 15 species. Meat Sci 68(1):137–144.

Ruusunen M, Puolanne E. 2005. Reducing sodium intake from meat products. Meat Sci 70:531–541.

Ruusunen M, Särkkä-Tirkkonen M, Puolanne E. 1999. The effect of salt reduction on taste pleasantness in cooked bologna type sausages. J Sens Stud 14:263–270.

Ruusunen M, Simolin M, Puolanne E. 2001a. The effect of fat content and flavor enhancers on the perceived saltiness of cooked bologna-type sausages. J Muscle Foods 12:107–120.

Ruusunen M, Tirkkonen MS, Puolanne E. 2001b. Saltiness of coarsely ground cooked ham with reduced salt content. Agric Food Sci Finland 10:27–32.

Samelis J, Aggelis G, Metaxopoulos J. 1993. Lipolytic and microbial changes during the natural fermentation and ripening of Greek dry sausages. Meat Sci 35(3):371–385.

Sanderson GW, Co H, González JG. 1971. Biochemistry of tea fermentation: the role of carotenes in black tea aroma formation. J Food Sci 36(2):231–236.

Sa nudo C, Enser ME, Campo MM, Nute GR, Maria G, Sierra I, Wood JD. 2000. Fatty acid composition and sensory characteristics of lamb carcasses from Britain and Spain. Meat Sci 54:339–346.

Sárraga C, Carreras I, Garcia-Regueiro JA, Guardia MD, Guerrero L. 2006. Effects of alpha-tocopherol acetate and beta-carotene dietary supplementation on the antioxidant enzymes, TBARS and sensory attributes of turkey meat. Br Poult Sci 47(6):700–707.

Sayas-Barberá E, Fernández-López J. 2000. El avestruz como animal de abasto en el siglo XXI. In: Rosmini MR, Pérez-Alvarez JA, Fernández-López J, eds., *Nuevas Tendencias en la Tecnología e Higiene de la Industria Cárnica*. Elche, Spain: Universidad Miguel Hernández, pp. 218–236.

Sayas-Barberá E, Fernández-López J, Pérez-Alvarez JA. 2008. Elaboración de embutidos crudo-curados. In: Pérez-Alvarez JA, Fernández-López J, Sayas-Barberá ME, eds., *Industrialización de Productos de Origen Animal*, vol. 1. Elche, Spain: Universidad Miguel Hernández, pp. 77–102.

Sebranek JG, Bacus AJN. 2007. Cured meat products without direct addition of nitrate or nitrite: What are the issues? Meat Sci 77:136–147.

Sentandreu MA, Stoeva S, Aristoy MC, Laib K, Voelter W, Toldrá F. 2003. Identification of small peptides generated in Spanish dry-cured ham. J Food Sci 68:64–69.

Shahidi F. 1989. Flavor of cooked meat. In: Teranishi R, Buttery RG, Shahidi F, eds., *Flavor Chemistry Trends and Developments*. Washington, DC: American Chemical Society, pp. 188–201.

Shu-Mei-Lai, Gray JI, Booren AM, Crackel RL, Gill JL. 1995. Assessment of off-flavor development in restructured chicken nuggets using hexanal and TBARS measurements and sensory evaluation. J Sci Food Agric 67(4):447–452.

Sitz BM, Calkins CR, Feuz DM, Umberger WJ, Eskridge KM. 2005. Consumer sensory acceptance and value of domestic, Canadian, and Australian grass-fed beef steaks. J Anim Sci 83(12):2863–2868.

Soriano A, García-Ruiz A, Gómez E, Pardo R, Galán FA, González-Vi nas MA. 2007. Lipolysis, proteolysis, physicochemical and sensory characteristics of different types of Spanish ostrich salchichon. Meat Sci 75:661–668.

Stahnke LH. 1995. Dried sausages fermented with *Staphylococcus xylosus* at different temperatures and with different ingredient levels: II. Volatile components. Meat Sci 41(2):193–209.

St Angelo AJ, Vercellotti JR, Legendre MG, Vinnett CH, Kuan JW, James JC, Dupuy HP. 1987. Chemical and instrumental analysis of warmed-over flavor in beef. J Food Sci 52:1163–1168.

Tai CY, Ho CT. 1997. Influence of cysteine oxidation on thermal formation of Maillard aromas. J Agric Food Chem 45(9):3586–3589.

Talon R, Chastagnac C, Vergnais L, Montel MC, Berdague JL. 1998. Production of esters by staphylococci. Int J Food Microbiol 45(2):143–150.

Tan SS, Aminah A, Zhang XG, Abdul SB. 2006. Optimizing palm oil and palm stearin utilization for sensory and textural properties of chicken frankfurters. Meat Sci 72:387–397.

Tang S, Kerry JP, Sheehan D, Buckley DJ. 2001. A comparative study of tea catechins and α-tocopherol as antioxidants in cooked beef and chicken meat. Eur Food Res Techol 213:286–289.

Tarladgis BG, Watts BM, Younathan MT, Dugan L. 1960. A distillation method for the quantitative determination of malonaldehyde in rancid foods. J Am Oil Chem Soc 37:44–48.

Taylor G, Andrews L, Gillespie J, Schupp A. 1997. A sensory panel evaluation of ratite meat. La Agric 40(1):18–19.

Terrell RN. 1983. Reducing the sodium content of processed meats. Food Technol 37(7):66–71.

Turk R. 1993. Metal free and low metal salt substitutes containing lysine. U.S. patent 5,229,161.

Turner BE, Larick DK. 1996. Palatability of sous vide processed chicken breast. Poult Sci 75(8):1056–1063.

Walter JM, Soliah L, Dorsett D. 2000. Ground ostrich: a comparison with ground beef. J Am Diet Assoc 100(2):244–245.

Warner RD, Dunshea FR, Ponnampalam EN, Cottrell JJ. 2005. Effects of nitric oxide and oxidation in vivo and post-mortem on meat tenderness. Meat Sci 71:205–217.

25

POULTRY MEAT COLOR

ALESSANDRA GUIDI AND L. CASTIGLIEGO
Department of Animal Pathology, Prophylaxis and Food Hygiene, University of Pisa, Pisa, Italy

INTRODUCTION

The visual appearance of meat and meat products is the first factor influencing the choice of the consumer and therefore the commercial success. As a consequence of the strong psychological impact on the "man in the street," the color of meat represents a considerable concern, not only for the buyer but also for the producer.

Handbook of Poultry Science and Technology, Volume 2: Secondary Processing, Edited by Isabel Guerrero-Legarreta and Y.H. Hui

To avoid or prevent undesirable chromatic alterations, a number of studies have been carried out either to clarify the basis of meat color origin or to better understand which factors may influence its change during processing, shelf life, or cooking. In this chapter we overview this important issue.

COLOR MEASUREMENT

Meat color measurement is very important for the following reasons: to acquire information on customers' appraisal, to better understand preferences and reactions to different shades of color, to investigate chromatic changes, and to determine its causes and evolution over time. Moreover, the possibility of using color measurements to predict the functional properties of poultry meat has been suggested by a number of studies. Unfortunately, color, as it is perceived by the human eye, represents a sensorial experience, and for this reason it is always subjective given that it is influenced by either physiological characteristics or by the psychological stance of the observer. In addition, the human visual system works more as a comparator than as an absolute sensor, so that even the background may substantially influence the overall perception of color. Essentially, color depends not only on the measured material, but also on the light source and the detector, be it human or instrumental, making different measurement responses as a consequence of different experimental conditions. This implies that any type of color evaluation must be done according to standardized conditions, preferably using specific instruments, which are now in widespread use for research or quality control purposes. In general, the most common procedures used for color measurements are visual assessment, spectrophotometry, and reflectance colorimetry.

The first trials for color were made using a panel of people who had to describe the object of the observation using predefined terms, each associated with a different shade of an established color. The responses to this test were too imprecise and varied. The comparative method was then introduced, by which sets of color range standards were developed, ordered on a numbered scale, and applied in a scoring system to carry out color matches for egg yolk, poultry skin, and meat, such as chicken shanks (Brown, 1930; Bird, 1943; Harms et al., 1971; Twining et al., 1971). The most famous and successful of these systems was the Roche color fan, which was reported by Vuilleumier in 1969. Visual standards were also developed for the poultry carcass and meat defects, which are often associated with chromatic variations (Daniels et al., 1989). Nowadays, panel assessment for meat color is still common and is still based on reference color standards and specific rules. Recommendations by meat science associations are used primarily to limit the influence of the environment or characteristics of the meat other than color (e.g., fat, texture) (AMSA, 1991). However, sensory panel assessment has some intrinsic limits and does not allow an objective response, which is instead based on instrumental measurements. The first attempts to overcome subjectivity were carried out using a spectrophotometer to measure the carotenoid content of the skin and the heme pigments of meat after chemical extraction (Day and Williams, 1958; Saffle, 1973). This method has often been used to correlate the quantity and form of myoglobin in meat (which is among the main factors responsible for meat color, as discussed further later) with the potential for meat color or color defects (Guidi et al., 2006).

Most research and quality control procedures contemplate the use of reflectance colorimeters, which express meat color in terms of color differences from a designated color standard. Yet despite the fact that this type of instrument ensures high levels of accuracy and reproducibility, it is designed for opaque surfaces, whereas meat is a translucent material and may scatter much of the light, reducing the reproducibility of the measurements and the comparison of color values (Uijttenboogaart, 1991). The most commonly used colorimetric scale is nowadays the CIE L^*, a^*, b^* (International Commission on Illumination, 1986), even though other color scales can be used, such as the Hunter L, a, b and YXZ space. The CIELAB color scale is organized in cube form, a sort of tridimensional space in which the color can be collocated and relies on three color parameters, expressed as L^*, a^*, and b^* (Figure 1). L^* expresses the lightness, with 0 associated with black (complete absorption of light) and 100 with white (complete reflection). a^* indicates the level of redness (or greenness, since the two colors are complementary), ranging from−60 (pure green) to 60 (pure red). Finally, b^* indicates yellowness (or blueness), also ranging from−60 (pure blue) and 60 (pure yellow). Hue angle is the arctangent of b^*/a^*, determined by rotation about the a^* and b^* axes and defines the color. The

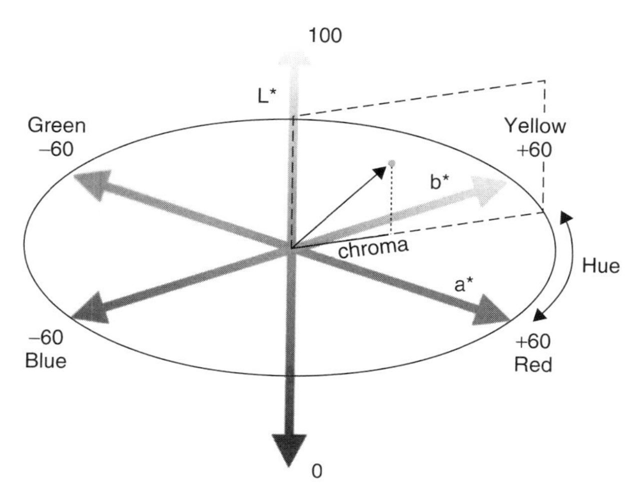

FIGURE 1 $L^*a^*b^*$ color space. The color of a surface is associated with a point in three-dimensional space with precise coordinates, related to each dimension (black to white; green to red; blue to yellow). Hue angle is the arctangent of b^*/a^*, and chroma (intensity) is the distance between the origin of the axes and the orthogonal projection of the point on the circle surface $(a^{*^2} + b^{*^2})^{-2}$.

color intensity (chroma) is defined as $(a^{*2} + b^{*2})^{-2}$. If a comparative analysis is performed, delta values (ΔL^*, Δa^*, Δb^*) indicate how much a sample and standard differ from one another. The total color difference (ΔE^*) may also be calculated, and it takes into account all the L^*, a^*, and b^* differences between samples and standards. The difference in chroma (ΔC^*) and in hue angle (ΔH^*) are described in a polar coordinate system.

Honikel (1998) specified a reference method for color measurements using the CIE scale, defining procedures for sample preparing, light source, geometry of illumination and viewing, observer angle, and all the parameters that have to be specified to ensure reproducibility of data. Despite instrumental measurement being more accurate and reliable than sensorial measurement, differences in sample presentation and measurement conditions make it difficult to compare absolute color values between laboratories. For instance, under practical industry conditions, breast meat size variation and background color (i.e., belt surfaces, packaging material, etc.) would be factors affecting color difference or absolute color measurements (Bianchi and Fletcher, 2002). Sandusky and Heath (1996) reported that broiler breast meat color measurements based on reflectance colorimetry could differ on the basis of sample thickness and background color. Bianchi and Fletcher (2002) obtained similar results in chicken and turkey breast meat. This suggests that the choice of the sample must be accurate, since part of the incident light is diffused when interacting with a wet meat surface and partly penetrates, to a different extent inside the tissues, and is reflected internally. Because of the translucency of the meat, the sample to be measured must be thick enough to avoid light reaching the background material and being reflected by it. Reflectance spectrophotometry also allows calculation of the proportion of the three major myoglobin pigments by means of measurements at the isobestic point (i.e., the wavelength with the same extinction coefficient for all the pigments) and specific equations (Stewart et al., 1965; Van den Oord and Wesdorp, 1971; Krzywicki, 1979).

Different color space values from instrumental color measurements have been suggested to correlate with panel evaluation and consumer scores and were associated with the subjective perception of a number of meat defects, such as paleness, darkness, and discoloration. Some studies showed that PSE (pale, soft, exudative)–like meats can be associated with specific color measurement responses as well as water-holding capacity (Barbut, 1993, 1998; McCurdy et al., 1996; Fletcher et al., 2000; Owens et al., 2000; Wilkins et al., 2000; Qiao et al., 2001). Relationships were also found between color and shelf life (Allen et al. 1997) or composition (Qiao et al., 2002) of broiler breast meat. Finally, lightness values have been shown to be useful as an indicator of poultry breast meat quality for further processing (McCurdy et al., 1996; Barbut, 1997; Owens et al., 2000). Reflectance colorimetry has been applied to a large number of experimentations, especially on yolk and meat, but it is less applicable for skin color evaluation, due to the small measurable area in relation to the whole sample and the nonhomogeneity of the skin surface. It was calculated that only about 0.03 to 0.13% of the entire surface of the broiler can be measured by a colorimeter (Fletcher, 1988).

An interesting alternative for color measurement is computer vision based on an analysis of digital camera images. The advantages over other methods can be found in the possibility of processing the acquired digital images to remove background fat and bone (Lu et al., 2000); several measurement systems can be used (e.g., CIE, Hunter, XYZ), and the values can be measured on a given area. Only a single observation could be representative for the color evaluation, and the surface variation of the myoglobin redox state can be assessed (O'Sullivan et al., 2003). Moreover, implementation of the data by statistical or neural network models can help to predict sensory color scores (Lu et al., 2000) since the camera takes measurements over the entire sample surface, resulting in a more representative result than that obtained by the point-to-point measuring of a colorimeter (O'Sullivan et al., 2003).

HOW MEAT COLOR IS DETERMINED

Many factors determine or influence the color of meat or, more properly, all its shades, making its final manifestation the effect of an intricate ensemble of causes. Some of them have been clearly identified, but a certain grade of uncertainty remains for others, due to the not always concordant results of the wide array of experimentation. This truth testifies to the great complexity and richness of the topic.

Skipping for a moment argument as to its bases, it can be asserted that the color of meat depends primarily on:

- The type and quantity of pigments present in the muscle
- The types of fibers composing the muscle and their spatial relationships, which determine the scattering grade of light and thus its deepness of penetration
- The intramuscular fats and surface dehydration, which confer different degrees of glossiness and thus affect light scattering and reflection

Meat Pigments and Their Chemistry

The pigment most responsible for the color of meat is myoglobin, although there is also a small amount of hemoglobin, which is very similar, but is present only in small amounts in well-slaughtered animals, and of cytochromes and ribonucleases, which play a minor role because of their low concentrations. Myoglobin is a single-chain globular protein of about 16,700 kDa, whose role is to store oxygen in the muscle once it is released by hemoglobin in the capillaries and spreads inside the myocites before entering the oxidative phosphorylation cycle to produce new energy. Myoglobin consists of a peptide portion (apoprotein) and an oxygen-binding prosthetic group, called heme, containing an iron atom in its core and placed in a hydrophobic pocket formed by the apoprotein (Figure 2). The iron is coordinated by four noncovalent bonds to pyrrole nitrogen of heme and by another bond to the proximal His 93 of the protein. A sixth coordination

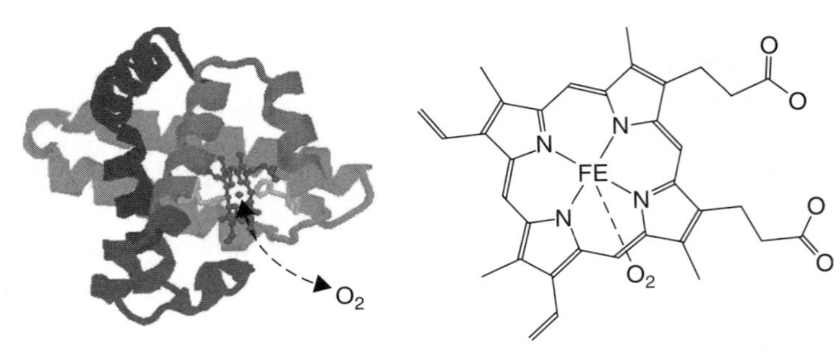

FIGURE 2 Schematic representation of myoglobin and heme, which is located in the pocket formed by the folding of the apoprotein. Proximal and distal histidines are yellow colored.

bond may potentially be formed by a number of small molecules, among them oxygen, which is capable of penetrating the pocket and interacting with the iron itself. Iron's property to attract electron pair–donating ligands, due to its electron deficiency, leads to the formation of bonds with such molecules, whose nature depends on the ease with which electrons are donated. The intact conformation of the pocket containing heme is important to guarantee stabilization of the iron's oxidation state, making the O_2 binding to the iron reversible (Morrison and Boyd, 1981) and allowing access to only a limited number of molecules, such as cyanide, nitric oxide, and carbon monoxide (Lycometros and Brown, 1973). A distal histidine (His 64) present in the pocket controls the spatial relations between heme and bound compounds, influencing the chemical dynamics related to color. The heme core can also absorb electromagnetic radiation in the visible range, yet the absorbance spectrum may vary according to the oxidation state of the iron and the type of bound molecule (Giddings and Markakis, 1972), conferring on the meat a great range of possible tonalities, from dark brown to bright red; some uncommon colors, such as green, can also appear, as in the case of bacterial contamination.

There are three main forms of myoglobin in muscles: deoxymyoglobin (Mb^{2+}), oxymyoglobin (Mb^{2+} O_2), and metmyoglobin (Mb^{3+} OH_2) (Stewart et al., 1965) (Figure 3), each conferring a different color to the meat. Under very low oxygen tension, iron in the form 2+ (Fe^{2+}) tends to interact ionically with H_2O when no other electron donors are present in the pocket (deoxymyoglobin), conferring a dark-red purple color to the tissue. On the other hand, at higher tension, the molecular oxygen tends to occupy the sixth coordination site and binds the iron, which maintains the lower oxidation state (Fe^{2+}): $Mb^{2+} + O_2 \rightleftharpoons Mb^{2+}O_2$. This complex determines the bright red color that is typical of fresh meat. The two forms are rapidly convertible, even if the reaction is influenced by the temperature, O_2 partial pressure, pH, and possible competitors for binding the iron. However, in meat, the conversion from oxymyoglobin to deoxymyoglobin may undergo more complex dynamics, characterized by a two-step reaction: $Mb^{2+}O_2 \rightarrow Mb^{3+}OH_2 \rightarrow Mb^{2+}$.

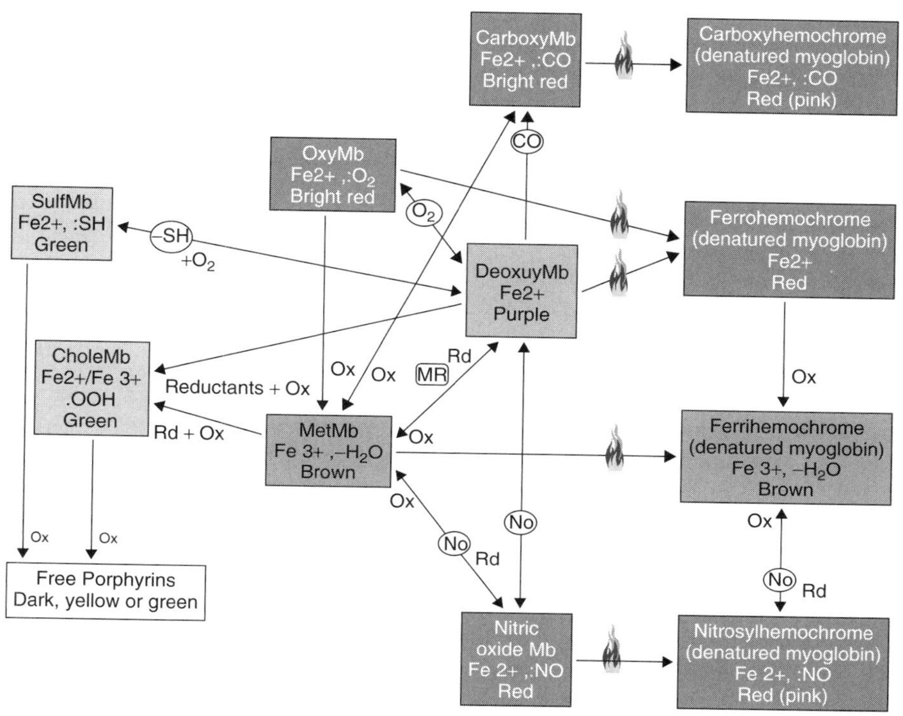

FIGURE 3 Major pathways of interconversion between different forms of myoglobin. Ox, oxidation; Rd, reduction; MR, metmyoglobin reductase. (Data from Lawrie, 1966; Forrest et al., 1975; Livingston and Brown, 1982; Varnam and Sutherland, 1995; DeMan, 1999.)

The oxygenation of deoxymyoglobin can frequently be observed when some parts of a piece of meat appearing darker than the rest are exposed to the atmosphere after being in prolonged contact with a gas-impermeable packaging film, which causes a temporary reduction in the local oxygen concentration. If the iron is oxidized further (Fe^{3+}), it loses its capability to bind molecular oxygen, but can form a different type of complex that interacts with other molecules. An important example is the formation of metmyoglobin, where iron binds a molecule of water and determines a shift in the meat's color toward brown. This is known as *meat discoloration*, and its extent is related to the proportion of metmyoglobin and the amount of the meat surface that has a high percentage of metmyoglobin. Such an occurrence is characterized by an autoxidation process that causes the formation of superoxide anion (O_2^-) according to the reaction $Mb^{2+}O_2 + H_2O + H^+ \rightleftharpoons Mb^{3+}OH_2 + HO_2$.

Superoxide anion can, in turn, be converted by spontaneous dismutation into hydrogen peroxide (H_2O_2), this also being a potent oxidant of MbO_2 (Tajima and Shikama, 1987). This process, whose molecular mechanisms are still a puzzle, was demonstrated to be dependent on the O_2 partial pressure, with a maximum around 2 torr (George and Stratmann, 1952; Gill and McGinnis, 1995), and it is favored by both acidic and basic pH (Shikama and Sugawara, 1978). In particular, in the meat context, the range of slightly acidic pH is of more interest, since postmortem muscle pH is supposed to decrease. Oxymyoglobin autooxidation to metmyoglobin can also be enhanced by temperature (Brown and Mebine, 1969) and an increase in free radicals (Giroux et al., 2001) in postmortem muscles for microbial growth or during meat processing. All the aforesaid effects occur to different degrees, depending on the type of protein involved (Shikama, 2006).

In general, the browning of meat produces a resolute rejection reaction in the consumer, despite the substantial harmlessness to health. In living muscle tissues, the enzymes metmyoglobin reductase, in the presence of the cofactor NADH and the coenzyme cytochrome *b*4, brings the Fe^{3+} back to the Fe^{2+} of normal myoglobin (Livingston et al., 1985). When tissues die and convert into meat, the functionality of their enzymes fades little by little and fails to contain the formation of metmyoglobin, which becomes irreversible. The condition of the postmortem muscle is not favorable for the reduction of metmyoglobin to deoxymyoglobin, due to progressive depletion of the reducing enzyme systems and cofactors (such as NADH). Moreover, at low-oxygen concentrations (deep portions or vacuum packaging), the formation of metmyoglobin has been widely reported (George and Stratmann, 1952; Gill and McGinnis, 1995), resulting in an accumulation of metmyoglobin (Young and West, 2001). However, the interconversion dynamics among the various forms of myoglobin also depends on other factors, such as pH, temperature, ions, and microbic metabolite concentration (Renerre et al., 1990; Osborn et al., 2003) and the presence of endogenous or exogenous antioxidants, such as tocopherols, ascorbic acid, and plant extracts, which can be added to the diet to preserve the meat oxidative stability during storage (Bartov and Kanner, 1996; Sheldon et al., 1997; Du et al., 2002; Young et al., 2003).

Undesirable color change from red to brown have also been observed in bone marrow, which, since it is rich in hemoglobin, undergoes oxidation–reduction reactions similar to those described for myoglobin. Lyon et al. (1976) demonstrated that meat and bone darkening of thigh pieces was related to pigment migration from the femur to tissue. Although many compounds can bind to iron other than O_2, only two molecules have some importance for meat: carbon monoxide (CO) and nitric oxide (NO), which, as well as O_2, can bind only Fe^{2+} but not Fe^{3+}, since the latter binds electrons of the donor molecules more tightly because of its higher charge, and is not able to donate back other electrons to form stabilizing bonds (Livingstone and Brown, 1982). Carbon monoxide has much more affinity for Mb^{2+} than O_2, so it can displace the latter and form a relatively stable compound called carboxymyoglobin (Mb^{2+} CO), which confers a bright red color to meat. Its importance derives from an increasing interest in packaging with a low CO level (Sørheim et al., 1999, Hunt et al., 2004).

Nitric oxide myoglobin (Mb^{2+} NO), which represents the pigment intentionally formed by cured-meat processing, also confers a red color to meat. Mb^{2+}

NO is formed by the degradation of nitrates into nitrites and further into nitric oxide and is more stable than Mb^{2+} O_2, but not enough to avoid the displacement of NO by O_2, which is normally far more abundant. In curing technology, meat and meat products can be cooked to prevent the loss of NO and form a stable pigment, the nitrosylhemochrome, which is responsible for the final pink color (Forrest et al., 1975) (Figure 3). This pigment is desirable in cooked red meat but undesirable in poultry meat and meat products. Sharper chromatic alterations can occur in the flesh of slaughtered animals after denaturation of the muscle proteins and reduction of the heme's core, as a consequence of their exposure to hydrogen sulfur and oxygen or to hydrogen peroxide and ascorbic acid, with the formation of sulfmyoglobin (reversible) and cholemyoglobin (irreversible), respectively, which determine the appearance of a green color. Such events can be caused by the growth of microorganisms, favored by tissue lesions in vivo. Further oxidation of cholemyoglobin may cause heme opening, forming the verdoheme (green in color) or the loss of iron and the formation of a chain of porphyrin rings (Lawrie, 1966; DeMan, 1999).

Every chemical process leading to the denaturation or destabilization of proteins, including myoglobins, may enhance the formation of pigment forms which can determine color alterations in the muscle. Besides the ability of the various myoglobin forms to catalyze lipid oxidation in an acidic environment, several studies have highlighted how transition elements, among them iron, can catalyze the decomposition of peroxides by generating free radicals, which in turn oxidize unsaturated fatty acids (Nawar, 1996; Mancuso et al., 1999). Lipid oxidation, also favored by low pH, may be correlated to myoglobin oxidation, which results in the formation of metmyoglobin (Lynch and Faustman, 2000). Thus, in an oxidizing and acidic environment, meats with high myoglobin and iron contents and rich in unsaturated fatty acid may show a tendency toward browning (Guidi et al., 2006).

Factors Affecting the Color of the Meat

As stated earlier, the great differences in color shades and intensity of meats of different origins is the complex synergism of many physical and biochemical characteristics of the muscle; among these the concentration of myoglobin, which plays a pivotal role, is influenced by such factors as:

- *animal species.* Poultry meat has a minor myoglobin content with respect to beef, ovine, or pork meat.
- *Strain.* Genetic selection implies differences in muscle growth rate, size, and biochemical characteristics.
- *Sex.* Females and gelded males have paler meat than that of whole males.
- *Age.* In relation to a different texture and a higher concentration of pigments, the color tends to intensify with aging.
- *Functional activity.* Species with an elevated motor activity have intensely pigmented muscles (e.g., game).

- *Anatomical district.* Within the same species, different muscles may present sharp variations. Muscles responsible for sustained activity, rich in oxidative fibers, such as those belonging to the leg in poultry, have a higher myoglobin content than that of muscles responsible for short-power movement, such as the pectoralis of nonflying birds. Moreover, the concentration may vary even within a few centimeters in the same muscle.

Some of such differences are reported in Table 1.

Other than these intrinsic factors, meat color is also affected by antemortem factors such as diet, which influences the chemical characteristics of meat, or animal management, which is a determinant for both long-term and preslaughter stress conditions in bred birds. Often, stress levels reflect on a decrease in pH in postmortem muscle and ultimate pH, which has been found to strongly influence many technological properties of meat, such as tenderness, water-holding capacity, cooking loss, juiciness, microbial stability, and raw and cooked meat color. pH is normally much lower in meat than in living muscle, due to the postmortem residual activities of enzymes, among which the glycolytic enzymes, which operate to demolish glycogen and its final transformation in lactic acid. The latter cannot be removed from tissues, due to the termination of circulation; it contributes to the process of pH lowering until its value is so low that it inhibits enzymatic activities or the glycogen reservoir is terminated. At the end of the process, pH begins to increase slightly again until the meat tenderization has been concluded. The final pH can be very different, again according to the animal species and type of muscle. Red muscles will be characterized by a higher pH, due to the presence of a less efficient glycolytic apparatus with respect to muscle rich in white fibers, whose role in supporting brief but intense contraction needs a larger amount of glycolitic enzymes. Ultimate pH and the rate of its lowering, together with other factors, such as temperature, influence chemical reactions and determine some of the structural characteristics of the muscles, which in turn strongly affect the color of the meat.

A number of studies have shown a strong relationship between final pH and the L^* value of breast meat in chicken (Allen et al., 1997; Barbut, 1997; Fletcher, 1999) and turkey (Barbut, 1996; McCurdy et al., 1996; Owens et al., 2000). In general, a paler appearance than normal is associated with lower values and a darker appearance with higher pH. Some authors have advanced a classification of chicken breast meat dividing it in three groups according to the level of lightness: darker than normal ($L^* < 46$), normal ($48 < L^* < 53$), and lighter than normal ($L^* > 53$), and showed that each group was associated with a precise pH, which averaged around 6.23, 5.96, and 5.81, respectively. The measure of initial L^* values, more than b^* values, was also found to have a strong relationship to the water-holding and emulsification capacities, suggesting that color extremes could be used to segregate meats with different functional properties (Qiao et al., 2001). It has clearly been demonstrated that excessively fast lowering of pH, associated with inadequate temperatures, determines the premature precipitation

TABLE 1 Comparative Data on Total Heme, Hemoglobin, and Myoglobin Content in Chicken Muscles

Muscle	Chicken Type	Total Heme (mg/g)	Hemoglobin (mg/g)	Myoglobin (mg/g)	Reference
Breast	Broiler	0.51[a]	0.12[b]	0.31[b]	Niewiarowicz et al. (1986)
Pectoralis profundus	Broiler	0.32[c,d]/0.44[d,e]	0.17[b,c]/0.28[b,e]	0.15[b,c]/0.16[b,e]	Fleming et al. (1991)
	Broiler	0.46[d]/0.51[a]			Pikul et al. (1982)
	Broiler	0.01[d]			Saffle (1973)
	Adult	0.08[d,f]/0.09[d,g]			Saffle (1973)
	?			0.84[b]	Enoki et al. (1988)
	NH[h]			1.05[b]	Pages and Planas (1983)
Pectoralis superficiales or profundus	Layer			0[i]	Nishida and Nishida (1985)
Leg/thigh	Broiler	1.83[a]	0.58[b]	1.17[b]	Niewiarowicz et al. (1986)
Biceps femoris	Broiler	0.59[c,d]/0.79[d,e]	0.38[b,c]/0.48[b,e]	0.21[b,c]/0.30[b,e]	Fleming et al. (1991)
	Broiler	1.77[d]/1.83[a]			Pikul et al. (1982)
	Broiler	0.39[d]			Saffle (1973)
	Adult[j]	1.12[d,f]/1.51[d,g]			Saffle (1973)
	NH			1.41[b]	Pages and Planas (1983)
	Young layer[k]			0.33[i]	Nishida and Nishida (1985)
Sartorius	Adult layer			2.53[i]	Nishida and Nishida (1985)
Adductor magnus	Adult layer			4.44[i]	Nishida and Nishida (1985)
Adductor longus	Adult layer			5.82[i]	Nishida and Nishida (1985)
Heart	Broiler[l]	10.8[d]	0.109[m]	0.04[f,m]	O'Brien et al. (1992a)
	Broiler		10.9[m]	0.044[m]	O'Brien et al. (1992b)
	Broiler	3.65[d]			Saffle (1973)
	NH			1.34[b]	Pages and Planas (1983)
	Young layer[k]			1.51[i]	Nishida and Nishida (1985)
	Adult layer			3.63[i]	Nishida and Nishida (1985)

Source: Kranen et al. (1999).

[a]Isobestic analysis (525 nm). [b]Analysis of absorbance of CO-heme, measured at two wavelengths. [c]Ice-slush-chilled carcasses. [d]Cyanometheme derivatives, measured at 540 nm. [e]Air-chilled carcasses. [f]Females. [g]Males. [h]NH, *Gallus gallus domesticus*, New Hampshire strain. [i]Immunodiffusion analysis. [j]6-week-old adult hens. [k]1-week-old layer. [l]1-day-old broilers. [m]Differential $(NH_4)_2SO_4$ precipitation.

of muscle proteins, which tend to aggregate, causing water loss and the detachment of myofibrils from the cell membranes. The muscle's open structure thus formed indicates greater surface reflectance and the pH promotes the oxidation of oxymyoglobin and deoxymyoglobin to metmyoglobin. These conditions together determine drip loss, soft consistency, and pale color, which characterize PSE meats, commonly found not only in pork, but also in poultry, especially in turkey (McKee and Sams, 1998; Owens et al., 2000). PSE meats are rejected by consumers and represent a growing concern in the industry. It is well know that PSE conditions are induced by acute preslaughtering stress, which causes an increase in tissue metabolism and a rapid consumption of glycogen, with a prompt decrease in pH associated with the high muscle temperature. When the glycogen reservoir has been depleted before slaughtering, as happens in very active animals or in animals kept at low temperatures or stressed over a long period, the ultimate pH of meat is usually higher, as observed in DFD (dark, firm, dry) cases. High postmortem pH determines a higher solubility of proteins, which bind a greater quantity of water. This condition produces a higher compactness of the fibers, resulting in a more tightly packed assembly. This closed structure of the muscle implies a lower diffusion of O_2 and favors the maintenance of myoglobin in its deoxy form. Moreover, it implies a lower scattering capability of the muscle, so that light is absorbed rather than reflected. These two factors determine the darker appearance of the meat (Adams and Moss, 2000; Warriss, 2000). Chen et al. (1991) observed DFD-like muscles in ducks that have been stressed by 10 min of forced exercise after feed deprivation. Dark chicken breasts have been found to have a higher pH, myoglobin, and iron content than those of normally colored samples. The establishment of cutoff values for the color a^* parameter has been shown to represent highly sensitive and specific diagnostic methods to differentiate normal from dark-colored carcasses, with a threshold of 2.72 at 97% sensitivity and 90% specificity (Boulianne and King, 1998).

With regard to the rate of pH decline, Debut et al. (2003) found that an accelerated rate (low pH 15 min after slaughter) was associated primarily with higher values of a^* in broilers, whereas no significant correlations were found with L^*. Yet other studies performed on turkey have not led to similar conclusions. Any significant variations in L^* and b^* of breast meat were found by Rathgeber et al. (1999) and Hahn et al. (2001) comparing turkeys characterized by a slow, a normal, or a rapid fall in pH, whereas McKee and Sams (1997), Pietrzak et al. (1997), and Wynveen et al. (1999) observed higher L^* and b^* values in breasts of birds with low pH a short time after slaughter. In the aforesaid studies, no relationship was observed between the initial rate of pH decline and the a^* value of the meat, while other studies revealed increased a^* associated with rapid glycolysis (Rathgeber et al., 1999; Hahn et al., 2001; Fernandez et al., 2002; Debut et al., 2003).

The negative influence on meat quality of preslaughter stress has been studied extensively, especially in beef and pork, but existing data on poultry show that detrimental effects may affect turkeys, ducks, and chickens to a considerable extent, even if commercial problems are mostly associated with turkey rather than with the broiler chicken industry, probably because such incidences occur more frequently in the former than in the latter (Kannan et al., 1997). The last hours before slaughtering represent a very critical period for the bird, since catching, crating, transportation, unloading, and hanging produce stress and can reduce meat quality and yield if performed improperly.

Kannan et al. (1997) observed that crating broilers for 1 h yielded lighter breast meat than did crating them for 3 h; it also showed that allowing the birds to rest for a period of 4 h between transport and slaughter reduced the plasma levels of corticosterone, which was found to be associated with an increase in the L^* values of the meat. Transportation stress has also been reported to increase the lightness of chicken meat. Cashman et al. (1989) assessed the ultimate pH and color of broiler meat and found that it was paler in birds transported for 2 h than in birds that were crated for only 10 minutes. On the other hand, Debut et al. (2003) and Owens and Sams (2000) observed a significant decrease in L^* values. Bianchi et al. (2006) found a negative correlation between the redness of chicken breast fillets and the length of transportation. Furthermore, the shortest holding time produced the highest L^* values and the lowest a^* values. These reports suggest that stress associated to transportation can also influence the color of broiler meat.

In addition, preslaughter heat stress has been reported to accelerate rigor mortis development, reduce water-holding ability, and increase paleness in poultry meat (Northcutt et al., 1994; McKee and Sams, 1997). By reducing the final pH and increasing the L^* of meat, acute heat stress represents one of the most detrimental preslaughter condition (Debut et al., 2003). This trend has been observed in breasts of chickens exposed to acute heat stress by Holm and Fletcher (1997) and Sandercock et al. (2001) and on breasts of turkeys subjected to chronic heat stress by McKee and Sams (1997), but not by others. In fact, Debut et al. (2003) observed an increase in L^* only in thighs of broilers going through heat stress, and Petracci et al. (2001) observed a slight decrease in a^* values in breast from chickens held in higher holding temperatures, suggesting that the influence of acute heat stress on meat quality could vary according to application conditions (duration or intensity), but also according to the genotypes and muscle used (Debut et al., 2003). However, thigh meat seems to be more sensitive than breast meat to environmental factors, as suggested by a number of studies, in which such factors appeared most dominant for some thigh characteristics (i.e., pH, color) in turkey (Le Bihan-Duval et al., 2003) and broilers (Kannan et al., 1997; Debut et al., 2003).

Interestingly, other factors provoking antemortem stress have been studied. For example, the activity of struggle on the shackle line revealed an important impact on meat characteristics; the most active birds had the highest initial rates of pH decline. Froning et al. (1978) and Ngoka and Froning (1982) compared turkeys free to flap on the shackle line with turkeys immobilized before death by anesthesia and observed that in the meat of the former, the initial rate of pH fall was accelerated and the a^* value was increased. Probably, the redder coloration of the meat was associated with the increase in pigments due to a higher blood inflow as a consequence of struggle (Ngoka and Froning, 1982). Similar effects have also been observed in chickens. The impact of struggle activity was found to differ according to the type of muscle, and a correlation was observed between duration of wing flapping and rate of initial pH decline, or the a^* value of the meat. Effects on breast muscle were found to be much more pronounced than in thigh muscle, probably due partly to its greater involvement in wing flapping and its content in white glycolytic fibers, which make it more sensitive to a fast rate of pH decline (Debut et al., 2003).

Strain also seems to play a fundamental role in meat color qualities. In a study by Debut et al. (2003), color differences were observed between different lines of chickens in normal or stressed preslaughter conditions, with fast growing line (FGL) birds having lighter breast and thigh meats than those of slow growing line (SGL) birds; heat- and transportation-induced stresses appeared as an additional source of variability among birds more than as the main cause of color variation. Selection for growth and muscle development should slowly modify meat color, according to some observations in broilers (Le Bihan-Duval et al., 1999; Bianchi et al., 2006), turkeys (Santá et al., 1991), and ducks (Baéza et al., 1997), which showed a significant decrease in color intensity in high performance compared to that of genotypes selected less.

Stunning is another antemortem procedure that can have some effect on meat color. In electric stunning, the current causes generalized contractions that can affect muscle characteristics and induce hemorrhages or broken bones if excessive current or too low frequencies are used (Veerkamp and DeVries, 1983; Gregory and Wilkins, 1989; Rawles et al., 1995). The effect of electric stunning on meat quality and carcass damage depends largely on the electric conditions used (e.g., voltage, frequency, stunning duration), and it has been suggested to have little direct effect on chicken breast (Papinaho and Fletcher, 1995; Craig and Fletcher, 1997) or turkey breast color (Owen and Sams, 1997), even if some authors reported a loss of lightness in breasts from both high-current- and low-voltage-stunned chickens and a gain of redness in low-voltage stunned birds (Craig et al., 1999) or a slight decrease of lightness and increase of yellowness according to the stunning duration in chicken breast (Young and Buhr, 1997).

An alternative to electrical stunning that has received attention is gas stunning, since it can reduce carcass damage (Raj et al., 1997; Kang and Sams, 1999a). This procedure may induce hypercapnic hypoxia (usually caused by a mix of CO_2/air) or anoxia, by depletion of oxygen with other gases. Other than stunning, gas killing can be induced via asphyxia by a higher concentration of CO_2 or argon or nitrogen or a mix of these gases. Yet these practices seem to affect meat color more than electric stunning. Because of the use of potentially different gas mixes or different types of electric stunning, it turns out to be more difficult to understand their effect on meat color. Rigor mortis development, which is known to affect ultimate pH and thus meat color, was reported to be accelerated with CO_2 stunning by some authors (Raj et al., 1997), whereas others observed no differences with electrical stunning (Kang and Sams, 1999a). Carbon dioxide stunning has been reported to cause a loss of redness in turkey breast and thigh meat (Raj et al., 1990), whereas gas argon killing has been reported to decrease darkness in broiler muscle with respect to electrical or CO_2 stunning (Fleming et al., 1991). Savenije et al. (2002) reported that in normally processed carcasses, CO_2/argon-stunned chickens had higher L^* values and lower a^* values than those of electrically stunned chickens, whereas chickens stunned with a $CO_2/O_2/N_2$ gas mixture had higher b^* values. Studies comparing breast fillets from CO_2-stunned and CO_2-killed chickens showed that the former were lighter than the latter after 24 h (Kang and Sams, 1999b). Northcutt et al. (1998) observed no significant differences in turkey breasts following 24 h after electrical or CO_2 stunning, nor did Poole and Fletcher (1998), who compared the effect of electrical stun and CO_2/argon killing in chicken breasts. Other than preslaughter stress factors, slaughtering modalities, processing procedures, chemical exposure during processing or packaging, storing conditions, and cooking temperatures may have considerable effects on meat and skin color.

COLOR OF THE SKIN

Since poultry meat is present on the market both skinned and unskinned, the color of the skin is another important factor to take in consideration. The xanthophylls, a particular group of carotenoids that accumulate in the epidermis, are the main determinant of skin color. With regard to consumer preference, which can be different according to the region of provenance, poultry skin can be made whiter or more yellow by varying the content and types of pigments in the diet, which can be of natural or synthetic origin, and confer a higher magnitude in b^* values to the skin color (Casta neda et al., 2005). Yet pigmentation of the skin depends strongly on genetic characteristics such as the ability to deposit carotenoid pigments in the epidermis to produce melanine. In the poultry industry, strains have been selected genetically for lower melanine production potential or a lower ability to deposit xanthophyll (Fletcher, 1989). The differences between white- and yellow-colored skin poultry depends primarily on the xanthophyll content in the epidermis. A high melanin content in dermis confers a dark color to poultry skin, which is shaded black when melanin is also found in the epidermis, blue if not, and green if xanthophylls are present in the epidermis.

Since skin color changes over time, computer-assisted vision grading and inspection systems have been developed for carcass and meat quality assessment and have also been applied to skin color monitoring. The postmortem change of skin color has not yet been well documented, but studies on broilers have shown that the most dramatic variation occurs within the first 4 h, after which changes are less pronounced up to 12 to 24 h. Skin color changing during storage (from 1 to 8 days) is variable and depends on processing and holding conditions. The changes have been identified primarily by increased lightness, regardless of scalding treatments or surface location, even if skin from semiscalded birds (50°C) undergoes less change than skin from subscalded birds (57°C). Areas with higher xantophyll deposition (associated with fat deposits) are most subject to color change (Petracci and Fletcher, 2002).

IRRADIATION

The use of ionizing radiation is a common practice in some countries for meat processing and prevention of foodborne diseases, and it is regarded as one of the most effective methods to eliminate pathogens in meat and poultry (Gants, 1996). The energy of the incident radiation on the meat may be absorbed by some electrons, which can leave their orbital and transfer part of this energy to other electrons, in a sort of cascade reaction, until there is enough energy for the orbital leaving (the Compton effect). This event may cause the scission of molecules; water, which is very abundant in meat, is highly subject to radiolysis, which brings about the formation of free radicals, such as hydrated electrons, hydrogen radicals, and hydroxyl radicals (Thakur and Singh, 1994). Radicals attack lipids and proteins and generate abnormal color and off-odors (Nanke et al., 1998; Ahn et al., 2001).

Irradiation can induce a number of different effects on myoglobin and on the molecules present in meat, whose alterations are responsible for the color change that occurs after treatment. Different final effects on color are related to the myoglobin concentration and its state before irradiation and also to substrate conditions, such as the pH and Eh values of the meat. Temperature and atmospheric composition of the packaging during irradiation and the animal species have also been shown to strongly influence the meat color after irradiation. In fact, instrumental hue measurement reveals different color change trends after storage in turkey, pork, and beef (Nanke et al., 1999; Kim et al., 2002). In particular, studies on turkey breasts showed that irradiation determines a dose-dependent increase in redness regardless of the type of packaging used (Miller et al., 1995; Nanke et al., 1998, 1999; Kim et al., 2002; Yan et al., 2006b), also perceivable by means of visual evaluation.

Also, the effects on the redness value of meat have been shown to be much greater than those provided by dietary supplementation of functional ingredients (Yan et al., 2006b). Similar results on the increase in redness after irradiation have been observed in chicken breast muscles (Lewis et al., 2002; Liu et al., 2003), in turkey breast patties (Nam and Ahn, 2003), and in sausages prepared with turkey thigh meat, thus testifying to a common trend in poultry, regardless of the muscle of origin or the type of processing (Du and Ahn, 2002), even though the magnitude of the effects may vary according to anatomical districts, as demonstrated in other species (Ahn et al., 1998). The red color of meat is often associated with its freshness, and irradiated turkey appears to be preferred by consumers (Lee et al., 2003; Yan et al., 2006a). Yet some studies have shown that the redness of meat was still higher in irradiated meat from turkey breast than in nonirradiated meat even after cooking, and the inside of the meat had a greater redness intensity than the surface. Color changes in irradiated meat after cooking are of greater concern because consumers can consider residual pink to be a sign of undercooking or contamination (Nam and Ahn, 2003). Nevertheless, aerobic storage may lead to a reduction in the color after cooking, as observed in irradiated broiler breast fillets (Du et al., 2002).

CO is a major radiolytic gas arising from irradiated foodstuffs (Pratt and Kneeland, 1972; Simic et al., 1979). It has been suggested that the red or pink color of irradiated turkey meat is due to the CO produced, which has a very strong affinity to heme pigments, forming carboxymyoglobin, which is more stable than oxymyoglobin (Kim and others, 2002; Nam and Ahn, 2002a). Irradiation also increases the reducing power of meat, which facilitates carbon monoxide–myoglobin complex formation. Furthermore, CO formation is dose-dependent (Nam et al., 2003). Nam and Ahn (2003) found an increased redness of vacuum-packaged turkey breast by irradiation that was stable even after 10 days of refrigerated storage. However, in some cases, the redness of aerobically packaged meat decreased significantly. This finding indicated that exposing irradiated meat to aerobic conditions was effective in reducing CO–heme pigment complex formation, probably due to the presence of oxygen, which accelerates the dissociation of Mb^{2+} CO (Grant and Patterson, 1991). Concentration of CO in precooked turkey was found to be similar regardless of the type of packaging, yet the concentration remained stable only in impermeable film-packaged samples (Nam and Ahn, 2002b).

However, the very importance of Mb^{2+} CO complexes and their evolution during storage must be considered even in relation to the evolution of redness values in time, which still seems to be controversial, and Mb^{2+} CO alone probably cannot explain all the irradiated meat color. In fact, in some other experiments a^* values were found to decrease after some storage time and at a faster rate at a higher dose of irradiation in aerobical packaging (Nanke et al., 1999), whereas no significant change was observed in vacuum packaging (Nanke et al., 1998). Similar results were reported for irradiated chicken after cooking (Du et al. 2002). On the contrary, Kim et al. (2002) found a slight change in vacuum packaging and no consistent trend in aerobical packaging, as observed by Liu et al. (2003) for chicken breast and Yan et al. (2006b) for ground turkey breast. Irradiation has also been performed in a nitrogen atmosphere. Some studies have shown the formation of a pink color after treatment on chickens with a reflectance similar to that associated with Mb^{2+} O_2 (Satterlee et al., 1972). The storage stability of the pigment formed and the irradiation atmosphere without oxygen caused others to ascribe the color to Mb^{2+} CO or Mb^{2+} NO (Millar et al., 1995). Furthermore, the combination of antioxidants with double packaging showed a synergistic effect in reducing the redness of irradiated meat (Nam and Ahn, 2003), whereas antioxidants should inhibit the radiolytic generation of CO (Grant and Patterson, 1991).

Finally, the effects of irradiation on lightness and yellowness seem to be markedly lower than the effect on redness, but the occurrence, the magnitude of possible changes of these parameters, and their development according to display time are still controversial as to details. However, the increase in brownness caused by a metmyoglobinlike pigment perceivable in other species could not be observed in turkey (Nanke et al., 1999).

PACKAGING

Packaging has been shown to influence meat color. The main reasons are related to the different degrees of permeability to the gases and moisture in the packaging film and the type of gas mix introduced in modified-atmosphere packaging (MAP). The function of the various types of packaging is not only to extend the shelf life of the products but also to maintain a good appearance, controlling transpiration, and preserving or enhancing the most desirable shades of meat color. In MAP it is important to find the best blend of gases to optimize color, its stability, and shelf life, limiting microbial growth and lipid oxidation. In general, the chemical properties of the various molecules that interact with meat constituents may induce different chemical reactions and produce different effects on the macroscopical appearance of the meat during its shelf life. Myoglobin, lipids, and metals are primarily involved in this type of reaction. The majority of data come from studies on red meat.

High-oxygen atmospheres, whose recommended gas mixture composition for poultry is 25% CO_2 and 75% O_2 (Parry, 1993), should promote pigment oxygenation, increasing the redness of the meat, but enhanced lipid oxidation and different results on color stability have been reported. Some authors observed a decrease in color deterioration and a stabilization in redness during storage in beef and pork meat (Gill, 1996; Jayasingh et al., 2002), but different levels of decreased a^* values at different times after packaging have been reported for ground turkey and chicken meat, probably as a consequence of myoglobin oxidation and the limited ability of poultry meat to form oxymyoglobin compared to beef and pork, due to high metmyoglobin-reducing activity and high O_2 consumption. In addition, the oxidation activity of high-O_2-content atmospheres on lipids leads to a shift from red to yellow (Saucier et al., 2000; Dhananjayan et al., 2006; Keokamnerd et al., 2008). Moreover, discoloration or localized color alterations have been reported in turkey thighs (Guidi et al., 2006).

In contrast, ultralow-oxygen atmospheres (usually N_2 as a inert filler, plus CO_2 at various concentrations) minimize lipid oxidation and growth of aerobic microorganisms, and their use has been shown to be associated with the greatest color stability during shelf life in both turkey (Santé et al., 1994; Dhananjayan et al., 2006) and chicken meat (Saucier et al., 2000; Keokamnerd et al., 2008). Generally, the prevalence of deoxymyoglobin in low-oxygen atmospheres confers a darker color to beef and pork meat. To overcome this drawback, low concentrations of CO can be added to the modified atmosphere, since it forms Mb^{2+} CO red stable complexes on the meat surface and improves color stability in beef (Hunt et al., 2004) and pork (Krause et al., 2003), but few data exist on poultry meat.

STORING

Generally, skin and meat color change dramatically in the first hours after slaughter. As shown by Petracci and Fletcher (2002), such changes have their highest magnitude during the first 4 h, while the carcasses are still in the processing plant, whereas later variations in color are less pronounced. Skin color shifts to increasing lightness, especially in the areas with greater xanthophyll deposition, in the feather tracts. Also, for meat the major changes involve particularly a gain in lightness, especially in breast cuts, during the first 2 days (Le Bihan-Duval et al., 1999; Alvarado and Sams, 2000; Owens and Sams, 2000; Owens et al., 2000), after which the values may to decrease again (Petracci and Fletcher, 2002).

Finally, freezing and thawing have been reported to cause a reduction in L^* values of pale broiler breast-filets and an increase in dark broiler breast filets (Galobart and Moran, 2004), but did not modify the values of lightness of normal filets (Lyon et al., 1976; Galobart and Moran, 2004).

COOKING

Heat has a strong effect on meat color, provoking the development of marked brownish hues. The factor most involved in determining the color of cooked meat are the different forms of myoglobin, which tend to denaturate and precipitate along with other proteins. Myoglobins begins to denaturate after 55°C and the process is accomplished primarily around 75 to 80°C (Varnam and Sutherland, 1995; Hunt et al., 1999). Cooking causes an increase in pH, which is likely to be the reason for the slower rate of protein denaturation as the temperature arises (Geileskey et al., 1998). According to the oxidation state of iron and the complex formed by myoglobin with other molecules, sensitiveness to heat varies. Deoxymyoglobin is less sensitive than oxymyoglobin and metmyoglobin (Van Laack et al., 1996; Hunt et al., 1999). Denaturation of oxymyoglobin and deoxymyoglobin with cooking leads to the formation of ferrohemochrome (Fe^{2+}), a red pigment that is rapidly oxidized to ferrihemochrome (Fe^{3+}); metmyoglobin is converted directly to ferrihemochrome, which results in higher amounts in cooked meat, conferring its typical brown color (Varnam and Sutherland, 1995) (Figure 3). Yet the final shade of the cooked meat depends on the ferrihemochrome/ferrohemochrome ratio, which is determined by either the type of cooking process (intensity and duration) or the initial relative concentration of the three main forms of myoglobin, even though under certain conditions meat can be enriched with other forms of myoglobin, as, for example, Mb^{2+} CO or Mb^{2+} NO, which are more stable than Mb^{2+} O.

Different types of muscles or animal species are associated with different structural and biochemical characteristics, which may cause a different color evolution during cooking and also a final hue, generally ranging from off-white to gray or brown. Differences may also be determined by breeding management, preslaughter stress, and postmortem processing. As with raw meat, this implies that many variables influence the color development of meat during cooking. One of the most important is pH. Experiments performed on chicken breasts showed that the darkness of raw meat, associated with higher pH, tends to persist after cooking. A linear relationship was found between L^* values of cooked meat and raw meat pH as well as L^* values of cooked meat and cooked meat pH, even though with a lower level of correlation, indicating that raw breast meat color and pH affect cooked breast meat color and pH, but that cooking reduces the degree of color variation (Fletcher et al., 2000).

An explanation of the association of pH with color development after cooking may be sought in the fact that a lower pH facilitates denaturation of the myoglobins, thus determining faster browning, due to the formation of ferrihemochrome, especially if subjected to heat treatment. However, different animal species are affected differently by pH. Experiments performed adjusting pork, beef, and turkey meat to different pH values cooked to reach different internal temperatures showed a common tendency to preserve the appearance of higher redness and a lower amount of denatured myoglobin, but at higher temperatures, turkey meat showed greater persistence of this tendency, with the percentage of denatured myoglobin at 83°C shifting from 75 to 95%, depending on pH, against 100% in other species at all pH levels (Trout, 1989). Nevertheless, even though some studies correlate the grade and direction of color variation after cooking with the content and state of myoglobin or pH level of raw meat, meat being a complex biochemical system, the final color of cooked meat, and raw meat as well, must be influenced by many other factors that are still unclear, and whose synergy must be taken into consideration. For example, high pH has been shown to lessen the effects on cooked meat color produced by other factors, such as fat content, freezing, and rate of thawing (Berry, 1998).

Beyond the aforesaid factors described above, others may be regarded as determinant, such as muscle fiber arrangement (e.g., DFD vs. PSE) and the denaturation processes of other meat proteins, including enzymes, which could also be affected by pH and to which myoglobin denaturation is intricately linked (King and Whyte, 2006). The authors of some studies, in which ground lamb myoglobin has been shown to denaturate more slowly at lower pH, suggested a possible explanation in the fact that other proteins have been denatured first, which otherwise would interact with myoglobin destabilizing it (Lytras et al., 1999). However, cooking at a high-temperature endpoint reduces red or bloody discoloration. Reduced discoloration with a low endpoint is also possible if the product is frozen before cooking (Smith and Northcutt, 2004).

Defects of Cooked Poultry Meat

One of the most recurring defects of cooked poultry meat is pinkness (or pinking), which can be observed even when the internal temperature exceeds 70°C. This defect is associated primarily with commercial problems, because of rejection by consumers, who perceive the product to be undercooked and unsafe to eat. Factors involved in pinking incidence are several and include the presence of specific pigments, genetics, feed and stress, processing methods, and incidental nitrate/nitrite contamination through diet, water supply, processing equipment, freezing, and storing (Howlonia et al., 2003a). Reduction of external contaminants in poultry industry plants has not significantly reduced the pinking occurrence, but assurance of the shortest processing time would be prudent. Other factors related to the chemical and physical properties of the meat, such as the presence of reducing agents, state and reactivity of pigments, and pH, may be involved in pinking (Howlonia et al., 2003a).

Other than undenatured oxymyoglobin or deoxymyoglobin and reduced hemochromes, pink defects are related primarily to the presence of nitrosylhemochrome, carbomonoxyhemochrome, or citochrome *c*. One part per million of sodium nitrite is enough for the pinking occurrence, but high pH and low oxidation–reduction potentials may affect the pigment activity, permitting its occurrence even at lower concentrations (Howlonia et al., 2003b). Compounds other than nitrites may promote pinking occurrence, such as tryphosphates or erythorbate, which have been used for simulating in situ conditions (Howlonia et al., 2004).

Another defect of cooked poultry meat is darkening, in which the tissue around the bone is discolored, attaining a burgundy or black appearance. This defect was first associated with frozen poultry (Spencer et al., 1961). The dark discoloration is seemingly determined by the infiltration of bone marrow onto the surrounding meat as a consequence of leaks in the bone. Bone marrow tends to get dark according to different cooking methods: If meat is frozen before cooking, darkening is more evident than if heated after cooking and freezing (Lyon and Lyon, 1986). Blast freezing enhances darkness and redness of raw and cooked bone-in broiler thighs, while removal of the femur before freezing decreases the redness of raw thighs (Lyon et al., 1976).

Finally, red discoloration of bone in fully cooked product is a defect often found in poultry but not yet investigated in depth. Smith and Northcutt (2003) reported that 11% of several different cooked chicken products available at retail were affected either severely or extensively by red discoloration. The defect tends to be localized, and it is associated more with bone darkening than with meat pinking, differing from the former by the dark red or bloody red hue. Marrow was determined to be the most important component in inducing red discoloration of breast meat (Smith and Northcutt, 2004). Discoloration varied among product types, as suggested by the different lightness and redness values, and it is affected by cooking methods and piece type, as shown in a study by Smith and Northcutt (2003), who reported that breasts were less discolored than thighs or drums and that thigh discoloration seemed to be redder than that of breasts or drumsticks.

REFERENCES

Adams M, Moss M. 2000. *Food Microbiology*. Cambridge, UK: Royal Society of Chemistry, p. 479.

Ahn DU, Olson DG, Jo C, Chen X, Wu C, Lee JI. 1998. Effect of muscle type, packaging and irradiation on lipid oxidation volatile production and color in raw pork patties. Meat Sci 49:27–39.

Ahn DU, Nam KC, Du M, Jo C. 2001. Volatile production of irradiated normal, pale soft exudative (PSE) and dark firm dry (DFD) pork with different packaging and storage. Meat Sci 57:419–426.

Allen CD, Russell SM, Fletcher DL. 1997. The relationship of broiler breast meat color and pH to shell-life and odor development. Poult Sci 76:1042–1046.

Alvarado CZ, Sams AR. 2000. Rigor mortis development in turkey breast muscle and the effect of electrical stunning. Poult Sci 79:1694–1698.

AMSA (American Meat Science Association). 1991. *Guidelines for Meat Color Evaluation*. Ames, IA: AMSA.

Baéza E, De Carville H, Salichon MR, Marché G, LeClercq B. 1997. Effect of selection, over three or four generations, on meat yield and fatness in Muscovy ducks. Br Poult Sci 38:359–355.

Barbut S. 1993. Colour measurements for evaluating the pale soft exudative (PSE) occurrence in turkey meat. Food Res Int 26:39–43.

Barbut S. 1996. Estimates and detection of the PSE problem in young turkey breast meat. Can J Anim Sci 76:455–457.

Barbut S. 1997. Problem of pale soft exudative meat in broiler chickens. Br Poult Sci 38:355–358.

Barbut S. 1998. Estimating the magnitude of the PSE problem in poultry. J Muscle Foods 9:35–49.

Bartov I, Kanner J. 1996. Effect of high levels of dietary iron, iron injection, and dietary vitamin E on the oxidative stability of turkey meat during storage. Poult Sci 75:1039–1046.

Berry B. 1998. Cooked color in high pH beef patties as related to fat content and cooking from the frozen or thawed state. J Food Sci 63:797–800.

Bianchi M, Fletcher DL. 2002. Effects of broiler breast meat thickness and background on color measurements. Poult Sci 81:1766–1769.

Bianchi M, Petracci M, Cavani C. 2006. The influence of genotype, market live weight, transportation, and holding conditions prior to slaughter on broiler breast meat color. Poult Sci 85:123–128.

Bird HR. 1943. Increasing yellow pigmentation in shanks of chickens. Poult Sci 22:205–208.

Boulianne M, King AJ. 1998. Meat color and biochemical characteristics of unacceptable dark-colored broiler chicken carcasses. J Food Sci 759–762.

Brown WL. 1930. *Some Effects of Pigmento Pepper on Poultry*. Georgia Experimental Station Bulletin 160.

Brown WC, Mebine LB. 1969. Autoxidation of oxymyoglobins. J Biol Chem 244:6696–6701.

Cashman PJ, Nicole CJ, Jones RB. 1989. Effect of stresses before slaughter on changes to the physiological, biochemical, and physical characteristics of duck muscle. Br Poult Sci 32:997–1004.

Casta neda MP, Hirschler EM, Sams AR. 2005. Skin pigmentation evaluation in broilers fed natural and synthetic pigments. Poult Sci 84:143–147.

Chen MT, Lin SS, Lin LC. 1991. Effect of stresses before slaughter on changes to the physiological, biochemical and physical characteristics of duck muscle. Br Poult Sci 32:997–1004.

Craig EW, Fletcher DL. 1997. A comparison of high current and low voltage electrical stunning systems on broiler breast rigor development and meat quality. Poult Sci 76:1178–1181.

Craig EW, Fletcher DL, Papinaho PA. 1999. The effects of antemortem electrical stunning and postmortem electrical stimulation on biochemical and textural properties of broiler breast meat. Poult Sci 78:490–494.

Daniels HP, Fris C, Veerkamp CH, De Vries AW, Wijnker P. 1989. *Standard Method of Classification of Broiler Carcasses*. Zeist, The Netherlands: Commodity Board for Poultry and Eggs.

Day EJ, Williams WP Jr. 1958. A study of certain factors that influence pigmentation in broilers. Poult Sci 37:1373.

Debut M, Berri C, Baéza E, Sellier N, Arnould C, Guémené D, Jehl N, Boutten B, Jego Y, Beaumont C, Le Bihan-Duval E. 2003. Variation of chicken technological meat quality in relation to genotype and preslaughter stress conditions. Poult Sci 82:1829–1838.

DeMan, JM. 1999. Color. In: *Principles of Food Chemistry*, 3rd ed. Gaithersburg, MD: Aspen Publishers, pp. 239–242.

Dhananjayan R, Han IY, Acton JC, Dawson PL. 2006. Growth depth effects of bacteria in ground turkey meat patties subjected to high carbon dioxide or high oxygen atmospheres. Poult Sci 85:1821–1828.

Du M, Ahn DU. 2002. Effect of antioxidants on the quality of irradiated sausages prepared with turkey thigh meat. Poult Sci 81:1251–1256.

Du M, Cherian G, Stitt PA, Ahn DU. 2002. Effect of dietary sorghum cultivars on the storage stability of broiler breast and thigh meat. Poult Sci 81:1385–1391.

Enoki Y, Morimoto T, Nakatani A, Sakata S, Ohga Y, Kohzuki H, Shimizu S. 1988. Wide variation of myoglobin contents in gizzard smooth muscles of various avian species. Adv Exp Med Biol 222:709–716.

Fernandez X, Santé V, Baéza E, LeBihan-Duval E, Berri C, Rémignon H, Babilé R, Le Pottier G, Millet N, Berge P, Astruc T. 2001. Post mortem muscle metabolism and meat quality in three genetic types of turkey. Br Poult Sci 42:462–469.

Fleming BK, Froning GW, Beck MM, Sosnicki AA. 1991. The effect of carbon dioxide as a preslaughter stunning method for turkeys. Poult Sci 70:2201–2206.

Fletcher DL. 1988. Methods of determining broiler skin pigmentation. In: *Proceedings of the National Pigmentation Symposium*, p. 37.

Fletcher DL. 1989. Factors influencing pigmentation in poultry. CRC Crit Rev Poult Biol 2:149–170.

Fletcher DL. 1999. Broiler breast meat color variation, pH, and texture. Poult Sci 78:1323–1327.

Fletcher DL, Qiao M, Smith DP. 2000. The relationship of raw broiler breast meat color and pH to cooked meat color and pH. Poult Sci 79:784–788.

Forrest JC, Aberle ED, Hedrick HB, Judge MD, Merkel RA. 1975. Principles of meat processing. In: *Principles of Meat Science*. New York: W.H. Freeman, p. 190.

Froning GW, Babji AS, Mather FB. 1978. The effect of preslaughter temperature, stress, struggle and anesthetization on color and textural characteristics of turkey muscle. Poult Sci 57:630–633.

Galobart J, Moran ET Jr. 2004. Refrigeration and freeze–thaw effects on broiler fillets having extreme L^* values. Poult Sci 83:1433–1439.

Gants R. 1996. Pathogen countdown. In: *Meat and Poultry*. Kansas City, MO: Sosland Publishing, pp. 26–29.

Geileskey A, King RD, Corte D, Pinto P, Ledward DA. 1998. The kinetics of cooked meat haemoprotein formation in meat and model systems. Meat Sci 48:189–199.

George P, Stratmann CJ. 1952. The oxidation of myoglobin to metmyoglobin by oxygen: 2. The relation between the first-order rate constant and the partial pressure of oxygen. Biochem J 51:418–425.

Giddings GG, Markakis P. 1972. Characterization of the red pigments produced from ferrimyoglobin by ionizing radiation. J Food Sci 37:361–364.

Gill CO. 1996. Extending the storage life of raw chilled meats. Meat Sci 43(Suppl 1):99–109.

Gill CO, McGinnis JC. 1995. The effect of residual oxygen concentration and temperature on the degradation of the color of beef packaged under oxygen-depleted atmospheres. Meat Sci 39:387–394.

Giroux M, Uattara B, Yefsah R, Smoragiewicz W, Saucier L, Lacroix M. 2001. Combined effect of ascorbic acid and gamma irradiation on microbial and sensorial characteristics of beef patties during refrigerated storage. J Agric Food Chem 49:919–925.

Grant IR, Patterson MF. 1991. Effect of irradiation and modified atmosphere packaging on the microbiological and sensory quality of pork stored at refrigeration temperatures. Int J Food Sci Technol 26:507–519.

Gregory NG, Wilkins LJ. 1989. Effect of stunning current on carcass quality in chickens. Vet Rec 124:530–532.

Guidi A, Castigliego L, Benini O, Armani A, Iannone G, Gianfaldoni D. 2006. Biochemical survey on episodic localized darkening in turkey: deboned thigh meat packaged in modified atmosphere. Poult Sci 85:787–793.

Hahn G, Malenica M, Müller WD, Taubert E, Petrak T. 2001. Influence of post-mortem glycolysis on meat quality and technological properties of turkey breast. In: *Proceedings of the 15th European Symposium on the Quality of Poultry Meat*, Kusadasi, Turkey. Ismir, Turkey: WPSA Turkish Branch, pp. 325–328.

Harms RH, Ahmed EH, Fry JL. 1971. Broiler pigmentation: factors affecting it and problems in its measurement. *Proceedings of the Maryland Nutrition Conference*, p. 81.

Holm CG, Fletcher DL. 1997. Antemortem holding temperatures and broiler breast meat quality. J Appl Poult Res 6:180–184.

Holownia K, Chinnan MS, Reynolds AE, Koehler PE. 2003a. Evaluation of induced color changes in chicken breast meat during simulation of pink color defect. Poult Sci 82:1049–1059.

Holownia K, Chinnan MS, Reynolds AE. 2003b. Pink color defect in poultry white meat as affected by endogenous conditions. J Food Sci 68:742–747.

Holownia K, Chinnan MS, Reynolds AE. 2004. Cooked chicken breast meat conditions related to simulated pink defect. J Food Sci 69:194–199.

Honikel KO. 1998. Reference methods for the assessment of physical characteristics of meat. Meat Sci 49:447–457.

Hunt MC, Sørheim O, Slinde E. 1999. Color and heat denaturation of myoglobin forms in ground beef. J Food Sci 64:847–851.

Hunt MC, Mancini RA, Hachmeister KA, Kropf DH, Merriman M, Del Duca G, et al. 2004. Carbon monoxide in modified atmosphere packaging affects color, shelf life, and microorganisms of beef steaks and ground beef. J Food Sci 69:45–52.

International Commission on Illumination. 1986. *Colorimetry*, 2nd ed. CIE Publication 15.2. Vienna, Austria: CIE Central Bureau.

Jayasingh P, Cornforth DP, Brennand CP, Carpenter CE, Whittier R. 2002. Sensory evaluation of ground beef stored in high oxygen modified tmosphere pack. J Food Sci 67:3493–3496.

Kang IS, Sams AR. 1999a. Bleed-out efficiency, carcass damage and rigor mortis development following electrical stunning or carbon dioxide stunning on a shackle line. Poult Sci 78:139–143.

Kang IS, Sams AR. 1999b. A comparison of texture and quality of breast fillets from broilers stunned by electricity and carbon dioxide on a shackle line or killed with carbon dioxide. Poult Sci 78:1334–1337.

Kannan G, Heath JL, Wabeck CJ, Souza MC, Howe JC, Mench JA. 1997. Effects of crating and transport on stress and meat quality characteristics in broilers. Poult Sci 76:523–529.

Keokamnerd T, Acton JC, Han IY, Dawson PL. 2008. Effect of commercial rosemary oleoresin preparations on ground chicken thigh meat quality packaged in a high-oxygen atmosphere. Poult Sci 87:170–179.

Kim YH, Nam KC, Ahn DU. 2002. Color, oxidation reduction potential and gas production of irradiated meat from different animal species. J Food Sci 61:1692–1695.

King NJ, Whyte R. 2006. Does it look cooked? A review of factors that influence cooked meat color. J Food Sci 71:31–40.

Kranen RW, Van Kuppevelt TH, Goedhart HA, Veerkamp CH, Lambooy E, Veerkamp JH. 1999. Hemoglobin and myoglobin content in muscles of broiler chickens. Poult Sci 78:467–476.

Krause TR, Sebranek JG, Rust RE, Honeyman MS. 2003. Use of carbon monoxide packaging for improving the shelf life of pork. J Food Sci 68:2596–2603.

Krzywicki K. 1979. Assessment of relative content of myoglobin, oxymyoglobin and metmyoglobin in the surface of beef. Meat Sci 3:1–10.

Lawrie R. 1966. *Meat Science*. Oxford, UK: Pergamon Press.

Le Bihan-Duval E, Millet N, Rémignon H. 1999. Broiler meat quality: effect of selection for increased carcass quality and estimates of genetic parameters. Poult Sci 78:822–826.

Le Bihan-Duval E, Berri C, Baéza E, Santé V, Astruc T, Rémignon H, Le Pottier G, Bentley J, Beaumont C, Fernandez X. 2003. Genetic parameters of meat technological quality traits in a grand-parental commercial line of turkey. Genet Sel Evol 35:623–635.

Lee EJ, Love J, Ahn DU. 2003. Effect of antioxidants on consumer acceptance of irradiated turkey meat. J Food Sci 5:1659–1663.

Lewis SJ, Velásquez A, Cuppett SL, McKee SR. 2002. Effect of electron beam irradiation on poultry meat safety and quality. Poult Sci 81:896–903.

Liu Y, Fan X, Chen YR, Thayer DW. 2003. Changes in structure and color characteristics of irradiated chicken breasts as a function of dosage and storage time. Meat Sci 63:301–307.

Livingston DJ, Brown WD. 1982. The chemistry of myoglobin and its reactions. Food Technol 35:244–252.

Livingston DJ, McLachlan SJ, La Mar GN, Brown WD. 1985. Myoglobin: cytochrome b_5 interactions and the kinetic mechanism of metmyoglobin reductase. J Biol Chem 260:15699–15707.

Lu J, Tan J, Shatadal P, Gerrard DE. 2000. Evaluation of pork color by using computer vision. Meat Sci 56:57–60.

Lycometros C, Brown WD. 1973. Effects of gamma irradiation on myoglobin. J Food Sci 38:971–977.

Lynch MP, Faustman C. 2000. Effect of aldehyde lipid oxidation products on myoglobin. J Agric Food Chem. 48:600–604.

Lyon BG, Lyon CE. 1986. Surface dark spotting and bone discoloration in fried chicken. Poult Sci 5:1915–1918.

Lyon CE, Townsend WE, Wilson RL Jr. 1976. Objective color values of non-frozen and frozen broiler breasts and thighs. Poult Sci 55:1307–1312.

Lytras GN, Geileskey A, King RD, Ledward DA. 1999. Effects of muscle type, salt and pH on cooked meat haemoprotein formation in lamb and beef. Meat Sci 52:189–194.

Mancuso JR, McClements DJ, Decker EA. 1999. Ability of iron to promote surfactant peroxide decomposition and oxidize alpha-tocopherol. J Agric Food Chem 47:4146–4149.

McCurdy RD, Barbut S, Qinton M. 1996. Seasonal effect on pale soft exudative (PSE) occurrence in young turkey breast meat. Food Res Int 29:363–366.

McKee SR, Sams AR. 1997. The effect of seasonal heat stress on rigor development and the incidence of pale, exudative turkey meat. Poult Sci 76:1616–1620.

McKee SR, Sams AR. 1998. Rigor mortis development at elevated temperatures induces pale exudative turkey meat characteristics. Poult Sci 77:169–174.

Miller SJ, Moss BW, MacDougall DB, Stevenson MH. 1995. The effect of ionizing radiation on the CIE Lab color co-ordinates of chicken breast meat as measured by different instruments. Int J Food Sci Technol 30:663–674.

Morrison RT, Boyd RN. 1981. Amino acids and protein. In: *Organic Chemistry*, 3rd ed. Boston: Allyn & Bacon, pp. 1132–1162.

Nam KC, Ahn DU. 2002a. Carbon monoxide-heme pigment is responsible for the pink color in irradiated raw turkey breast meat. Meat Sci 60:25–33.

Nam KC, Ahn DU. 2002b. Mechanisms of pink color formation in irradiated precooked turkey breast meat. J Food Sci 67:600–607.

Nam KC, Ahn DU. 2003. Use of double packaging and antioxidant combinations to improve color, lipid oxidation, and volatiles of irradiated raw and cooked turkey breast patties. Poult Sci 82:850–857.

Nam KC, Min BR, Park KS, Lee SC, Ahn DU. 2003. Effects of ascorbic acid and antioxidants on the lipid oxidation and volatiles of irradiated ground beef. J Food Sci 68:1680–1685.

Nanke KE, Sebranek JG, Olson DG. 1998. Color characteristics of irradiated vacuum-packaged pork, beef, and turkey. J Food Sci 63:1001–1006.

Nanke KE, Sebranek JG, Olson DG. 1999. Color characteristics of irradiated aerobically packaged pork, beef and turkey. J Food Sci 64:272–278.

Nawar WW. 1996. Lipids. In: *Food Chemistry*. New York: Marcel Dekker.

Ngoka DA, Froning GW. 1982. Effect of free struggle and preslaughter excitement on color of turkey breast. Poult Sci 61:2291–2293.

Niewiarowicz A, Pikul J, Czajka P. 1986. Gehalt an Myoglobin und Hämoglobin in Fleisch verschiedener Geflügelarten. Fleischwirtsch 66:1281–1282.

Nishida J, Nishida T. 1985. Relationship between the concentration of myoglobin and parvalbumin in various types of muscle tissues from chickens. Br Poult Sci 26:105–115.

Northcutt JK, Buhr RJ, Young LL. 1998. Influence of preslaughter stunning on turkey breast muscle quality. Poult Sci 77:487–492.

O'Brien PJ, O'Grady M, McCutcheon LJ, Shen H, Nowack L, Horne RD, et al. 1992a. Myocardial myoglobin deficiency in various animal models of congestive heart failure. J Mol Cell Cardiol 24:721–730.

O'Brien PJ, Shen H, McCutcheon LJ, O'Grady M, Byrne PJ, Ferguson HW, et al. 1992b. Rapid, simple and sensitive microassay for skeletal and cardiac muscle myoglobin and hemoglobin: use in various animals indicates functional role of myohemoproteins. Mol Cell Biochem 112:45–52.

Osborn HM, Brown H, Adams JB, Ledward D. 2003. High temperature reduction of metmyoglobin in aqueous muscles extracts. Meat Sci 65:631–637.

O'Sullivan MG, Byrne DV, Martens H, Gidskehaug GH, Andersen HJ, Martens M. 2003. Evaluation of pork colour: prediction of visual sensory quality of meat from instrumental and computer vision methods of colour analysis. Meat Sci 65:909–918.

Owens CM, Sams AR. 1997. Muscle metabolism and meat quality of pectoralis from turkeys treated with postmortem electrical stimulation. Poult Sci 76:1047–1051.

Owens CM, AR Sams. 2000. The influence of transportation on turkey meat quality. Poult Sci 79:1204–1207.

Owens CM, Hirschler EM, McKee SR, Martinez-Dawson R, Sams AR. 2000. The characterisation and incidence of pale, soft, exudative turkey meat in a commercial plant. Poult Sci 81:579–584.

Pages T, Planas J. 1983. Muscle myoglobin and flying habits in birds. Comp Biochem Physiol 74A:289–294.

Papinaho PA, Fletcher DL. 1995. Effect of stunning amperage on broiler breast muscle rigor development and meat quality. Poult Sci 74:1527–1532.

Parry RT. 1993. *Principles and Applications of Modified Atmosphere Packaging of Food*. Glasgow, UK: RT Parry, pp. 1–18.

Petracci M, Fletcher DL. 2002. Broiler skin and meat color changes during storage. Poult Sci 81:1589–1597.

Petracci M, Fletcher DL, Northcutt JK. 2001. The effect of holding temperature on live shrink, processing yield, and breast meat quality of broiler chickens. Poult Sci 80:670–675.

Pietrzack M, Greaser ML, Sosnicki AA. 1997. Effect of rapid rigor mortis on protein functionality in pectoralis major muscle of domestic turkeys. J Anim Sci 75:2106–2116.

Pikul J, Niewiarowicz A, Pospieszna H. 1982. Gehalt an Hämfarbstoffen im Fleisch verschiedener Geflügelarten. Fleischwirtsch 62:900–905.

Poole GH, Fletcher DL. 1998. Comparison of a modified atmosphere stunning-killing system. Poult Sci 77:342–347.

Pratt GB, Kneeland LE. 1972. *Irradiation Induced Head-Space Gases in Packaged Radiation-Sterilized Food*. Technical Report 72–55-FL. Natick, MA: U.S. Army Natick Laboratories.

Qiao M, Fletcher DL, Smith DP, Northcutt JK. 2001. The effect of broiler breast meat color on pH, moisture, water holding capacity, and emulsification capacity. Poult Sci 80:676–680.

Qiao M, Fletcher DL, Northcutt JK, Smith DP. 2002. The relationship between raw broiler breast meat color and composition. Poult Sci 81:422–427.

Raj ABM, Grey TC, Audsely AR, Gregory NG. 1990. Effect of electrical and gaseous stunning on the carcase and meat quality of broilers. Br Poult Sci 31:725–733.

Raj ABM, Wilkins LJ, Richardson RI, Johnson SP, Wotton SB. 1997. Carcase and meat quality in broilers either killed with a gas mixture or stunned with an electric current under commercial processing conditions. Br Poult Sci 38:169–174.

Rathgeber BM, Boles JA, Shand PJ. 1999. Rapid postmortem pH decline and delayed chilling reduce quality of turkey breast meat. Poult Sci 78:477–484.

Rawles D, Marcy J, Hulet M. 1995. Constant current stunning of market weight broilers. J Appl Poult Res 4:109–116.

Renerre M. 1990. Review: factors involved in discoloration of beef meat. Int J Food Sci Technol 25:613–630.

Saffle RL. 1973. Quantitative determination of combined hemoglobin and myoglobin on various poultry meat. J Food Sci 38:968–970.

Sandercock DA, Hunter RR, Nute GR, Mitchell MA, Hocking PM. 2001. Acute heat stress-induced alterations in blood acid–base status and skeletal muscle membrane integrity in broiler chickens at two ages: implications for meat quality. Poult Sci 80:418–425.

Sandusky CL, Heath JL. 1996. Effect of background color, sample thickness, and illuminant on the measurement of broiler meat color. Poult Sci 75:1437–1442.

Santé V, Bielicki G, Renerre M, Lacourt A. 1991. Post mortem evolution in the pectoralis superficialis muscle from two turkey breeds: relationship between pH and colour changes. In: *37th International Congress of Meat and Technology*, vol. 1. Kulmbach, Germany: Federal Center for Meat Research, pp. 465–468.

Santé V, Renerre M, Lacourt A. 1994. Effect of modified atmosphere packaging on color stability and on microbiology of turkey breast meat. J Food Qual 17:177–195.

Santé VS, Lebert A, Le Pottier G, Ouali A. 1996. Comparison between two statistical models for prediction of turkey breast meat colour. Meat Sci 43:283–290.

Satterlee LD, Brown WD, Lycometros C. 1972. Stability and characteristics of the pigment produced by gamma irradiation of metmyoglobin. J Food Agric 31:213–217.

Saucier L, Gendron C, Gariepy C. 2000. Shelf life of ground poultry meat stored under modified atmosphere. Poult Sci 79:1851–1856.

Savenije B, Schreurs FJ, Winkelman-Goedhart HA, Gerritzen MA, Korf J, Lambooij E. 2002. Effects of feed deprivation and electrical, gas, and captive needle stunning on early postmortem muscle metabolism and subsequent meat quality. Poult Sci 81:561–571.

Sheldon BW, Curtis PA, Dawson PL, Ferket PR. 1997. Effect of dietary vitamin E on the oxidative stability, flavor, color, and volatile profiles of refrigerated and frozen turkey breast meat. Poult Sci 76:634–641.

Shikama K. 2006. Nature of the FeO_2 bonding in myoglobin and hemoglobin: a new molecular paradigm. Prog Biophys Mol Biol 91:83–162.

Shikama K, Sugawara Y. 1978. Autoxidation of native oxymyoglobin: kinetic analysis of the pH profile. Eur J Biochem 91:407–413.

Simic MG, Merritt C Jr, Taub IA. 1979. Fatty acids, radiolysis. In: Pryde EH, ed., *Fatty Acids*. Champaign, IL: American Oil Chemists' Society, pp. 457–477.

Smith DP, Northcutt JK. 2003. Red discoloration of fully cooked chicken products. J Appl Poult Res 12:515–521.

Smith DP, Northcutt JK. 2004. Induced red discoloration of broiler breast meat: effect of blood, bone marrow and marination. Int J Poult Sci 3:248–252.

Sørheim O, Nissen H, Nesbakken T. 1999. The storage life of beef and pork packaged in an atmosphere with low carbon monoxide and high carbon dioxide. Meat Sci 52:157–164.

Spencer JV, Sauter EA, Stadelman WJ. 1961. Effect of freezing, thawing and storing broilers on spoilage, flavor and bone darkening. Poult Sci 40:918–920.

Stewart MR, Zipser MW, Watts BM. 1965. The use of reflectance spectrophotometry for the assay of raw meat pigments. J Food Sci 30:464–469.

Tajima G, Shikama K. 1987. Autooxidation of oxymyoglobin. J Biol Chem 262:12603–12606.

Thakur BR, Singh RK. 1994. Food irradiation: chemistry and applications. Food Rev Int 10:437–473.

Trout G. 1989. Variation in myoglobin denaturation and color of cooked beef, pork, and turkey meat as influenced by pH, sodium chloride, sodium tripolyphosphate, and cooking temperature. J Food Sci 54:536–544.

Twining PV Jr, Bossard EH, Lund PG, Thomas OP. 1971. Relative availability of xanthophylls from ingredients based on plasma level and skin measurements. In: *Proceedings of the Maryland Nutrition Conference*, p. 90.

Uijttenboogaart TG. 1991. Colour of fresh and cooked poultry meat. In: *Workshop on Welfare, Hygiene and Quality Aspects of Poultry Processing*, University of Bristol, pp. 16–17.

Van den Oord AHA, Wesdorp JJ. 1971. Analysis of pigment in intact beef samples. J Food Sci 6:1–13.

Van Laack R, Berry B, Solomon M. 1996. Effect of precooking conditions on color of cooked beef patties. J Food Prot 59:976–983.

Varnam A, Sutherland J. 1995. *Meat and Meat Products*. London: Chapman & Hall, p. 430.

Veerkamp CH, DeVries AW. 1983. Influence of electrical stunning on quality aspects of broilers. In: Eikelenboom G, ed., *Stunning of Animals for Slaughter*. Boston: Martinus Nijhoff, pp. 197–212.

Vuilleumier JP. 1969. The 'Roche yolk color fan': an instrument for measuring yolk color. Poult Sci 48:767–779.

Warriss P. 2000. *Meat Science: An Introductory Text*. London: CABI Publishing, p. 310.

Wilkins LJ, Brown SN, Phillips AJ, Warriss PD. 2000. Variation in the color of broiler breast fillets in the UK. Br Poult Sci 41:308–312.

Wynveen EJ, Bowker BC, Grant AL, Demos BP, Gerrard DE. 1999. Effects of muscle pH and chilling on development of PSE-like turkey breast meat. Br Poult Sci 40:253–256.

Yan HJ, Lee EJ, Nam KC, Min BR, Ahn DU. 2006a. Effects of dietary functional ingredients and packaging methods on sensory characteristics and consumer acceptance of irradiated turkey breast meat. Poult Sci 85:1482–1489.

Yan HJ, Lee EJ, Nam KC, Min BR, Ahn DU. 2006b. Dietary functional ingredients: performance of animals and quality and storage stability of irradiated raw turkey breast. Poult Sci 85:1829–1837.

Young LL, Buhr RJ. 1997. Effects of stunning duration on quality characteristics of early deboned chicken fillets. Poult Sci 76:1052–1055.

Young O, West J. 2001. Meat colour. In: Hui YH, Nip WK, Rogers R, Young O, eds., *Meat Science and Applications*. New York: Marcel Dekker, pp. 39–70.

Young JF, Stagsted J, Jensen SK, Karlsson AH, Henckel P. 2003. Ascorbic acid, α-tocopherol, and oregano supplements reduce stress-induced deterioration of chicken meat quality. Poult Sci 82:1343–1351.

26

REFRIGERATED POULTRY HANDLING

ESTHER SENDRA-NADAL, ESTRELLA SAYAS BARBERÁ, AND JUANA FERNÁNDEZ LÓPEZ
Departamento de Tecnología Agroalimentaria, Escuela Politécnica Superior de Orihuela, Universidad Miguel Hernández, Orihuela, Alicante, Spain

INTRODUCTION

Refrigerated is the preferred preservation method for poultry by consumers, so is the most common form of presentation of poultry meat. The measurable basis for defining fresh or frozen is temperature, either of the product itself or of the environmental chamber containing the product. Consumers often perceive that soft

Handbook of Poultry Science and Technology, Volume 2: Secondary Processing, Edited by Isabel Guerrero-Legarreta and Y.H. Hui

flesh equals fresh and that hard flesh equals frozen, and that soft flesh is of better quality. Refrigeration aims to maintain the original characteristics of fresh poultry, which are largely dependent on animal genetics, age and management (especially feeding), stress, slaughter conditions, and postmortem treatment (refrigeration and deboning). As is well known, refrigeration decreases the reaction rate of the chemical and biochemical reactions, modifies the microbial ecology of the system, and is usually combined with other preservation technologies, such as the use of chemicals, physical treatments (irradiation, pressurization), packaging, and modification of the surrounding atmosphere. Quality factors generally associated with chilled storage of poultry include changes in flavor, texture, and microbial profile. The main critical factors affecting the effectiveness of refrigeration as a preservation method are the quality and treatments applied to the raw material, the temperature of storage and the avoidance of temperature fluctuations, the integrity of the packaging material, and the proper handling of the food.

QUALITY INDICATORS OF REFRIGERATED POULTRY

The principal quality parameters of meat products are those related to sensory perception (color, texture, taste, and flavor) and safety (microbial populations and chemicals). One of the main defects occurring during refrigerated storage of poultry is lipid oxidation: the oxidation of unsaturated fatty acids, and the main volatile originated by oxidation is hexanal. Poultry meat is susceptible to rancidity due to a high polyunsaturated fat content, together with the presence of heme, nonheme iron, and phospholipids. The oxidative stability may be increased primarily by increasing antioxidants in the diet of the livestock, mainly α-tocopherol. Poultry meat oxidation seems to be related to the development of a characteristic termed warmed-over flavor (WOF), which is an issue in further-processed and ready-to-eat poultry products. Several compounds have been investigated for the reduction of WOF in poultry as sodium tripolyphosphate, sodium ascorbate monophosphate, sodium lactate, calcium chloride, calcium acetate, and calcium gluconate, which reduced the WOF problem but did not fully prevent other off-flavor development (McKee, 2007a). Sheldon et al. (1997) reported that when feeding supplementary vitamin E to turkey, color scores and lipid stability of refrigerated turkey increased and the incidence of very pale meat was reduced, enhancing the maintenance of quality parameters of fresh turkey breast.

The color of poultry meat is also the topic of Chapter 25. As is well known, the color of poultry meat may be taken as an indicator of meat characteristics and pH; L^* values, especially, may be related to PSE meat (Pérez-Alvarez, 2006). Several researchers have suggested classifying poultry carcasses according to L^* values in order to optimize meat functionality. Galobart and Moran (2004) described that L^* values for refrigerated poultry fillets decreased as storage proceeds for 24 to 48 h postmortem, indicating drip losses; further, L^* decreases were related to meat shrinkage. Regarding poultry color, the most usual color defects observed during refrigerated storage are black spots (which are due to the growth of molds and bacteria), blue, pink or red, and greenish bruise (which is due to damaged muscles and rupture of blood vessels originated by improper catching or shanking) and green iridescence (caused by microstructural diffraction) (Totosaus et al., 2007). The color of poultry meat under the skin is not influenced by refrigerated storage (Totosaus et al., 2007).

The absence of off-odors and slime is also a determinant quality indicator of poultry meat. Both defects are due to the growth of microorganisms and are discussed later in the section on microbial quality. The texture of refrigerated poultry is related primarily to rigor, temperature, and pH (events related to rigor have are reviewed extensively in another chapter). Chicken muscle pH and shear values decrease as chilling prior to deboning increases from 0 to 8 h; a holding time of at least 4 h prior to excision is recommended (McKee, 2007a).

FACTORS AFFECTING REFRIGERATED POULTRY MEAT QUALITY

The main critical factors affecting the effectiveness of refrigeration as a preservation method are the quality and treatments applied on the raw material, the temperature of storage and the avoidance of temperature fluctuations, the integrity of the packaging material, and proper handling of the food. The shelf life of poultry and ostrich depends on several factors: mainly initial bacterial loads, storage temperature, and the gaseous environment around the products (Fernández-López et al., 2008). The average shelf life of refrigerated poultry meat is 2 to 7 days; the most common causes of deterioration are pathogen growth, microbial proliferation (bacteria and yeasts), and rancidity; and the most critical environmental factors involved in deterioration are oxygen, temperature of storage, and light exposure.

Carcass Decontamination

In some cases, retail chicken microbial loads are unacceptable, as they are higher than the maximum limits established in the guidelines for poultry meat (Alvarez-Astorga et al., 2002). Several decontamination treatments have been proposed to reduce poultry microbial loads, including chemical (e.g., organic and inorganic acids, chlorine compounds, organic preservatives, oxidizers such as ozone), physical (e.g., water rinse, high pressure, irradiation, ultraviolet light), and microbiological (e.g., lactic acid bacteria, bacteriocins, microbial parasites) (Pérez-Chabela, 2007). Treatments applied to poultry prior to package also affect poultry shelf life. Under appropriate conditions, the technologies applied to carcasses may reduce mean microbiological counts by approximately 1 to 3 log CFU/cm^2 (Sofos and Smith, 1998).

Several organic compounds have been tested to reduce the microbial load of fresh poultry, the use of chemicals for poultry preservation is reviewed elsewhere in the book. As an example, Patterson et al. (1984) reported that dipping poultry in potassium sorbate followed by storage temperatures between 1 and 2°C extended poultry meat quality (assessed by panelists) to 6 weeks. They also tested lactic acid, which succeeded in preserving the microbial quality of poultry when stored at 1°C, but the color and appearance of poultry was described as "gray and unattractive." Other compounds, such as acetic acid, hydrogen peroxide, chlorine, and sodium tripolyphosphate, have also been tested to increase poultry shelf life (Yang et al., 1998). Zuckerman and Abraham (2002) successfully tested Microgard (a mixture of bactericins and organic acids) and nisin in extending shelf life and inhibiting *Listeria monocytogenes* on kosher poultry, which is especially prone to have high microbial loads, due to the scalding, defeathering, and salting in a koshering operation.

In recent years much attention has been focused on extracts from herbs and spices as antimicrobials, especially due to their content of tannins and essential oils (Viuda et al., 2007). Gulmez et al. (2006) compared the effect of 10-min surface wash treatments on the bateriological quality and shelf life of broiler meat: sterile distilled water (DW), 8% water extract of sumac (*Rus coriaria* L.) (WES), and 2% lactic acid (LA). Shelf life was 7 and 14 days for meat treated with DW and WES, respectively, whereas the LA-treated meat did not spoil after 14 days of cold storage (3°C). Nevertheless, an undesirable pale color and an acidulous odor occurred in the LA-treated meat. Berrang (2001) evaluated the presence and levels of *Campylobacter*, coliforms, *Escherichia coli*, and total aerobic bacteria recovered from broiler parts with and without skin, and observed that no trends were evident when comparing bacterial populations recovered from store-bought skin-on and skin-off products. However, removal of skin from partially processed broiler carcass may be useful in lowering the level of contamination carried forward in the plant.

Additives: Marination

Marination by addition of polyphosphate and sodium chloride has been reported to improve moisture absorption and water-holding capacity synergistically while reducing cooking loss and drip loss of poultry meat. These solutions favor the release of soluble proteins and increase meat tenderness, thus enhancing the binding properties of poultry meat; however, the effect of phosphates in poultry shelf life is not clear (Allen et al., 1998). Inconsistent observations regarding color and pH changes due to marination have been reported by several authors, although L^* increase and a^* and b^* decrease, together with pH increase, are the most repeated observations (Allen et al., 1998). These authors compared the difference between light- and dark-colored fillets in pH, marination pickup, and shelf life and found that dark, high-pH broiler breast meat had reduced shelf life but higher marination pickup. They also reported no effect of marination on poultry tenderness, and that initial and tumbled L^* values correlated positively with drip loss and cooking loss.

Handling of Chilled Poultry

U.S. Department of Agriculture (USDA) regulations require that poultry carcasses be chilled to 7°C within 2 h of processing. The most usual temperature for the product to be sold "fresh" is 4°C. Temperature evolution during processing strongly affects the color of poultry. For example, higher temperatures during antemortem holding and product holding prior to deboning and storage and delays in postmortem chilling are associated with lighter meat colors (McKee, 2007). Slow chilling rates lead to meat discoloration due to cell disruption and blood migration. Chilling poultry prior to rigor mortis, when ATP is still present, causes meat toughening, due to the process of *cold shortening*, which dramatically affects poultry quality. Poor chilling conditions may lead to PSE (pale, soft, exudative) conditions in normal glycolysis carcasses. Poultry cuts including bones must be chilled within the shortest time, as they are susceptible to quality changes as temperature fluctuates due to the low thermal transference rate of bones. Leg quarters are the most susceptible parts, due to their content in dark meat and bones. In the present chapter we are focused on refrigeration from the perspective of a preservation method, whereas postmortem treatment and changes are reviewed extensively in other chapters.

Mielnik and others (1999) compared air chilling with evaporative air chilling (water spray followed by blowing cold air) on the quality of fresh chicken carcasses. Cooling efficiency and total heat loss were significantly higher for evaporative air chilling. Chicken chilled in cold air lost considerably more weight than chicken cooled by evaporative chilling (1.8% difference). After evaporative chilling, the chicken carcasses had a lighter color and more water on the back and under the wings. Regarding color, spraying with water prevented discoloration and improved chicken appearance. The shelf life of the chicken stored at 4 and −1°C were not affected significantly by the chilling method. Moisture content, cooking loss, pH, odor, and flavor were not affected either by the chilling method. The shelf life of chicken stored at 4°C (8 days) and −1°C (13 days) was not affected significantly by the chilling method, but by the storage temperature.

Storage and Packaging

The most popular methods to increase the shelf life of refrigerated poultry are the effective implementation of a HACCP (hazard analysis and control of critical points) plan (including a careful maintenance of the cold chain) and the combination with other preservation technologies, such as packaging and irradiation, which are reviewed extensively in other chapters. Regarding packaging, the most common materials used for meat are plastics, combining polymers with high oxygen barrier properties together with polymers with good humidity barrier and sealing properties, such as polyethylene and polypropylene. Microperforated film has been tested successfully for preserving raw meat during refrigerated storage; the inclusion of natural antioxidants in the polymer layers has also been tested, together with vacuum or MAP packaging, also with good results in preventing oxidation. The use of time–temperature indicators in the packages could help in the assurance of good maintenance of the cold chain. An effective temperature control is critical for populations of Enterobacteriaceae, *Clostridia*, and spoilage bacteria.

Regarding refrigerated poultry packaging, the poultry is packed commonly in polystyrene trays and wrapped in film or bulk packed in polyethene-lined cardboard boxes (Totosaus and Kuri, 2007). There are a wide variety of packaging materials, which allows a combination of materials to obtain a composite with unique properties: oxygen permeability, humidity, hardness, and stability as well as impression and sealing properties, heat resistance, market requirements, and reasonable costs. The most common materials used for fresh meat are stretchable and shrinkable films, absorbent pads, trays, trays with a transparent sealed film on top, bags for whole birds, pouches for leg quarters and breasts, thermoform roll stock, and chub films for ground poultry. The absorbent layer is sometimes built in within the tray and helps in controlling free water and enhancing product appearance.

The packaging may be nonpreservative or preservative. Nonpreservative packaging protects only from cross-contamination and water losses. Preservative packaging also modifies the environment to modify or restrict microbial growth. The main modification is the gas atmosphere: vacuum or modified atmosphere, together with the use of barrier films and proper sealing. The composition of the atmosphere around the product determines the color of meat and the nature of spoilage that develops. In vacuum packaging, oxygen is reduced to less that 1% and is effective in extending the shelf life of poultry. Modified-atmosphere packaging (MAP) consists of packing the food product in gas-barrier materials when the environment has been changed or modified to inhibit the action of spoilage agents. The objective is to maintain the quality of a perishable product or to extend its shelf life. The most common gases used are nitrogen, carbon dioxide, and oxygen. Nitrogen is an inert gas used to prevent package collapse or replace oxygen. The shelf life of poultry and ostrich meat, offal, and meat products packaged in MAP depends on gas composition (Fernández-López et al., 2004, 2006, 2008), initial carcass contamination, storage temperature, film permeability, and headspace volume in the package. The combined use of MAP and decontamination systems as the addition of several short-chain fatty acids may be useful in the extension of the shelf life of refrigerated poultry.

When methods to reduce microbial load are applied there is a shift in the microbial ecology of poultry meat. Vacuum packaging or carbon dioxide fluxing, and the consequent reduction in oxygen levels, inhibits gram-negative psychrotrophs such as *Pseudomonas* but induces the growth of anaerobes or facultative anaerobes such as *Lactobacillus*, which have been reported to be the predominant spoilage microorganism in poultry packaged under vacuum (Totosaus and Kuri, 2007). The initial microbiota of meat is mesophilic and can reach 10^2 to 10^4 bacteria/cm^2 (Dainty and Mackey, 1992). When meat is refrigerated, psychrotrophic bacteria develops; usually, *Pseudomonas* spp. prevail ($<5^\circ$C) and may constitute up to 50 to 90% of the overall microbial populations. Enterobactiaceae prevail under conditions of poor refrigeration

(10°C) and spoil the meat, and under an anaerobic environment lactic acid bacteria prevail, as they are more tolerant to CO_2 than the other two groups (Dainty and Mackey, 1992). The use of CO_2 affects microbial growth by extending the lag phase and increasing the generation time. A minimum concentration of 20 to 30% is necessary to have an inhibitory effect.

According to Bailey and others (1979), at 2°C, the shelf life of poultry stored conventionally (on air) is about 14 days, storage in 65% carbon dioxide may extend poultry shelf life to 19 days, whereas 20% carbon dioxide may extend it to 18 days. Paterson and others (1984) reported that at 1°C and treated with either carbon dioxide or nitrogen, breast portions had a shelf life of 42 days, and leg/thigh portions, of 35 days, assessed by microbial load (10^7CFU/cm^2).

Saucier and others (2000) tested two MAP conditions for ground poultry meat and evaluated meat shelf life. The authors concluded that from a microbial standpoint, poultry meat was better preserved in the presence of oxygen and high levels (>60%) of CO_2, but maintained better appearance and took longer to discolor in the absence of oxygen and moderate CO_2 content (20%). Results indicate that an appropriate gas mixture that can maintain a desirable color offers no guarantee with respect to the microbial profile of the product. MAP may help in increasing poultry meat shelf life, but it is most common packaging used on trays wrapped either in plastic foil or with a sealed cover. Poultry meat has a quick turnover time in retail marketing, and the extra costs for packaging materials and equipment may be a problem. Charles et al. (2006) studied the effects of three packaging systems on the natural microbiota, color, and sensory characteristics of chicken breast meat. Packages tested were (1) a Styrofoam tray with poly(vinyl chloride) overwrap and an absorbent pad, (2) the same type of tray and wrap but without the absorbent pad, and (3) a Fresh-R-Pax container equipped with an absorbent liner-gel system. Samples were stored at 1.2°C for 8 days. In general, *Pseudomonas* spp. and psychrotrophic counts increased as storage time increased for all packaging systems. Total phsychrotrophic counts reached log 7 and at least log 8 CFU/g after 6 and 8 days, respectively, for all breast meat samples in all packaging systems, and the detection of off-odor was most evident after 6 and 8 days of storage. Color parameters were not affected by the packaging system used. Although the absorbent pad did not control microbial growth, it maintained aesthetic appeal by absorbing all visible moisture released from the meat during storage.

Temperature is a critical factor that affects poultry shelf life during cold storage. Sawaya et al. (1993) studied the effect of temperature on vacuum and conventionally packaged poultry quality by panelist assessment of odor: When vacuum-packaged poultry was stored at 4°C, samples were rated as unacceptable after 17 storage days, at 7°C unacceptability was reached after 14 days, and at 9°C, at 10 days. If samples were conventionally packaged, unacceptability was reached 3 to 4 days sooner than vacuum-packaged samples. Color changes, as greening and slime formation appear some days after off-odors appear.

Vainionpää and others (2004) evaluated the effect of temperature on the preservation of broiler chicken cuts packed under MAP (80% carbon dioxide, 20%

nitrogen) and observed that maintenance of the cold chain was the most critical factor determining poultry shelf life. Attending to sensory quality, higher temperature (7.7°C) caused poultry rejection at 5 days of storage, moderate temperature (6.6°C) extended rejection time to 9 days, and a further temperature decrease (5.5°C) extended rejection to 12 storage days.

Listeria monocytogenes growth was inhibited by storage at 1°C, regardless of the atmosphere and inhibited by environments containing carbon dioxide when chicken breasts where stored at 6°C (Hart et al., 1991). Bailey et al. (2000) studied the effect of different refrigeration temperatures on the microbiological profile of chicken. They reported that at day 0, mesophilic bacteria counts were about 4.6 log CFU/mL (sampled by rinsing chicken halves with 100 mL of phosphate-buffered saline water) and increased by 2 logs after 7 days on carcasses held at 4°C, psychrotrophic counts were about 3.6 and increased during the initial 7 days of storage by about 3.9, 1.9, and 1.4 logs, respectively, on carcasses held at 4, 0, and −4°C, coliform counts were about log 2.2/mL and declined to about log 1.5/mL or less by day 7 for all storage temperatures tested; *E. coli* counts were about log 2/mL and were reduced by about 1 log and salmonellae counts were about log 1.5 on salmonellae-positive carcasses and did not change at any storage temperature.

The main cause of quality changes during refrigerated storage of poultry meat is temperature fluctuation. Fresh samples are lighter (higher L^* value) than frozen stored samples. Poultry meat stored at 4°C showed higher a^* values (redness) than lower refrigeration temperatures or freezing temperatures up to −12°C, but were not different from the −18°C pieces (Lyon and Lyon, 2002). However, cooking negated color differences attributed to chilling conditions.

MICROBIAL QUALITY OF REFRIGERATED POULTRY: FOOD SAFETY

Food safety and shelf life are the major microbial concerns regarding poultry meat production. Raw poultry is often highly contaminated, and poultry products are often involved in outbreaks of foodborne illness. Improper handling or cross-contamination of food and temperature abuse during transport, preparation, or storage are common causes of the high microbial loads of poultry meat (McKee, 2007b). Although it is not possible to assure total absence of pathogens, several measures of control are available and can enhance the microbial quality of poultry meat: separation of flocks (according to sanitary conditions, management, and vaccination plans), proper slaughter plants and operations, carcass decontamination, optimum refrigeration conditions, accurate maintenance of the refrigeration chain, and the implementation of a balanced and operational HACCP system (Bolder, 2007).

Slaughter practices involve a shift from a mesophilic environment at the start of processing to a psychrotrophic environment at the end of the production cycle. Poultry meat spoilage is due largely to gram-negative, psychrotrophic bacteria such as *Pseudomonas* spp., *Achromobacter* spp., and *Micrococcus* spp.

(Dawson and Spinelli, 2007). Barnes and Thornley (1966) reported that the bacteria on broiler meat immediately after processing were micrococci (50%), gram-positive rods (14%), flavobacteria (14%), Enterobacteriaceae (8%), and unidentified organisms (5%). After the samples were stored at 1°C for 10 to 11 days, the bacterial microbiota changed to predominantly psychrotrophs, including 90% *Pseudomonas* spp., 7% *Acinetobacter*, and 3% Enterobacteriaceae.

The most common pathogens in poultry are, in order of prevalence, *Campylobacter* spp., *Listeria* spp., *E. coli*, and *Salmonella* spp.; and the most common spoilage microorganisms are *Pseudomonas* spp., associated with the spoilage of refrigerated poultry stored under aerobic conditions; *Lactobacillus* spp., associated with the spoilage of refrigerated poultry stored under microaerophilic or anaerobic conditions; and proteolytic or lipolytic yeats such as *Candida zeylanoides* and *Yarrowia lipolytica* (McKee, 2007b). High microbial loads are associated with off-odors, related mainly to sulfur-containing compounds, as well as sliminess on poultry. Both defects become noticeable when microbial levels reach 10^6 to 10^8 CFU/cm^2.

Poultry spoilage has been defined as "strong off-odor," and it was associated with microbial loads of 10^6 CFU/cm^2, *Lactobacillus* being more than 90% of the microbiota. Controlled spoilage at 3°C resulted in odors described subjectively as "sulfur," "dishrag," "ammonia," "wet dog," "skunk," "dirty socks," "rancind fish," and "canned corn" (Russell et al., 1995). These odors were attributed primarily to *Shewanella putrefaciens, Pseudomonas fluorescens*, and *Pseudomonas fragi*. Off-odors belonging to the family of sulfur compounds are usually associated with *P. fluorescens* and *Pseudomonas putida*. Sliminess of poultry meat is associated with *Pseudomonas* populations over 10^6 CFU/g (Totosaus and Kuri, 2007).

Bailey et al. (2000) determined the effect of different refrigeration and freezer temperatures on the microbial profile of chicken. Mesophilic bacteria increased in refrigerated poultry, psychrotrophic bacteria increased in poultry held at refrigerated but not at subfreezing temperatures, coliforms and *E. coli* decreased at all refrigerated and frozen conditions tested, and salmonellae did not change appreciably at any storage temperature.

Coleman et al. (2003) studied the effect of the microbial ecology of poultry products on the growth of several pathogens, especially regarding the mathematical model to be used in future research. The initial probability of contamination was established based on USDA data: more than 90% of broiler samples positive for *Campylobacter* spp., about 12% positive for *Listeria monocytogenes*, and about 20% for *Salmonella* spp. The authors also assumed the fact reported by Thomas and Wimpenny (1996): temperature dependence of shifts in dominance of the mesophilic pathogen *Salmonella* with psychrotrophic nonpathogen *Pseudomonas*. At 30°C, mesophiles inhibit the psychrotroph, whereas at 20°C the psychrotrophic strains inhibit the mesophile's growth. The indigenous microbiota appear to have a strong competitive advantage over pathogens in numerical dominance and faster growth rates at refrigeration temperatures. They also took into account the *Jameson effect*, a theory that the total population density of the food system might suppress the growth of all microbial populations present. Their final recommendations include taking good care of real time–temperature abuse conditions and running additional research to bridge gaps in knowledge of growth kinetics between traditional culture broth and food matrices to assist in risk analysis and test the effects of clustering (nonhomogeneous distribution) of pathogens in foods. Regarding data on times in transit and temperatures at the beginning and end of transport, the authors reported that in the United States the main time in transit for shoppers was about 1 h (standard deviation 26 min, range 13 min to 6 h and 20 min). The temperature range was difficult to interpret but had a clear seasonal dependency; increased meat temperatures after transport correlated with higher ambient air temperature.

CONCLUSIONS

Main quality concerns related to refrigerated poultry are meat safety and shelf life; lipid oxidation is also important but to a lesser extent. Although it is not possible to assure complete absence of pathogens, several measures of control are available, and if properly applied can enhance the microbial quality of poultry meat: separation of flocks (according to sanitary conditions), proper slaughter plants and operations, carcass decontamination, proper packaging, optimum refrigeration conditions and accurate maintenance of the refrigeration chain, and the implementation of a balanced and operational HACCP system. Temperature fluctuations during refrigerated storage seem to be the main cause of quality loss in refrigerated poultry meat.

REFERENCES

Allen CD, Fletcher DL, Northcutt JK, Russell SM. 1998. The relationship of broiler breast meat color to meat quality and shelf-life. Poult Sci 77:361–366.

Alvarez-Astorga M, Capita R, Alonso-Calleja C, Moreno B, García-Fernández. 2002. Microbiological quality of retail chicken by-products in Spain. Meat Sci 62:45–50.

Bailey JS, Reagan JO, Carpenter JA, Schuler GA, Thomson JE. 1979. Types of bacteria and shelf life of evacuated carbon dioxide-injected and ice-packed broilers. J Food Prot 42(3):218–221.

Bailey JS, Lyon BG, Lyon CE, Windham WR. 2000. The microbial profile of chilled and frozen chicken. J Food Prot 63:1228–1230.

Barnes EM, Thornley MJ. 1966. The spoilage flora of eviscerated chickens stored at different temperatures. J Food Technol 1:113–119.

Berrang ME, Ladely SR, Buhr RJ. 2001. Presence and level of *Campylobacter*, coliforms, *Escherichia coli* and total aerobic bacteria recovered from broiler parts with and without skin. J Food Prot 64(2):184–188.

Bolder N. 2007. Microbial challenges of poultry meat production. World's Poult Sci Assoc 63(9):401–411.

Charles N, Williams SK, Rodrick GE. 2006. Effect of packaging systems on the natural microflora and acceptability of chicken breast meat. Poult Sci 85:1798–1801.

Coleman ME, Sandberg S, Anderson SA. 2003. Impact of microbial ecology of meat and poultry products on predictions from exposure assessment scenarios for refrigerated storage. Risk Anal 23(1):215–228.

Dainty RH, Mackey BM. 1992. The relationship between the phenotypic properties of bacteria from chill-stored meat and spoilage process. J Appl Bacteriol 73:103S–114S.

Dawson PL, Spinelli N. 2007. Poultry meat flavour. In: Nollet LML, ed., *Handbook of Meat, Poultry and Seafood Quality*. Oxford, UK: Blackwell Publishing, pp. 439–453.

Fernández-López J, Sayas-Barberá E, Sendra E, Pérez-Alvarez JA. 2004. Quality characteristics of ostrich liver pate. J Food Sci 69(2):SNQ85–SNQ91.

Fernández-López J, Yelo A, Sayas-Barberá E, Sendra E, Navarro C, Pérez-Alvarez JA. 2006. Shelf life of ostrich (*Struthio camelus*) liver stored under different packaging conditions. J Food Prot 69(8):1920–1927.

Fernández-López J, Sayas-Barberá E, Muñoz T, Sendra E, Navarro C, Pérez-Alvarez JA. 2008. Effect of packaging conditions on shelf-life of ostrich steaks. Meat Sci 78:143–152.

Galobart J, Moran ET. 2004. Refrigeration and freeze-thaw effects on broiler fillets having extreme L^* values. Poult Sci 83:1433–1439.

Gulmez M, Oral N, Vatansever L. 2006. The effect of water extract of suma (*Rhus coriaria* L.) and lactic acid on decontamination and shelf life or raw broiler wings. Poult Sci 85(10):1466–1471.

Hart CD, Mead GC, Norris AP. 1991. Effects of gaseous environment and temperature on the storage behaviour of *Listeria monocytogenes* on chicken breast meat. J Appl Bacteriol 70:40–46.

Lyon BG, Lyon CE. 2002. Color of uncooked and cooked broiler leg quarters associated with chilling temperature and holding time. Poult Sci 81:1916–1920.

McKee L. 2007a. General attributes of fresh and frozen poultry meat. In: Nollet LML, ed., *Handbook of Meat, Poultry and Seafood Quality*. Oxford, UK: Blackwell Publishing, pp. 429–437.

McKee L. 2007b. Microbial and sensory properties of fresh and frozen poultry. In: Nollet LML, ed., *Handbook of Meat, Poultry and Seafood Quality*. Oxford, UK: Blackwell Publishing, pp. 487–496.

Mielnik MB, Dainty RH, Lundby F, Mielnik J. 1999. The effect of evaporative air chilling and storage temperature on quality and shelf life of fresh chicken carcasses. Poult Sci 78:1065–1073.

Patterson JT, Gillespie CW, Hough B. 1984. Aspects of the microbiology of vacuum- and gas-packaged chicken, including pre-treatments with lactic acid and potassium sorbate. Br Poult Sci 25:457–465.

Pérez-Alvarez JA. 2006. Color de la carne y productos cárnicos. In: Hui YH, Guerrero I, Rosmini MR, eds., *Ciencia y Tecnología de Carnes*. Mexico City, Mexico: Limusa, pp. 161–198.

Pérez-Chabela ML. 2007. Shelf-life of fresh and frozen poultry. In: Nollet LML, ed., *Handbook of Meat, Poultry and Seafood Quality*. Oxford, UK: Blackwell Publishing, pp. 467–474.

Russell SM, Fletcher DL, Cox NA. 1995. Spoilage bacteria in fresh broiler chicken carcasses. Poult Sci 74(12):2041–2047.

Saucier L, Gendron C, Gariépy C. 2000. Shelf life of ground poultry meat stored under modified atmosphere. Poult Sci 79(11):1851–1856.

Sawaya WN, Abu-Ruwaida AS, Hussain AJ, Khalafawi MS, Dashti BH. 1993. Shelf-life of vacuum-packaged eviscerated broiler carcasses under simulated market storage conditions. J Food Safety 13:305–321.

Sheldon BW, Curtis PA. Dawson PL, Ferket PR. 1997. Effect of dietary vitamin E on the oxidative stability, flavor, color and volatile profiles of refrigerated and frozen turkey breats meat. Poult Sci 76:634–641.

Sofos JN, Smith GC. 1998. Nonacid meat decontamination technologies: model studies and commercial applications. Int J Food Microbiol 44:171–188.

Thomas LV, Wimpenny JW. 1996. Competition between *Salmonella* and *Pseudomonas* species growing in and on agar, as affected by pH, sodium chloride concentration and temperature. Int J Food Microbiol 29(2–3):361–370.

Totosaus A, Kuri V. 2007. Packaging of fresh and frozen poultry. In: Nollet LML, ed., *Handbook of Meat, Poultry and Seafood Quality*. Oxford, UK: Blackwell Publishing, pp. 475–485.

Totosaus A, Pérez-Chabela ML, Guerrero I. 2007. Color of fresh and frozen poultry. In: Nollet LML, ed., *Handbook of Meat, Poultry and Seafood Quality*. Oxford, UK: Blackwell Publishing, pp. 455–466.

Vainionpää J, Smolander M, Alakomi HL, Ritvanen T, Rajamaki M, Rokka M, Ahvenainen R. 2004. Comparison of different analytical methods in the monitoring of the quality of modified atmosphere packaged broiler chicken cuts using principal component analysis. J Food Eng 65:273–280.

Viuda-Martos M, Ruiz-Navajas Y, Fernández-López J, Pérez-Alvarez JA. 2007. Antifungal activities of thyme, clove and oregano essential oils. J Food Safety 27(1):91–101.

Yang Z, Li Y, Slavik M. 1998. Use of antimicrobial spray applied with an inside–outside birdwasher to reduce bacterial contamination on pre-chilled chicken carcasses. J Food Prot 61:829–832.

Zuckerman H, Abraham RB. 2002. Quality improvement of kosher chilled poultry. Poult Sci 81:1751–1757.

PART V

ENGINEERING PRINCIPLES, OPERATIONS, AND EQUIPMENT

27

BASIC OPERATIONS AND CONDITIONS

M.C. PANDEY AND AMARINDER S. BAWA
Defence Food Research Laboratory, Siddartha Nagar, Mysore, India

Handbook of Poultry Science and Technology, Volume 2: Secondary Processing, Edited by Isabel Guerrero-Legarreta and Y.H. Hui

INTRODUCTION

Production of good-quality meat products from a live bird involves a series of efficiently performed specific tasks carried out in a sanitary manner. Broiler farms are often quite large and raise several million birds each year. Chicks are highly susceptible to many diseases, so broiler producers must practice rigid husbandry with respect to temperature and humidity control, sanitation, and feeding practices. The principal types of poultry are chicken, turkey, ducks, geese, and ostriches. Broilers or fryers are generally preferred for processing into flesh or frozen chicken, where tenderness is essential. Modern plants are efficient, with continuous-line facilities in which birds are moved from operation to operation via monorail. The operation can be partially mechanized and highly efficient in large plants if the birds are remarkably uniform with respect to size, shape, weight, and other characteristics. It is common to raise a 2.3-kg broiler in just six weeks with a feed conversion of 1.8 kg of feed per kilogram of bird. In other words, a 2.3-kg broiler is raised from a chick on just about 4.1 kg of feed. This is one reason why chicken may be purchased at a lower price, on an edible-weight basis, than beef, which has a lower feed-conversion ratio (Potter and Hotchkiss, 1996).

Defects resulting in carcass downgrading and product losses are caused by diseases, damage to birds before slaughter or by maladjusted equipment, and manual error during processing. The type of management also affects poultry processing and meat quality. Broilers reared under a low-temperature regime (i.e., 12°C) during the last 4 weeks of the rearing period were significantly more tender and had a higher flavor score than those reared under a higher-temperature regime (i.e., 28°C). Tenderness, juiciness, and flavor intensity were related to the sexual maturity of the birds (Touraille and Ricard, 1981). Seemann (1981) reported that older and male birds had higher yields than younger and female birds. He concluded that fattening of modern broilers for a longer time would result in higher yields: higher breast, breast meat, and lower wing and back percentages.

Birds should be processed and stored at the proper temperature to ensure that the risk of foodborne illness is minimized. They should be taken off feed 8 to 10 h prior to slaughter to reduce the amount of material in the digestive tract, and care should be taken not to use birds that have such symptoms as lumps or spots of any size on the surface of the liver, any measurable quantity of fluid in the body cavity, fat in a poorly fleshed bird that is orange rather than yellow or white, any intestinal organs that are abnormally large, breast meat with the same coloration as that of thigh and leg meat, and meat showing white steaks or an area of abnormal enlargement. Birds with defects such as bruises, blisters, and skin can be processed as wholesome carcasses by removing the damaged tissue.

Broilers are caught for loading at night, as at that time they are easier to catch, struggle less, and settle down in crops faster, and in summer the weather is cooler at night. Reduction of animal preslaughter stress, rapid carcass chilling, and variations in processing methods to regenerate and protect functional properties of proteins (particularly, water-binding capacity) are required to minimize the incidence of these defects in poultry meat (Lesiow and Kijowski, 2003). The loading schedule should be arranged such that the birds arrive at the plant an hour before they are to be unloaded for slaughter. During loading and unloading, loads of poultry are weighed on public scales near the processing plant. Shrinkage of 3 or 4% is permitted from place to place. Shrinkage above the maximum level is borne by the seller or hauler. In some cases, broilers are paid for on the basis of weights as delivered to the plant.

The objectives of a centralized poultry-processing plant of any scale of operation is to produce material at a price that customers can afford and that is hygienic, wholesome, and attractive, consistent in appearance and quality, with a realistic shelf life. The appearance of the product involves showing its fat and bone. These factors are controlled by breeding and selection of carcasses at the processing plant.

RECEIVING

Depending on distance from the slaughterhouse, poultry should be taken off their feed and water 1 to 4 h before they are loaded and taken for slaughter. This ensures that the birds are significantly empty and their feces are dry. If the period is extended to, say, 10 h, the feces becomes more fluid and the chance of cross-contamination between birds during transportation is increased. In the tropics it is essential that birds are not overcrowded and thus liable to overheat. Larger birds should be allowed more space than smaller birds.

The transport vehicle should be parked in the shade. The vehicle used will depend on the number of birds to be carried and the distance. Adequate ventilation will reduce the transportation stress of the livestock. Close-sided vehicles are unsuitable. The crates should be kept in the shade during transport, which should be carried out in the cool of the day. Early morning transport is recommended. Actual movement of the vehicle is important in reducing the transport stress of young birds. Birds should not be subjected to excessive vibration acceleration or to breaking at speed.

On arrival at the abattoir, crates should be carefully unloaded from the transport vehicle in the reception area. After unloading, poultry should be kept for a minimum time before slaughter and should be left in crates under cover until required. Empty crates should be returned to the wash area, and the transport vehicle should be cleaned and disinfected before it is removed from the compound.

STUNNING

Except where slaughter is performed ritually, regulations insist that birds be stunned prior to killing and bleeding. Almost all birds are rendered unconscious through stunning and allowed painless humane killing. In religious practices such as kosher and halal, stunning is not performed, but it is very helpful when working with turkeys or geese because of their large size. Stunning immobilizes a bird to increase the killing efficiency. It accelerates the bleeding rate, relaxes the muscles holding the feathers, thereby making defeathering very easy, and contributes to the overall meat quality. Generally, turkey wing and tail feathers are pulled from the carcass immediately after stunning. A variety of methods are used to stun birds: electrical stunning, stunning gun, and modified-atmospheric stunning.

Electrical stunning has been the most common method used to immobilize poultry for slaughter in poultry-processing plants (Bilgili, 1992). Electrical stunning is performed by the use of large water baths which can stun up to 12,000 birds/h. The head of the bird is dipped into the saline water and an electric current is used to render the birds temporarily unconscious (1 to 2 min, time enough for cutting and bleeding to death). The current applied will be of low voltage and amperage (20 V, 10 to 20 mA for 3 to 5 s for broilers and 20 to 40 mA for 10 to 12 s for turkeys, depending on the size of the bird). Careful control of the current is very important in this process. When a high-voltage, high-frequency (50 to 60 Hz) current is used, wing hemorrhages, red skin, poor defeathering, broken bones, and unaccepted blood spots may result (Gregory and Wilkins, 1989; Walther, 1991).

In the late 1970s and early 1980s, reports from Europe began to indicate problems with electrical stunning of poultry. Electrical stunning was shown to be unreliable, with approximately one-third of the birds emerging from the stunner dead, while another one-third were unstunned (Heath, 1984). When a stun gun is used for larger birds, it is critical to set it properly to achieve maximum efficiency. Modified-atmosphere stunning is now in use. An article in *Poultry International* (Anon., 1997) recommended considerating the possible benefits (including the meat quality benefits) of the use of controlled-atmosphere killing as an alternative to conventional stunning methods. Both single-stage anoxic and two-stage anesthetic systems are in use. In a single-stage anoxic system, the bird enters a tunnel filled with a mixture of gases, which typically contain argon and either carbon dioxide or nitrogen, by means of a moving conveyor. In two-stage systems, birds are anesthesized in the first stage by an atmosphere rich in oxygen and carbon dioxide, then rendered reversibly unconscious in a separate atmosphere high in carbon dioxide. In this operation, there is no chance of a bird entering the killing or scalding process while still alive. The concentration of carbon dioxide required must be controlled carefully; otherwise, the bird will be killed. A higher concentration of CO_2 is required for to stun male birds than to stun female birds (Drewniak et al., 1955). The gas stunner reduced the incidence of blood spots on breast fillets and tenders while lightening the breast meat color (Nunes, 1994). Practical limits for stunning turkeys varied from 73 to 75% concentration. A modified atmosphere virtually eliminates the bone breakage and blood spots seen with electrical stunning and increases the quality and yield benefits.

In a study by Hoen and Lankhaar (1999), anoxia generated through the use of argon or a mixture of argon and carbon dioxide or hypercapnic hypoxia appeared to be very promising. Tests revealed that meat tenderness and drip loss improved. Blood spots, especially those on the thighs and breasts caused by stunning and hanging, disappeared altogether.

SLAUGHTERING

Slaughter refers specifically to the killing of animals for food. Traditional slaughter practices have dealt primarily with factors that ensure both the wholesomeness and quality of the meat (Fletcher, 1999). Almost all birds are rendered unconscious through stunning prior to killing. Some exceptions are made for religious meat processing (e.g., kosher, halal). Various methods have been adopted for the killing of birds so that they can be bled easily. *Modified kosher-killed birds* have their jugular vein severed such that the windpipe (trachea) and esophagus remain uncut. If a bird's head is not in the correct position for neck cutting, the trachea and esophagus are severed. Usually, birds are placed on cones on the killing stand, the head is held in the left hand, and a cut is made by placing the blade of the knife just behind and below the earlobe. The bleed cut can be made manually or automatically. In *automatic cutting* the carotid vein and artery are usually cut on one side of the neck only and the trachea and esophagus are not damaged (Mead, 2004). If done correctly, the bird will bleed rapidly and fully from the severed carotid artery. Another method that has been used, in which birds are pierced through the brain, consists of severing the veins in the roof of the mouth, but this method of slaughter has been discontinued. *Neck dislocation killing* is also performed in some poultry plants. For this operation, 24-h fasting is desirable before slaughter. Average-sized birds are chosen for this, and the legs are held in the left hand and the neck is held just below the head between the first and second fingers of the right hand. A bird's neck is "stretched" upward and the head is bent sharply backward to break the vertebrae. The bird is then placed in a cone, the blood vessel is cut using a knife, and the bird is allowed to bleed. With turkeys and larger birds, it may be necessary to suspend the bird by the hocks after stunning to allow both hands to be available to apply sufficient pressure to the head and neck to perform killing.

The age at slaughter has been found to influence organoleptic quality factors. Electrical stimulation (ES) of a carcass can follow slaughter. ES is a process in which an electrical current is applied to an animal carcass shortly after slaughter to stimulate muscular contraction and postmortem metabolic activity. Postmortem electrical stimulation of meat carcasses was developed in the 1950s and became widely used by the red meat industry in the 1970s (Chrystall and Devine, 1985). Postmortem stimulation has been shown both to accelerate rigor mortis and to result in microstructural changes that lead to more tender meat (Fletcher, 1999).

BLEEDING

As soon as a cut is made, the head is twisted slightly to see that the blood gushes out from the cut, and broilers are bled as they pass through a bleeding tunnel designed to collect blood to reduce wastewater biochemical oxygen demand and total nitrogen concentrations. Of the three principal methods of killing, the modified kosher method is widely practiced to obtain good bleeding. Bleeding must take place for a minimum of 90 s. On average, broilers are held in the tunnel for 45 to 125 s for bleeding, with an average time of 80 s; a turkey is held for 90 to 210 s, with an average of 131 s. Blood loss approaches 70% in some plants but is generally between 34 and 50% during the killing operations. If you allow a bird to bleed for a very short time, the result will be poor picking. The U.S. Department of Agriculture (USDA) recommends a bleeding time in the range 55 to 133 s. In work on factors that influence bleeding, Davis and Coe (1954) reported that debraining before cutting the carotid arteries and jugular veins was of little value during the first 20 s of bleeding, other than to immobilize the bird. The blood is collected in troughs and transported to a rendering facility through vacuum, gravity, or pump systems.

SCALDING

After bleeding, birds are conveyed for scalding. Scalding is performed to relax feather follicles and facilitate the removal of feathers (Keener et al., 2004) using hot water. The heat breaks down the protein that is holding the feathers in place. The secret of good scalding is control of the temperature of the scald water, to obtain product consistency. The scald water can be heated by direct steam injection, low-pressure steam, or hot water circulation through integral heating panels mounted in the scald tank. The temperature of the scald water is monitored and controlled electronically. Immersion scalding and spray scalding are the two types used. Virtually all plants use scald tanks (long troughs of hot water) for immersing the bled birds, because of the higher water use and inconsistent feather removal associated with spray scalding. In operation, a scald tank is used with a continuous inflow of water, sufficient to replace that lost in removing wet birds. To ensure both full penetration of the feathers and the desired water temperature, scalders are agitated by means of impellers or air injection. The temperature used for the scalding water is determined by the market. Mainly hard, sub, and semi are scalding used, the principal difference among the three being the scalding temperature. Pool et al. (1954) observed that the force required to remove feathers from turkey carcasses decreased as the scalding temperature decreased.

Poultry immersed in water heated to 71 to 82°C for 30 to 60 s is considered to be *hard scalded* (Mountney, 1976). It is easier to remove the feathers from carcasses scalded at this temperature, but there is greater danger of removing skin portions in the defeathering machines and of discoloration of the carcass due to uneven moisture loss. The resulting carcass will be red in appearance.

Hard scalding is used primarily for waterfowl because it is the only satisfactory way to release feathers, and the skin of waterfowl does not discolor as readily as it does in other species of poultry.

Subscalding is performed at a water temperature of 58 to 60°C for 30 to 75 s. These carcasses have the outer layer of the skin broken down, but the flesh is not affected as it is in hard scalding. This results in easy removal of feathers and uniform coloration of the meat. Klose et al. (1961) reported that scalding resulted in a reduction in feather pulling force, ranging from 30% at 50°C to over 95% at 60°C. Klose and Pool (1954) suggested that turkeys scalded at 60°C were acceptable for frozen storage if proper moisture control was maintained.

Semiscalding, often called *soft* or *slack scalding*, is carried out at a temperature of 50 to 55°C for 90 to 120 s. This loosens feathers without skin damage. The chief advantage of this method is that it leaves the skin intact and permits more diverse methods of chilling and packing. Many tests over the years have proven that water with a low mineral content (0° hardness) permits a shorter scalding period and results in substantially less need for pinning by hand. Addition of detergent or a specific water softener improved the penetration of water through feathers.

DEFEATHERING

During the past few years, several studies have been undertaken to determine some of the factors that affect feather-release mechanisms. The quality of the defeathering is related to the scald, and it has been observed that the force required to remove feathers from carcasses decreased as the scalding temperature was increased. The feathers usually come out easily and can be removed by hand if scalding is done correctly. However, hand picking is time consuming. Turkeys and laying hens are more difficult to defeather; and waterfowl feathers are especially difficult to remove. Klose et al. (1961) reported that it required more force to pull feathers from males than from females and that fasting for 8 h increased the force required to remove feathers.

In poultry-processing industries, feathers are removed using electrically or hydraulically driven automatic pluckers with mounted rubber fingers. Feather removal is most successful when pluckers are installed close to the scalders so that a bird's body temperature remains high during feather removal. There are two possible mechanized methods of removing feathers. The first is to hand-hold each bird against a revolving plucker which has a number of protruding rubber fingers. As the fingers rub against the bird, their abrasive action removes the feathers much as a person's fingers do when a bird is picked by hand. The bird is rolled across the drum so that all parts are exposed to the fingers long enough to remove the feathers but not damage the skin. A manual drum picker can process up to 200 to 300 birds an hour, while a good operator can defeather a bird in 10 to 12 s if the bird has been effectively slack- or hard-scalded.

When the capacity exceeds 300 birds, it is more economical to use an automatic drum or spin picker type of machine. All of the various types of machines have a tub in which the birds are placed. The machine has either a rotating disk at the bottom or a drum in the center that rotates the birds in the tub while rubber fingers in the walls of the tub and in the rotating disk or drum rub the feathers off. The feathers removed are then washed down from pluckers into a channel running underneath the machines. The pickers can be a major source of carcass bruising, wing breakage, and broken hocks, especially if the rubber fingers are worn or not positioned correctly.

The pinfeathers that survive a conventional defeathering machines are usually removed by *pinning*, either by hand using a pinning knife or by dipping in a hot-wax bath. Hot wax consists of a mixture of wax, gum, and fat, and the temperature of the bath is maintained at about 55°C (Gerrard, 1964). Generally, the carcasses are dropped into the wax bath while suspended on shackles by the head and feet, then are removed and dipped a second time. The wax is then made to set by immersion in cold water. When the wax is hardened, it is peeled off in large pieces, pulling out the small pinfeathers. The wax is then heated to kill the microbial load, to reduce contamination, and is reused. *Singeing* is performed to remove hairlike projections called *filoplumes* and involves passing the bird through a sheet of flame as it moves along the conveyor line. Small-scale processors utilize a propane torch to burn them off, being careful not to burn the skin. Commercial poultry breeds have white feathers that do not leave stains.

EVISCERATION

Evisceration involves a thorough cleaning of the carcass in a multistep process that begins with removing the neck and head and opening the body cavity. The viscera are pulled out but are kept attached to the birds until they are inspected for evidence of disease by a veterinary inspector. This is generally carried out in a cold room. Both manual methods and automatic machines are employed for the process; however, manual techniques are preferred and in widespread use because of limitations in using machines. Automatic machines for evisceration are usually species specific; the species involved and uniformity in size are very important for proper operation. In manual methods the entire process is performed on a table, and the bird is passed along from one operator to another until it is finished.

During the evisceration process, carcasses can easily become contaminated with fecal material, especially if improper actions take place. Maximum care has to be taken during the removal of viscera. Full intestines are more easily cut or torn during evisceration, and the contents may leak onto the carcass during extraction. On the other hand, when birds go without feed for too long before processing (more than 14 h), the intestinal lining is lost and intestines may break during extraction.

In a manual evisceration process, a cut is usually made around the vent, taking care not to puncture the intestines. The opening must be big enough so that a hand may be placed inside the carcass. Assuming that the operator is right-handed, the carcass is held firmly with the left hand and two or three fingers of the right hand are inserted through the incision in the abdomen. The three middle fingers (depending on the convenience of the operator), extended, slide past the viscera until the heart is reached; then with a loose grip on the organs, everything is pulled out carefully with a gentle twisting action.

Automation of the evisceration process varies depending on plant size and operation (Childs and Walters, 1962). A fully automated line can eviscerate approximately 6000 broilers per hour. A variety of equipment is available, varying as to location and manufacturer. Many parts of the operation can be performed manually, especially with turkeys. Birds entering an automated evisceration area are rehung by their hocks on shackles on a conveyor line that runs directly above a wet or dry offal collection system. A bird's neck is disconnected by breaking the spine with a blade that applies force just above the shoulders. As the blade retracts, the neck falls downward and hangs by the remaining skin while another blade removes the preen gland from the tail. This gland produces a type of oil that birds use for grooming, whose odor humans find unpleasant. Next, a venting machine cuts a hole with a circular blade around the anus for the extraction of viscera. The evisceration machine immobilizes the bird and passes a clamp through the abdominal opening to grip the visceral package and pull out slowly.

Next, the viscera are separated from the bird and the edible components (i.e., heart, liver, neck, spleen, kidneys, gizzard) are harvested carefully. The feet can be washed thoroughly and used for stock. The inedible offal, such as intestines, sex organs, lungs, and the waste removed from the gizzard harvesting process, is usually shipped to a rendering facility. Once the evisceration is complete, the birds are washed thoroughly. Nozzles are used to spray water both inside and outside the carcass. High-pressure nozzles are designed to eliminate the majority of remaining contaminants on both carcass and conveyor line. Once the washing is over, the meat must be chilled to slow down harmful microbial growth.

Dietary factors affect abdominal fat markedly and influence evisceration yield, shelf life, and water uptake during chilling. Fat loss through evisceration pollutes water during processing and chilling (Essary and Dawson, 1965; Hemm et al., 1967).

GIBLET HARVESTING

The viscera are removed from birds that have passed inspection and are pumped into a harvesting area where edible viscera are separated from inedible viscera. A giblet harvester is used to collect the edible viscera. The heart and liver are removed from the remaining viscera by cutting or pulling them loose. As soon as they removed, the gallbladder is cut or pulled from the liver, and at the same time the pericardial sac and arteries are removed from the heart. More edible tissues are lost by cutting than by pulling (Mountney, 1976). The gizzard is split, its contents washed away, its hard lining peeled off, and it is washed thoroughly.

CUTTING

Depending on market focus, a large percentage of carcasses are cut into portions for retail sale, for use in canteens and catering outlets, and as a raw material for a wide range of fast-food products. Cutting includes cutting the carcass into halves, quarters, legs, thighs, wings, breasts, and drumsticks, or complete deboning. Whole legs and leg quarters can be cut into thighs and drumsticks. Common cut-up configurations consist of eight pieces (wings, breast, thighs, and drumsticks). The wings can also be cut into drumettes, with the remaining racks as a by-product. The instruments used for the process include cleavers, knives, secateurs, and automatic machines.

CHILLING

During processing, both scalding and chilling exert an influence on the quality of the meat. Different methods of chilling have been the subject of a number of investigations (Brant, 1963). After cutting, the carcass temperature must be lowered quickly to prevent microbial growth (Thomson et al., 1966). According to the USDA, the internal temperature of the carcass must be lowered to 4°C within 4 h for 1.8- to 2-kg broilers, 6 h for 2- to 4-kg broilers, and 8 h for 3.6- to 4-kg broilers or turkey. Chilling of carcasses can be done through one of three methods, according to convenience. *Air chilling* involves passing the carcass through rooms with air circulating in and around the carcass at between −7 and 2°C for 1 to 3 h, but wingtips and neck flaps should not be frozen. *Wet chilling* is done by immersing the carcass in cold running tap water, crushed ice, and slush ice. Most poultry plants use two chilling tanks in series, a prechiller and a main chiller. The direction of water flow is from the main chiller to the prechiller, opposite the direction of carcass movement. Because water and ice are added only to the main chiller, the water in the prechiller is somewhat warmer than that in the main chiller. Most plants chlorinate chiller water to reduce potential carcass microbial contamination. Dry-air chilling provides a better alternative to chlorinated water chilling than unchlorinated water chilling. Surface drying has repeatedly been shown to inhibit microbial growth and, in some cases, to induce significant reductions in microbial loads (Burton and Allen, 2002). Slush ice and water chill faster than crushed ice alone because with water the carcasses are completely immersed in the cooling medium. Slush ice agitated with compressed air or a circulating pump or in an on-the-line chiller can also be used. Agitation makes the water a very effective washer and often cleans off any remaining contaminants. *Combination chilling* comprises a shortened immersion or drag-through wet-chill system followed by a shortened air-chill tunnel. Giblets are chilled similarly to the carcasses, although the chilling systems used are separate and smaller. The advantages and disadvantages of various chilling systems have been discussed by Ziolecki (1990) and Vranic et al. (1991).

Weight change with water uptake during chilling predisposes an increase in drip and cooking losses, although the taste, tenderness, and juiciness of the meat remain unaffected. A number of workers have observed losses in protein and ash contents in broiler carcasses subjected to a water-chilling process (Pippen and Klose, 1955; Fris Jensen and Bøgh-Sørensen, 1973; Zenoble et al., 1977).

PACKING

After a carcass is properly chilled, it is ready for packing. Chickens are sold in both fresh and frozen forms. Quick packing in a cold room is recommended once the carcass has been taken from the chiller and before it warms up. A product for retail sale can be packed in a number of ways. Ice packing is preferred for meat and includes receiving, storing, setting up, and distributing the boxes. Portions to be sold frozen are individually quick frozen, weighed, and packed in preprinted polyethylene bags. They are then placed into a cardboard or stainless steel carton for dispatch. The cardboard carton has a polyethylene laminate inside to prevent the cardboard from taking up moisture from the product. In the packaging of poultry products, it is necessary to print the required consumer information on the packaging material: weight, date of packaging, and expiry date. Recent developments have drawn increased attention to deboned poultry packed using the modified-atmosphere technique. Products utilizing this method have a prolonged shelf life and give both the processor and the retailer increased operating flexibility. Small processors also package in individual bags, but they usually shrink-wrap them. For this they use bags designed to evacuate the air on dipping in hot water. Vacuum packing is also preferred for poultry products to obtain a shelf life of 5 to 12 days. In large plants, meat is packed in dry tray packs or bulk ice packs. In dry tray packing, the meat is packed in overwrapped trays. Only the top quarter inch of meat is frozen (crust frozen), to insulate the product. This provides a shelf life of 21 days when stored at −2°C. In bulk ice packing, meat is packed in boxes of 18 to 32 kg with ice on top and drainage holes in the boxes. The shelf life of bulk ice-packed meat is 7 days at 4°C. Generally, 14 to 16 kg of crushed ice is placed on top of the birds in each box. Most processors now supply carcasses iced or dry packed with CO_2 (Hale et al, 1973).

STORAGE

Facilities installed in a poultry-processing plant include a chill room, a blast freezer, and a freezer. This allows maximum flexibility for product preservation. On-farm processors store carcasses for their customers in refrigerators or freezers. Loading birds that have not been chilled to 4°C in a refrigerator or freezer may drive up the temperature and allow microbial growth on the carcasses. In large-scale industries, the packaged poultry is taken into the chill room and stored overnight before dispatch the following day. Frozen-packed meat in boxes is stored in the freezer room, which operates at a temperature of about −20°C. The product enters at a temperature of −40°C on the outside of the pack and −10°C on the inside. The boxes attain equilibrium during storage and are rested about 24 h before dispatch. The shipment room should be maintained at a temperature of 0°C, and poultry sold frozen is cooled to approximately −18°C. Storage and shipping temperatures have also been reported to be important determinants for moisture retention (Bigbee and Dawson, 1963; Thomson et al., 1966).

CONCLUSIONS

As the poultry industry has become more involved in processing, the importance of strict process control to maintain consistent meat quality is increasing. Therefore, basic operations and conditions for poultry processing are very important for obtaining good output. The important operations in poultry processing are receiving, stunning, slaughtering, bleeding, scalding, defeathering, evisceration, giblet harvesting, cutting, chilling, storage, and transportation. Receiving the bird at the slaughterhouse should be without transport stress and bruises. Although there are different methods of stunning, electrical stunning is the most widely used. Poultry should be processed under the appropriate temperature, humidity, and hygienic conditions. The bleed cut of the bird during slaughtering can be done manually or automatically. Bleeding time for broilers, turkeys, and waterfowl differs by species, and the USDA recommends a bleeding time range of 55 to 133 s. Scalding is done using hot-water troughs. Temperature control during scalding holds the key to carcass quality. Defeathering of the carcass is done using manual and automatic methods. Both manual and automatic drum pickers are used widely in poultry processing. The pinfeathers that survive the defeathering removed by a wax treatment. Manual evisceration is preferred over an automatic method, due to some of the limitations of the latter. Following evisceration the carcasses undergo giblet harvesting, cutting, and washing. Edible and inedible viscera are separated in giblet harvesting. The carcasses are cut into a variety of pieces according to demand. The washed carcasses are chilled quickly to retard the growth of microorganisms and are stored under recommended temperature and humidity conditions to retain the quality of the product.

REFERENCES

Anon. 1997. European perspective on poultry stunning. Poult Int 36(8):Poultry Processing Worldwide Suppl., 2 pp.

Bigbee DG, Dawson LE. 1963. Some factors that affect change in weight of fresh chilled poultry: 1. Length of chill period, chilling medium and holding temperature. Poult Sci 42:457–462.

Bilgili SF. 1992. Electrical stunning of broilers—basic concepts and carcass quality implications: a review. J Appl Poult Res 1:135–146.

Brant AW. 1963. Chilling poultry - A review. Poult Process Market 69(5):14.

Burton CH, Allen VM. 2002. Air-chilling poultry carcasses without chlorinated water. Poult Int 41(4):32, 34, 36, 38.

Childs RE, Walters RE. 1962. *Methods and Equipment for Evicerating Chickens*. USDA Marketing Research Report 549. Washington, DC: U.S. Department of Agriculture.

Chrystall BB, Devine CE. 1985. Electrical stimulation: its early development in New Zealand. In: *Advances in Meat Research*, vol. 1, *Electrical Stimulation*. Westport, CT: AVI Publishing, pp. 73–119.

Davis LL, Coe ME. 1954. Bleeding of chicken during killing operations. Poult Sci 33:616–619.

Drewniak EE, Bausch ER, Davis LL. 1955. *Carbon-Dioxide Immobilization of Turkeys Before Slaughter*. USDA Circular 958. Washington, DC: U.S. Department of Agriculture.

Essary EO, Dawson LE. 1965. Quality of fryer carcasses as related to protein and fat levels in the diet: 1. Fat deposition and moisture pick up during chilling. Poult Sci 44:7.

Fletcher DL. 1999. Symposium: recent advances in poultry slaughter technology. Poult Sci 78:277–281.

Fris Jensen J, Bøgh-Sørensen L. 1973. The effect of chinnilng on the nutritive and organoleptic quality of broiler meat. In: *Proceedings of the 4th European Poultry Conference*, London, 1972. Edinburgh, UK: British Poultry Science Ltd., p. 359.

Gerrard F. 1964. *Meat Technology*. Leonard Hill Publishing, pp. 243–254.

Gregory NG, Wilkins LJ. 1989. Effect of stunning current on carcass quality in chickens. Vet Rec 124:530–532.

Hale KK, Thompson JC, Toledo RT, White HD. 1973. *An Evaluation of Poultry Processing*. Special Report. University of Georgia College Experiment Station Committee on Poultry and the Poultry Processing Industry. Athens, GA: University of Georgia College of Agriculture.

Heath GBS. 1984. The slaughter of broiler chickens. World's Poult Sci J 40:151–159.

Hemm E, Childs RE, Mercuri AJ. 1967. Relationship between fats in broiler finisher rations and fats in chiller water from broiler processing. Poult Sci 8:23–33.

Hoen T, Lankhaar J. 1999. Controlled atmosphere stunning of poultry. Poult Sci 78(2): 287–289.

Keener KM, Bashor MP, Curtis PA, Sheldon BW, Kathariou S. 2004. Comprehensive review of *Camphylobacter* and poultry processing. Compr Rev Food Sci Food Saf 3:105–109.

Klose AA, Pool MF. 1954. The effect of scalding temperature on the quality of stored frozen turkeys. Poult Sci 33:280–289.

Klose AA, Mecchi EP, Pool MF. 1961. Observations on factors influencing feather release. Poult Sci 40:1029–1036.

Lesiow T, Kijowski J. 2003. Impact of PSE and DFD meat on poultry processing: a review. Pol J Food Nutr Sci 12/53(2):3–8, 56.

Mead GC. 2004. *Poultry Meat Processing and Quality*. Cambridge, UK: Woodhead Publishing, pp. 95–107.

Mountney GJ. 1976. *Poultry Products Technology*, vol. 2. Westport, CT: AVI Publishing, pp. 141–154.

Newell, GW, Shaffner CS. 1950. Blood loss by chicken during killing. Poult Sci 33:274–279.

Nunes FG. 1994. CO_2 stunning of broilers: a quality-boosting reality. Misset World Poult 10(11):17, 19–20.

Pippen EL, Klose AA. 1955. Effects of water chilling on flavor of chicken. Poult Sci 34:1139.

Pool MF, Mecchi EP, Lineweaver H, Klose AA. 1954. The effect of scalding temperature on the processing and initial appearance of turkeys. Poult Sci 33:274–279.

Potter NN, Hotchkiss JH. 1996. Meat, poultry and eggs. In: *Food Science*, vol. 5. New Delhi, India: CBS Publishers and Distributors, pp. 333–336.

Seemann G. 1981. The influence of age, sex and strain on yield and cutting of broilers. In: *Proceedings of the 5th European Symposium on the Quality of Poultry Meat*, Apeldoorn, The Netherlands, May 17–23, pp. 21–23.

Thomson JE, Mercuri AJ, Kinner JA, Sanders DH. 1966. Effect of time and temperature of commercial continuous chilling of fryer chickens on carcass temperatures, weight and bacterial counts. Poult Sci 45:363–369.

Touraille C, Ricard EH. 1981. Relationship between sexual maturity and meat quality in chickens. In: *Proceedings of the 5th European Symposium on the Quality of Poultry Meat*, Apeldoorn, The Netherlands, May 17–23, pp. 259–261.

USEPA (U.S. Environmental Protection Agency). 2002. *Meat and Poultry Products Industry Overview*. Washington, DC: USEPA.

Vranic V, Nedeljkovic L, and Veljic Z. 1991. Current methods for poultry meat chilling and their future improvements. Tehnol Mesa 32:109–114.

Walther JH. 1991. Minimizing product loss in the hang, stun and kill areas. In: *Proceedings of the 26th Poultry Health and Condemnations Meeting*, University of Delaware, Newark, DE, pp. 160–163.

Zenoble OC, Bowers JA, Cunningham FE. 1977. Eating quality and composition of spent hens processed with or without immersion chilling. Poult Sci 56:843.

Ziolecki J. 1990. How processing and storage affect carcass appearance. Poult Int 29:52, 54, 56.

INTERNET RESOURCES

http://www.fao.org/docrep/003/t0561e/T0561E05.html

http://ceplacer.ucdavis.edu/files/46823.pdf

http://www.jfequipment.com/upload/PoultryBulletin-August2007.pdf

http://www.knasecoinc.com/documents/systemsguide_2003.pdf

http://pubs.caes.uga.edu/caespubs/pubcd/b1156-w.html

http://www.hyfoma.com/en/content/food-branches-processing-manufacturing/meat-fish-shrimps/poultry-slaughtering/

28

POULTRY-PROCESSING EQUIPMENT

JOSÉ JORGE CHANONA-PÉREZ, LILIANA ALAMILLA-BELTRÁN, ERNESTO MENDOZA-MADRID, JORGE WELTI-CHANES, AND GUSTAVO F. GUTIÉRREZ-LÓPEZ
Departamento de Graduados en Alimentos, Escuela Nacional de Ciencias Biológicas, Instituto Politécnico Nacional, México D.F., México

Handbook of Poultry Science and Technology, Volume 2: Secondary Processing, Edited by Isabel Guerrero-Legarreta and Y.H. Hui

RAW POULTRY PROCESSING AND PRODUCTION SCALES

According to required production levels, processes have been classified into three categories: on farm, small scale, and large scale. Table 1 is a comparison of these production levels. In general, the three production levels are important in technological applications and commercial products. For the on-farm case, a growing number of small producers are raising poultry outdoors on pasture, processing the birds on the farm, and selling the meat directly to customers at the farm or at a farmers' market. In many countries it is permissible to process birds on the farm, and each year they are sold directly to consumers with no inspection. Some of these small producers go further and build government-licensed processing plants to supply regional or niche markets. Specialty "religious kill" is often done in small plants. *Kosher* is the term for Jewish slaughter and *halal* for Muslim slaughter (Fanatico, 2003). In many countries small-scale producers cover the internal demand for poultry products in local or rural markets. Consolidation in the meat-processing industry has left very few small plants that will do custom poultry processing. Large plants generally do not process for small producers, as they cannot keep track of a small batch of birds and are not able to make money on small-volume orders. A large plant's output is directed mainly to urban zones and exporting markets.

TABLE 1 Comparison of Poultry Process Types According to Production Level

	Production Level		
Feature	On-Farm	Small	Large
Size	Outdoor or shed facility	2000–3000 ft^2	150,000 ft^2
Equipment	Manual	Manual/mechanical	Fully automated
Cost	Less than $15,000	Less than $500,000	$25,000,000
Labor	Family	Family/hired	Hired
Capacity	50–100 birds per day	200–5000 birds per day	250,000 birds per day
Operation	Seasonal; 1–30 processing days per year	Seasonal or year-round; 50-plus processing days per year	Year-round; process daily
Marketing	Product sold fresh, sometimes frozen; whole birds	Fresh and frozen, whole and parts	Mainly cut-up, sold fresh, further-processed
Notes	Independent operation; labor-intensive; low-risk; usually not inspected, direct sales	Independent or part of a collaborative group; requires good markets and grower commitments	Part of an integrated operation, including grow-out, processing, and marketing

Source: Adapted from Fanatico (2003).

Poultry processing comprises three main segments or steps (Zeidler and Curtis, 2002):

1. *Dressing*. The birds are placed on a moving line, killed, and defeathered.
2. *Eviscerating*. Viscera are removed, the carcasses are chilled, and the birds are inspected and graded.
3. *Further processing*. The largest portion of the carcasses are cup up, deboned, and processed into various products, which are packed and stored chilled and frozen.

The most common commodity produced in poultry slaughterhouses is the whole bird, which may also be processed further into various products based on the type of poultry meat (e.g., from simple cuts to ready-to-eat meals). Figure 1 provides a simplified flow diagram of the processing of fresh poultry, which includes further-processed subproducts and service operations.

BASIC EQUIPMENT FOR RAW POULTRY PROCESSING

A simplified diagram of equipment for poultry processing on a small scale (capacity of 350 chickens/hour, building size of 115 m^2) is shown in Figure 2, and a simplified equipment layout for processing fresh poultry on a large scale is shown in Figure 3. According to Mountney and Parkhurst (1995), overall, considering on farm, small scale, and large scale, the main steps in processing fresh poultry are discussed below.

Assembling and Transportation

Assembling and transporting steps of poultry processing includes loading and carrying the broilers from farm to the plant. Nowadays, catching methods could be classified in various stages, including transporting birds from the house to trucks by vacuum lines, special conveyor systems, enlarging and redesigning coops and trucks, and totally mechanized systems for production, harvesting, and transporting birds (Silverside and Jones, 1998).

Weighing

Currently, bulk weighing is used almost exclusively in weighing birds. Transportation of birds (generally on trucks) includes weighing empty and loaded trucks at a public scale. Then the trucks proceed to the farm and the broilers are loaded into crates that are left on the trucks. The load of poultry is weighed again at the public balance nearest the processing plant (Mountney and Parkhurst, 1995).

Hanging

In the hanging step the loaded crates of broilers are moved off the trucks onto roller conveyors and pushed to the processing line. Several people are involved

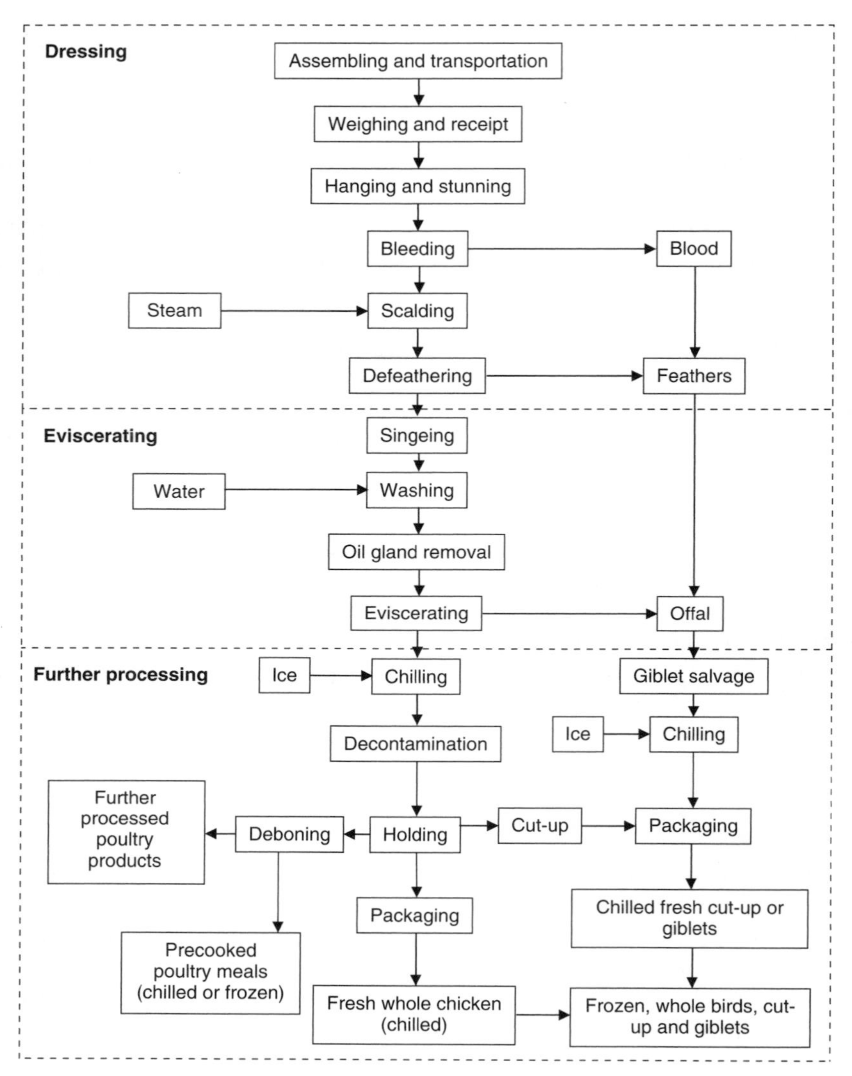

FIGURE 1 Simplified flow diagram for processing fresh poultry (processing, subproducts, and services). (Adapted from Zeidler and Curtis, 2002.)

in removing the birds from crates and shackling them for slaughter (Mountney and Parkhurst, 1995).

Slaughtering: Stunning and Bleeding

Slaughtering includes the stunning and bleeding operations. Often, birds are bled without stunning. An electrical shocker is frequently used as a stunning method. This operation prevents struggling and also relaxes the muscles that hold the

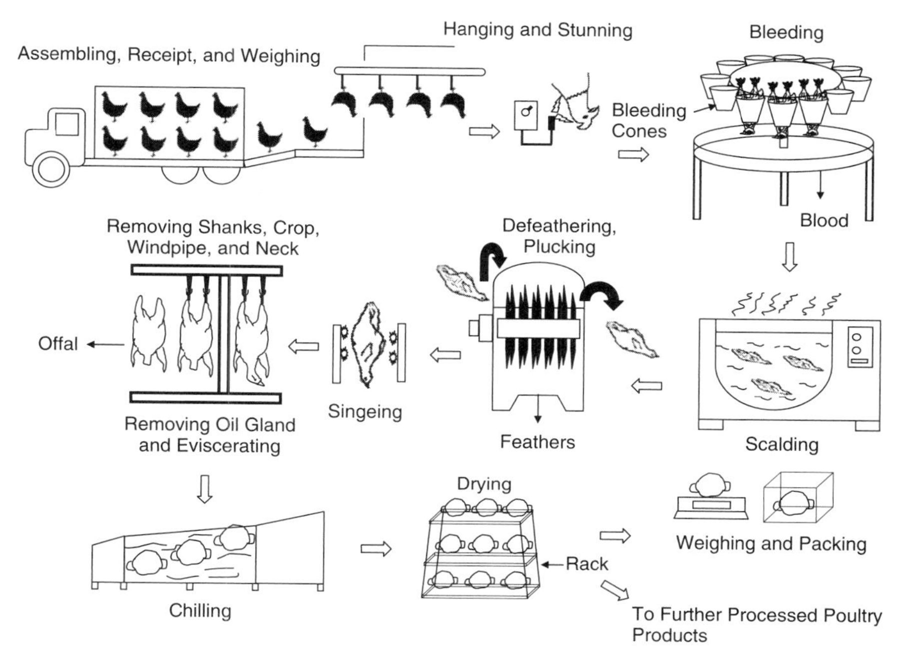

FIGURE 2 Equipment for small-scale processing of fresh poultry. (Adapted from Silverside and Jones, 1998.)

feathers. Another advantage of the stunning operation may be the bleeding reduction rate. In some systems, birds are removed from the crates and hung on an overhead shackle, where they are stunned by a low-voltage system before being placed in bleeding cones. Generally, to carry out the operation, a bleed knife is used to cut the blood vessels of the neck. Practice is required to use this technique (Silverside and Jones, 1998). Usually, there are three methods of cutting a bird's neck so it can bleed in a convenient manner. One method is *modified kosher*, in which birds are killed by severing the jugular vein just below the jowls so that the windpipe and esophagus remain uncut. Another method is *decapitation*; even though it can be considered a slaughter method, it is rarely used. A third method, *piercing the brain* of broilers, consists of severing the veins in the roof of the mouth. This method is essentially no longer used. Of the three methods of slaughter, modified kosher is the most widely used in modern processing operations, due to the fact that it is easier to obtain good bleeding and leaves the head and neck intact for use when suspending the carcass for later eviscerating operations. Birds should be allowed to bleed for $1\frac{1}{2}$ to 2 min before dressing is begun (Fanatico, 2003).

Scalding

On-farm processors use a single tank of hot water, usually scalding one to four birds at a time. Small processors remove the birds from the killing cones for

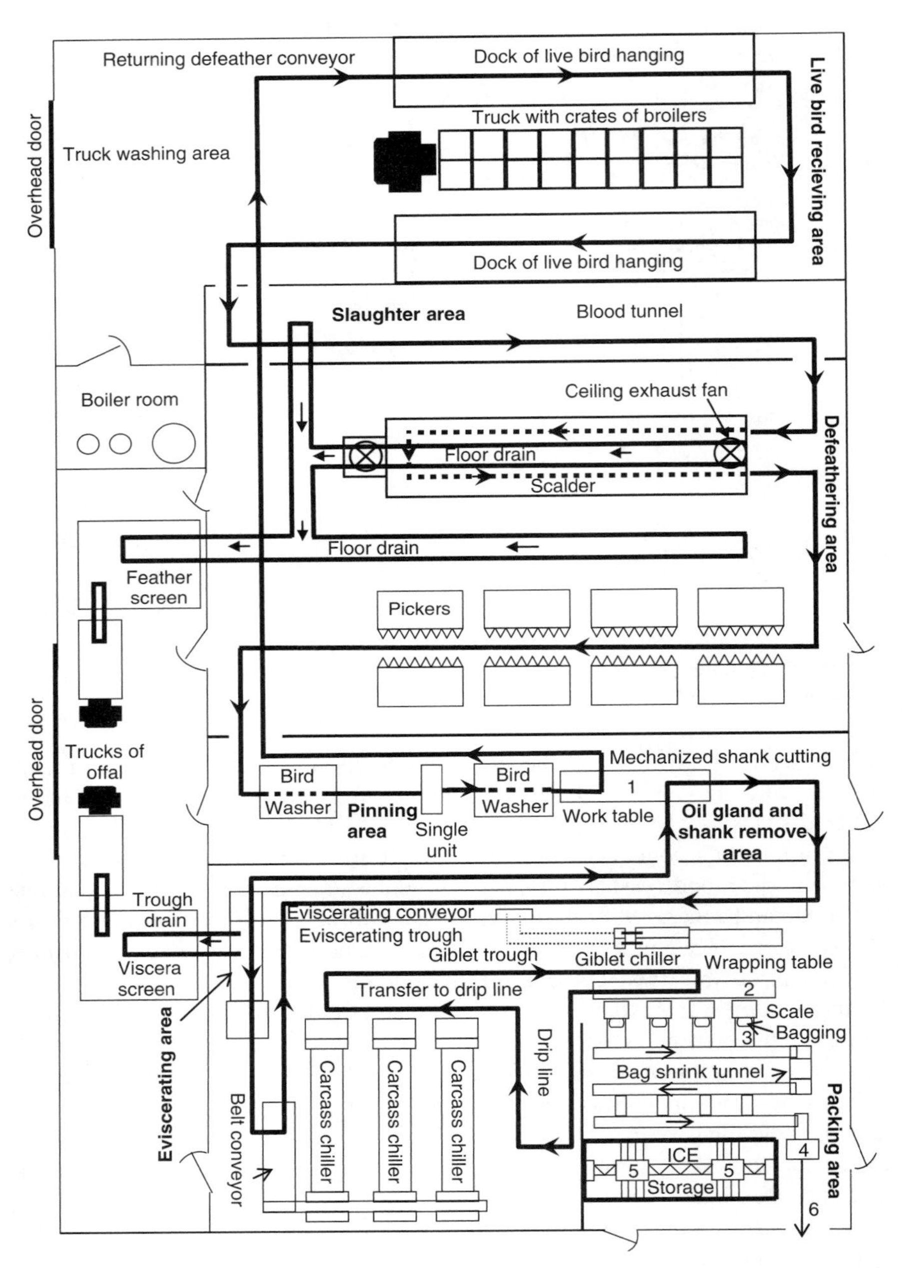

FIGURE 3 Equipment layout for processing fresh poultry. 1, Area of transfer from defeathering conveyor to evisceration conveyor; 2, transfer to scales; 3, bag vacuum; 4, on-line scale; 5, ice machine; 6, toward shipping dock. (Adapted from Zeidler and Curtis, 2002.)

scalding. In large plants, the birds stay on the shackles. The overhead rail moves the birds from the stunning/bleeding area and lowers them into a scalding tank. The speed of the rail ensures that the birds are scalded for the right length of time and are agitated sufficiently in the water (Fanatico, 2003).

There are three methods of scalding, classified based on the water temperature and residence time. In the hard-scalded method, broilers are immersed in water and heated from 71.1 to 82°C for 30 to 60 s. When the carcasses are scalded in water of 58.8 to 60°C for 30 to 75 s, the process is considered subscalding. Semiscalding, often called soft or slack scalding, is carried out at 50.5 to 54.4°C for 90 to 120 s. After the birds have passed through the scalding tank, they are removed from the overhead conveyors for plucking (Mountney and Parkhurst, 1995).

Defeathering

The quality of the pick is related to the scald. The force required to remove feathers from carcasses decreases as the scalding temperature increases; the scalding temperature is more important than the scalding time. If the scald water is too cool, the feathers won't loosen; if it is too hot, the skin will tear in the picker. But if it is just right, the feathers usually come out more easily and can even be removed by hand. However, hand picking is time consuming. To process a large number of birds, it is necessary to count with a mechanical picker. It is also possible to remove the feathers by abrasion; these machines can pick a bird and clean it in about 30 s (and sometimes break the wings). Some on-farm processors skin the birds instead of removing the feathers. A drum picker, a cylinder with rubber fingers around the exterior, defeathers one bird at a time. Sometimes small processors carry the birds to the picker. Large plants use continuous in-line pickers that look like a tunnel with rubber fingers (Fanatico, 2003; Zeidler and Curtis, 2002).

Singeing

After picking and pinning, carcasses are singed to remove hairlike appendages called *filoplumes*. Each carcass passes through a sheet of flame as it moves along the conveyor line (Mountney and Parkhurst, 1995).

Removing Shanks and Oil Glands

After feather removal, the heads, oil glands, and shanks are removed. This can be carried out in the dressing area, after which the carcasses are washed. These operations can be carried out simultaneously (Silverside and Jones, 1998). Shanks and heads can be removed by knives, saws, and manual or mechanical shears. It has been observed that the most efficient method of removing the oil gland is by suspending carcasses by the hocks rather than by the neck. With this method some skilled operators could achieve a rate of 40 birds per minute (Mountney and Parkhurst, 1995).

Evisceration

Methods of eviscerating poultry vary considerably not only among different areas and for different species of poultry, but also among different scales of processing. Eviscerating is a complex process that involves several operations, such as opening the body cavity and removing viscera, decapitating, processing giblets, and removing lungs, crop, windpipe, and neck, among others organs. Manual evisceration involves cutting around the vent, opening the body, and drawing out the organs. Inedible viscera or guts (e.g., intestines, esophagus, spleen, reproductive organs, lungs) are removed. The crop is loosened so that it will come out with the guts. The kidneys are difficult to remove, so are left inside the carcass. Instead of shackles, on-farm processors usually eviscerate on a flat surface (stainless steel for easy cleaning, or a disposable plastic sheet). On-farm processors and small plants eviscerate manually with scissors, knife, or a handheld vent-cutter gun with a circular blade, and draw out the guts by hand. Large plants use automated machines that scoop out the guts; high-speed lines eviscerate 2000 to 8000 birds per hour. These automated lines are usually designed for one specie, and uniformity in size is very important for proper operation. Descriptions of operations, methods, and eviscerating equipment are widespread; many detailed studies are available in the literature (Mountney and Parkhurst, 1995; Fanatico, 2003), where more aspects of evisceration can be reviewed.

Chilling

The carcass temperature must be lowered rapidly to prevent microbial growth. Soaking the carcass in chilled water is the most common method of chilling poultry. Poultry products may be chilled to -3.5°C or frozen to -3.5°C or less). Means of refrigeration include ice, mechanically cooled water or air, dry ice (carbon dioxide sprays), and liquid nitrogen sprays. Continuous chilling and freezing systems with various means of conveying the product are common (Zeidler and Curtis, 2002).

CHILLING EQUIPMENT

On-farm processors use large plastic tubs filled with cold water and ice. Sometimes they have two tubs, the first used to remove the initial body heat and the second to chill the carcass. Carcasses usually stay in the water for about 1 h. Small-plant processors use food-grade plastic or stainless steel bins filled with ice; a slush forms as the ice melts. Water chilling is used in large plants. Carcasses are removed from shackles and put in large chill tanks filled with cold water. About $\frac{1}{2}$ gallon of water is required per carcass for the initial tank of water (makeup water).

Some chillers hold more than 300,000 gallons of water. They are either a through-flow type with paddles or rakes, or a countercurrent type with augers to move birds. In consequence, several types of chilling equipment can be used to reduce the carcass temperature to the levels required for packing. The most important equipment in chilling operations are continuous-immersion slush ice chillers, which are fed automatically from the end of the evisceration conveyer line and have replaced slush ice chilling tanks, a batch process. In general, tanks are used only to hold chilled carcasses in an iced condition prior to cutting up, or to age prior to freezing (Fanatico, 2003). The following types of continuous chillers are used in poultry processing:

- *Continuous slush ice chiller*. The simplest chiller configuration is the slush ice chiller; this equipment is illustrated in Figure 4, where the carcasses are pushed by a continuous series of power-driven rakes.
- *Continuous drag chiller*. Suspended carcasses are pulled through containing agitated cool water and ice slush. Typical drag poultry chillers are shown in Figure 5. Their counterflow design moves birds gently in the opposite direction of water flow, creating efficient chilling by ensuring that the coldest birds stay in constant contact with the coldest water. The configuration of rodded paddles allows adequate water flow and improves the heat transfer (Morris and Associates, n.d.).

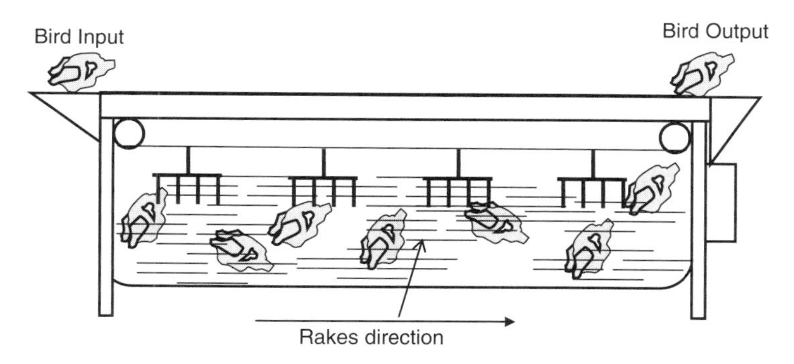

FIGURE 4 Slush ice chiller.

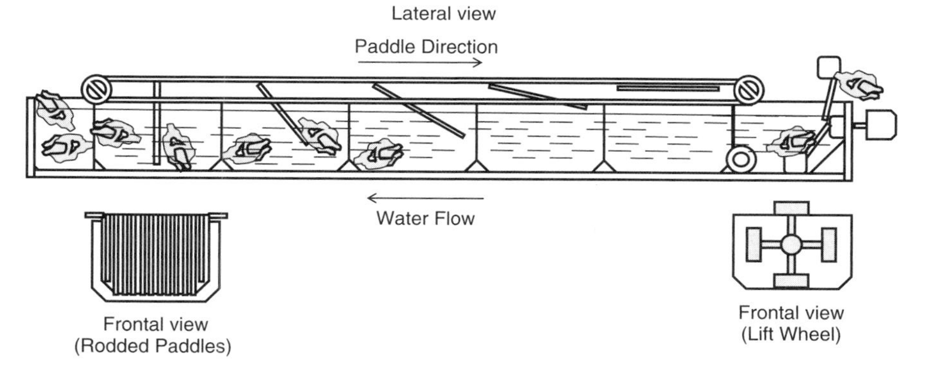

FIGURE 5 Drag poultry chiller (counterflow).

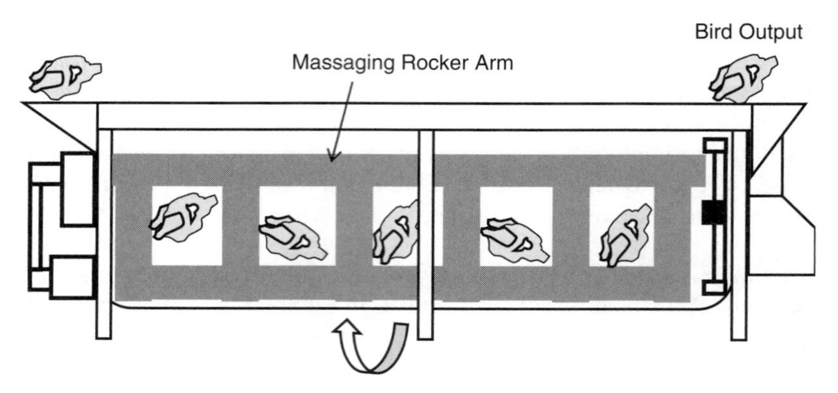

FIGURE 6 Poultry chiller rocker vat system.

Vat Systems

Carcasses are conveyed by the recirculating water flow, and agitation is accomplished by an oscillating, longitudinally oriented paddle. Carcasses are removed from the tanks automatically by continuous elevators (Morris and Associates). A chiller-type rocker vat system is shown in Figure 6.

Auger Chillers

Concurrent Tumble Systems Figure 7 is a diagram of a concurrent tumble system or auger chiller. In this equipment, free-floating carcasses pass through horizontally rotating drums suspended in successive tanks of cool water and ice slush. Movement of the carcasses is regulated by the flow rate of recirculated water in each tank (Zeidler and Curtis, 2002). The design of this type of equipment can increase the water flow and lower the overall average water temperature for a more consistent product.

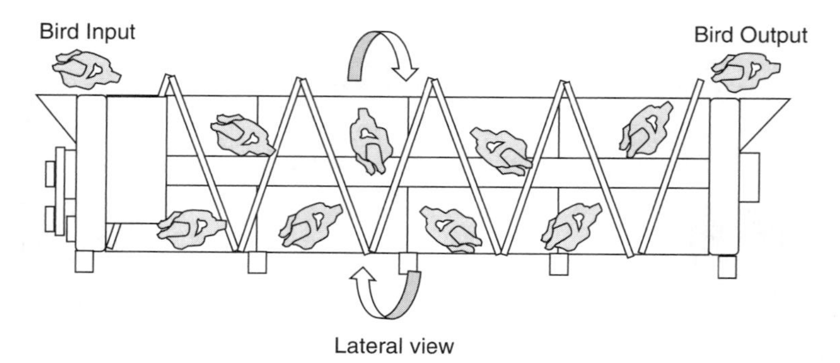

FIGURE 7 Auger poultry chiller.

Counterflow Tumble Chillers A counterflow tumble chiller is essentially the same as a concurrent tumble system except that the water circulates opposite to the direction of the carcasses. The carcasses are carried through tanks of cool water and ice slush by horizontally rotating drums with helical flights on the inner drum surfaces.

Other chilling methods are water-spray chilling, air-blast chilling, carbon dioxide snow, and liquid-nitrogen spray. These methods are alternatives for bacterial contamination reduction in carcasses by chilling immersion. However, these chilling methods have some limitations, as liquid water has a much higher heat transfer coefficient than any gas at the same temperature; consequently water immersion is more rapid and efficient than gas chilling. Water-spray chilling requires larger amounts of liquid than are required by immersion chilling (Reid, 1993; Heber et al., 2000). Air-blast, carbon dioxide, and nitrogen chilling could promote surface dehydration. In consequence, air chilling without packaging could cause from 1 to 2% of moisture loss, while water immersion chilling permits 4 to 15% moisture uptake (Zeidler and Curtis, 2002).

Air or Gas Chilling

Air chilling of poultry is commonly practiced in Europe, Canada, and Brazil. Air chilling takes longer than water chilling, usually at least 2 h. As an example of this air-chilling equipment, a one-tiered evaporative air chiller is shown in Figure 8. In air-blast chilling and evaporative chilling, heat is conduced partly by air-to-carcass contact and partly by evaporation of moisture from the carcass surface. Air chilling takes place in an insulated room or tunnel in which the temperature is kept between 20 and 35°F by coolers in the ceiling. Air is blown from nozzles directly into the cavity of each bird or around it. An overhead track conveys the carcasses into the room. It saves labor to keep the birds on the shackles for chilling since there is no need to handle them, but sometimes they are removed and put in baskets or on racks. To prevent the formation of an

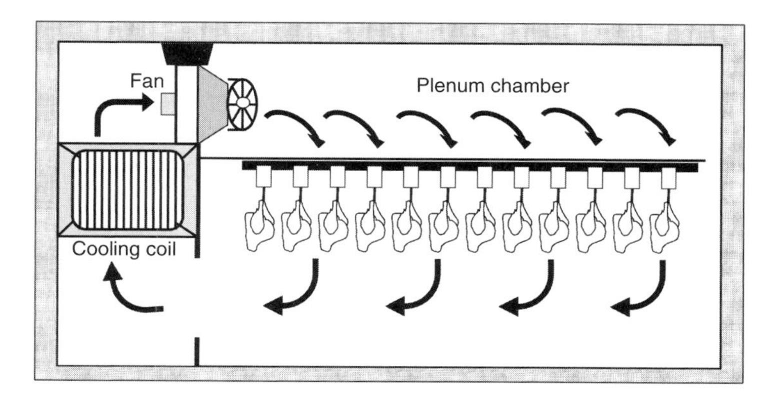

FIGURE 8 One-tiered evaporative air chiller.

upper layer of carcasses from dripping on the lower layer, the birds are usually not stacked. Heightened humidity or a water spray prevents the carcass from drying out. Evaporative chilling is a type of air chill in which water is sprayed on the carcass; water absorbs heat during evaporation. Air-chilling equipment requires more space and uses more energy than water-chilling equipment and costs more; however, the water use is low. Both types of chilling are effective; the choice depends on water availability, the market, and other factors (Heber et al., 2000; Zeidler and Curtis, 2002).

Spray Chilling

Spray chilling involves atomization of cold water on the carcasses. This method has the disadvantage of consuming more water than immersion chilling and has the characteristic of having lower heat transfer efficiency than immersion chilling, but has the advantage that microbes transfer between carcasses is unlikely. If the surface of a carcass freezes as part of the chilling process, the bacterial load may be reduced as much as 90% (Zeidler and Curtis, 2002).

Cryogenic Chilling

In cryogenic chilling, the heat transfer medium is nitrogen or carbon dioxide, which is liquefied in large plants and shipped to poultry-processing plants at low temperature in well-insulated pressure vessels. Cryogenic chillers are generally used in long insulated tunnels through which the carcasses are conveyed on a continuous belt. Figure 9 illustrates a typical example of cryogenic chilling. In a liquid-nitrogen chiller, nitrogen is sprayed into the chiller and evaporates from the carcasses. Cryogenic chilling can be found as straight belt, multitier, spiral belt, and immersion design. Key attributes such as high heat transfer rate, low investment costs, and rapid installation and startup are especially attractive for chilling operations (Reid, (1993, 1998)). Some freezing of the outer layer (crust freezing) usually occurs and the temperature is allowed to settle. Some plants use a combination of continuous water immersion chilling to reach 2 to 5°C and a cryogenic gas tunnel to reach −2°C (Zeidler and Curtis, 2002). In another method, liquid carbon dioxide is normally stored under pressure; when the liquid is released to the atmosphere, 50% becomes dry-ice snow and 50% vapor, both at −70°C. As a result of these unusual properties of carbon dioxide, the chiller design can vary widely (Reid, 1993).

PACKAGING

Before packaging, other operations are carried out, such as drying, weighing, and inspecting. After a carcass is chilled properly, it is ready to be packed. In other cases, when a carcass is not to be packed as a whole carcass, it can be used in other operations, called further-processed poultry (Figures 1 and 3). Packing of

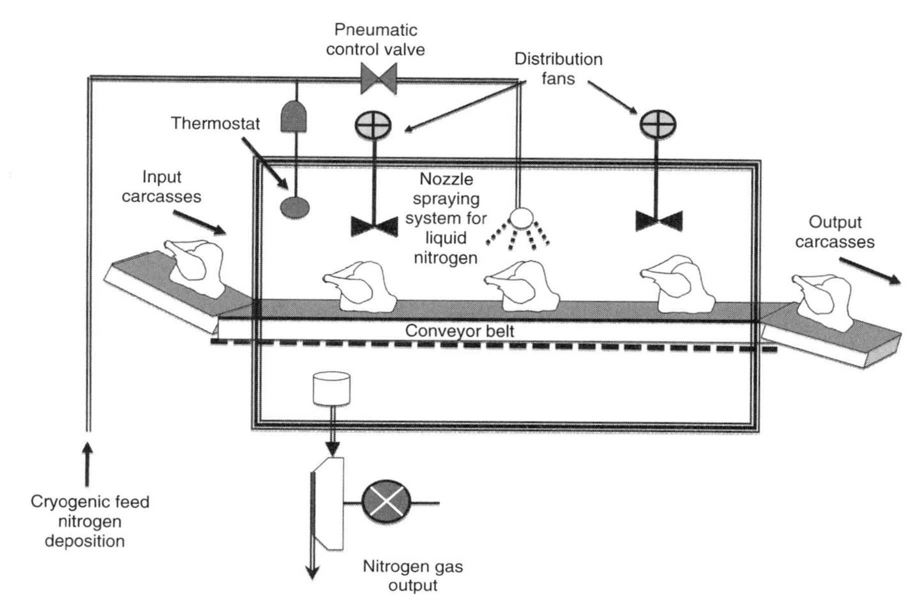

FIGURE 9 Cryogenic chiller.

poultry products is done for both aesthetic and utilitarian proposes (Reid, 1998). Packing provides a number of functions in addition to purely utilitarian ones and also helps to assure consumers of high-quality products. Utilitarian functions derived from packaging are an assemblage of a number of small units into one larger, easier-to-handle unit protected from physical damage, dehydration, oxygen and other gases, and protection from odors, microorganisms, dirt, filth, insects, and other contaminants.

Changes in packing methods and materials are so rapid that the best sources of information on this subject are the companies that fabricate films and packages and distribute the materials. Most packaged poultry is now tray packed for either frozen or chilled distribution. All-plastic packages and automated packaging lines using plastic film have been engineered. The majority of chilled product poultry is packaged as a whole carcass, cut-up parts, deboned, or ground at the processing plant. Individual portions are not only cut up and wrapped, but each individual package is weighed, priced, and printed with the store's label and bar code for automated checkout.

Several types of packages are commonly used for handling chilled poultry: wooden wire-bound crates, corrugated paraffin-lined containers, polyethylene-coated cardboard containers, and plastic containers. Other packages in common use are plastic and wax-impregnated fiberboard containers. Sometimes pads are used in the bottom to absorb moisture. Overwraps such as films, paperboard cartons, and carton liners are used for tray packing. Various transparent films are used to prepackage poultry: among them Mylar, a polyester film; CryOvac;

L, a shrinkable irradiated polyethylene film; Saran S, a poly(vinylidene chloride); Pliofilm, a rubber hydrochloride; and cellophane (Mountney and Parkhurst, 1995).

For packing frozen poultry, different packaging could be used, such as plastic bags, which generally are heat shrinkable; waxed cardboard boxes overwrapped and heat-sealed with waxed paper or cellophane; or fiber boxes laminated with aluminum. Microwave and aluminum containers and edible coatings are also used as packaging.

FURTHER PROCESSING

Further processing includes not only cut-up and deboned (see Figure 10) products but also portioned, formed, cooked, cured, smoked, and brined products. Formed products are made by reducing the particle size of the meat, adding ingredients for flavor or functionality, tumbling to increase brine penetration, and forming with a stuffer or mold. Some products are also breaded and cooked. Curing and smoking are ancient ways of preserving meat that also contribute to flavor. Curing uses nitrites as a preservative. Smoking can be done without nitrites (Fanatico,

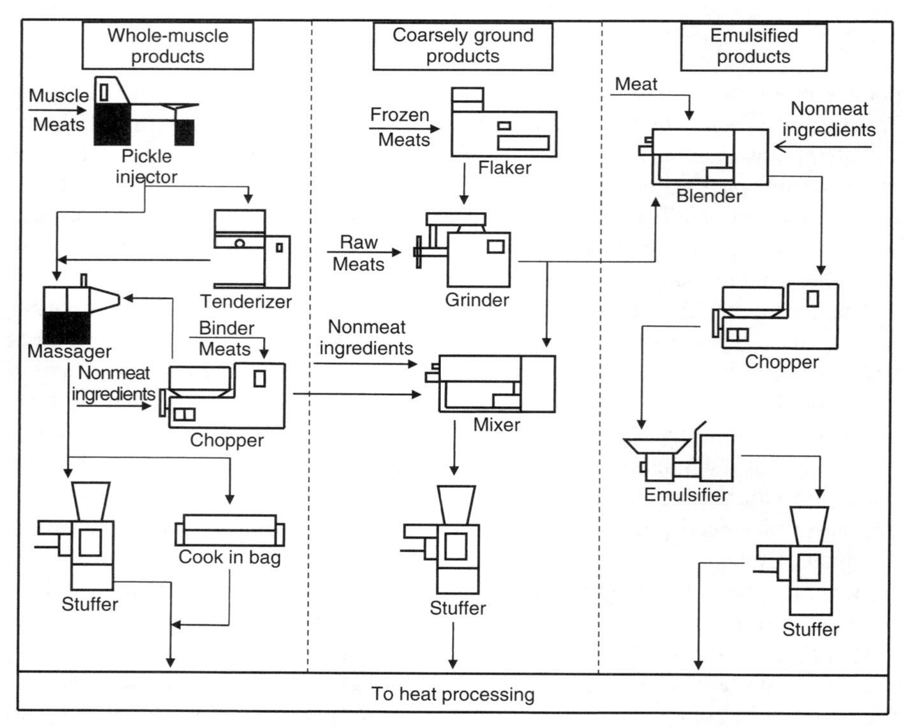

FIGURE 10 Meat product processing flowchart. (Adapted from Zeidler and Curtis, 2002.)

2003). Formed products or poultry meat products include the following three basic types (Zeidler and Curtis, 2002):

1. *Whole-muscle products,* such as nuggets, patties, Buffalo wings, and schnitzels. These have pieces that can still be recognized as meat.
2. *Coarsely ground products,* such as ground poultry meat, deli rolls, loaves, and meatballs. The pieces of meat have been chopped and are smaller; breast meat or deboned meat and skin are used.
3. *Emulsified products,* such as hot dogs, sausages, and bologna. For these products the pieces of meat are very small and when mixed with fat and water may not be recognizable as meat.

Figure 10 is a simplified flowchart to describing the elaboration of the products listed above. Various unit operations can be used to further processing. The following types of equipment used for further processing of poultry products are also used in red meat facilities (Zeidler and Curtis, 2002):

- *Size reduction and mixing machines*. The following unit operations could be used: grinding, flaking, chopping, mixing and tumbling, and injection.
- *Shaping forms and dimension*. These machines establish the form, size, and desired mass of size-reduced poultry meats, and the unit operations that could be used are stuffing forming, molding, and coating.

Cooking Techniques Many meat products are produced as ready-to-eat meals that only need warming or are eaten cold. These products are fully cooked in the plant by various methods. Other products are produced as ready-to-cook and so skip the cooking step at the plant. The unit operations that could be used are smoking/cooking, continuous hot-air cooking, cooking in a water bath, frying, microwave heating, and rotisserie roasting.

IMAGE PROCESSING FOR POULTRY PRODUCT INSPECTION

In the poultry industry, some quality evaluation are still performed manually by trained inspectors, but this is tedious, laborious, costly, and inherently unreliable, due to its nature. Increased demands for objectivity, consistency, and efficiency have necessitated the introduction of computer-based image-processing techniques (Cheng-Jin and Da-Wen, 2003). Machine vision is a noninvasive technology that provides automated production processes with vision capabilities when the majority of inspection tasks are highly repetitive, and their effectiveness depends on the efficiency of the human inspectors. A number of investigators have demonstrated various applications using machine vision techniques for the agricultural and food industries, particularly in grading and inspection (Chung-Chieh et al., 2005). In all machine vision systems, image acquisition is the main step to features evaluation of food material. A very intensive field of research in image acquisition is the development of sensors. In recent years there have been attempts to develop nondestructive, no-invasive sensors for assessing the composition and quality of food products. Various sensors, such as charge-coupled-device (CCD) cameras, ultrasound, magnetic resonance imaging (MRI), computed tomography (CT), electric tomography (ET), and x-ray, are used widely to obtain images of poultry products. For example, CCD cameras have been used to classify poultry carcasses (Park et al., 2002). Also, ultrasound techniques were applied to the measurement of breast meat and the evaluation of carcass merit (Grashorn and Komender, 1990; Cheng-Jin and Da-Wen, 2003). MRI instruments have been used in the estimation of poultry breast meat yield (Davenel et al., 2000). Examples of CT applications are broiler measurements of in vivo breast meat amount yield (Cheng-Jin and Da-Wen, 2003) and detection of bone fragments in deboned poultry (Tao and Ibarra, 2000). X-ray imaging has been used for many years to find foreign bodies; a source of x-rays passes through the product, moving on a conveyor belt to a sensor underneath the conveyor belt, which converts the x-ray signal into a digital signal. This digital signal corresponds to the x-ray absorption image of the product, and it is this image that is processed to make a decision as to whether or not the product contains a foreign object. A schematic diagram of a basic x-ray imaging system is shown in Figure 11; this system has been used to inspect poultry products. Graves (2003) has reported an automatic inspection system for poultry meat using x-rays coupled to machine vision. This BoneScan machine system is used specifically for bone detection in poultry products.

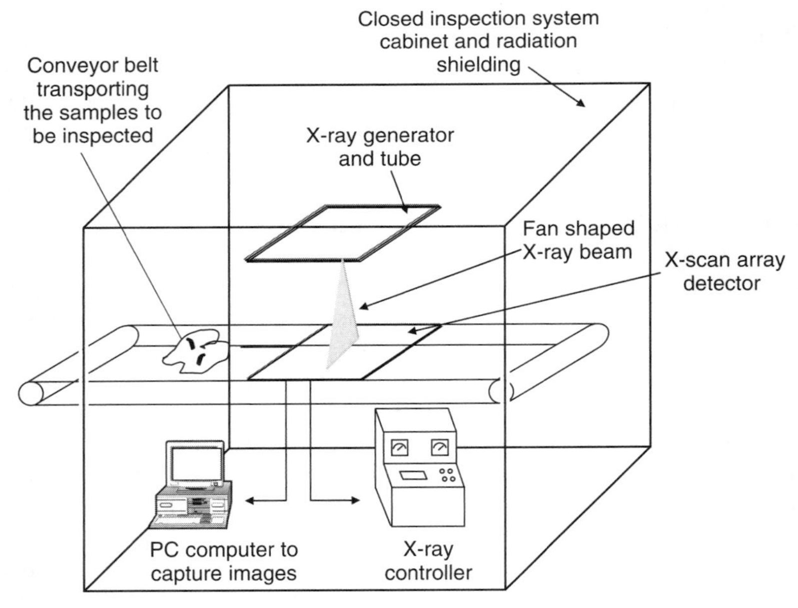

FIGURE 11 X-ray imaging system. (Adapted from Graves, 2003.)

REFERENCES

Cheng-Jin D, Da-Wen S. 2004. Recent developments in the applications of image processing techniques for food quality evaluation. Trends Food Sci Technol 15: 230–249.

Chun-Chieh Y, Kuanglin C, Yud-Ren C. 2005. Development of multispectral image processing algorithms for identification of wholesome, septicemic, and inflammatory process chickens. J Food Eng 69: 225–234.

Davenel A, Seigneuring F, Collowet G, Rémignon H. 2000. Estimation of poultry breast-meat yield: magnetic resonance imaging as a tool to improve the positioning of ultrasonic scanners. Meat Sci 56: 153–158.

Fanatico A. 2003. *Small Scale Poultry Processing*. ATTRA Publication IP231. National Center for Appropriate Technology. http://attra.ncat.org/attra-pub/PDF/poultryprocess.pdf.

Grashorn MA, Komender P. 1990. Ultrasonic measurements of breastmeat. Poult Int 29: 36–40.

Graves M. 2003. X-ray bone detection in further processed poultry production. In: *Machine Vision for the Inspection of Natural Products*. New York: Springer-Verlag, pp. 421–449.

Heber J, Löndahl G, Persson P, Rynnel L. 2000. Freezing systems for the food industry. In: Francis FJ, ed., *Encyclopedia of Food Science and Technology*, vol. 2, 2nd ed. New York: Wiley, pp. 1121–1137.

Morris and Associates, Inc. (n.d.). http://www.morris-associates.com/page/poultry_industry.

Mountney JG, Parkhurst RC. 1995. *Poultry Products Technology*, 3rd ed. New York: Food Products Press, pp. 171–186.

Park B, Lawrence KC, Windham WR, Chen YR, Chao K. 2002. Discriminant analysis of dual-wavelength spectral images for classifying poultry carcasses. Comput Electron Agric 33: 219–231.

Reid D. 1993. Basic physical phenomena in the freezing and thawing of plant and animal tissues. In: Mallett CP, ed., *Frozen Food Technology*. London: Blackie Academic & Professional/Chapman & Hall, pp. 1–19.

Reid D. 1998. Freezing preservation of fresh foods: quality aspects. In: Taub IA, Singh P, eds., *Food Storage Stability*. Boca Raton, FL: CRC Press, pp. 387–397.

Silverside D, Jones M. 1998. *Small-Scale Poultry Processing*. Rome: FAO. http://www.fao.org/docrep/003/t0561e/t0561e00.htm.

Tao Y, Ibarra JG. 2000. Thickness-compensated x-ray imaging detection bone fragments in deboned poultry-model analyis. Trans ASAE 44: 1005–1009.

Zeidler G, Curtis PA. 2002. Poultry products. In: *2002 ASHRAE Refrigeration Handbook*. Atlanta, GA: American Society of Heating, Refrigerating, and Air-Conditioning Engineers, pp. 17.1–17.14.

29

THERMAL PROCESSING

ISABEL GUERRERO-LEGARRETA
Departamento de Biotecnología, Universidad Autónoma Metropolitana, México D.F., México

Y.H. HUI
Science Technology System, West Sacramento, California

Handbook of Poultry Science and Technology, Volume 2: Secondary Processing, Edited by Isabel Guerrero-Legarreta and Y.H. Hui

INTRODUCTION

Heat treatment is probably the cheapest and most efficient preservation method used for poultry meat and products. The main objective of heat treatment is to ensure the destruction of microorganisms present in the substrate and to inactivate enzymes that promote deterioration. Depending on further processing and the expected shelf life, heat treatments have a severity rate, usually calculated according to the microorganism most likely to colonize the food or the most dangerous strain from a sanitation point of view. Heat treatment is also related to sensory attribute improvement, or to develop specific physical properties, such as in luncheon meats or in sausages to change from a semifluid to a solid material during canning due to gelation.

HEAT TRANSFER MECHANISMS

Thermal processing is basically an operation in which heat flows from a hot element, the heating medium, to a cold element, the food. As it is a dynamic process, heat flux is proportional to the driving force and inverse to the flow resistance. Heat transfer obeys one of the following mechanisms: conduction, convection, or radiation.

When heat is transferred by *conduction*, it is transported within a body due to molecular vibrations, following Fourier's law:

$$q = k\frac{A\Delta T}{L} \tag{1}$$

where A is the area, ΔT the temperature difference, L the material thickness, and k the thermal conductivity of the material (food). This mechanism occurs in solids, such as pieces of canned food or batters gelling inside a can (Mittal and Blaisdell, 1984). Conduction heating also depends on thermal conductivity of the material (k). In foods, k is very low; Mittal and Usborme (1985) reported that, on average, for meat $k = 1.464$ kg · cal/h · m^2. Due to the fascicular structure of the striated muscle, conductivity in meats also depends on the heat flow direction; if it is perpendicular to the muscle fibers, it is 1.72 kJ/h · m · K at 78% relative humidity and 0°C; whereas at the same conditions, if the flow is parallel to the muscle fibers, conductivity is 1.76 kJ/h · m · K (Pérez and Calvelo, 1984).

Siripona and others (2007) reported that the average thermal conductivity for white and dark poultry meat is 0.5093 and 0.4930 W/m · K, respectively; the average specific heat values of white and dark poultry meats are 3.521 and 3.654 kJ/kg · K, respectively. Note the difference between these values and those for stainless steel, where $k = 59.47$ kJ/h · m · K (Green and Maloney, 1997).

Convection occurs in fluids; its driving force is movement due to density differences when a fluid is heated or cooled. It follows Newton's law:

$$q = hA\Delta T \tag{2}$$

where A is the area, ΔT the temperature difference, and h depends on the flow properties, surface type, and flow velocity of the heating medium. h varies widely; for gases in a natural convection regime $h = 2.5$ to 25 kcal/h $\cdot$ m^2 $\cdot$ K; for water in a forced-convection regime $h = 500$ to 5000 kcal/h $\cdot$ m^2 $\cdot$ K; for condensing steam $h = 5000$ to 15,000 kcal/h $\cdot$ m^2 $\cdot$ K (Green and Maloney, 1997).

Heat transfer in food canning, such as in soups, sauces, and brines, occurs by this mechanism. The heat flow direction is from the heating medium (hot water or steam) through a barrier (the can) to a cold fluid within the can (the food). Heat diffusion is faster if an external force is applied, such as can rotation, decreasing the temperature difference to a minimum (Welti-Chanes et al., 2003).

In some products, the heat mechanism changes during heating from convection to conduction, due to changes in flow properties. For example, in canned luncheon meats, based on meat emulsions, a gel forms once the food inside the can is heated; therefore, the heating rate varies.

In the *radiation* mechanism, heat is transmitted by electromagnetic waves from a hot body and absorbed by a cold body. Although this mechanism is seldom used in food processing, it is commonly used in food preparation, just before consumption, in homes, hotels, restaurants, and so on. Infrared and microwave are radiation systems; waves are absorbed by food and transformed into heat.

THERMAL PROCESSING PARAMETERS

As stated before, a heat treatment's first aim is to destroy pathogens, spoilage microorganisms, and enzymes. In fact, theoretical considerations for microbial destruction are also valid for enzyme inactivation. In general, the strict anaerobe *Clostridium botulinum* is taken as the target microorganism due to its pathogenicity; however, other target microorganisms are *Bacillus stearothermophilus, B. thermoacidurans, B. macerans*, and *B. polymyxa* (Guerrero-Legarreta, 2004), in addition to specific pathogens most likely to colonize a specific food. In the case of raw poultry meat and poultry products, these are *Clostridum prefringens, Salmonella* spp., *Staphylococcus* spp., and *Campylobacter* spp. Microbial inactivation calculations are based on how long a food's shelf life must be extended. The main criteria for thermal destruction are:

1. All spores and viable cells able to growth and produce toxins must be eliminated, taking as a calculation basis *C. botulinum*, the most dangerous microorganism from the public health point of view, which also produces a relatively thermostable toxin.
2. Spoilage microorganisms must be reduced to a limit that ensures food quality for a given time. From a commercial point of view, a food can be considered sterile if it is free of *B. stearothermophilus* or *C. perfringens*.

Sporulated thermophiles are also of consideration if the food will be stored at high temperatures; this is the case of tropical preserves (Manev, 1984). Heat treatment conditions that destroy *C. botulinum* and *Clostridium sporogenes* result in a thermostable food with considerably long shelf life and without the need of another preservation process. Inactivation of either pathogen or spoilage-causing microorganisms is calculated by heat penetration. Vegetative cells are destroyed at temperatures slightly higher than optimum growth temperatures, whereas spores can survive at higher temperatures (Zamudio, 2006).

Traditionally, process calculations assume that as heat applications involve the destruction of at least one microbial enzyme, vegetative cells and spores are inhibited according to a first-order reaction rate equation (Baranyi and Roberts, 1995) even though Han (1975) and Peleg (2006) stated that there is evidence that bacterial spore inactivation, including *C. botulinum* spores, does not follow first-order kinetics. Peleg (2006) asserts that the exponential inactivation rate depends on the spores' previous thermal history, which is not considered in the exponential inactivation rate equations, which follow a log-linear Arrhenius model. However, the author concluded that canning operations are generally a safe procedure, due to overprocessing.

To calculate thermal process parameters, it is necessary to take into account the food's chemical composition as well as the initial microbial population and its heat sensitivity. The final microbial load expected and further storage must also be considered. The main thermal processing parameters are the following (Guerrero-Legarreta, 2006a):

- *D-value*. Microbial destruction follows a log-linear rate; 90% of the microorganisms are destroyed in a given time at constant temperature. This interval, specific for each microorganism, is called *decimal reduction time* or *D*-value; it represents the number of minutes necessary to destroy 90% of a given microbial species at a given temperature. For instance, $D_{110^\circ C}$ for *C. sporogenes* (i.e., to reduce 10^5 to 10^4 cells if heated at 110°C) is 10 min ($D_{110^\circ C} = 10$ min) (ICMSF, 1980).
- *z-value*. The *z*-value is the temperature increase necessary to obtain a $\frac{1}{10}$ reduction in *D*-values. For example, the *z*-value for *C. botulinum* type A is 10°C; this means that the same destruction is achieved at 131°C in 0.02 min, at 121°C in 0.2 min, and at 111°C in 2 min (ICMSF, 1996). *z*-Values increase with microbial heat resistance. Known *z*-values are used to compare heating times at a given temperature.
- *F-value*. The *F*-value is sum of all the destructive effects acting on a microbial population. *F*-values make it possible to compare thermal treatments among different foods. The relationship between *D* and *F*, taking the initial and final microbial cell concentrations into consideration, is

$$F = D(\log a - \log b) \tag{3}$$

where *a* is the initial microbial cell concentration and *b* is the final microbial cell concentration.

Because commercial sterility must be achieved in every part of a food product, process evaluation is based on the point that takes the longest time to reach the process temperature, that is, the *cold point*. In conduction mechanisms, the geometric center is also the cold point. In convection mechanisms it is located along a vertical axis, around one-third of the distance from the container's bottom end (Guerrero-Legarreta, 2001). However, Siripona et al. (2007) calculated the slowest heating point and optimum cooking time of whole chicken cooking in hot water at different temperatures and reported that the temperature of chicken did not significantly affect the thermal properties. As the F-value is the effect of every part of the process, a simple but precise method to calculate the lethal effect during heating and cooling operations is to record with thermocouples the temperature in the cold point of a food or food container, and calculate the corresponding F-value. The addition of all F-values will be the total F-value for the process (Guerrero-Legarreta, 2006b).

MICROBIAL INACTIVATION

To calculate a specific process or to modify process parameters for a given product, information on the heat resistance of the most abundant microorganism or the most likely microorganism to contaminate the food is needed. It is also important to know the temperature history of the meat, from slaughter to reception in the processing plant, due to native microflora that may proliferate under improper or temperature abuse conditions. To calculate the heat process severity required for a product, it is also necessary to define the shelf life required and the conditions under which the product will be handled. For example, half-preserves, stored for up to 12 months at less than 10°C require only $F_c = 0.4$; in this case, nonsporulated *Streptococcus faecium, S. faecalis*, and psychrophiles, and *Bacillus* and *Clostridium* spores are destroyed. At the other extreme, $F_c = 12$ is required for tropical preserves to remain safe and edible for up to one year at 40°C; in this case, a process that inactivates sporulated thermophiles, such as *Bacillus* and *Clostridium*, is needed (Manev, 1984; Stiebing, 1992). Populations of *C. botulinum* types A and B are the calculation basis for D-values at 121.1°C and 0.21 min. Under the $12D$ concept, foods should be heat treated at $F = 2.5$, called *botulinum heating*, to ensure that *C. botulinum* is practically absent (Mathlouthi, 1986).

The most frequently found pathogens in poultry meat and products, and their main characteristics, are:

- *C. perfringens* contaminates poultry meat and meat products, especially stews, gravies, and pies. This organism is found in the waste of animals and humans and often in raw meat and in soil. It thrives in airless conditions and survives ordinary cooking (Christchurch City Council, 1998). The D-value for *C. perfrinegns* is 47.40 to 57.58 min at 55°C (Juneja, 2006).
- *Salmonella* spp. contaminate meat and meat products, especially poultry. *Salmonella* is often present in the waste of humans and animals (especially rodents and poultry) (Christchurch City Council, 1998). Its D-value is 15.5 min at 55°C (Murphy et al., 2003).

- *Staphylococcus* spp. contaminate moist protein foods, primarily meat, poultry, eggs, and fish products. These bacteria may come from infected sores, nasal secretions, and skin (perspiration and hair). The toxins survive ordinary cooking (Christchurch City Council, 1998). The *S. aureus* D value at 55°C is 13 to 21.7 min (Kennedy et al., 2005).
- *Campylobacter* spp. contaminate meat and meat products, especially poultry. Bacteria are often present in the waste of humans and animals (especially domestic animals and poultry) (Christchurch City Council, 1998). This illness is infectious and can be spread to other people. The D-value at 55°C is 2.12 to 2.25 min (Food Safety Authority of Ireland, 2007).

ENZYME INACTIVATION

Another objective in heat processing is enzyme inactivation. It depends on the same factors as those affecting the microbial inactivation rate, as it depends on the destruction of at least one enzyme involved in any metabolic pathway. The protein moiety is denatured by heat, affecting the secondary and tertiary protein structure. However, some isoenzymes are heat resistant and can cause off-odors or off-flavors. Heat-processing calculations are carried out considering the most resistant enzyme. In general, enzyme inactivation and spore destruction take place at the same time between 130 and 145°C (Dziezak, 1991).

EFFECT ON SENSORY CHARACTERISTICS

In addition to producing a safe, shelf life–extended food free of pathogens, heat processing is aimed at improving sensory characteristics and digestibility. Although heat treatments are of various severity levels, all of them result in improved palatability compared to that of raw meat or uncooked products. When meats are cooked above 70°C for 5 min or longer, proteins begin to denature and coagulate; at higher temperatures protein molecular structure is altered completely, leading to the liberation of some amino acids and producing characteristic flavors; carbohydrates caramelize and react with amino compounds; fats generate odor-related compounds. Ngadi and Ikediala (1999) studied the thermal denaturation of chicken-drumstick proteins and found that specific heat capacity was related to the state of thermal denaturation of the chicken muscle protein. Murphy et al. (2001) studied the effect of air convection oven cooking on chicken breast patty moisture loss, product yield, and soluble proteins. The authors found that moisture loss in the cooked products increased with increasing final product temperature and oven air temperature, whereas soluble proteins decreased with increasing final product temperature. They concluded that soluble proteins might be used as an indicator for the degree of cooking.

HEAT TREATMENTS AND PROCESSING EQUIPMENT

As stated earlier, heat treatment severity depends on factors such as type and amount of contaminating microorganisms, food composition, expected shelf life, and further storage conditions. It can be as mild as scalding, aimed to inhibit several heat-labile enzymes and to clean the food surface. At the other extreme are sterilization processes, including commercial sterilization, where a long shelf life can be attained.

Scalding

Scalding is carried out by treating the food with hot water or steam for a given time, which depends on the process objectives, either enzyme inactivation or partial cooking. In general, this is a continuous process, although it can be a batch operation. Batch processing consists of food immersion in water at 90 to 100°C; continuous scalding is carried out in conveyors where the food is steam treated. This type of process is seldom applied to meats, although it can be used for such meat products as sausage and bologna.

Pasteurization

Pasteurization is carried out at temperatures below 100°C, usually in scalding tanks. Pasteurization kills part but not all of the viable cells in foods; therefore, it is used with foods that will receive additional preservation methods to minimize microbial growth. In most cases, the aim of pasteurization is to destroy pathogens, but spoilage microorganisms can survive, making it necessary to apply other, less severe preservation methods, such as refrigeration, addition of chemical preservatives, and packaging. Pasteurization and commercial sterilization are basically the same process. They differ in severity, and therefore the extent of microbial destruction varies (Masana and Rodríguez, 2006). The easiest pasteurization method to use for meat and meat products is a water bath; the packaged product is placed in stainless steel tanks, hot water is then applied, followed by cold water for rapid cooling. The continuous process is carried out in conveyors that move the product through water tanks (Hanson, 1990).

Cooking

Probably the thermal treatment applied most often to meat and meat products is oven cooking. In the case of sausages and similar products, they are first stuffed in impermeable casings; heat is transferred from the heating medium (e.g., hot air, steam, smoke) to the product, and product humidity is transferred to the heating medium. Therefore, a combined heat and mass transfer mechanism takes place: heat transfer from the heating medium to and within the product, and between the product and the heating medium, and mass transfer as water and nutrient diffusion.

Cooking can be carried out as oven cooking, grilling, roasting, frying, boiling, and steam cooking; the way heat is applied depends on the cooking type. Dry heat at more than 100°C is employed in oven cooking, grilling, and roasting; boiling and steaming are carried out by placing the food in water. Dry heat is less efficient than humid heat in inactivating vegetative cells or spores (Mathlouthi, 1986).

Due to the fact that the heat driving force depends on temperature differences, the larger the difference, the higher the heat flux; the heating rate is determined by the temperature difference between the product surface and the cold point. Convection and conduction are the leading mechanisms; conduction occurs by direct contact between food particles. This is the leading mechanism inside the meat or meat product, starting from the surface in a transient state; that is, the temperature in any point of the product changes continuously. Conversely, a convection mechanism takes place from the heating medium (air or steam) toward the product surface, due to density gradients as a result of temperature variation; heating is more efficient if forced convection is applied. If the transfer coefficient in the product surface is low, such as in the case of air-free convection (2.5 to 25 kcal/h $\cdot$ m^2 $\cdot$ K), the limiting factor is the convection from the heating medium to the product surface; if the convection coefficient is high, as in condensing steam (5000 to 15,000 kcal/h $\cdot$ m^2 $\cdot$ K), the limiting factor is the conduction rate within the product (Hanson, 1990).

Murphy et al. (2001) reported that in cooking chicken breast patties, moisture loss in the cooked products increased with increased final product temperature and oven air temperature, and also that increasing humidity increased the heat transfer coefficient and therefore reduced the cooking time, oven temperature, and internal temperature, and increased the air humidity and product yield.

Heat transfer mechanisms during the cooking of certain meat products may change during processing: for example, meat products containing large amounts of starch or protein gelling within the can or the casing, such as luncheon meats or finely comminuted sausages. During gelling, proteins or carbohydrates interlink, trapping water in the network formed, behaving as particles in suspension (Li-Chan et al., 1985). At initial heating stages, the dominant mechanism is convection caused by a density gradient; as particles interlink, heat is transferred by vibration between molecules, changing the mechanisms to conduction. As a result, the heating rate changes, and the heating time should be calculated to avoid overheating.

Cooking equipment varies in operational principles, but the most widely used is the force convection batch or continuous oven; the smallest industrial equipment can load approximately 180 kg, whereas the largest one can process up to 25,000 kg. In a continuous oven, the product is loaded in a conveyor moving though one or more cooking zones; in some cases the product is also transported into a cooling section. In traditional ovens the heating medium is forced air, steam, or water; transference coefficients vary accordingly. Boiling water has a high coefficient (1500 to 20,000 kcal/h $\cdot$ m^2 $\cdot$ K), favoring a high heat transfer rate by convection from the heating medium to the product surface. However, most ovens use hot air as the heating medium; here, the heat transfer rate is low

(2.5 to 25 kcal/h · m^2 · K), although it can be increased in a forced convection regime, available in most ovens (10 to 100 kcal/h · m^2 · K) (Hanson, 1990).

Smoking

Smoking is also a variation of cooking; in this case meat products such as finely or coarsely comminuted sausages, or whole birds, are placed in moisture-permeable casings, acting as a barrier between the heating medium and the meat; heat is transferred from the hot air in the smokehouse, together with chemicals in the smoke, to the casing surface; from there, smoke components diffuse though the casing and into the meat (Müller, 1990).

Frying

Li (2005) discussed mechanisms involving water in chicken nugget frying. According to this author, water in a frying food migrates from the center to the surface. The water migration toward the product inner part causes the formation of a dry surface layer, characteristic of fried foods. Water evaporation from the surface of a frying food also removes heat and inhibits surface burning. Subsurface water moves heat from the surface toward the center of the product. Heat transfer to the product's inner part results in product cooking.

The normal temperature range for food frying is 160 to 190°C (Hanson, 1990). At this temperature range the products develop a highly acceptable color, producing a crisp texture; oil absorption is between 8 and 25%. Lower frying temperatures produce lighter color and increased oil absorption. High-temperature frying produces thinner crusts, which cook faster than the inner part, with less oil absorption.

According to Ngadi et al. (2006), oven cooking results is higher mass transfer characteristics than deep frying. These authors studied the effect of frying or baking chicken nuggets on mass transfer characteristics. Moisture loss profiles in the breading and core portions of the product were significantly different in fried and oven-baked products; there was a rapid initial moisture loss from the breading portion of deep-frying nuggets, but moisture loss took 15 min in oven baking. Values were considerably lower for deep frying than for baking: moisture diffusivity was 20.93×10^{-10} to 29.32×10^{-10} m^2/s for deep fat frying and 1.90×10^{-10} to 3.16×10^{-10} m^2/s for oven baking; activation energies were 8.04 and 25.7 kJ/mol for deep fat frying and oven baking, respectively. Li (2005) also reported that chicken nugget frying in oils with higher degrees of hydrogenation resulted in products with a lighter and harder (more crispy) texture. Products fried in nonhydrogenated oil absorbed more oil but also retained more moisture than did samples fried in hydrogenated oil. The rate of change in the color parameters of oils was observed to increase with increasing frying time and degree of hydrogenation.

Tangduangdee et al. (2003) considered the thermal denaturation of actin as the quality index of the meat-based product being fried. They report the cause to be between the predicted and observed results, corresponding to actin kinetic parameters determined experimentally by differential scanning calorimetry, with the assumption of a single-step irreversible reaction.

Microwave Cooking

Microwaves are part of the electromagnetic spectrum, having frequencies from 300 MHz to 300 GHz: that is, wavelengths between 1 mm and 1 m. For domestic microwave ovens, 915 to 2450 MHz is the only wavelength range legally allowed, the latter value being the most commonly used. Microwaves generate heat due to a rapid dipole change in water molecules. When a food material is exposed to microwaves, water dipoles change their alignment at a fast rate (5×10^9 times/s); the resulting friction generates heat. Similar to direct heating, microbial destruction is based on enzyme inactivation and protein denaturation. However, overheating can occur in localized areas, due to energy absorption at higher levels than average (due to food heterogeneity); on the other hand, cold spots can develop. Therefore, care must be taken, as heterogeneous microbial inactivation may occur. In general, microwave heating is used only in very thin products such as bacon (Guerrero-Legarreta, 2006a).

Most of the same factors that affect traditional cooking also influence microwave cooking. However, the moisture and salt content of foods being subjected to microwave cooking play an important role. This is due to the nature of the electric field involved in causing molecular friction. Internal temperatures should be relied on to assure proper cooking (USDA–AFDO, 1999).

Canning

A *sterile product* is defined as one in which no microorganism is present. However, *sterilization* is not an accurate term for food heat treatment, since the sterility criterion cannot be applied to a food product. Therefore, foods are commercially sterile, microbiologically inactive, or partially sterile. Commercially sterile foods are merchandised in hermetic containers to prevent recontamination; therefore, strict aerobic vegetative cells cannot grow (Guerrero-Legarreta, 2006a).

Canning and aseptic processing are the two basic methods used for commercial sterilization. In aseptic processing, the food is heated up to time–temperature conditions of commercial sterilization, placed in a container, and sealed. Canning is used primarily for fluids and fluids containing small particles in suspension; operational efficiency is reduced considerably when larger particles are processed, due mainly to pumping. Direct steam injection has been used for milk sterilization in a process known as *uperization*; it is efficient only if further refrigeration is applied (Thumel, 1995).

The objective of canning is to destroy certain microbial populations (vegetative cells and spores) and/or enzymes that promote spoilage or are harmful to human health. Problems solved by canning, from a sanitation point of view, are prevention of vegetative cells and spores' ability to grow and produce toxins and to eliminate or inhibit microbial development. Time–temperature relationship depends on microbial heat resistance, related primarily to a specific microorganism of a given food. In canned poultry meat or meat products, process calculations are based on the destruction of *C. perfringens, Salmonella* spp., *Staphylococcus* spp., and *Campylobacter* spp. (Christchurch City Council, 1998), although for long-shelf-life products, botulinum heating is employed ($F = 2.5$).

The canning process consists basically of four operations: food preparation (i.e., cleaning, selection, size reduction, scalding, etc.); can, pouch, or jar filling; air exhaustion; and sealing and thermal process (heating and cooling).

1. *Filling*. Heat penetration depends on the solid–liquid distribution in a container. For canned sausages arranged side by side, convection–conduction takes place; solid material loosely packed heats faster that does tightly packed material. Approximately 30% of a can must be filled with brine to improve the heat transfer rate; in addition, 0.5% of the total container volume must be left as headspace.
2. *Exhaustion and sealing*. Meat and meat products react easily with oxygen, affecting several chemical components, mainly pigments and fats. Therefore, air exhaustion from the food and headspace is necessary; exhaustion removes air from the meat tissue, favoring heat penetration. Small air bubbles may cause insufficient heat treatment and therefore insufficient sterilization. If large solid pieces are canned, exhaustion during filling and sealing is enough. Conversely, can filling with raw batters incorporates air bubbles; therefore, vacuum filling is necessary. Air exhaustion also reduces the risk of can blowing or deformation. Exhaustion is achieved by heating, mechanical air removal, or vapor injection. Heating (75 to 95°C) induces steam to replace air in the headspace; cans are closed and sealed immediately. Once cooled down, vacuum is promoted by vapor condensation. Alternatively, cans are transported in a conveyor where they are heated at 85 to 95°C, and 90% or more of the air is removed from the headspace, depending on the residence time and temperature of the exhaustion tunnel (Mathlouthi, 1986).
3. *Thermal processing*. Thermal processing consists of two cycles: heating and cooling. Time–temperature relationships are calculated according to microbial destruction and enzyme inactivation criteria, as described earlier.

Canning is carried out by various processes; batches in still retorts is the oldest method, commonly used in small to medium-sized operations. The food, placed in cans, pouches, or glass jars, is loaded into a retort that is fed with steam; once the time–temperature conditions satisfy the processing parameters (D, z, F), the product is cooled using cold water. Care must be taken to reduce the retort pressure gradually to avoid deformation or breaking of containers. Continuous retorts are generally used in large operations. In hydraulic seal retorts, cans enter the system through a pressure seal and go through the retort in a helix-shaped conveyor, taking the material through heating–cooling zones.

REFERENCES

Baranyi J, Roberts TA. 1995. Mathematics of predictive food microbiology. Int J Food Microbiol 26:199–218.

Christchurch City Council. 1998. *A Food Safety for Food Workers Information Source*. Christchurch, New Zealand: CCC.

Dziezak JD. 1991. Enzymes in food and beverage processing. Food Technol 45(1):77–85.

Food Safety Authority of Ireland. 2007. http://www.fsai.ie/publications/factsheet/factsheet_campylobacter.pdf. Accessed Sept. 11, 2008.

Green DW, Maloney JO. 1997. *Perry's Chemical Engineers' Handbook*. New York: McGraw-Hill.

Guerrero-Legarreta I. 2001. Meat canning technology. In: Hui YH, Shorthose R, Young O, Koohmaraie M, Rogers P, eds., *Meat Science and Applications*. New York: Marcel Dekker, pp. 521–535.

Guerrero-Legarreta I. 2004. Canning. In: Jensen WK, Devine C, Dikeman M, eds., *Encyclopaedia of Meat Sciences*. London: Academic Press, pp. 139–144.

Guerrero-Legarreta I. 2006a. Procesamiento térmico. In: Hui YH, Guerrero-Legarreta I, Rosmini M, eds., *Ciencia y Tecnología de Carnes*. Mexico City, Mexico: Noriega Editores, pp. 437–461.

Guerrero-Legarreta I. 2006b. Thermal processing of meat. In: Hui YH, Castell-Perez E, Cunha LM, Guerrero-Legarreta I, Liang HH, Lo YM, Marshall DL, Nip WK, Shahidi F, Sherkat F, Winger RJ, Yan KL, eds., *Handbook of Food Science, Technology and Engineering*. Boca Raton, FL: Taylor & Francis, Chap. 162.

Han YW. 1975. Death rates of bacterial spores: non-linear survivors' curves. Can J Microbiol 21:1464–1467.

Hanson RE. 1990. Cooking technology. In: *Proceedings of the Reciprocal Meat Conference*, pp. 109–115.

ICMSF (International Commission on Microbiological Specifications for Foods). 1980. *Microbial Ecology of Foods*, vol. 1, *Factors Affecting Life and Death of Microorganisms*. New York: Academic Press.

ICMSF. 1996. *Microorganisms in Foods*, vol. 5, *Microbiological Specifications of Food Pathogens*. London: Blackie Academic & Professional.

Juneja VK. 2006. Delayed *Clostridium perfringens* growth from a spore inocula by sodium lactate in sous-vide chicken products. Food Microbiol 23(2):105–111.

Kennedy J, Blair IS, McDowell DA, Bolton DJ. 2005. An investigation of the thermal inactivation of *Staphylococcus aureus* and the potential for increased thermotolerance as a result of chilled storage. J Appl Microbiol 99(5):1229–1235.

Lan YH, Novakofski J, McCusker RH, Brewer MS, Carr TR, McKeith FK. 1995. Thermal gelation of pork, beef, fish, chicken and turkey muscles as affected by heating rate and pH. J Food Sci 60(5):936–940, 945.

Li Y. 2005. Quality changes in chicken nuggets fried in oils with different degrees of hydrogenation. M.Sc. thesis, Department of Bioresource Engineering, Macdonald Campus, McGill University, Montreal, Quebec, Canada.

Li-Chan E, Nakai S, Wood DF. 1985. Relationship between functional (fat binding, emulsifying) and physicochemical properties of muscle proteins: effects of heating, freezing, pH and species. J Food Sci 50(4):1034–1040.

Manev G. 1984. *La Carne y Su Elaboración*, vol. II. Havana, Cuba: Editorial Científica y Técnica, pp. 308–402.

Masana MO, Rodríguez R. 2006. Ecología microbiana. In: Hui YH, Guerrero-Legarreta I, Rosmini M, eds., *Ciencia y Tecnología de Carnes*. Mexico City, Mexico: Noriega Editores, pp. 293–336.

Mathlouthi M. 1986. *Food Packaging and Preservation: Theory and Practice*. London: Elsevier Applied Science Publishers.

Mittal GS, Blaisdell JL. 1984. Heat and mass transfer properties of meat emulsions. Lebensm-Wiss Technol 17(2):94–98.

Mittal GS, Usborne WR. 1985. Moisture isotherms for uncooked meat emulsions of different compositions. J Food Sci 50:1576–1579.

Müller WD. 1990. The technology of cooked cured products. Fleischwirt Int 1:36–41.

Murphy RY, Johnson ER, Duncan LK, Clausen EC, Davis MD, March JA. 2001. Heat transfer properties, moisture loss, product yield, and soluble proteins in chicken breast patties during air convection cooking. Poult Sci 80:508–514.

Murphy RY, Duncan LK, Beard BL, Driscoll KH. 2003. *D* and *z* values of *Salmonella, Listeria innocua*, and *Listeria monocytogenes* in fully cooked poultry products. J Food Sci 68(4):1443–1447.

Ngadi MO, Ikediala JN. 1999. Heat transfer properties of chicken-drum muscle. J Sci Food Agric 78(1):12–18.

Ngadi M, Dirani K, Oluka S. 2006. Mass transfer characteristics of chicken nuggets. Int J Food Eng 2(3). http://www.bepress.com/ijfe/vol2/iss3/art8. Accessed Sept. 11, 2008.

Peleg, M. 2006. It's time to revise thermal processing theories. Food Technol 60(7):92.

Pérez MGR, Calvelo A. 1984. Modeling the thermal conductivity of cooked meat. J Food Sci 49:152–156.

Siripona K, Tansakula A, Mittal GS. 2007. Heat transfer modeling of chicken cooking in hot water. Food Res Int 40(7):923–930.

Stiebing A. 1992. Tratamiento por calor. In: Wirth F, ed., *Tecnología de Embutidos Escaldados*. Zaragoza, Spain: Editorial Acribia, pp. 171–190.

Tangduangdee C, Bhumiratana S, Tia S. 2003. Heat and mass transfer during deep-fat frying of frozen composite foods with thermal protein denaturation as quality index. Sci Asia 29:355–364.

Thumel H. 1995. Preserving meat and meat products: possible methods. Fleischwirt Int 3:3–8.

USDA–AFDO (U.S. Department of Agriculture–Association of Food and Drug Officials). 1999. Cooking and Cooling of Meat and Poultry Products. Distance Learning Training Course, Bethesda, MD.

Welti-Chanes J, Velez-Ruiz JF, Barbosa-Cánovas GV. 2003. *Transport Phenomena in Food Processing*. Boca Raton, FL: CRC Press.

Zamudio M. 2006. Microorganismos patógenos y alternantes. In: Hui YH, Guerrero-Legarreta I, Rosmini M, eds., *Ciencia y Tecnología de Carnes*. Mexico City, Mexico: Noriega Editores, pp. 337–370.

30

PACKAGING FOR POULTRY PRODUCTS

S.N. Sabapathi and Amarinder S. Bawa
Defence Food Research Laboratory, Siddartha Nagar, Mysore, India

Handbook of Poultry Science and Technology, Volume 2: Secondary Processing, Edited by Isabel Guerrero-Legarreta and Y.H. Hui

INTRODUCTION

There is a continuous search for improved methods for transporting food products. With increasing urbanization, the problems associated with the keeping quality of fresh flesh foods such as red meat and chicken carcasses have become more accentuated. The flesh foods industry is an important sector of the world food industry and ranks among the top five agricultural commodities. Large-animal slaughter and processing facilities have developed in areas where livestock production is highly concentrated, and such areas are surprisingly not far away from centers of dense population (Hicks, 2002). There is a major world trade in fresh and preserved flesh foods. In all these situations, packaging has a key role to play in protecting the product from extrinsic environmental factors and ensuring the required shelf life for a food in the particular market (Smith et al., 1990).

IMPORTANCE OF FLESH FOOD PACKAGING

Food packaging is an integral part of food processing and a vital link between the processor and the eventual consumer for the safe delivery of a product through the various stages of processing, storage, transport, distribution, and marketing. All over the world, consumers are showing greater awareness of food packaging, as it provides a clue to the quality, quantity, and hygienic standards of a product. A very important aspect of flesh food preservation is suitable packaging. The main purpose of packaging is to protect flesh foods from microbial contamination, light, oxygen, or any physical damage or chemical changes. The selection of packaging material has to be done very carefully to protect the various physicochemical properties, such as the nature of pigments, sensory attributes, and microflora (Charles et al., 2006). The purpose is to retard or prevent the main deteriorative changes and make products available to consumers in the most attractive form. However, the initial quality of the flesh foods has to be very good because packaging can only maintain the existing quality or delay the onset of spoilage by controlling the factors that contribute to it (Farber, 1991). The product is therefore protected for a limited period, determined by the system used. Thus, flesh foods need a specialized package profile, depending on the type of processing as well as on conditions of storage and distribution.

Literature on packaging requirements for meat and meat products is plentiful (Inns, 1987). Although the packaging requirements for fresh dressed chicken (carcasses) are seemingly similar to those for fresh meat, physiological and biological factors make the requirements unique and challenging. Therefore, the packaging requirements for poultry products are discussed separately.

PACKAGING OF POULTRY MEAT

Although most poultry meat is sold in the form of whole oven-ready carcasses, there is increasing demand for cut-up portions and a variety of other further-processed products, both raw and cooked. Raw poultry meat is a perishable commodity of relatively high pH (5.7 to 6.7) which readily supports the growth of microorganisms when stored under chill or ambient conditions. The shelf life of such meat depends on the combined effect of certain intrinsic and extrinsic factors, including the numbers and types of psychrotropic spoilage organisms present initially, the storage temperature, and muscle pH and type (red or white), as well as the type of packaging material used and the gaseous environment of the product (Gill, 1990). The main pathogenic organisms associated with poultry and poultry products are *Salmonella* spp., *Staphylococcus aureus*, and *Clostridium perfringens*. Therefore, most studies on the extension of shelf life using CO_2 in modified atmospheres have focused on the suppression of spoilage organisms rather than the survival and growth of pathogens (Russell et al., 1996).

Further, raw poultry meat is high in protein, low in calories, and easy to chew and digest, but poultry fat is unsaturated and is very prone to the development of oxidative rancidity. The shelf life of poultry varies according to the type of processing, nature of the processing environment, initial flora, and postslaughter treatment. Poultry meat is packaged immediately after the dressing operations are over. Unpacked refrigerated storage may result in surface dehydration or freezer burn, characterized by surface discoloration, tough texture, and diminished juiciness as well as flavor loss (Fletcher, 1999). Poultry is usually packed as whole dressed poultry, cut-up poultry, and poultry organs. Dressed poultry has a shelf life of 5 to 7 days under refrigerated storage conditions (0 to 5°C).

MODIFIED-ATMOSPHERE PACKAGING

It has been known for over 100 years that the preservative effect of chilling can be greatly enhanced when it is combined with control or modification of the gas atmosphere surrounding the food. The normal composition of air by volume is 78% nitrogen, 21% oxygen, 0.9% argon, 0.3% carbon dioxide, and traces of nine other gases in very low concentrations. Generally, the atmosphere is changed by increasing or decreasing the concentration of O_2, and/or by increasing the concentration of CO_2. Several terms are used to describe changes in the gas atmosphere inside individual packages of food. *Controlled-atmosphere packaging* (CAP) is, strictly speaking, the enclosure of food in a gas-impermeable package inside which the gaseous environment with respect to CO_2, O_2, N_2, water vapor, and trace gases has been changed and is controlled selectively (Brody, 1989). Using this definition, there are no CAP systems in commercial use. Modified-atmosphere packaging (MAP) is the enclosure of food in a package in which the atmosphere inside the package is modified with respect to CO_2, O_2, N_2, water vapor, and trace gases. This modification is generally achieved using one of two processes: by removing air and replacing it with a controlled mixture of gases, a procedure generally referred to as *gas flush packaging*, or by placing the food in a gas-impermeable package and removing the air, a procedure known as *vacuum packaging*. In vacuum packaging, elevated levels of carbon dioxide of from 10 to 20% can be produced by microorganisms as they consume residual oxygen, or by respiring produce (Sofos, 1995).

Principal Factors in MAP

The principal factors in a successful MAP operation are the choice of gas or gas mixture and its effect on a product, the use of suitable packaging material; and the packaging machine. Overriding all of these is close control of temperature throughout the packaging, distribution, and retailing of MAP foods.

Temperature One of the major concerns with MAP foods is temperature abuse: holding food at temperatures above chill temperatures, such that the growth of pathogens is accelerated. In addition, because the biostatic effects of CO_2 are temperature dependent, a rise in temperature during storage could permit the growth of microorganisms that had been inhibited by CO_2 at lower temperatures. If O_2 were present in the package, growth of aerobic spoilage organisms during periods when the food was at nonrefrigerated temperatures would alert consumers to temperature abuse due to the appearance of undesirable odors, colors, or slime (Christensen, 1983). However, the absence of O_2 will favor the growth of anaerobic microorganisms (including *Clostridium botulinum)* over aerobic spoilage organisms. It should be noted that both aerobic and anaerobic pathogens can grow at temperatures as low as 4°C and produce toxin without any sensory manifestation of food deterioration.

Choice of Gas The choice of gas or gas mixture to replace air depends largely on the nature of the food and its principal mode(s) of deterioration. Microbial growth and oxidation are commonly the two major deteriorative modes, and thus the concentration of oxygen is frequently reduced, and in some cases, removed completely. Carbon dioxide inhibits the growth of a wide range of microorganisms (Avery et al., 1996). In aerobic systems, atmospheres containing 20 to 30% CO_2 are used (greater concentrations have little additional inhibitory effect on spoilage floras), and in anaerobic systems, atmospheres of 100% CO_2 may be used.

Carbon dioxide is highly soluble in water and oils and will therefore be absorbed by the food until equilibrium is attained. Nitrogen is used to purge air from a package to achieve a sufficiently low level of oxygen to prevent aerobic microbial spoilage. It also frequently functions as a filler gas in MAP to reduce the concentration of other gases in the package and to keep the package from collapsing as CO_2 dissolves into the product. Since containers for gas packaging are comparatively good gas barriers, the internal atmosphere will be modified by the food during storage. The relative volumes of gas and food are therefore important in determining the progress of the changes in concentration of gases during storage, and cognizance must also be taken of the high solubility of CO_2 compared to the relatively low solubility of CO_2 and N_2 in foods (Church and Parson, 1995). The presence of CO_2 is important because of its biostatic activity against many spoilage organisms which grow at chill temperatures. In general, the inhibitory effects of CO_2 increase with decreasing temperature, due primarily to the increased solubility of CO_2 at lower temperatures; dissolution of CO_2 in water lowers the pH and consequently slows reaction rates (Fletcher et al., 2000). The overall effect of CO_2 is to increase both the lag phase and the generation time of spoilage microorganisms; however, the specific mechanism for the bacteriostatic effect is not known.

It is important to note that whereas CO_2 inhibits some types of microorganisms, it has no effect on others. Furthermore, to be an effective biostat, it must dissolve into the aqueous portion of the product. Although the growth of anaerobic pathogens will be inhibited by the presence of O_2, the shelf life of the food will not necessarily be extended. Atmospheres enriched with CO_2 have been advocated for extending the shelf life of poultry products.

Choice of Packaging Material The choice of packaging material is an important factor in any MAP operation. A low water vapor transmission rate, together with a high gas barrier, must generally be achieved. Virtually all MAP packages are based on thermoplastic polymers. A point that should be remembered is that all packages made purely from such materials allow some gas transmission, even at chill temperatures. Thus, over the relatively long storage times for which many MAP foods are held, there will be diffusion of gases through the package walls. The comparative dearth of gas permeability data for thermoplastic polymers at chill temperatures and high relative humidities makes prediction of the extent of such gas transport tenuous.

The packaging material also needs to have the mechanical strength to withstand machine handling and subsequent storage, distribution, and retailing. Materials in use are laminations or coextrusions of polyethylene with polyester or nylon, with or without the addition of a high barrier layer of vinylidene chloride/vinyl chloride copolymer or ethylene/vinyl alcohol (EVA) copolymer, depending on the barrier required.

Choice of Packaging Machinery The packaging machinery requirements are obviously related to the method of packaging employed: thermoforming or pillow packaging. The thermoforming method involves the use of a rigid or semirigid base material which is thermoformed into a tray. The pillow wrap or horizontal form–fill–seal machine employs a single reel of flexible packaging material which is formed into a tube and the two edges heat-sealed.

The most comprehensive study of the use of CO_2-enriched atmospheres for extending the shelf life of poultry meat (chicken portions) had established that the shelf life of MAP chicken carcasses or pieces was two to three times more than air-packed counterparts. Halved ready-to-cook chicken carcasses stored under carbon dioxide had a shelf life of 26 days and frozen (IQF) turkey strips stored in MAP up to 84 days had the highest sensory scores in respect to aroma, appearance, and structure (Bohnsack et al., 1988; Yan et al., 2006). Increased levels of carbon dioxide were synergistic, with the lethal effect of irradiation of fresh minced chicken increasing its shelf life (Heath et al., 1990). It was necessary to employ a carbon dioxide concentration of 60 to 80% for a shelf life of 28 days for ground chicken meat (Baker et al., 1986), whereas the same concentration of carbon dioxide extended the shelf life of chicken quarters to 35 days at 2°C (Hotchkiss, 1988). The shelf life of 70% carbon dioxide and 30% nitrogen for MA-packed chicken carcasses stored at 2, 4, 7, and 9°C was 25, 21, 12, and 8 days, respectively, compared with 20,15, 8, and 8 days, respectively, for 30% carbon dioxide and 70% nitrogen MAP carcasses. The inhibitory effect of MAP on the growth of Enterobacteriaceae and the production of spoilage metabolites such as free fatty acids and extract release volume was more pronounced at lower temperatures (i.e., at 2 and 4°C), and the effect was negligible at 7 and 9°C (Kakouri and Nychas, 1994).

Treatment of chicken broiler leg and breast meat with potassium sorbate (PS) and lactic acid (LA) caused a reduction in total bacterial counts under refrigerated storage conditions. The shelf life of poultry meat treated with PS and LA and vacuum packed was 30 days compared to 18 and 6 days for vacuum-packed control and unsealed control samples, respectively (Lin et al., 1989). Pretreatment with LA increased the shelf life of broiler carcasses 6 to 7 days; when stored in modified atmospheres (70% carbon dioxide + 5% oxygen + 25% nitrogen), the shelf life was extended to 35 days at 7°C and > 36 days at 4°C compared with 13 days and 22 days for control carcasses (Cosby et al., 1999)

Whereas the number and types of microorganisms found on stored poultry are important factors when determining shelf life, the real determinant is the sensory quality of the raw and cooked products. Unfortunately, most published studies have not included sensory evaluation. Hotchkiss et al. (1985) evaluated the quality of raw and cooked poultry that had been stored under a modified atmosphere and refrigeration for up to 5 weeks. The result of their study indicated that MAP (80% CO_2) poultry would be quite acceptable to consumers for up to 4 to 6 weeks, depending on the storage temperature. It was observed that commercial poultry processors may not be getting as long a shelf life because of the difficulties inherent in controlling the packaging process and temperature under production conditions.

As to the safety of MAP chicken, possible problem organisms would be *Camplobacter jejuni*, which may be able to survive better in a MAP product, and *Listeria monocytogenes* and *Aeromonas hydrophila*, which may, because of the extended storage lives of MAP products, have additional time to grow to potentially high numbers. Although *C. perfrigens* may be able to survive better in some MAs than in air, it would not be able to grow at the chill temperatures commonly used for MAP products. Thus, it is unlikely to be a health hazard in a MAP product unless the product is temperature abused, because high numbers of the organism must be ingested to cause illness (Stadelman, 1995).

A wide range of manufactured poultry products has been developed, including rolls, roasts, burgers, and sausages. However, in most cases very little information is available on either the keeping quality or the influence on shelf life of particular gas mixtures or packaging materials.

PACKAGING TECHNIQUES

Vacuum Packaging

The vacuum packaging of poultry carcasses, cuts, and other manufactured products can extend the shelf life, provided that the product is held under chill conditions. During storage at 1°C in either O_2-permeable film or vacuum packs, extensions in shelf life from 16 to 25 days in the case of breast fillets (pH 5.9 to 6.0) and from 14 to 20 days for drumsticks (pH 6.1 to 6.3) were observed for vacuum-packaged products. However, deleterious flavor changes that tended to precede the development of off-odors in vacuum packs of both types of muscle were observed. The ideal materials for vacuum packaging of poultry carcasses are laminates of polyester/polyethylene, polyamide/polyethylene, poly(vinyl dichloride) (PVDC) copolymer film, and nylon/EVA.

Active Packaging

Sometimes, certain additives are incorporated into the polymeric packaging film or within packaging containers to modify the headspace atmosphere and to extend shelf life. This is referred to as *active packaging* (Stupak et al., 2003). The concept of active packaging has been developed to rectify the deficiencies in passive packaging. For example, when a film is a good barrier to moisture but not to oxygen, the film can still be used along with an oxygen scavenger to exclude oxygen from the pack. Similarly, carbon dioxide absorbents and emitters, ethanol emitters, and ethylene absorbents can be used to control oxygen levels inside the MA pack. The appropriate absorbent materials are placed alongside the food. By their activity, they modify the headspace of the package and thereby contribute to extending the shelf life of the contents.

Thus, the MAP system is a dynamic one in which respiration of the packaged product and gas permeation through the packaging film take place simultaneously. During respiration, the packaged product takes oxygen from the package atmosphere, and the carbon dioxide produced by the product is given away to the package atmosphere. This results in the depletion of oxygen and accumulation of carbon dioxide within the package. Consequently, the composition of the package air changes. Initially (i.e., soon after sealing of the package), the composition of package air remains nearly same as that of the ambient air. As the concentration of oxygen in the package air is reduced, that of carbon dioxide increases, and oxygen and carbon dioxide concentration gradients begin to develop between the package atmosphere and the ambient atmosphere. This decrease in respiration and increase in gas permeation continues until equilibrium is reached. At equilibrium, the rate of oxygen permeating (ingress) becomes equal to the rate of oxygen consumption (respiration), and the rate of carbon dioxide permeating (exgress) becomes equal to the rate of carbon dioxide evolution (respiration). Thus, oxygen consumed during respiration is replaced simultaneously by the ingress of oxygen. Similarly, an equal amount of carbon dioxide evolved by the packaged produce permeates out of the package. As a result, the air composition remains constant. This state is known as an *equilibrium* or *steady state*. The attainment of an equilibrium state depends on proper design of the MA package.

OVERWRAPS

Packaging of whole dressed chicken halves or cut-up parts are done in plastic films such as polyethylene, polypropylene, PVDC, rubber hydrochloride, or nylon-6. These are films of 150- to 200-gauge thickness. Polyethylene is the most widely used packaging material because of its low cost and easy availability. Sheets of this thermoplastic film can be fabricated into bags and a dressed eviscerated bird inserted into a bag. Sometimes, an individual dressed bird is wrapped in waxed paper or parchment paper before bagging. The problem of body fluid accumulation is avoided by putting an absorbent pad or blotter on the back of each bird to soak up the liquid. The bag is then heat sealed or twist-tied or clipped shut.

Trays with Overwraps

Small whole dressed chicken, broilers, and roasted chickens are placed in a polystyrene foam tray and overwrapped with transparent plastic film. A blotter underneath absorbs the excessive meat juice accumulated. Chicken thus wrapped has a shelf life of 7 days at 4°C in a refrigerator.

Shrink-Film Overwraps

Many thermoplastic films, such as polyethylene, polypropylene, and polyvinylidene, can be biaxially oriented to stay stretched at ambient temperature. Dressed chicken is overwrapped with such films and passed through a hot-air tunnel or dipped in a water tub maintained at 90°C for a few seconds to effect shrinkage of the film.

MODERN TRENDS IN MEAT AND POULTRY PACKAGING

With the entry of new types of flesh foods in the market, high-quality alternative packaging materials are also emerging. Consumers are becoming more discerning in their choice of food products and there is a trend toward a shift from traditional food items and eating habits. Some of the recent trends in the flesh foods packaging industry are covered here.

Retortable Flexible Pouch

A retort pouch is a flexible package into which a food product is placed, sealed, and sterilized at temperatures between 110 and 140°C. The finished product is commercially sterile, shelf stable, and does not require refrigeration. A retortable pouch is made of a laminate of three layers held together by an adhesive. The outer layer, made of polyester, polyamide, or oriented polypropylene, provides support and physical strength to the composite.

The middle layer of aluminum foil acts as a barrier against water vapor, gases, and light. The inner layer of polyethylene, polypropylene, or poly(vinyl chloride) provides heat sealability and food contact. The different laminates used for a retort pouch are polyester/aluminum foil/modified high-density polyethylene or polyester/aluminum foil/polypropylene–ethylene copolymer. Like a metal can, a retortable pouch can be sterilized by heat and has the advantage of lower cooking time, as it has a thinner profile than that of a metal can (Srivatsa et al., 1993; Sabapathi et al., 2001). They do not require storage at refrigerator temperatures and like canned food, are shelf stable. These types of packages also require less storage space and are lighter in weight.

Roast-in Bags

A roast-in bag is an oven-stable vacuum skin package that can be used to cook meat at a temperature up to 204°C. It is fabricated from poly(ethylene terephthalate) (PET) film, due to PET's unusual properties of not becoming brittle with age, long shelf life, resistance to most chemicals and moisture, and dimensional stability.

Microwave Packages

Convenience foods fall into two categories, frozen and retortable. The current trend in frozen food is dual ovenability: products that can be heated in a microwave oven or in a conventional oven. Shelf-stable retortable foods are better suited to microwave heating. Owing to the growing importance of microwave ovens, other materials are overtaking conventional aluminum trays.

When selecting thermoplastics for dual-ovenable packages, the critical properties to be considered are dimensional stability up to 200 to 250°C, good impact strength at freezer temperatures, and microwavability. Heat-resistant plastic trays made of materials such as polyester, polypropylene, nylon, and polycarbonate can be used in combination or as a monolayer. The trays are closed with heat-sealable lidding materials and are overwrapped, shrink-wrapped, or sealed inside a microwave bag. Thus, due to their flexibility to form different shapes and sizes on in-line formation, rigid plastic trays dominate microwavable packaging.

Polyester-coated paperboard cartons are also used as microwave packages. These containers can be formed on a conventional tray-making-, carton-forming, or folded-carton-making machine. Crystallized polyester containers are also very popular for microwave packaging as well as for conventional oven cooking. These trays are very stiff and can be sealed by a high-speed tray sealing machine with transparent or nontransparent lidding material. They are easy to handle, sturdy, attractive, cost-competent, and can be compartmentalized for multicomponent food items. They are self-serving and reusable.

CONCLUSIONS

With continually growing demand for processed, packed, convenient ready-to-eat and ready-to-serve meat products, a variety of specialized package profiles are available, depending on the type of processing techniques and storage conditions. From fresh meat to cured meat, from pork to poultry, the main purpose of packaging is to make the products available to the customers in most attractive form while maintaining the quality of the contents. Plastics are used in every form of packaging: trays, overwraps, shrink films, MAP, and retort packaging. Plastics in the form of laminates, plain films, overwraps, and so on, play a major role in imparting barrier properties and aesthetics to the packaging medium. Overall, the use of plastics in meat and poultry packaging is one of the most important factors contributing to the growth of the processing industry today.

REFERENCES

Avery SM, Rogers AR, Bell RG. 1996. J Food Sci Technol 30:725.

Baker RC, Qureshi RA, Hotchkiss JH. 1986. Poult Sci 65:729.

Bohnsack U, Knippel G, Hopke HU. 1988. Fleischwirtschaft 68:1553.

Brody AL, ed. 1989. Controlled modified atmosphere. In: *Vacuum Packaging of Foods*. Trumball, CT: Food and Nutrition Press.

Charles N, Williams SK, Rodrick GE. 2006. Poult Sci 85:1798.

Christensen CM. 1983. J Food Sci 48:787.

Church IJ, Parson AL. 1995. J Sci Food Agric 67(2):143.

Cosby DE, Harrison MA, Toledo RT, Craven SE. 1999. J Appl Poult Res 8:185.

Farber JM. 1991. J Food Prot 54:58.

Fletcher DL. 1999. Poult Sci 78:1323.

Fletcher DL, Qiao M, Smith DP. 2000. Poult Sci 79:784.

Gill CA. 1990. Food Control 2:74.

Heath JL, Owens SL, Tesch S, Hannah KW. 1990. Poult Sci 69:313.

Hicks A. 2002. Aust J Technol 6(2):89.

Hotchkiss JH. 1988. Food Technol 42(9):55.

Hotchkiss JH, Baker RC, Qureshi RA. 1985. Poult Sci 64:333.

Inns R. 1987. Modified atmosphere packaging. In: Paine FA, ed., *Modern Processing, Packaging and Distribution Systems for Food*. Glasgow, UK: Blackie & Son, Chap. 3.

Kakouri A, Nychas GJE. 1994. J Appl Bacteriol 76:163.

Lin CF, Gray JI, Asghar A, Buckley DJ, Booren AM, Flegal CJ. 1989. J Food Sci 54:1457.

Russell SM, Fletcher DL, Cox NA. 1996. Poult Sci 75:2041.

Sabapathi SN, Ramakrishna A, Srivatsa AN. 2001. Indian Food Ind 20(3):67.

Smith JP, Ramaswamy HS, Simpson BK. 1990. Trends Food Sci Technol 11:111.

Sofos N, Smith GC, Williams SN. 1995. J Food Sci 60:1179.

Srivatsa AN, Ramakrishna A, Gopinathan VK, Nataraju S, Leela RK, Jayaraman KS, Sankaran R. 1993. J Food Sci Technol 30:429.

Stadelman W. 1995. *Egg Science and Technology*. Binghamton, NY: Haworth Press.

Stupak P, Miltz J, Sonneveld K, Bigger SW. 2003. J Food Sci 68(2):408.

Yan HJ, Lee EJ, Nam KC, Min BR, Ahn DU. 2006. Poult Sci 85:1482.

PART VI

CONTAMINANTS, PATHOGENS, ANALYSIS, AND QUALITY ASSURANCE

31

CONTAMINATION OF POULTRY PRODUCTS

Marcelo L. Signorini
Consejo Nacional de Investigaciones Científicas y Técnicas, Instituto Nacional de Tecnología Agropecuaria, Estación Experimental Rafaela, Departamento de Epidemiología y Enfermedades Infecciosas, Provincia de Santa Fe, Argentina

José L. Flores-Luna
Food Safety Management Systems (Consultant), México D.F., México

Handbook of Poultry Science and Technology, Volume 2: Secondary Processing, Edited by Isabel Guerrero-Legarreta and Y.H. Hui

INTRODUCTION

Foodborne disease has emerged as an important and growing public health and economic problem in many countries during the last two decades. Frequent outbreaks caused by new pathogens or the use of antibiotics in animal husbandry and the transfer of antibiotic resistance to humans are just a few examples (Rocourt et al., 2003). Nevertheless, it is possible that these outbreaks are alone the most visible aspect of a much more widespread and persistent problem. Foodborne diseases not only have a significant impact on the health and well-being of the population, but also have economic consequences for individuals and their families, in the provision of medical attention, and in the reduction of economic productivity (Keene, 2006).

Access of countries to international food markets will continue, depending on their ability to fulfill the regulatory requirements of importing countries. The creation and maintenance of a demand for their products on world markets presupposes confidence by importers and consumers in the integrity of their food safety systems (FAO, 2004). Based on the fact that food hazards can be introduced in the food chain at the very beginning and are able to continue growing at any point along the chain, food safety systems are being developed increasingly on a farm-to-table basis as an effective way of reducing foodborne diseases (FAO, 2004).

Since the mid-1990s there has been a transition toward risk analysis based on better scientific knowledge of foodborne diseases and their causes. This approach offers a prevention basis for risk management measures, such as implementation of a hazard analysis and critical control point (HACCP) system by food organizations, as well as regulatory measures for national and international food safety programs (Hoornstra and Notermans, 2001).

MICROBIAL HAZARDS

Food poisoning incidents usually arise when the causative organisms, initially present in low numbers, are allowed to multiply during manufacture, distribution, preparation, or storage of foods. Factors that contribute to the problem are well known and include inadequate cooking, lack of refrigeration, and cross-contamination of cooked items. In most developed countries, poultry meat is frequently contaminated with *Salmonella* and *Campylobacter* spp., the organisms responsible for many cases of human enteritis, and other pathogens may also be present (e.g., *Listeria monocytogenes* was isolated from 60% of raw poultry examined in the United Kindom in 1988) (Mead, 1993).

Foodborne disease comprises a broad group of illnesses. Among them, gastroenteritis is the most frequent clinical syndrome and can be attributed to a wide range of microorganisms, including bacteria, viruses, and parasites. Different degrees of severity are observed, from a mild disease that does not require medical treatment to a more serious illness requiring hospitalization, with long-term disability and death. The outcome of exposure to foodborne diarrheal pathogens depends on a number of host factors, including preexisting immunity, the ability to elicit an immune response, nutrition, age, and nonspecific host factors (Rocourt et al., 2003).

As the human population increases and megacities grow, there is greater risk that infectious diseases will evolve, emerge, or spread readily among populations. There are new conditions for the emergence of pathogens (Mead, 1993; Rocourt et al., 2003; Gilchrist et al., 2007):

1. *Changes in animal husbandry*. Modern intensive animal husbandry practices introduced to maximize production seem to have led to the emergence and increased prevalence of zoonoses (diseases transmissible from animals to humans), such as *Salmonella* serovars and/or *Campylobacter* in herds of all the most important production animals (especially poultry and pig). The transmission of foodborne pathogens in poultry production is strongly influenced by the intensive nature of present systems for breeding, growing, and poultry processing (tends to spread microbial contamination).
2. *Increase in international trade*. This has three main consequences: (a) rapid transfer of microorganisms from one country to another; (b) increasing time between processing and consumption of food, leading to augmented opportunities for contamination and time–temperature abuse of products, and hence a risk of foodborne illness; and (c) a population is more likely to be exposed to a higher number of different strains/types of foodborne pathogens.
3. *Increase in susceptible populations*. Advances in medical treatment have resulted in a rising number of elderly and immunocompromised people, mostly in industrialized countries.
4. *Increase in international travel*. 90% of salmonellosis in Sweden, 71% of typhoid fever cases in France, and 61% of cholera cases in the United States are attributed to international travelers.
5. *Change in lifestyle and consumer demands*. While dining in restaurants and salad bars was relatively rare 50 years ago, today it is a major source of food consumption in a number of OECD countries. As a result, an increasing number of outbreaks are associated with food prepared outside the home.

Countries with reporting systems have documented significant increases in the incidence of foodborne disease. Nontyphoid salmonellosis is the only foodborne disease reported in all countries, with an annual reported incidence rate ranging from 6.2 to 137 cases per 100,000 population. Campylobacteriosis appears to be one of the most frequent bacterial foodborne diseases in many countries, with reported annual incidence rates up to 95 cases per 100,000 population. For other bacterial foodborne disease, reported annual incidence rates are lower, between 0.01 and 0.5 case per 100,000 population for listeriosis, between 0.01 and 1.6 cases per 100,000 population for botulism (OECD countries) (Zhao et al., 2001; Rocourt et al., 2003).

Foods most frequently involved in outbreaks in OECD countries are meat and meat products, poultry, and eggs and egg products, with the likely implication that these foods are associated with *Salmonella* and *Campylobacter*. Case–control studies confirmed the same food sources for sporadic cases: raw and undercooked eggs, eggs containing food and poultry for salmonellosis; and poultry for campylobacteriosis (Rocourt et al., 2003).

The three main foodborne pathogens associated with poultry (*Salmonella, Campylobacter*, and *Listeria* spp.) are usually carried asymptomatically in the intestines of infected birds. It is well known that good husbandry hygiene is essential in controlling the spread of avian pathogens, and the same principles are relevant to agents of foodborne disease. What makes it particularly difficult to control is the ubiquity of the organisms and the insidious nature of most flock infections. In addition, only regular testing of flocks can determine whether control measures are effective. There can be little doubt that any effective control of foodborne pathogens in poultry production must be multifactorial and heavily dependent on measures to limit live bird infection. The next stage in this objective may well require stringent husbandry hygiene, even for broiler flocks, but such an approach requires the support of prophylactic treatment for chicks at a time when susceptibility to infection is high. Protection of them by early introduction of a mature intestinal microflora (competitive exclusion) is becoming well established as part of the strategy against food-poisoning salmonellas and in the future may be extended to cover other pathogens as well (Mead, 1993).

The most difficult problem to control in poultry processing is that of cross-contamination, which can arise from aerosols, process water, and contact between carcasses and equipment or operators' hands. Also, line speeds are such that there is little or no opportunity to sanitize implements after one bird has been dealt with and before another is ready (Mead, 1993). The stages in processing that are most often associated with transmission of foodborne pathogens are scalding, plucking, and evisceration. The need to loosen the feathers by immersing birds in a water bath leads to large numbers of organisms being released into the water, approximately 10^9 from each bird entering the tank. Thus, there is ample opportunity for cross-contamination, especially when the water is maintained at 50 to 53°C, as it must be for birds that will be air-chilled and sold fresh to avoid subsequent discoloration of the skin. During the next stage, which is mechanical defeathering, microorganisms are disseminated via the aerosols produced and through contamination of the flexible rubber fingers that scour the surface of each carcass. Since the atmosphere inside the machines is both warm and moist, microbial growth can occur and cause further contamination of the birds as they pass through. A particular problem arises with strains of *Staphylococcus aureus*, which colonize equipment and tend to survive there for long periods of time. Automatic evisceration equipment often causes fecal contamination of carcasses because of gut breakage. This is a consequence of natural variations in bird size and the inability of such machines to adjust automatically to size variation. The spread of fecal material will transmit any enteric pathogens, such as *Salmonella* and *Campylobacter*. Because birds must remain whole throughout the processing operation, the abdominal cavity is a site that is particularly difficult to clean effectively following evisceration (Mead, 1993).

Campylobacter spp.

Campylobacteriosis is typically self-limiting, with symptoms rarely lasting more than 10 days; however, it can be fatal in more vulnerable population. The Guillain–Barré syndrome, a sequel of campylobacteriosis, is a subacute polyneuropathy affecting motor, sensory, and autonomic nerves (cranial nerves may also become involved) that supply the limbs and respiratory muscles. It has a mortality of approximately 10%, and recovery is often incomplete, delayed, or both. The other sequela is the reactive arthritis, which may cause pain and incapacitation for several weeks to months in approximately 1% of *Campylobacter jejuni* cases (Potter et al., 2003; Price et al., 2007).

Campylobacteriosis is one of the most important foodborne disease in the world. For example, in the Netherlands, the incidence of campylobacteriosis is estimated to be 80,000 cases per year in a population of 16 million (Nauta et al., 2006). In Finland, the number of cases of campylobacteriosis in 2005 was 4002 (Karenlampi et al., 2007). In Norway in 1988 an incidence rate of 13.3 cases per 100,000 population was reported (Kapperud et al., 1992). In Iceland, the incidence of domestically acquired human campylobacteriosis peaked in 1999 at 117.6 cases per 100,000 persons. An estimated 2.5 million cases of *Campylobacter* infections occur each year in the United States, and 80% of these cases have been found to be the result of foodborne transmission, with a cost estimated at between $1.3 billion and $6.2 billion (Potter et al., 2003; Bhaduri and Cottrell, 2004). In Switzerland, where 92 cases per 100,000 inhabitants were reported, *Campylobacter* spp. are the leading cause of bacterial zoonoses (Ledergerber et al., 2003). In Denmark 78 cases per 100,000 inhabitants were reported, and it was the most frequent foodborne zoonosis (Heuer et al., 2001).

Broiler chickens are generally regarded as one of the main sources of campylobacteriosis (Harris et al., 1986a; Kapperud et al., 1992; Van Gerwe et al., 2005; Cortez et al., 2006; Idris et al., 2006). Two major species of the genus *Campylobacter* that occur in the poultry industry are *C. jejuni* and *C. coli*. Humans can thus be exposed to *Campylobacter* by the consumption of improperly heated broiler or other foods that are cross-contaminated with *Campylobacter* during food preparation with broiler meat (Slader et al., 2002; Potter et al., 2003; Bhaduri and Cottrell, 2004; Karenlampi et al., 2007). One risk assessment model in broiler meat suggested that human exposure and illness are predominantly the consequence of *Campylobacter* present on the exterior at the entrance of the processing plant, not of *Campylobacter* in the chickens' feces (Nauta et al., 2006).

The prevalence of broiler flocks colonized with *Campylobacter* spp. varies among countries, ranging from 5% of flocks to more than 90%. Once a flock is exposed, the bacteria spread rapidly through the flock, and most of the birds become colonized and remain so until slaughter (Heuer et al., 2001; Newell et al., 2001; Zhao et al., 2001; Ledergerber et al., 2003; Newell and Fearnley, 2003;

Luangtongkum et al., 2006; Price et al., 2007). Sampling of broiler carcasses and domestic human cases showed that 85% of *Campylobacter* isolates in humans had genetic sequences identical to those of isolates from broilers. Due to the difficulties in eliminating contamination carcasses in slaughter plants, the control of *Campylobacter* in broiler flocks and subsequent production of birds free from colonization at slaughter is essential for preventing human cases (Van de Giessen et al., 1992; Guerin et al., 2007).

In general, the factors associated with an increased risk of *Campylobacter* were increasing median flock size, spreading manure on a farm in the winter, and increasing the number of broiler houses on the farm. For each additional house on the farm, the risk of *Campylobacter* colonization increased by approximately 6 to 14%, possibly as a consequence of the introduction of bacteria into the house from the environment, the increased movement of farmworkers between houses, or difficulty in maintaining strict hygiene or biosecurity practices (Guerin et al., 2007). Some studies showed that a farm using official water sources had approximately one-third to one-half the risk of *Campylobacter* as that of farms using nonofficial untreated sources. Van Gerwe (2005) reported that after introduction of *Campylobacter* in a flock, each broiler will infect on average 1.04 new broilers per day. Various studies (Karenlampi et al., 2007) suggested that different genotypes may be more prevalent in different geographical areas. Colonization with *Campylobacter* is not limited to *Campylobacter* sources within a broiler facility; the immediate external environment has also been shown to be an important source of *Campylobacter* for colonization. Once a flock becomes colonized with *Campylobacter*, these organisms can be pumped into the environment via tunnel ventilation systems (Price et al., 2007).

With an unusually high minimum growth temperature of around 30°C and a requirement for low-oxygen conditions, *Campylobacter* are unlikely to multiply on either carcasses or processing equipment. However, relatively high numbers can be introduced into a processing plant on the skin and feathers and in the intestines of carrier birds. Invariably, this results in widespread contamination of processing equipment, working surfaces, and process water, so that control of product contamination is extremely difficult in the case of these organisms (Mead, 1993).

Since *Campylobacter* is much more sensitive than many other types of bacteria to adverse environmental conditions, it might be expected that the organism would rapidly die out during processing and that scalding in particular would eliminate surface contamination. In practice, levels on the skin are reduced during scalding at 58°C but not at lower temperatures. Nevertheless, it appears that *C. jejuni* is more heat-resistant when attached to poultry skin, and even at 60°C many of the skin-associated cells may remain viable (Mead, 1993).

Different studies observed that the evisceration, knives, and the pooling of edible carcasses and parts in tubs contribute to overall bacteria contamination (Harris et al., 1986b). With poultry carcasses, the exposure to cold air markedly reduced levels of *Campylobacter* on the breast surface but not on the neck skin or inside the abdominal cavity, where sufficient moisture was retained. Thus, despite their apparent fragility, *Campylobacter* are largely able to survive the effects of processing (Daud et al., 1978; Mead, 1993; Newell et al., 2001; Bhaduri and Cottrell, 2004). The presence of viable *C. jejuni* after refrigerated and frozen storage is significant given that ingestion of only 500 *C. jejuni* cells has resulted in illness in human experimental infections (Bhaduri and Cottrell, 2004).

Salmonella spp.

Salmonella may cause gastroenteritis in people of all ages and severe invasive disease in infants, elderly persons, and immunocompromised persons. During the past two decades, the incidence of zoonotic foodborne *Salmonella* infections in industrialized countries has increased progressively. In addition, the frequency of antimicrobial resistance and the number of resistance determinants in *Salmonella* has risen markedly. Food animals, and especially poultry and their products (approximately 50% of the foodborne outbreaks were the result of the consumption of poultry products surface-contaminated with *Salmonella*), are considered to be the main source of human salmonellosis (Bello-Pérez et al., 1990; Infante et al., 1994; Boonmar et al., 1998a; Mokgatla et al., 1998). *S. enteritidis* poisoning is even associated with contaminated grade A eggs, and *S. typhimurium* is a major foodborne pathogen regularly associated with poultry (Mokgatla et al., 1998). *S. enteritidis* was detected in the chicken feces collected on farms; the contamination of chicken meat with *S. enteritidis* might be due to the contamination of intestinal contents through the equipment in slaughterhouses (Boonmar et al., 1998b). Nontyphoidal *Salmonella* infections are an important public health problem worldwide (Zaidi et al., 2006; Kim et al., 2007).

Zaidi et al. (2007) reported the emergence and rapid dissemination of multidrug-resistant *S. typhimurium* in food animals and humans in Mexico. Some studies (Boonmar et al., 1998b; Zaidi et al., 2006). suggest that some of the sporadic human *Salmonella* infections are due to the consumption of contaminated broiler chicken meat because the bacteria have similar genotypes. A risk assessment suggests that the ability of *S. enteritidis* to survive food processing and/or cause disease is almost seven times greater than that of *S. typhimurium* and 17 and 50 times greater than that of *S. infantis* and *S. dublin*, respectively (Hald et al., 2004).

Three main sources of salmonellae have been identified as contamination of processed carcasses: contaminated feed, cross-contamination during feed, and contamination during transport from the farm to the slaughterhouse. Feed has been identified as a source of salmonellae for growing broiler flocks, but some studies suggest that contamination of the feathers of growing broilers may be more common than intestinal carriage of salmonellae, and for that reason it is possible that salmonellae entering a processing plant on the feathers of incoming birds may be a more important source of contamination (Morris and Wells, 1970; Rigby et al., 1980).

Cooked foods continue to be important vehicles in human salmonellosis in many countries, and the occurrence of strains of *Salmonella* that appear to have greater tolerance to heat highlights the need for the food industry to continue to challenge the safety of food production systems. It may be that muscle attachment, which may occur naturally during the preparation of comminuted meat products, could permit greater survival during subsequent cooking. This may be a possible explanation for the involvement of cooked foods in outbreaks and cases of infection with *Salmonella* spp. (Humphrey et al., 1997).

Escherichia coli

Escherichia coli strains are part of the normal anaerobic microflora of the intestinal tracts of humans and animals of warm blood. *E. coli* strains cause diarrhea and are categorized in specific groups based on their virulent properties, mechanisms of pathogenicity, clinical syndromes, and various serogrups. Some of these categories are enteropathogenic strains, enterotoxigenic strains, enteroinvasive strains, those of diffuse adherence, enteroagregatives, and enterohemorragics.

Listeria monocytogenes

The genus *Listeria* includes different species of gram-positive bacteria with different forms of bacilli. The species *Listeria monocytogenes* provokes severe diseases such as meningitis and meningoencephalitis in human beings. Another syndrome associated with this microorganism is listeriosis in pregnant women, which generally causes an abortion or a premature septic birth, as well as the syndrome of meningitis in the newborn child. *L. monocytogenes* presents a different type of problem because it is one of the few foodborne pathogens that is capable of growth under chill temperature. Once introduced into the processing plant, this relatively hardy organism is likely to grow on any suitable wet surface, thus increasing the chances of carcass contamination (Mead, 1993). Despite the high incidence of this microorganism in the food, a low incidence of listeriosis has been observed.

Staphylococcus spp.

The genus *Staphylococcus* is a group of bacteria that cause a wide range of human and animal diseases and they are divided into positive or negative coagulase, according to their ability to clot rabbit plasma. *S. aureus* is the most notable member of the coagulase-positive staphylococci and is considered the most virulent of the staphylococci. In contrast, coagulase-negative staphylococci are normally considered benign organisms that are part of the normal flora. However, in recent years the number of coagulase-negative organisms implicated in human and animal disease has risen dramatically (DeBoer et al., 2001). On the average 20% of all foodborne outbreaks is a consequence of the consumption of food contaminated with enterotoxins produced by bacteria of the genus *Staphylococcus*, and principally for the species *S. aureus*.

S. aureus produces some extracellular compounds such as the hemolisins, enterotoxins, coagulase, nucleases, and lipases. The enterotoxins produce symptoms of the poisoning staphylococci, and many of them have an important role in the pathogenicity of some other diseases. This poisoning is characterized by nausea, vomiting, abdominal cramps, and occasionally, diarrhea without the presence of fever; general discomfort and headache are also possible.

CHEMICAL HAZARDS

Contamination of food may occur through environmental pollution of the air, water, and soil, such as with toxic metals, polychlorinated biphenyls, and dioxins. Other chemical hazards, such as naturally occurring toxicants, may arise at various points during food production, harvest, processing, and preparation. Because the period of time between exposure to chemicals and the effect is usually long, it is difficult to attribute disease caused by long-term exposure to chemicals in a specific type of food. This is one of the reasons why, in contrast to biological hazards, the protection of public health from chemical hazards has long largely employed the risk assessment paradigm. Exposure to chemicals in foods can result in acute and chronic toxic effects ranging from mild and reversible to serious and life threatening. These effects include cancer, birth defects, and damage to the nervous, reproductive, or immune systems (Rocourt et al., 2003).

A significant portion of human cancers may relate to dietary factors, including both exogenous and endogenous mutagens. Of the exogenous factors, certain metals and certain pesticides (both naturally produced and those manufactured by industry), *N*-nitroso compounds, heterocyclic amines, and polycyclic aromatic hydrocarbons are all probable human carcinogens (Rocourt et al., 2003). The dioxins belong to the aromatic polyhalogenated hydrocarbon group of chemical compounds. The dioxins are a group of chemical toxic compounds manufactured by industry that accumulate in the lipidic tissue of animals and humans. The Belgian dioxin crisis (January 1999), which probably entailed a higher exposure to dioxin through its polychlorinated biphenyl content than through its polychlorinated dibenzodioxin/polychlorinated dibenzofuran, should be considered a potentially important public health event (van Larebeke et al., 2001).

Antibiotic resistance is increasing among most human pathogens. The many bacteria resistant to multiple antibiotics in particular present a heightened concern. In some cases there are few or no antibiotics available to treat resistant pathogens (Gilchrist et al., 2007). Increased antibiotic resistance can be traced to the use and overuse of antibiotics. Much of that use occurs in human medicine, although antibiotic overuse in animals is problematic and the magnitude of the problem is unknown. The Union of Concerned Scientists has estimated that 11.2 million kilograms of the antibiotics used annually in the United States are administered to livestock as growth promoters. This compares with their estimate of 1.4 million kilograms for human medical use. This prolonged use of antibiotics, especially at low levels, presents a risk of not killing the bacteria while promoting their resistance by selecting for resistant populations. The resistance genes can pass readily from one type of bacteria to another. Consumers of meat become colonized through insufficient cooking. Ultimately, these genes may pass into pathogens, and diseases that were formerly treatable will be capable of causing severe illness or death (Zaidi et al., 2003; Gilchrist et al., 2007; Price et al., 2007). Antimicrobial-resistant *Campylobacter* may be persistent contaminants of poultry products for years after on-farm antimicrobial use has ended (Ishihara et al., 2006; Luangtongkum et al., 2006; Price et al., 2007). The antimicrobial resistance rates vary significantly in different production types, because, in general, conventionally raised broilers and turkeys harbor more antimicrobial-resistant bacteria than do organically raised broilers and turkeys (Ledergerber et al., 2003; Luangtongkum et al., 2006).

FOOD SAFETY PROGRAMS AND REGULATIONS

Traditional Programs and Regulations

Food safety programs and regulations were introduced to avoid consumption of tainted foods and promote fair trade. Regulations became more relevant as knowledge about etiology and other factors affecting food safety developed. But it is now evident that in the development of food safety regulations, it is practically impossible to foresee all possible conditions that could generate unsafe foods (Elbert, 1981). Worldwide, small and less developed businesses (SLDBs) present diverse and sometimes very complex challenges to ensuring food safety. These challenges are similar, irrespective of geographical location: lack of resources and lack of technical expertise. Large food businesses supplying export markets are more capable of complying requirements perhaps because the adoption of these is sometimes a basic requirement in major international food markets. Then the dilemma consists of the development of regulations for general applications or specific regulations for large and small LDBs, with the unavoidable disagreement of large businesses when they are concurrent in the same market.

During the process of inspection there are always subjective criteria implied in the assessment of compliance with regulations. The inspection has other limitations, such as the time spent on the premises. Nevertheless, inspections are relevant in those operations in which prevalent physical conditions are critical or sanitary practices violate elementary regulatory requirements. In these cases, basic sanitary problems can be solved through traditional inspection to prevent foodborne diseases, but this does not mean a recognition that traditional inspections have severe limitations as a mean to assure food safety (Fernández, 2000).

The use of microbial testing should assume that microbial contamination is usually sporadic and the distribution of a pathogenic agent does not follow a regular pattern of distribution, and that the exactness and precision of present microbiological methods are low (Mossel and Drake, 1990). Sampling and testing of final products have severe limitations, such as limited representatives of samples, small number of samples tested, high cost, limitations to investigating all possible pathogens present in the sampled product, and a limited number of laboratories to sample and test all pertinent samples. Also, microbiological analyses usually take several days to complete.

Evolution of Food Safety Programs

Mandatory microbiological criteria apply to those products and/or points of the food chain where no other more effective tools are available, because a preventive approach offers more control than microbiological testing, since the effectiveness of microbiological examination to assess the safety of foods is limited. Where the use of microbial criteria is appropriate, they should be product-type specific and applied only at the point of the food chain specified in the regulation. Microbiological criteria should be developed for a particular food to define the acceptability of a product or a food lot, based on the absence or presence or the number of microorganisms, including parasites, and/or the quantity of their toxins or metabolites per unit of mass, volume, area, or lot. The principles for the establishment and application of microbiological criteria for foods should consider purpose, microbiological methods, microbiological limits, sampling plans, and methods and handling and reporting (CAC, 1997)

The Recommended International Code of Practice, General Principles of Food Hygiene, developed within the Codex Alimentarius Commission (CAC, 2003a) lay a firm foundation for ensuring food hygiene. The principles include primary production, design, and facilities of the establishment, control of operation, maintenance, and sanitation, personal hygiene, transportation, product information, and awareness and training. They are used in conjunction with specific codes of hygienic practice such as the *Code of Practice for the Hygiene of Meat* (CAC, 2005a) and guidelines on microbiological criteria.

Prerequisite programs originated from food regulations and voluntary food industry programs. Good manufacturing practices (GMPs) were established to help define for the food industry minimal sanitary conditions for safe processing of food products. The NACMCF 1997 HACCP Guidelines list 11 prerequisite programs in its Appendix A: facilities, supplier control, specifications, production equipment, cleaning and sanitation, personal hygiene, training, chemical control, receiving, storage, and shipping, traceability and recall, and pest control (Surak and Wilson, 2007).

The HACCP system, which is science based and systematic, identifies specific hazards and measures for their control to ensure the safety of food. HACCP is a tool to assess hazards and establish control systems that focus on prevention rather than relying mainly on end-product testing. Any HACCP system is capable of accommodating change, such as advances in equipment design, processing procedures, or technological developments. HACCPs can be applied throughout the food chain from primary production to final consumption and its implementation should be guided by scientific evidence of risks to human health. In addition to enhancing food safety, implementation of HACCPs can provide other significant benefits. In addition, the application of HACCP systems can aid inspection by regulatory authorities and promote international trade by increasing confidence in food safety (CAC, 2003b).

INTERNATIONAL REFERENCE STANDARDS

World Organization for Animal Health

The need to control animal diseases at the global level led to the creation of the Office International des Epizooties through an international agreement signed on January 25, 1924. In May 2003 the Office became the World Organization for Animal Health but kept its historical acronym, OIE. The OIE is the intergovernmental organization responsible for improving animal health worldwide. As of January 2008 it had a total of 172 member countries and territories. OIE standards are recognized by the WTO as a reference for international sanitary rules (www.oie.int).

The organization is placed under the authority and control of an international committee consisting of delegates designated by the governments of all member countries. Each member country undertakes to report the animal diseases that it detects on its territory. The OIE then disseminates information to other countries, which can take the necessary preventive action. This information also includes diseases transmissible to humans and intentional introduction of pathogens. Information is sent out immediately or periodically, depending on the seriousness of the disease. This objective applies to both naturally occurring diseases and those caused deliberately.

The OIE collects and analyses the latest scientific information on animal disease control. This information is then made available to the member countries to help them to improve the methods used to control and eradicate these diseases. The OIE develops normative documents relating to rules that member countries can use to protect themselves from the introduction of diseases and pathogens without setting up unjustified sanitary barriers. The main normative works produced by the OIE are the *Terrestrial Animal Health Code*, the *Manual of Diagnostic Tests and Vaccines for Terrestrial Animals*, the *Aquatic Animal Health Code*, and the *Manual of Diagnostic Tests for Aquatic Animals*.

Codex Alimentarius Commission

The Codex Alimentarius Commission was created in 1963 by the Food and Agriculture Organization (FAO) and the World Health Organization (WHO) to develop food standards, guidelines, and related texts, such as codes of practice under the Joint FAO/WHO Food Standards Programme. The main purposes of this program are to protect the health of consumers and to ensure fair trade practices in the food trade, and to promote coordination of all food standards work undertaken by international governmental and nongovernmental organizations: www.codexalimentarius.net/web/index_en.jsp). Different sets of standards that arose from national food laws and regulations inevitably gave rise to trade barriers that were of increasing concern to food traders. The advantages of having universally agreed food standards for the protection of consumers were recognized by international negotiators during the Uruguay

Round. It is not surprising, therefore, that the World Trade Organization's (WTO's) SPS Agreement and TBT Agreement, within the separate areas of their legal coverage, both encourage the international harmonization of food standards. Importantly, the SPS Agreement cites Codex's food safety standards, guidelines, and recommendations for facilitating international trade and protecting public health. The standards of Codex have also proved an important reference point for the dispute settlement mechanism of the WTO (www.wto.org/english/thewto_e/coher_e/wto_codex_e.htm),

The Codex Alimentarius is a collection of standards, codes of practice, guidelines, and other recommendations supported by committees with representatives of almost all countries of the world. The development of a new standard or other text follows a stepwise process. Starting from the project proposal and until the standard, guideline, or other text is published by the Codex Secretariat, decisions are made by consensus. The CAC supports its decisions in the joint FAO/WHO expert committees, which are independent of the commission (and the commission's subsidiary bodies), although their output contributes significantly to the scientific credibility of the commission's work. Some texts are specifically relevant for the safety of poultry products, such as *Maximum Residue Limits for Veterinary Drugs in Foods* (CAC, 2006a), *Code of Practice for the Prevention and Reduction of Dioxin and Dioxin-like PCB Contamination in Foods and Feeds* (CAC, 2006b), *Code of Practice to Minimize and Contain Antimicrobial Resistance* (CAC, 2005b), as well as the above-mentioned *Code of Practice for the Hygiene of Meat* (CAC, 2005a).

REGIONAL FOOD SAFETY REGULATIONS

Australia and New Zealand

On December 20, 2002, the Australia New Zealand Standards Code (FSANZ) became the uniform code applying to both countries, with some exceptions reserved for certain food standards not to apply in New Zealand. The industry had two years before the code came into effect to decide on complying with either the old regulations or the new code, but not a mixture (Food Standards Australia New Zealand Act, 1991). The primary function of FSANZ is to develop, modified, and review standards in the Australia New Zealand Food Standards Code and, where appropriate, make recommendations to governments about these standards. Other functions are to oversee matters of food surveillance, to develop food safety education initiatives, to coordinate food recalls, and to develop policies on imported foods (Food Legal Consultancy, 2008).

FSANZ requires that food industries have a food safety plan (e.g., a HACCP system) before starting a business. Some industries may operate either voluntary industry compliance schemes or have contractually imposed standards. These additional standards may address issues such as food composition and hygienic specifications, traceability, and transport requirements. Any body or person may apply to FSANZ for the development or variation of food standards. If FSANZ decides to proceed with the application, it must make a full assessment. However, the authority has the power on its own initiative to develop, modify, or review standards.

While the FSANZ code was expected by the federal government to introduce more uniformity in food safety practices and food standards throughout Australia and New Zealand, in most states and territories actual enforcement of food law was dealt with by the local municipal councils, taking into account their different priorities and key food industry operators.

European Union

In 2000 the European Union (EU) published its White Paper on Food Safety (EC, 2000), setting out a legislative action plan for a proactive new food policy. Key elements in the new approach were the establishment of a framework regulation; the establishment of the European Food Safety Authority (EFSA), an independent body providing scientific advice to the legislators; the development of specific food and feed safety legislation, including a major overhaul of the existing hygiene legislation; and the creation of a framework for harmonized food controls. The EU developed a "farm-to-fork" approach covering all sectors of the food and feed chain, with traceability as a basic concept. Application of the "precautionary principle" as described in the February 2000 Commission Communication on the Precautionary Principle (EC, 2000b) is also an important concept in the EU's approach. The overall aim of the radical revision of the EU's food safety rule has been to create a single hygiene regime covering food and food operators in all sectors, together with effective instruments to manage food safety and any possible food crises throughout the food chain. A regulatory package for food hygiene went into force on January 1, 2006.

Food producers bear primary responsibility for the safety of food through the use of a food safety programmes and procedures based on HACCP principles or an HACCP system. It was also determined that community rules should not apply either to primary production for private domestic use, or to the domestic preparation, handling, or storage of food for private domestic consumption. However, it was considered appropriate to protect public health through national law, in particular because of the close relationship between the producer and the consumer. These and other general requirements for primary production, technical requirements, registration/approval of food business, and national guides to good practice were included in the regulation (EC, 2004a).

Specific hygiene rules for food of animal origin (approval of establishments, health and identification marking, imports, food chain information) were included. Requirements were determined for poultry meat in the transport of live animals to the slaughterhouse, for slaughterhouses, for cutting plants, for slaughter hygiene, for hygiene during and after cutting and deboning, for slaughter on the farm, for minced meat, and for meat preparations and mechanically separated meat products and by-products (EC, 2004b). The regulatory package also detailed rules for the organization of official controls on products of animal origin (methods to verify compliance of general and specific hygiene rules and animal by-products regulation 1774/2002) (EC, 2004c).

Also, rules for the production, processing, distribution, and importation of products of animal origin entered into force on January 1, 2005. The purpose of this regulation is to ensure that proper and effective measures are taken to detect and control *Salmonella* and other zoonotic agents at all relevant stages of production, processing, and distribution, particularly at the level of primary production, including in feed, in order to reduce their prevalence and the risk they pose to public health (EC, 2002). Additional regulations have been issueed to ensure that proper and effective measures are taken to detect and control *Salmonella* and other zoonotic agents at all relevant stages of production, processing, and distribution, particularly at the level of primary production, including in feed, to reduce their prevalence and the risk they pose to public health (EC, 2003).

Mercosur

On March 26, 1991, Argentina, Brazil, Paraguay, and Uruguay signed the Agreement of Asuncion, creating the Southern Common Market, Mercosur. The institutional organizational structure of Mercosur rests on nine main groups. The Common Market Group's Task Force No. 3, Technical Regulations and Conformity Assessment, includes the Commission on Food (www.mercosur.int/msweb/).

Each country has its own food safety legislation. The poultry regulation is applied independent of other countries. When it comes to standards applicable to Mercosur, the Food Commission, composed of representatives of all member countries, harmonize standards, which are adopted and implemented by each country through different mechanisms. For example, Argentina has the Argentine Food Code, which comprises all food safety regulations applicable domestically. When Mercosur emits a regional standard, it becomes part of Argentinean law.

The food safety national authorities are: for Argentina, the Ministry of Health and the Ministry of Economy and Production through the Secretary of Agriculture, Livestock, Fisheries and Food; for Brazil, the Ministry of Health through the National Sanitary Surveillance Agency; for Paraguay, the Ministry of Public Health and Social Welfare through the National Institute of Nutrition and Food and the Ministry of Industry and Commerce; and for Uruguay, the Ministries of Public Health and Industry, Energy, and Mining.

The regulations of each country (for both domestic and foreign trade) focus on building and equipment construction and design, good manufacture practices, standardized operational procedures of sanitization and HACCP, postmortem inspection and antemortem examination, mandatory rules for establishments manufacturing specific meat products, definition of products and specific design requirements, use of additives, classification and definition of birds, mandatory waste product management, packaging and labeling, health certificates, transport of food products from animal origin, and the regime of hearing and penalties. Most regional standards are concerned with food additives and flavors, packaging materials, labeling and specifications of identity, and quality of foodstuffs. They are consistent with international guidelines set by the Codex Alimentarius (www.puntofocal.gov.ar/mercosur_sgt_alimentos.htm).

North America

Under federal acts (Meat Inspection Act. 1985; Poultry Products Inspection Act, 21 U.S.C. 451; Federal de Health Animal, 2007; Ley General de Salud, 2007) and regulations, the Canadian Food Inspection Agency (CFIA), the Food Safety and Inspection Service (FSIS) of the U.S. Department of Agriculture, and the Mexican authorities of agriculture, the Mexican National Service of Agifood Health, Safety and Quality (SENASICA), and health, the Federal Commission for the Protection Against Sanitary Risk (COFEPRIS), inspect poultry products sold in interstate and foreign commerce, including imported products, to ensure that poultry products are safe, wholesome, and correctly labeled and packaged. The CFIA, FSIS, and SENASICA verify that poultry products leaving federally inspected establishments or being imported are safe and wholesome. The CFIA also monitors registered and unregistered establishments for labeling compliance. The FSIS monitors meat and poultry products after they leave federally inspected plants and state inspection programs, which inspect poultry products sold only within the state in which they were produced. In 1990, Canada issued the meat inspection regulation, which established facility and equipment design requirements for the registration of establishments and the design and implementation of prerequisite programs, other control programs, and HACCP for the licensing of registered establishment operators. The Food Safety Enhancement Program (FSEP) is the CFIA approach to encourage and support the development, implementation, and maintenance of HACCP systems in all federally registered establishments (*Food Safety Enhancement Program Manual*, 2006). It is consistent with international trends, emphasizing cooperation between government and industry, whose members are ultimately responsible for the safety of their own products.

In 1996, the United States, through the FSIS, implemented HACCP and pathogen reduction final rules applicable to meat and poultry processors to develop and implement working HACCP plans for their products. Establishments are also required to develop and implement written sanitation standard operating procedures (SSOPs) and regular microbial testing to verify the adequacy of establishments' process controls for the prevention and removal of fecal contamination and associated bacteria (CFR, 1996).

In Mexico since 1994, federally inspected establishments have to comply with regulations for facilities and equipment (Mexican Official Standard, 1994a) and the sanitary processing of meat (Mexican Official Standard, 1994b). Derived from the issue of a new animal health federal act in 2007, SENASICA would have the authority to regulate good hygienic practices in primary production and to consolidate the inspection of SSOPs and eventually HACCPs in federally inspected establishments. Hygienic practices in municipal slaughterhouses are inspected by COFEPRIS and state health regulatory agencies. Inspection and enforcement were strengthened in 2004, when a new regulation was issued in view of the less developed infrastructure of these establishments (Mexican Official Standard, 2004).

China

In response to circumstances in China, in August 2007 the Information Office of the State Council of the Chinese government published a white paper entitled China's Food Quality and Safety. This document addresses the quality and safety of foods in general. It reviews activities made to build and improve a supervisory system and mechanism for food safety, strengthened legislation, and the setting of relevant standards (Information Office of the State Council of the People's Republic of China, 2007).

The Food Safety Regime aimed at guaranteeing food safety, improving food quality, and regulating food imports and exports through the Law on the Quality and Safety of Agricultural Products, Food Hygiene Law, Law on Import and Export Commodity Inspection, Law on Animal and Plant Entry and Exit Quarantine, Frontier Health and Quarantine Law, and Law on Animal Disease Prevention, among others. Specific administrative regulations in this regard include those for strengthening safety supervision and administration of food and other products; for administration of production licenses; for certification and accreditation; for import and export commodity inspection; for animal and plant entry and exit quarantine; for veterinary medicine; and for feedstuffs and feed additives.

So far, China has promulgated more than 1800 national standards concerning food safety and more than 2900 standards for the food industry, among which 634 national standards are mandatory. To solve the problems of overlapping food safety standards and poor organization, China has sorted out and reviewed many national standards, industrial standards, local standards, and enterprise standards, and worked out plans to enact over 280 national standards. It has also worked hard to promote and enforce these standards, and it urges food-producing enterprises to abide by them strictly.

REFERENCES

Bello-Pérez LA, Ortiz-Dillanes M, Pérez-Memije E, Castro-Domíneguez V. 1990. Salmonella en carnes crudas: un estudio en localidades del Estado de Guerrero. Salud Publ Mexico 32:74–79.

Bhaduri S, Cottrell B. 2004. Survival of cold-stressed *Campylobacter jejuni* on ground chicken and chicken skin during frozen storage. Appl Environ Microbiol 70(12): 7103–7109.

Boonmar S, Bangtrakulnonth A, Pornrunangwong S, Marnrim N, Kaneko K, Ogawa M. 1998a. *Salmonella* in broiler chickens in Thailand with special reference to contamination of retail meat with *Salmonella enteritidis*. J Vet Med Sci 60(11):1233–1236.

Boonmar S, Bangtrakulnonth A, Pornrunangwong S, Terajima J, Watanabe H, Kaneko KI, Ogawa M. 1998b. Epidemiological analysis of *Salmonella enteritidis* isolated from humans and broiler chickens in Thailand by phage typing and pulsed-field gel electrophoresis. J Clin Microbiol 36(4):971–974.

CAC (Codex Alimentarias Commission). 1997. *Principles for the Establishment and Application of Microbiological Criteria for Foods*. CAC/GL-21-1997.

CAC. 2003a. *Recommended International Code of Practice: General Principles of Food Hygiene*. CAC/RCP 1-1969, Rev. 4–2003.

CAC. 2003b. Annex to CAC/RCP 1–1969 (Rev. 4–2003).

CAC. 2005a. *Code of Practice for the Hygiene of Meat*. CAC/RCP 58/2005.

CAC. 2005b. *Code of Practice to Minimize and Contain Antimicrobial Resistance*. CAC/RCP 61–2005.

CAC. 2006a. *Maximum Residue Limits for Veterinary Drugs in Foods*. CAC/MRL 02–2006. Updated as at the 29th Session of the Codex Alimentarius Commission (July 2006).

CAC. 2006b. *Code of Practice for the Prevention and Reduction of Dioxin and Dioxin-like PCB Contamination in Foods and Feeds*. CAC/RCP 62–2006, p. 1 of 11.

CFR. 1996. Pathogen Reduction-Hazard Analysis and Critical Control Point (HACCP) Systems final rule. 61 CFR 38806, July 25.

Cortez ALL, Carvalho ACFB, Scarcelli E, Miyashiro S, Vidal-Martins AMC, Burger KP. 2006. Survey of chicken abattoir for the presence of *Campylobacter jejuni* and *Campylobacter coli*. Rev Inst Med Trop 48(6):307–310.

Daud HB, McMeekin TA, Olley J. 1978. Temperature function integration and the development and metabolism of poultry spoilage bacteria. Appl Environ Microbiol 36(5):650–654.

DeBoer LR, Slaughter DM, Applegate RD, Sobieski RJ, Crupper SS. 2001. Antimicrobial susceptibility of staphylococci isolated rom the faeces of wild turkeys (*Meleagris gallopavo*). Lett Appl Microbiol 33:382–386.

Ebert H. 1981. Government involvement in the food industry. Dairy Food Sanitat 1: 458–459.

EC (European Commission). 2000a. *White Paper on Food Safety*. COM(1999) 719 final. Brussels, Jan. 12. http://ec.europa.eu/dgs/health_consumer/library/pub/pub06_en.pdf.

EC. 2000b. *Communication from the Commission on the Precautionary Principle*. COM(2000) 1 final. Brussels, Jan. 2. http://eur-lex.europa.eu/smartapi/cgi/sga_doc?smartapi!celexplus!prod!CELEXnumdoc &lg=en&numdoc=52000DC0001.

EC. 2002. Regulation 2002/99/EC.

EC. 2003. Regulation 2003/2160/EC.

EC. 2004a. Regulation 2004/852/EC.

EC. 2004b. Regulation 2004/853/EC.

EC. 2004c. Regulation 2004/854/EC.

FAO (Food and Agriculture Organization). 2004. Garantía de la inocuidad y calidad de los alimentos. Directrices para el fortalecimiento de los sistemas nacionales de control de los alimentos. In: *Estudio FAO Alimentación y Nutrición*. Rome: FAO.

FAO 2006. *FAO/WHO Guidance to Governments on the Application of HACCP in Small and/or Less-Developed Food Businesses*. FAO Food and Nutrition Paper 86. Rome: FAO.

FDA (U.S. Food and Drug Administration). 1986. *Current Good Manufacturing Practice in Manufacturing, Packing, or Holding Human Food*. 21 CFR 110.

Federal de Health Animal. 2007. *Diario Oficial de la Federación*. July 25, p. 39.

Fernández E. 2000. *Microbiologáa e Inocuidad de los Alimentos*. Querétaro, Mexico: Universidad Autónoma de Querétaro.

Food Legal Consultancy. 2008. http://www.foodlegal.com.au/.

Food Safety Enhancement Program Manual. 2006. Last amendment July 6, 2006. http://www.inspection.gc.ca/english/fssa/polstrat/haccp/haccpe.shtml.

Gilchrist MJ, Greko C, Wallinga DB, Beran GW, Riley DG, Thorne PS. 2007. The potential role of concentrated animal feeding operations in infectious disease epidemics and antibiotic resistance. Environ Health Perspect 15(2):313–316.

Guerin MT, Martin W, Reiersen J, Berke O, McEwen SA, Bisaillon JR, Lowman R. 2007. A farm-level study of risk factors associated with the colonization of broiler flocks with *Campylobacter* spp. in Iceland, 2001–2004. Acta Vet Scand 49:18–29.

Hald T, Vose D, Wegener HC, Koupeev T. 2004. A Bayesian approach to quantify the contribution of animal-food sources to human salmonellosis. Risk Anal 24(1):255–269.

Harris NV, Weiss NS, Nolan CM. 1986a. The role of poultry and meats in the etiology of *Campylobacter jejuni/coli* enteritis. Am J Public Health 76:407–411.

Harris NV, Thompson D, Martin DC, Nolan CM. 1986b. A survey of *Campylobacter* and other bacterial contaminants of pre-market chicken and retail poultry and meats, King Country, Washington. Am J Public Health 76:401–406.

Heuer OE, Pedersen K, Andersen JS, Madsen M. 2001. Prevalence and antimicrobial susceptibility of thermophilic *Campylobacter* in organic and conventional broiler flocks. Lett Appl Microbiol 33:269–274.

Hoornstra E, Notermans S. 2001. Quantitative microbiological risk assessment. Int J Food Microbiol 66:21–29.

Humphrey TJ, Wilde SJ, Rowbury RJ. 1997. Heat tolerance of *Salmonella typhimurium* DT104 isolated attached to muscle tissue. Lett Appl Microbiol 25:265–268.

Idris U, Lu J, Maier M, Sanchez S, Hofacre CL, Harmon BG, Maurer JJ, Lee MD. 2006. Dissemination of fluorquinolone-resistant *Campylobacter* spp. within an integrated commercial poultry production system. Appl Environ Microbiol 72(5):3441–3447.

Infante D, de Nouera C, León AJ, Catari M, Herrera AJ, Valdillo P. 1994. Aislamiento de *Salmonella* en canales de pollos. Vet Trop 19:91–99.

Information Office of the State Council of the People's Republic of China. 2007. China's food quality and safety, 2007. http://www.gov.cn/english/2007-08/17/content_720346.htm. Accessed June 6, 2008.

Ishihara K, Yano S, Nishimura M, Asai T, Kojima A, Takahashi T, Tamura Y. 2006. The dynamics of antimicrobial-resistant *Campylobacter jejuni* on Japanese broiler farms. J Vet Med Sci 68(5):515–518.

Kapperud G, Skjerve E, Bean NH, Ostroff SM, Lassen J. 1992. Risk factors for sporadic *Campylobacter* infections: results of a case–control study in southeastern Norway. J Clin Microbiol 30(12):3117–3121.

Karenlampy R, Rautelin H, Schonberg-Norio D, Pauli L, Hanninen ML. 2007. Longitudinal study of finnish *Campylobacter jejuni* and *C. coli* isolates from humans, using multilocus sequence typing, inlcuding comparison with epidemiological data and isolates from poultry and cattle. Appl Environ Microbiol 73(1):148–155.

Keene WE. 2006. Lessons from investigation of foodborne diseases outbreaks. JAMA 281(19):1845–1847.

Kim A, Lee YJ, Kang MS, Kwag SI, Cho JK. 2007. Dissemination and tracking of *Salmonella* spp. in integrated broiler operation. J Vet Sci 8(2):155–161.

Ledergerber U, Regula G, Stephan R, Danuser J, Bissig B, Stark KDC. 2003. Risk factors for antibiotic resistance in *Campylobacter* spp. isolated from raw poultry meat in Switzerland. BMC Public Health 3:39–47.

Ley General de Salud. 2007. Last amendment, DOF, July 19. http://www.cofepris.gob.mx/mj/documentos/leyes/LGS.pdf.

Luangtongkum T, Morishita TY, Ison AJ, Huang S, McDermott PF, Zhang Q. 2006. Effect of conventional and organic production practices on the prevalence and antimicrobial resistance of *Campylobacter* spp. in poultry. Appl Environ Microbiol 72(5):3600–3607.

Mead GC. 1993. Problems of producing safe poultry: discussion paper. J R Soc Med 86:39–42.

Meat Inspection Act. 1985. c. 25 (1st Suppl.). http://laws.justice.gc.ca/en/M-3.2.

Mexican Official Standard 1994a. NOM-008-ZOO-1994. Last amendment, DOF, Feb. 18, 1999.

Mexican Official Standard 1994b. NOM-009-ZOO-1994. DOF, Nov. 16, 2004. Last amendment, DOF, July 31, 2007.

Mexican Official Standard 2004. NOM-194-SSA1-2004. DOF Sept. 18.

Mokgatla RM, Brozel VS, Gouws PA. 1998. Isolation of *Salmonella* resistant to hypochlorous acid from a poultry abattoir. Lett Appl Microbiol 27:379–382.

Morris GK, Wells JG. 1970. *Salmonella* contamination in a poultry-processing plant. Appl Microbiol 19(5):795–799.

Mossel DDA, Drake DM. 1990. Processing food for safety and reassuring the consumer. Food Technol 44(12):63–67.

NACMF (U.S. National Advisory Committee on Microbiological Criteria for Foods). 1997. *Hazard Analysis and Critical Control Point Principles and Application Guidelines*. Washington, DC: U.S. Food and Drug Administration, App. A.

Nauta MJ, Jacobs-Reitsma WF, Havelaar AH. 2006. A risk assessment model for *Campylobacter* in broiler meat. Risk Anal 26(6):1–17.

Newell DG, Fearnley C. 2003. Sources of *Campylobacter* colonization in broiler chickens. Appl Environ Microbiol 69(8):4343–4351.

Newell DG, Shreeve JE, Toszeghy M, Domingue G, Bull S, Humphrey T, Mead G. 2001. Changes in the carriage of *Campyobacter* strains by poultry carcasses during processing in abattoirs. Appl Environ Microbiol 67(6):2636–2640.

Potter RC, Kaneene JB, Hall WN. 2003. Risk factors for sporadic *Campylobacter jejuni* infections in rural Michigan: a prospective case–control study. Am J Public Health 93(12):2118–2123.

Poultry Products Inspection Act. 21 U.S.C. 451.

Price LB, Lackey LG, Vailes R, Silbergeld E. 2007. The persistence of fluoroquinolone-resistant *Campylobacter* in poultry production. Environ Health Perspect 115(7):1035–1039.

Rigby CE, Pettit JR, Baker MF, Bentley AH, Salomons MO, Lior H. 1980. Flock infection and transport as sources of salmonellae in broiler chickens and carcasses. Can J Comp Med 44:328–337.

Rocourt J, Moy G, Schlundt J. 2003. *The Present State of Foodborne Disease in OECD Countries*. Geneva, Switzerland: World Health Organization.

Slader J, Domingue G, Jorgensen F, McAlpine K, Owen RJ, Bolton FJ, Humphrey TJ. 2002. Impact of transport crates reuse and of catching and processing on *Campylobacter* and *Salmonella* contamination of broiler chickens. Appl Environ Microbiol 68(2):713–719.

Surak J, Wilson S. 2007. *The Certified HACCP Auditor Handbook*. Milwaukee, WI: ASQ Press.

Van de Giessen A, Mazurier SI, Jacobs-Reitsma W, Jansen W, Berkers P, Ritmeester W, Wernars K. 1992. Study on the epidemiology and control of *Campylobacter jejuni* in poultry broiler flocks. Appl Environ Microbiol 58(6):1913–1917.

Van Gerwe TJWM, Bouma A, Jacobs-Reitsma WF, van den Broek J, Klinkenberg D, Stegeman JA, Heesterbeek JAP. 2005. Quantifying transmission of *Campylobacter* spp. among broilers. Appl Environ Microbiol 71(10):5765–5770.

van Larebeke N, Hens L, Schepens P, Covaci A, Baeyens J, Everaert K, Bernheim JL, Vlietinck R, De Poorter G. 2001. The Belgian PCB and dioxin incident of January–June 1999: exposure data and potential impact on health. Environ Health Perspect 109(3):265–273.

Zaidi MB, Zamora E, Díaz P, Tollefson L, Fedorka-Cray PJ, Headrick ML. 2003. Risk factors for fecal quinolone-resistant *Escherichia coli* in Mexican children: antimicrob Agents Chemother 47(6):1999–2001.

Zaidi MB, McDermott PF, Fedorka-Cray P, Leon V, Canche C, Hubert SK, Abbott J, León M, Zhao S, Headrick M, Tollefson L. 2006. Nontyphoidal *Salmonella* from human clinical cases, asymptomatic children, and raw retail meats in Yucatan, Mexico. Clin Infect Dis 42(1):1–9.

Zaidi MB, Leon V, Canche C, Perez C, Zhao S, Hubert SK, Abbott J, Blickenstaff K, McDermott PF. 2007. Rapid and widespread dissemination of multidrug-resistant *bla*_CMY - 2 *Salmonella typhimurium* in Mexico. J Antimicrob Chemother 60:398–401.

Zhao C, Ge B, de Villena J, Sudler R, Yeh E, Zhao S, White DG, Wagner D, Meng J. 2001. Prevalence of *Campylobacter* spp., *Escherichia coli*, and *Samonella* serovars in retail chicken, turkey, pork and beef from the Greater Washington, D.C., area. Appl Environ Microbiol 67(12):5431–5436.

32

MICROBIAL ECOLOGY AND SPOILAGE OF POULTRY MEAT AND POULTRY MEAT PRODUCTS

Elina J. Vihavainen and Johanna Björkroth
Department of Food and Environmental Hygiene, Faculty of Veterinary Medicine, University of Helsinki, Helsinki, Finland

CONCEPT OF A SPECIFIC FOOD SPOILAGE ORGANISM

When a food is spoiled, sensory changes make it unacceptable for human consumption. Spoilage may appear as physical damage, chemical changes (oxidation,

Handbook of Poultry Science and Technology, Volume 2: Secondary Processing, Edited by Isabel Guerrero-Legarreta and Y.H. Hui

color changes) or the appearance of off-flavors and off-odors. In poultry meat and meat products, water activity and nutritional content is high, and thus spoilage changes result from bacterial growth and metabolism. The repertoire of spoilage changes depend on the type of bacterial population developing and its metabolic activities. Off-odors and off-flavors are very typical bacterial spoilage changes, but slime formation, color and texture changes, and gas formation may also take place.

Bacterial populations in spoiled food consist of strains responsible for the sensory changes and those not having played any role in spoilage. Usually, the prevailing microbial group is responsible for the spoilage changes, but this is not necessarily the fact. Little is known about interactions of microbes during the development of the spoilage bacterial population. By the term *specific spoilage organism* (SSO) we refer to those microbes contributing actively to the development of spoilage changes.

The ability of a pure culture to produce the metabolites associated with the spoilage of a particular product is considered as the spoilage potential of a microorganism. A combination of microbiology, sensory analyses, and sometimes also chemistry is needed to determine which microorganism(s) are the SSOs of a particular food product. Prediction of the shelf life of a product is challenging because bacterial levels as total counts or even the level of a SSO as such cannot be used as an indicator of the sensory quality of a product. The level of SSOs can be used to predict the remaining shelf life of a product under conditions where the SSO is developing as expected. Therefore, sensory analysis remains important and product-specific data are needed for shelf-life predictions.

Extrinsic and intrinsic factors associated with a product select which microbes are SSOs. The meat industry relies on modified-atmosphere packaging (MAP) and cold storage to meet the demands of modern food logistics. This changes the order of significance of the prevailing SSOs from aerobic gram-negative bacteria such as *Pseudomonas* to anaerobic and facultatively anaerobic bacteria possessing fermentative metabolism. These bacteria belong primarily to psychrotrophic lactic acid bacteria (LAB), and to some extent Enterobacteriaceae.

MICROBIAL ECOLOGY IN A POULTRY MEAT ENVIRONMENT

Poultry Meat as a Growth Medium

Poultry meat is a good growth medium for various microorganisms, including those with relatively complex nutritional requirements. As mentioned earlier, poultry meat has a high water content and it contains variable nutrients, including carbohydrates (such as glycogen, glucose, glucose-6-phophate, ribose), amino acids, nucleotides, minerals, and B vitamins to support the growth of diverse microbial populations. In general, the microbial populations on meat preferentially utilize glucose as a source of energy (Nychas et al., 1998). Following glucose depletion, microorganisms continue to metabolize other low-molecular-weight compounds, such as lactate and glycolytic intermediates, and finally, amino acids. Although meat is rich in protein, breakdown of muscle proteins rarely appears until very late stages of spoilage (Nychas et al., 1998).

Factors in Selecting the Spoilage Population

To obtain safe poultry products with a reasonable storage life, perishable products must be stored chilled. During storage, refrigeration selects for psychrotrophic microorganisms, present initially only as a minor component (less than 10%) of the microbial population (Borch et al., 1996). Refrigeration extends the lag phase and the generation time of microorganisms, reducing the overall microbial growth and delaying the onset of spoilage. However, even a minor increase in temperature may stimulate microbial growth and have a critical effect on quality (Smolander et al., 2004). In addition to the cold chain, the type of packaging, the formulation and type of the product, and the number and type of initial spoilage bacteria determine the composition of the dominant bacterial population.

Among the intrinsic factors of poultry meat, pH and the availability of glucose and other simple sugars affect the development of the spoilage population and the rate of microbial growth and spoilage. The pH of the meat is highly dependent on the amount of glycogen in the muscle; in breast muscle, the postmortem glycolysis will lead to the accumulation of lactate and a reduction in pH to about 5.7 to 5.9. In contrast, the muscles in poultry legs have very low initial glycogen concentration and therefore a pH of 6.2 or above. The pH is also higher in poultry skin, with a pH as high as 6.6 to 7.2. Skin-on poultry cuts and high-pH meat and meat products may also support the development of different SSOs than those associated with spoilage skinless breast fillets.

Spoilage Changes Typical for Poultry Meat and Poultry Meat Products

The process of poultry spoilage is quite similar to that in beef and pork. The early signs of microbial spoilage are typically detectable when the bacteria have reached a population of 10^7 to 10^8 CFU/g (Borch et al., 1996; Stanbridge and Davis, 1998). The composition of the spoilage population determines the characteristic off-odors, with sour, acid, buttery, dairy, sweet, or fruity associated with early signs of poultry spoilage. At the later stages of spoilage, offensive putrid odors often develop as the microbes switch to utilize amino acids and produce compounds such as ammonia, hydrogen sulfide and other sulfides, and foul-smelling amines (such as putrescine and cadaverine). The repertoires of compounds formed during amino acid catabolism are dependent on the microbial species, the amino acids available, and the redox potential of the meat. In poultry cuts with high pH and low glucose content, glucose depletion leads to initiation of amino acid utilization, and putrefactive odors appear earlier than in breast fillet.

In addition to off-odors and off-flavors, formation of slime, gas, and purge are typical defects in spoilage of poultry. Slime formation is related to accumulation of exopolysaccharides produced by many spoilage microbes, whereas gas distension and loosening of packages is caused by the formation of carbon dioxide (CO_2). Purge accumulation is associated with SSOs with fermentative metabolism and is related to a reduction of pH and subsequent decrease in the water-holding capacity of the meat or meat product.

Spoilage of Aerobically Stored Poultry

Psychrotrophic, aerobic, or facultative anaerobic, particularly gram-negative bacteria, develop rapidly on poultry either stored in air or packaged using gas-permeable film. At low temperature, *Pseudomonas* spp., mainly species of *P. fragi, P. fluorescens*, and *P. lundensis*, are frequently found to dominate the spoilage populations (Stanbridge and Davis, 1998). These bacteria contribute to poultry spoilage by producing compounds such as ammonia, dimethyl sulfide, and nonvolatile amines, including putrescine and cadaverine. These off-odors result primarily from post-glucose utilization of amino acids and are detectable at microbial populations above 10^8 CFU/cm^2.

Shewanella putrefaciens is another gram-negative organisms associated with spoilage of aerobically stored poultry. The defects caused by this organism are attributed to the formation of malodorous substances, including ammonia and hydrogen sulfide. Condition favoring the growth of *S. putrefaciens* on cold-stored poultry include a pH above 6.0, and therefore *S. putrefaciens* mainly is important in spoilage of poultry leg cuts (pH 6.2 to 6.4) or skin-on poultry cuts rather than skinless breast fillets (pH 5.6 to 5.9) (Barnes and Impey, 1968; McMeekin, 1977).

As minor components, spoilage populations often include cold-tolerant members of the family Enterobacteriaceae (e.g., *Hafnia* spp., *Serratia* spp., *Enterobacter* spp.), *Acinetobacter* spp., LAB, and *Brochothrix thermosphacta*. Since the growth rates of these microbes are typically lower than those of *Pseudomonas* spp. or *S. putrefaciens*, their role in spoilage of aerobically stored products is less important.

Spoilage of Modified-Atmosphere and Vacuum-Packaged Poultry

Vacuum and modified-atmosphere (MA) packaging, combined with cold storage, prevent fast-growing aerobic spoilage organisms and extend the shelf life of products. The compositions of gas atmospheres employed in poultry packaging are tailored to each product. Typically, increased levels of CO_2 are needed in the MA to suppress the growth of aerobic spoilage organisms, whereas nitrogen is used to balance an MA package. For the MA of certain case-ready products, inclusion of high levels of oxygen (60 to 80%) may be necessary to reduce the discoloration of meat during storage.

Packaging of poultry under CO_2-enriched atmospheres favors the growth of psychrotrophic microorganisms with fermentative metabolism and leads to the development of spoilage populations dominated by LAB (Borch et al., 1996). The LAB frequently identified from spoiled MA-packaged poultry belong to the genera *Lactobacillus, Lactococcus, Leuconostoc*, and *Carnobacterium* (Holzapfel,

1998; Björkroth et al., 2005; Vihavainen et al., 2007). Compared with the strong growth of *Pseudomonas* spp. in aerobically stored meats, growth of LAB populations on packaged poultry is more moderate. However, during the shelf life of the product, LAB populations often reach high levels and may cause spoilage. Typically, defects described in LAB spoilage include sour and acid off-odors and off-flavors, which result from the accumulation of organic acids formed during fermentative metabolism. These defects are undesirable but less offensive than the putrid and ammonia odors characteristic of aerobic spoilage of meat. In addition to off-odors, heterofermentative LAB such as *Leuconostoc* spp. can produce large amounts of CO_2 and lead to distension of the package. Rapid gas formation in MA-packaged marinated skinless poultry meat strips has been attributed to *Leuconostoc gasicomitatum* (Björkroth et al., 2000). In marinated poultry products, the marked accumulation of gas may be the only indication of spoilage, as the marinade masks the potential sour or acid odors.

Psychrotrophic members of the family Enterobacteriaceae, such as *Serratia liquefaciens* and *Hafnia alvei*, also play an important role in spoilage of packaged poultry, particularly if the products are exposed to temperature abuse or stored at temperatures above 6°C (Smolander et al., 2004). In glucose depletion, Enterobacteriaceae may produce ammonia, hydrogen sulfide, and other volatile sulfides, as well as malodorous amines (cadaverine, putrescine) which impart the disagreeable off-odors. Additionally, formation of hydrogen sulfide may induce discolorations, due to its reactions with muscle color pigment.

Spoilage of Heat-Processed Poultry Products

Heat treatment (65 to 75°C) is effective in inactivating the majority of the vegetative cells on the surface of meat and meat products. However, depending on the heat treatment and the type and formulation of the product, certain heat-resistant bacteria may survive in the core of a product. Spoilage of refrigerated heat-processed poultry products is often a consequence of the growth of psychrotrophic organisms which contaminate the products after heat treatment. If the processing includes portioning or slicing, both the heat-resistant microorganisms surviving at the core and the postheat contaminant are easily distributed over the entire product. Processed products are often packaged under vacuum or anoxic MA with high levels of CO_2. These conditions favor the growth of psychrotrophic, anaerobic, and facultative anaerobic bacteria. The initial bacterial numbers following heat treatment are very low. However, during the relatively long shelf life of heat-processed poultry products, SSOs, if present, grow, giving rise to undesirable changes.

In processed poultry products, psychrotrophic LAB, primarily *Lactobacillus sake, L. curvatus, Leuconostoc* spp., and *Weissella* spp., frequently cause spoilage. These organisms are mainly present in packaged products as postheat contaminants, although certain species, such as *Weissella viridescens*, may survive the cooking process. Gas formation leading to gross distension or loosening of the vacuum package is due to heterofermetative LAB, mainly *Leuconostoc carnosum,*

L. gelidum, and *L. mesenteroides*, which form large amounts of CO_2 (Yang and Ray, 1994; Samelis et al., 2000). In addition, LAB often produce characteristic sour or acidic off-odors and off-flavors and reduced pH. The decline in pH affects the water-binding capacity of the product and may lead to an accumulation of purge. However, in products containing high levels of phosphates, a minor pH decline and little purge formation are often detected, even if the spoilage LAB populations have reached high levels. Processed products containing sucrose may also show slime formation resulting from *Leuconostoc* spp. synthesizing dextran (exopolysaccharide) from sucrose (Samelis et al., 2000).

Other organisms often involved in spoilage include *Clostridium* spp. and members of the family Enterobacteriaceae. In vacuum-packaged processed products, Enterobacteriaceae, particularly *S. liquefaciens*, are reported to cause a strong ammonia off-odor. Similarly with LAB, Enterobacteriaceae are typically post-cooking contaminants, whereas the spores of *Clostridium* spp. survive the heat treatment and grow in the processed product. In noncured, "cooked-in-bag" turkey breast, *Clostridium* spp. has been described to cause spoilage characterized by strong hydrogen sulfide odor and gas accumulation (Kalinowski and Tompkin, 1999). Also, pink discolorations in the interior of products have been attributed to *Clostridium* spp. (Meyer et al., 2003).

TRACING SPECIFIC SPOILAGE BACTERIA IN POULTRY SLAUGHTERING AND MEAT PROCESSING

Specific spoilage bacteria are not usually fecal contaminants. In many cases the natural habitat of the SSOs is not even known. Gram-negative spoilage bacteria are mainly associated primarily with soil, water, sewage, and plant material, and psychrotrophic LAB have been considered to originate primarily from plant material and other environmental sources. Studying SSO contamination as a part of the environmental contamination at a processing facility is challenging, since SSO levels may be very small compared to all other bacteria.

Gram-Negative Spoilage Bacteria in Poultry at Various Slaughtering Stages

To examine the effect of commercial processing and refrigerated storage on spoilage bacteria in the native microflora of broiler carcasses, Hinton et al. (2004) conducted trials on prescalded, picked, eviscerated, and chilled carcasses. The levels of psychrotrophs on processed carcasses stored at 4°C for 7, 10, or 14 days were enumerated using iron agar, *Pseudomonas* agar, and STAA agar. Bacterial isolates were identified based on numerical analysis of fatty acid profiles. Spoilage bacteria occurred in processed carcasses significantly less than in carcasses entering the processing line. *Acinetobacter* and *Aeromonas* spp. were the primary isolates recovered from carcasses taken from the processing line. During refrigerated storage, a significant increase in the bacterial levels was detected, and *Pseudomonas* spp. were the species predominantly recovered. Bacterial cross-contamination of carcasses seemed to occur during all stages of processing. Although poultry processing was found to decrease carcass contamination with psychrotrophic spoilage bacteria, significant bacterial cross-contamination occurred during processing.

Lues et al. (2007) studied the microbial composition of the air in various areas of a high-throughput chicken-slaughtering facility. They sampled six processing areas over a four-month period and monitored the influence of environmental factors. The highest counts of coliforms and *Pseudomonas* were 4.9×10^3 CFU/cm^3 and 7.0×10^4 CFU/m^3, respectively, recorded in the initial stages of processing, comprising the receiving–killing and defeathering areas. The counts decreased toward the evisceration, air-chilling, packaging, and dispatch areas. The highest counts of all airborne microorganisms were found in the receiving–killing and defeathering areas.

Tracing Spoilage LAB at a Broiler Chicken Slaughterhouse and Adjacent Processing Premises

Molecular typing methods have been used to trace spoilage LAB in a broiler chicken slaughterhouse and an adjunct processing plant. To show which LAB species are the SSOs in modified-atmosphere-packaged nonmarinated broiler chicken products, Vihavainen et al. (2007) enumerated and identified LAB in products at the end of the producer-defined shelf lives. Identification was done using numerical analysis of 16S and 23S rRNA gene *Hin*dIII RFLP patterns and a database containing patterns of approximately 300 type and reference strains. To reveal how spoilage-associated LAB were connected subsequently with poultry and processing environment contamination, broiler chicken handled during the early stages of slaughter and air from processing phases related to carcass chilling, cutting, and packaging were sampled. Isolates were collected from these samples and identified to the species level using the database. A total of 447, 86, and 122 isolates originating from broiler products, broiler carcasses, and processing plant air, respectively, were identified.

The LAB counts in late-shelf-life products varied from 10^4 to 10^8 CFU/g. *Carnobacterium divergens* and *C. maltaromicum* (*piscicola*) were two prevailing species in the developing spoilage LAB populations, forming 63% of the LAB isolated. Other major LAB species detected in the products belonged to the genera *Lactococcus, Leuconostoc*, and *Lactobacillus*. The broiler carcasses handled at the beginning of the slaughtering line did not contain SSOs detected in the late-shelf-life broiler products. However, they were recovered from the production plant air adjacent to the final cutting and packaging stages. According to these results, the incoming broiler chickens were not the major source causing psychrotrophic spoilage LAB contamination of the products, whereas finding them in the air suggested that the contaminated processing environment played a major role in product contamination. LAB associated with the developing spoilage population of MAP broiler meat were likely to be introduced to the products from the environment during late processing operations rather than being indigenous bacteria associated with the microbiome of broiler chicken.

According to these results, preventive hygiene-controlling acts should not be directed only at the handling of broiler chickens at the abattoir level, since psychrotrophic spoilage LAB contamination was evident in the air associated with the final product manufacturing and packaging phases.

REFERENCES

Barnes EM, Impey CS. 1968. Psychrophilic spoilage bacteria of poultry. J Appl Bacteriol 31(1):97–107.

Björkroth KJ, Geisen R, Schillinger U, Weiss N, De Vos P, Holzapfel WH, Korkeala HJ, Vandamme P. 2000. Characterization of *Leuconostoc gasicomitatum* sp. nov., associated with spoiled raw tomato-marinated broiler meat strips packaged under modified-atmosphere conditions. Appl Environ Microbiol 66:3764–3772.

Björkroth J, Ristiniemi M, Vandamme P, Korkeala H. 2005. *Enterococcus* species dominating in fresh modified-atmosphere-packaged, marinated broiler legs are overgrown by *Carnobacterium* and *Lactobacillus* species during storage at 6°C. Int J Food Microbiol 97:267–276.

Borch E, Kant-Muermans ML, Blixt Y. 1996. Bacterial spoilage of meat and cured meat products. Int J Food Microbiol 33:103–120.

Hinton A, Cason JA, Ingram KD. 2004. Tracking spoilage bacteria in commercial poultry processing and refrigerated storage of poultry. Int J Food Microbiol 91:155–165.

Holzaphel WH. 1998. The gram-positive bacteria associated with meat and meat products. In: Davies A, Board RG, eds., *The Microbiology of Meat and Poultry*. London: Blackie Academic & Professional, p. 64.

Hutchison ML, Walters LD, Mead GC, Howell M, Allen VM. 2006. An assessment of sampling methods and microbiological hygiene indicators for process verification in poultry slaughterhouses. J Food Prot 69:145–153.

Kalinowski RM, Tompkin RB. 1999. Psychrotrophic *Clostridia* causing spoilage in cooked meat and poultry products. J Food Prot 62:766–772.

Lues JFR, Theron MM, Venter P, Rasephei MHR. 2007. Microbial composition in bioaerosols of a high-throughput chicken-slaughtering facility. Poult Sci 86:142–149.

McMeekin TA. 1977. Spoilage association of chicken leg muscle. Appl Environ Microbiol 33:1244–1246.

Meyer JD, Cerveny JG, Luchansky JB. 2003. Inhibition of nonproteolytic, psychrotrophic *Clostridia* and anaerobic sporeformers by sodium diacetate and sodium lactate in cook-in-bag turkey breast. J Food Prot 66:1474–1478.

Nychas GJE, Drosinos EH, Board RG. 1998. Chemical changes in stored meat. In: Davies A, Board RG, eds., *The Microbiology of Meat and Poultry*. London: Blackie Academic & Professional, pp. 288–326.

Samelis J, Kakouri A, Rementzis J. 2000. The spoilage microflora of cured, cooked turkey breasts prepared commercially with or without smoking. Int J Food Microbiol 56:133–143.

Smolander M, Alakomi H-L, Ritvanen T, Vainionpää J, Ahvenainen R. 2004. Monitoring of the quality of modified atmosphere packaged broiler chicken cuts stored in different temperature conditions: A. Time–temperature indicators as quality-indicating tools. Food Control 15:217–229.

Stanbridge LH, Davis AR 1998. The microbiology of chill-stored meat, 1998: chemical changes in stored meat. In: Davies A, Board RG, eds., *The Microbiology of Meat and Poultry*. London: Blackie Academic & Professional, pp. 174–219.

Tuncer B, Sireli UT. 2008. Microbial growth on broiler carcasses stored at different temperatures after air- or water-chilling. Poult Sci 87:793–799.

Vihavainen E, Lundström HS, Susiluoto T, Koort J, Paulin L, Auvinen P, Björkroth KJ. 2007. Role of broiler carcasses and processing plant air in contamination of modified-atmosphere-packaged broiler products with psychrotrophic lactic acid bacteria. Appl Environ Microbiol 73:1136–1145.

Yang R, Ray B, 1994. Prevalence and biological control of bacteriocin-producing psychotropic leuconostocs associated with spoilage of vacuum-packaged processed meats. J Food Prot 57:209–217.

33

CAMPYLOBACTER IN POULTRY PROCESSING

Marja-Liisa Hänninen

Department of Food and Environmental Hygiene, Faculty of Veterinary Medicine, University of Helsinki, Helsinki, Finland

Handbook of Poultry Science and Technology, Volume 2: Secondary Processing, Edited by Isabel Guerrero-Legarreta and Y.H. Hui

INTRODUCTION

The genus *Campylobacter* includes 17 validly named species. The most important *Campylobacter* species in human infections is *C. jejuni*, which covers more than 90% of campylobacteriosis cases. A minority of the infections are caused either by *C. coli* (up to 5% of the cases) or some other *Campylobacter* spp. These bacteria are rather fastidious and are not usually able to grow outside the intestine of their host under natural conditions. They are microaerophilic; their optimum growth temperature is approximately 42°C and they do not grow at 30°C (Humphrey et al., 2007).

Human campylobacteriosis is an important enteric infectious disease affecting both industrialized and less developed countries throughout the world. In many countries campylobacteriosis is a notifiable disease. For example, the number of cases reported has been increasing in the European Union (EU) countries since the beginning of the 1990s (EFSA, 2006). Since 1998, the number *Campylobacter* cases reported annually to the local national public health institutes in Finland, Sweden, and Norway have been higher than those of salmonellosis cases. The annual incidence rates in Europe are approximately from 50 to more than 100 cases in 100,000 (EFSA, 2006). In the United States, much lower incidences have been reported (Jones et al., 2007). Most human infections occur as single cases or small family outbreaks; epidemics are uncommon and are associated primarily with drinking water and unpasteurized milk (Humphrey et al., 2007). Travel outside one's country is the most important risk factor associated with campylobacteriosis, and more than half of the patients in the Nordic countries acquire the illness when traveling abroad (EFSA, 2006).

To explain the relative importance of various sources and routes of transmission, more than 20 case–control studies have been performed. In most of these studies, undertaken in 1982–2002 throughout the world (e.g., in the United States, Canada, Australia, New Zealand, England, Sweden, Norway, and Finland), eating or handling chicken has been a significant risk factor for campylobacteriosis (Carrique-Mas et al., 2005) Specific risk factors associated with poultry have included eating undercooked chicken meat at a barbecue or restaurant. Some studies have found contrasting results, suggesting that contact with poultry meat at home may be a protective factor as well (Adak et al., 1995).

Birds, including poultry, are often colonized by *Campylobacter* spp., especially *C. jejuni* and *C. coli*. The high body temperature of the birds, approximately 42°C, and high population density of birds in commercial breeding facilities support the high colonization frequencies found in studies on *Campylobacter* and poultry. A flock is most often contaminated by campylobacters during the rearing period at the farm, and at the age of slaughter up to 100% of the birds can be colonized (Berndtson et al., 1996; Allen et al., 2007). In some studies, flocks were identified in which only 10 to 40% of the birds were colonized (Nauta et al., 2007; EFSA, 2006). In some regions (e.g., in northern European countries), a distinct seasonal variation in the prevalence of positive flocks has been found. The peak season is June–September, and the highest number of positive flocks occurs in August. In countries having a high percentage of positive flocks, no distinct seasonal variation may be visible (EFSA, 2006). The colony-forming units (CFU) of *Campylobacter* in the ceca of birds at slaughter are high: 10^6 to 10^8 per gram of cecal material (Allen et al., 2007; Nauta et al., 2007), suggesting that the potential *Campylobacter* load of a *Campylobacter*-positive flock is enormous. For example, if a flock with 10,000 birds is 100% colonized and has 10^7 CFU/g of fecal material, the estimated *Campylobacter* load could be 10^{11} to 10^{12} per day.

Why *Campylobacter* and Chicken?

The results of human case–control and food chain studies of *Campylobacter* as well as simultaneous increase both in the consumption of chicken meat and the number of reported campylobacteriosis cases suggest that poultry is one of the most important risk factors for human infections (Carrique-Mas et al., 2005; Wingstrand et al., 2006; Humphrey et al., 2007). Chicken has been the focus of attention in monitoring studies, development of risk assessment procedures, and as a target of potential intervention strategies to decrease the risk of human infections (FAO–WHO, 2002; Rosenquist et al., 2003; Nauta et al., 2007). Indirect evidence for the significance of chicken meat in human infections is available from Belgium, where chicken meat consumption was down during the dioxin crisis, and during this period, the incidence of campylobacteriosis decreased by 40% compared with the incidence in previous years (Wellinga and Van Loock, 2002). Some epidemiological studies have shown that exposure to fresh chicken meat is a risk factor, but exposure to frozen chicken meat is not (Wingstrand et al., 2006). Many countries routinely monitor campylobacters in chicken fecal samples at the farm 1 to 2 weeks before slaughter or take cecal samples at the slaughterhouse and/or meat samples, and report the results (EFSA, 2006). For example, in the EU, monitoring for *Campylobacter* is required by the EU's Zoonoses Directive (European Commission 2003/99/EC).

The contamination frequency and level of chicken meat products at retail are dependent on several factors associated with contamination, decontamination, and cross-contamination at various steps throughout the production chain (Figures 1 and 2). The process is dynamic, and contamination, decontamination, and cross-contamination of carcass surfaces all affect the end result. Parameters that need to be considered include (1) prevalence within the flock at the farm, (2) contamination of carcasses in transport, (3) contamination/decontamination and cross-contamination at various steps in the slaughter process, (4) meat cutting, (5) storage of raw chicken products at retail, and (6) handling of chicken meat at home or in a restaurant kitchen (Figure 1). A schematic quantitative view on *Campylobacter* counts at various stages of chicken meat processing is shown in Figure 2. Various studies and risk assessments showed that the most important site of contamination is at the farm (Hartnett, 2001; FAO–WHO, 2002; Rosenquist et al., 2003).

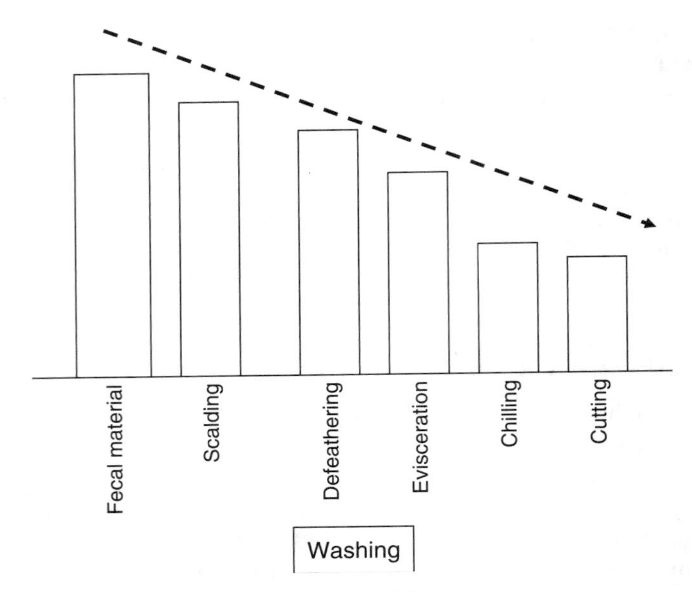

FIGURE 1 Schematic description of *Campylobacter* counts on carcasses in various stages of the slaughtering process.

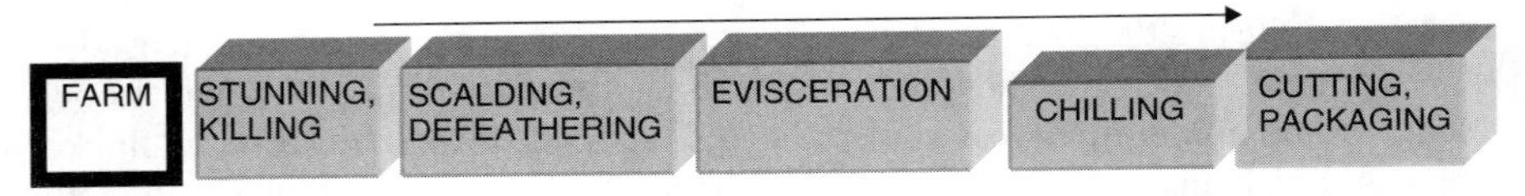

FIGURE 2 Transmission of *Campylobacter* in the chicken meat production chain from farm through processing line.

Detection and Counting of *Campylobacter* on the Processing Line

Many methodological approaches are used in studies of the prevalence of *C. jejuni* and *C. coli* at various stages in the processing chain. Similarly, quantitative approaches vary widely. Thus, differences both in sampling and cultivation methods in estimations of *Campylobacter* counts need to be recognized when the results of different studies are compared. Food inspection services and regulatory agencies need to consider which method is most suitable for their purposes. For the presence or absence testing of fecal samples, direct cultivation on a selective medium is commonly used (Berndtson et al., 1996; Stern et al., 2003). Enrichment procedures either for fecal material or meat samples are used if the counts are suspected to be low and only presence or absence results are needed. However, for risk assessment of the survival rate and fate of campylobacters throughout the process, quantitative counts are needed. Samplings performed on the processing line can handle whole carcasses, whole carcass rinses, swabbing the carcass surface, neck skin samples, skin and meat samples, and free weep fluid from a carcass (Musgrove et al., 2003). In addition to plate counts, the most probable number method can be applied (Ritz et al., 2007; Stern et al., 2007). The complexity of detection and quantification methods of campylobacters in the chicken meat processing chain is even increased when molecular methods, polymeraze chain reaction (PCR), and quantitative real-time PCR are employed.

SCALDING

In scalding, the carcasses are densely immersed in scalding water tanks, and fecal material contaminates the exterior of the carcasses intensively. The scalding water temperatures used vary from 50 to 52°C (soft scald) to 56 to 68°C (hard scald). The time–temperature combinations that can be used without denaturing the carcass skin do not decrease the *Campylobacter* contamination level of carcasses (Nauta et al., 2007). The counts on the carcass exterior before scalding are already very high on delivery at the processing plant: approximately $\log_{10}7$ (Figure 2); Stern and Pretanik, 2006; Nauta et al., 2007). The results of experiments testing the potential for decreasing contamination of the carcass exterior (e.g., using 2.5% lactic acid or 10% trisodium phosphate in the scalding water) have shown a maximum 1 $\log_{10}$ unit decrease in the counts (Havelaar et al., 2007). The scalding tank water is an important stage for cross-contamination both within a flock and between flocks. This cross-contamination can be especially important if a *Campylobacter*-negative flock is slaughtered after a *Campylobacter*-positive flock. Equipment, which could reduce fecal leakage in the scalding, has been estimated to decrease up to 80% of the consumers' risk of infection (Havelaar et al., 2007).

DEFEATHERING

Defeathering has a minor effect on *Campylobacter* counts on carcasses because counts are already high in contaminated flocks (Figure 2). The feather follicles in the skin at this stage are opened and may lead to movement of *Campylobacter* cells inside the follicles, which may decrease the wash-off effect of washing carcass surfaces (Cason et al., 2004). Cross-contamination is important at this stage.

EVISCERATION

Evisceration performed properly does not affect *Campylobacter* counts extensively. But ruptures of the viscera can occur in evisceration, and this may lead to extensive fecal contamination of a carcass (Rosenquist et al., 2006). In these cases, improvement in the evisceration process may lead to better hygienic quality and a lower *Campylobacter* contamination level. The FAO–WHO Expert

Group on Risk Assessment of *Campylobacter* in Chicken (FAO–WHO, 2002) concluded that reducing surface contamination after evisceration can have a significant impact on reducing the risk of exposure. It suggests that in the slaughtering process, the most significant intervention measures can be used in the process stages after evisceration.

CHILLING

In Europe, air chilling is used more often than water chilling, which is a common practice, for example, in the United States. Chilling decreases the counts of *Campylobacter* on carcasses by approximately 1 $\log_{10}$ unit (Rosenquist et al., 2006; Allen et al., 2007; Berrang et al., 2007) The effect of air or water-immersion chilling showed similar reductions in *Campylobacter* counts (Rosenquist et al., 2006; Huezo et al., 2007).

Counts of *Campylobacter* After Chilling

Variable counts of *Campylobacter*, ranging from $\log_{10}3$ to $\log_{10}6$, were found involving whole-carcass rinse-water samples within a flock. In counts of the carcasses, 18% of the samples were in the range $\log_{10}$ 2.70 to 4.99 and 20% were in the range $\log_{10}$ 5.0 to 6.99 (Jorgensen et al., 2002). A mean 2.6-$\log_{10}$ unit decrease in counts on carcass exteriors between incoming chickens and a carcass in the chilling room showed that even the variation between flocks and slaughtering processes can be large (Berrang et al., 2007; Huezo et al., 2007; Nauta et al., 2007).

CAMPYLOBACTER IN THE PROCESSING ENVIRONMENT

When a *Campylobacter*-positive flock is slaughtered, the equipment and air along the processing line from the scalding room to the evisceration room are contaminated by *Campylobacter*, which may no longer be detectable in the chilling area (Berndtson et al., 1992; Allen et al., 2007). The heavy contamination that occurs when a *Campylobacter*-positive flock is on the slaughtering line suggests that equipment and air are sources of cross-contamination and that special attention to cleaning is required. Contaminated equipment surfaces and aerosols in the air can constitute a significant occupational risk of campylobacteriosis for people working on the slaughtering line (Wilson, 2004).

Logistic Slaughtering

Logistic slaughtering, in which *Campylobacter*-positive flocks are slaughtered at the end of the working day, has been applied in several countries to prevent cross-contamination in the slaughtering process (Norström et al., 2007). A recent Dutch modeling study showed that the impact of logistic slaughter on the number of contaminated carcasses is marginal (Havelaar et al., 2007; Nauta et al., 2007). The counts of *Campylobacter* on carcasses contaminated in the slaughtering process are also lower than on those coming from a colonized flock (Allen et al., 2007).

Microbial Intervention Strategies on *Campylobacter* on the Processing Line

Washing of carcasses with inside–outside bird washers at various stages of the slaughtering process (Figure 1), starting after defeathering, is the most efficient way to remove fecal contamination and simultaneously to decrease *Campylobacter* counts on carcass surfaces and in the peritoneal cavity. Efficient washing before chilling can decrease counts by approximately 1 log (Bashor et al., 2004; Rosenquist et al., 2006; Berrand et al., 2007). Chemical decontamination applied at different stages in the process has been tested and used in some establishments. Chemicals can be added to the carcass washing water, brushes, cabinet sprays, or dip tanks (Bashor et al., 2004; Northcutt et al., (2005, 2007)). The most common antimicrobial treatment used in washing water has been sodium hypochlorite. It is rather inexpensive and easy to use, but the efficiency varies due to inefficient dosage or inactivation by organic material. The pH and temperature of the water are also important for the antimicrobial activity of chlorine. The odor and formation of organic chlorine by-products are disadvantages for the use of chlorine. Many countries also have a common policy of not using any chemical treatment for foods sold fresh. Chlorine dioxide and monochloramine are examples of other chlorine compounds tested. In a study (Northcutt et al., 2007), carcasses were treated by spray washing with acidified electrolyzed water or sodium hypochlorite (NaOCl), both of which decreased the counts by approximately 2 $\log_{10}$ units. Several other chemicals have been tested as well; for example, trisodium phosphate (10%) and acidified sodium chlorite in combination with washing reduced *Campylobacter* counts by an additional 0.5 log compared with washing as such. A limited number of experiments have tested lactate (2.5%) in chilling sprays or washing of carcasses (Nauta et al., 2007).

Other procedures known to reduce bacterial loads include irradiation and ultraviolet light. Irradiation of meat by high-energy gamma rays decreases significantly the counts of all types of bacterial contaminants, including *Campylobacter*, on the surface (Farkas, 1998). This treatment is not permitted in some areas (e.g., in the EU countries). In the United States the U.S. Department of Agriculture (USDA) approved a maximum dose of 3 kGy to control foodborne pathogens (Nauta et al., 2007). Crust freezing is a rapid-freezing procedure applied for a short period to the carcass surface. The procedure was able to decrease the campylobacter count by 0.9 log (James et al., 2007). All these measures reduce campylobacters but do not eliminate them from the final product completely and may have a negative impact on the organoleptic properties of a product.

EFFECT OF FREEZING ON *CAMPYLOBACTER* COUNTS AND REDUCTION OF RISK

In many countries chicken meat is sold as frozen carcasses or pieces. Studies have shown that the freezing process and frozen storage decrease *Campylobacter* counts by approximately 1 to 2 $\log_{10}$ units (Sandberg et al., 2005; Havelaar et al., 2007; Ritz et al., 2007). The greatest decrease is associated with the freezing process. Ritz et al., (2007) found that the actual decrease can be minor, because some methods used in the analysis may not detect injured cells. In risk assessment models a 2-$\log_{10}$-unit decrease in counts has been estimated to decrease the consumer risk twofold (FAO–WHO, 2002). Therefore, some countries have initiated an intervention strategy in which meat from *Campylobacter*-positive flocks is frozen at -20°C before distribution to consumers. For example, such a strategy has been used in Norway since 2001 (Norström et al., 2007). A similar strategy has also been used in Iceland (Stern et al., 2003). An additional risk for cross-contamination exists when frozen meat is thawed, and the thawing water may contaminate surfaces and other foods. The global market for frozen chicken meat is large, and therefore attention to *Campylobacter* and frozen chicken meat is required.

CUTTING

The consumer demand for foods that are ready for use at home has led to increased consumption of fresh or frozen chicken parts, such as drumsticks, breast fillets with or without skin, deboned fillets, skinless meat slices or ground chicken or turkey meat prepared from whole carcasses or fillets. These products are easily and rapidly prepared as foods at home. The largest impact of cutting on *Campylobacter* counts is in products with or without skin. Since most *Campylobacter* contamination is located on skin, skinless products have counts several log units lower than the respective meat products with skin (Havelaar et al., 2007). Cutting as such has no remarkable effect on *Campylobacter* counts on product surfaces.

PACKAGING

Selling of prepackaged meat products has increased from 42% in 1990 to 76% in 2001. Supermarkets mainly sell prepackaged meat. Most fresh meat products (pieces) at retail in Finland are packed in a modified gas atmosphere containing approximately 50/50 nitrogen and CO_2. Packaging in a modified gas atmosphere has no significant impact on the survival of campylobacters (Perko-Mäkelä et al., 2000). In a recent case–control study, a significant risk of acquisition of campylobacteriosis in children 0 to 6 months of age was riding in a shopping cart next to meat packages while shopping with a family member (Fullerton et al., 2007).

REFRIGERATED STORAGE OF CHICKEN MEAT PRODUCTS

The sellout time for prepackaged chicken meat products can vary from a few days to 10 days for products packaged in a modified gas atmosphere. Refrigeration reduced *Campylobacter* counts from 0.25 to 0.35 $\log_{10}$ unit at 4°C (Perko-Mäkelä et al., 2000; Havelaar et al., 2007) under experimental conditions. In practice, the temperature may fluctuate in distribution to retail establishments as well under refrigeration at shops.

MESSAGES OF RISK ASSESSMENT MODELS

To develop the most efficient intervention strategy for reducing the numbers of campylobacteriosis cases associated directly with consumption of chicken meat, analysis and modeling of the chicken meat production chain have been the focus of activity. During recent years, several countries have performed risk assessments of *Campylobacter* in broiler chickens: in Canada (Frazil et al., 1999, 2003), the UK (Hartnett, 2001), Denmark (Rosenquist et al., 2003), and the Netherlands (Havelaar et al., 2007; Nauta et al., 2007). FAO–WHO (2002) has performed a risk assessment based on expert consultations and found several gaps in our knowledge of how to perform full risk assessment. The most significant gap was lack of quantitative data, leading to a need to rely on assumptions, thus increasing the uncertainties in risk assessment. Local chicken meat production conditions also vary widely. One common conclusion of the experts at the 2002 FAO–WHO meeting was that at high contamination levels a twofold reduction in the risk of human campylobacteriosis would result if a *Campylobacter* contamination level reduces twofold, but at low contamination levels reduction

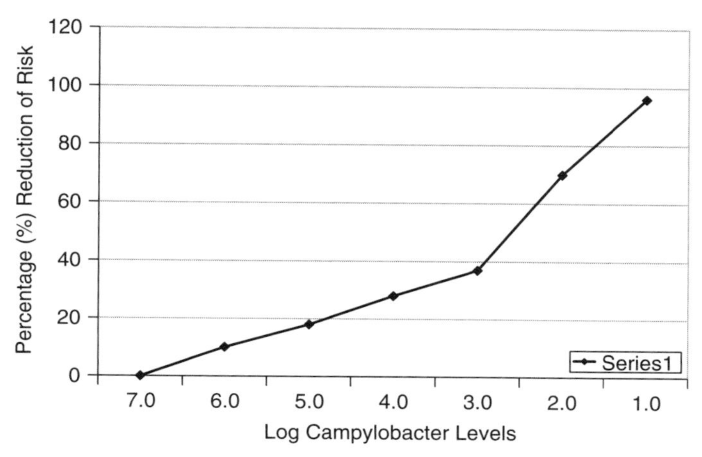

FIGURE 3 Model for reduction of risk of infection and levels of *Campylobacter* on carcasses. (From A.M. Frazil in FAO–WHO, 2002.)

would decrease the risk by more than 50% (Figure 3). The reduction in risk was counted mainly as a reduction in exposure, that is, in the number of *Campylobacter*-positive meals. Consumption habits, as well as the variety of chicken meat products on the market in different countries, suggest that the relative importance of direct consumption of meat or cross-contamination from chicken meat products is variable.

In conclusion, local customized risk assessment processes are needed to understand consumer risk, while a common risk assessment models can be used as a framework. This approach requires cooperation between chicken meat–producing companies, farmers, researchers, and risk assessment experts, as well as risk management and regulatory authorities.

TURKEY AND OTHER POULTRY

Most of the results and discussion on campylobacters in poultry are focused on broiler chicken, because its production is the most intensive branch in the poultry industry, and most of the studies and results on *Campylobacter* are based on chicken. The results can be applied to turkey as well, but differences in rearing, slaughtering, and so on, may require their own risk assessment procedures.

REFERENCES

Adak GK, Cowden JM, Nicholas S, Evans HS. 1995. The Public Health Laboratory Service national case–control study of primary indigenous sporadic cases of *Campylobacter* infection. Epidemiol Infect 115(1):15–22.

Allen VM, Bull SA, Corry JE, Domingue G, Jørgensen F, Frost JA, Whyte R, Gonzalez A, Elviss N, Humphrey TJ. 2007. *Campylobacter* spp. contamination of chicken carcasses during processing in relation to flock colonisation. Int J Food Microbiol 113(1):54–61.

Bashor MP, Curtis PA, Keener KM, Sheldon BW, Kathariou S, Osborne JA. 2004. Effects of carcass washers on *Campylobacter* contamination in large broiler processing plant. Poult Sci 83(7):1232–1239.

Berndtson E, Danielsson-Tham ML, Engvall A. 1996. Campylobacter incidence on a chicken farm and the spread of *Campylobacter* during the slaughter process. Int J Food Microbiol 32(1–2):35–47.

Berrang ME, Bailey JS, Altekruse SF, Pater, P, Shaw WK Jr, Meinersmann RJ, Fedorka-Cray PJ. 2007. Prevalence and numbers of *Campylobacter* on broiler carcasses collected at rehang and postchill in 20 U.S. processing plants. J Food Prot 70(7):1556–1560.

Carrique-Mas J, Andersson Y, Hjertqvist M, Svensson A, Torner A, Giesecke J. 2005. Risk factors for domestic sporadic campylobacteriosis among young children in Sweden. Scand J Infect Dis 37(2):101–1.

Cason JA, Hinton A Jr, Buhr R. 2004 Impact of feathers and feather follicles on broiler carcass bacteria. Poult Sci 83(8):1452–1455.

EFSA (European Food Safety Authority). 2006. *Campylobacter*. The Community Summary Report on Trends and Sources of Zoonoses, Zoonotic Agents, Antimicrobial Resistance and Foodborne Outbreaks in the European Union in 2005. EFSA J 94:88–103.

FAO–WHO. 2002. Risk assessment of *Campylobacter* spp. in broiler chickens and *Vibrio* spp. in seafood. Report of a Joint FAO/WHO Expert Consultation, Bangkok, Thailand, Aug. 2002. http://www.who.int/foodsafety/publications/micro/aug2002.pdf.

Farkas J. 1998. Irradiation as a method for decontaminating food: a review. Int J Food Microbiol 44(3):189–204.

Frazil AM, Lowman R, Stern N, Lammerding A. 1999. Quantitative risk assessment model for *Campylobacter jejuni* in chicken. Abstract CF10. 10th International Workshop on CHRO.

Frazil AM, Lammerding A, Ellis A. 2003. A quantitative risk assessment model for *Campylobacter jejuni* on chicken. http://www.who.int/foodsafety/micro/jemra/assessment/campy/en/.

Fullerton KE, Ingram LA, Jones TF, Anderson BJ, McCarthy PV, Hurd S, Shiferaw B, Vugia D, Haubert N, Hayes T, et al. 2007. Sporadic *Campylobacter* infection in infants: a population-based surveillance case–control study. Pediatr Infect Dis J 26(1):19–24.

Hartnett E, 2001. Human infection with *Campylobacter* spp. from chicken consumption: a quantitative risk assessment. Ph.D. dissertation, University of Strathglyde, Glasgow, UK.

Havelaar AH, Mangen MJ, de Koeijer AA, Bogaardt MJ, Evers EG, Jacobs-Reitsma WF, van Pelt W, Wagenaar JA, de Wit GA, van der Zee H, Nauta MJ. 2007. Effectiveness and efficiency of controlling *Campylobacter* on broiler chicken meat. Risk Anal 27(4):831–844.

Huezo R, Northcutt JK, Smith DP, Fletcher DL, Ingram KD. 2007. Effect of dry air immersion chilling on recovery of bacteria from broiler carcasses. J Food Prot 70(8):1829–1834.

Humphrey T, O'Brien S, Madsen M. 2007. Campylobacters as zoonotic pathogens: a food production perpective. Int J Food Microbiol 47(3):237–257.

James C, James SJ, Hannay N, Purnell G, Barbedo-Pinto C, Yaman H, Araujo M, Gonzalez ML, Calvo J, Howell M, Corry JE. 2007. Decontamination of poultry carcasses using steam or hot water in combination with rapid cooling, chilling or freezing of carcass surfaces. Int J Food Microbiol 114(2):195–203.

Jones TF, Scallan E, Angulo FJ. 2007. FoodNet: overview of a decade of achievement—a review. Foodborne Pathog Dis 4(1):60–66.

Jorgensen F, Bailey R, Williams S, Henderson P, Wareing DR, Bolton FJ, Frost JA, Ward L, Humphrey TJ. 2002. Prevalence and numbers of *Salmonella* and *Campylobacter* spp. on raw, whole chickens in relation to sampling methods. Int J Food Microbiol 76(1–2):151–164.

Musgrove MT, Cox NA, Berrang ME, Harrison MA. 2003. Comparison of weep and carcass rinses for recovery of *Campylobacter* from retail broiler carcasses. J Food Prot 66(9):1720–1723.

Nauta MJ, Jacobs-Reitsma WF, Havelaar AH. 2007. A risk assessment model for *Campylobacter* in broiler meat. Risk Anal 27(4):845–861.

Northcutt JK, Smith DP, Musgrove MT, Ingram KD, Hinton A Jr. 2005. Microbiological impact of spray washing broiler carcasses using different chlorine concentrations and water temperatures. Poult Sci 84(10):1648–1652.

Northcutt J, Smith D, Ingram KD, Hinton A Jr, Musgrove M. 2007. Recovery of bacteria from broiler carcasses after spray washing with acidified electrolyzed water or sodium hypochlorite solutions. Poult Sci 86(4):2239–2244.

Norström M, Johnsen G, Hofshagen M, Tharaldsen H, Kruse H. 2007. Antimicrobial resistance in *Campylobacter jejuni* from broilers and broiler house environments in Norway. J Food Prot 70(3):736–738.

Perko-Mäkelä P, Hänninen M-L, Koljonen M. 2000. Survival of *Campylobacter jejuni* in marinated chicken. J Food Saf 20:209–216.

Ritz M, Nauta MJ, Teunis PF, van Leusden F, Federighi M, Havelaar AH. 2007. Modelling of *Campylobacter* survival in frozen chicken meat. J Appl Microbiol 103(3):594–600.

Rosenquist H, Nielsen NL, Sommer HM, Norrung B, Christensen BB. 2003. Quantitative risk assessment of human campylobacteriosis associated with thermophilic *Campylobacter* species in chickens. Int J Food Microbiol 86(1):87–103.

Rosenquist H, Sommer HM, Nielsen NL, Christensen BB. 2006. The effect of slaughter operations on the contamination of chicken carcasses with thermotolerant *Campylobacter*. Int J Food Microbiol 108(2):226–232.

Sandberg M, Hofshagen M, Ostensvik O, Skjerve E, Innocent G. 2005. Survival of *Campylobacter* on frozen broiler carcasses as a function time. J Food Prot 68(8):1600–1605.

Slavik M, Kim J-W, Pharr MD, Raben DP, Tsai S, Lobsinger CM. 1994. Effect of trisodium phosphate on *Campylobacter* attached to post-chill chicken carcasses. J Food Prot 57:324–326.

Stern NJ, Pretanik S. 2006. Counts of *Campylobacter* spp. on U.S. broiler carcasses. J Food Prot 69(5):1034–1039.

Stern NJ, Hiett KL, Alfredsson GA, Kristinsson KG, Reiersen J, Hardardottir H, Briem H, Gunnarsson E, Georgsson F, Lowman R, et al. 2003. *Campylobacter* spp. in Icelandic poultry operations and human disease. Epidemiol Infect 130(1):23–32.

Stern NJ, Georgsson F, Lowman R, Bisaillon JR, Reiersen J, Callicott KA, Geirsdóttir M, Hrolfsdóttir R, Hiett KL. 2007. Campy-on-Ice Consortium: Frequency and enumeration of *Campylobacter* species from processed broiler carcasses by weep and rinse samples. Poult Sci 86(2):394–399.

Vellinga A, Van Loock F. 2002. The dioxin crisis as experiment to determine poultry-related *Campylobacter enteritis*. Emerg Infect Dis 8(1):19–22.

Wingstrand A, Neimann J, Engberg J, Nielsen EM, Gerner-Smidt P, Wegener HC, Mølbak K. 2006. Fresh chicken as main risk factor for campylobacteriosis, Denmark. Emerg Infect Dis 12(2):280–285.

Wilson IG. 2004. Airborne *Campylobacter* infection in a poultry worker: case report and review of the literature. Comm Dis Public Health 7(4):349–353.

34

MICROBIOLOGY OF READY-TO-EAT POULTRY PRODUCTS

CAROL W. TURNER

Department of Family and Consumer Sciences, New Mexico State University, Las Cruces, New Mexico

INTRODUCTION

The increasing availability of ready-to-eat (RTE) foods reflects consumer demand, and food-processing methods and packaging techniques are changing to meet these needs (Woteki and Kineman, 2003). Sales of minimally processed products are increasing with about half of the food dollar spent on RTE products (Frazoo, 1999). The U.S. food manufacturing system is very diverse. In 2000, more than 16,000 food-processing firms were responsible for producing over 40,000 products in the United States (Harris, 2002). The size and complexity of the food industry, coupled with the rapid changes in food products, creates a challenge to improved food safety.

Handbook of Poultry Science and Technology, Volume 2: Secondary Processing, Edited by Isabel Guerrero-Legarreta and Y.H. Hui

The United States enjoys one of the safest food supplies in the world. To ensure this, the U.S. Department of Agriculture's Food Safety and Inspection Service (USDA–FSIS) implemented a series of safeguards to protect against foodborne disease. These safeguards include in-plant procedures to reduce dangerous foodborne pathogens from entering the food supply. The Foodborne Diseases Active Surveillance Network (FoodNet) was established in 1996 as part of the Centers for Disease Control and Prevention's (CDC) Emerging Infections Program (Allos et al., 2004). Public health authorities in the United States have set target national health objectives to be met by 2010, which include 0.25, 6.8, and 1.0 cases per 100,000 people for *Listeria monocytogenes, Salmonella*, and *Escherichia coli*, respectively (USDHHS, 2000). The establishment of national health objectives in the United States is valuable; however, food safety objectives should be debated and agreed upon internationally.

Globally the incidences of reported cases of foodborne illness are increasing. The reported yearly incidence of listeriosis ranges from 0.3 to 7.5 cases per million people in Europe and 3 cases per million people in Australia (EC, 2003). There are a number of reasons for this increase: better detection methods, increased tourism and travel, an increase in the number of people dining out, bacterial resistance to antibiotics, changing food production methods, population shifts, poor hygiene practices, a lack of training in food safety, media coverage, and public awareness of their rights as to how to report these incidents (Kramer and Scott, 2004). Listeriosis is observed primarily in industrialized countries. What is not known is whether these differences in incidence rates between developed and less developed countries reflect true geographical differences, differences in food habits and food storage, or differences in diagnosis and reporting practices.

Although the food supply continues to be safe, the number of food recalls in the United States has increased in recent years due in part to a renewed focus on food safety and security by the U.S. government. Most major meat processors in the United States have been involved in a recall at some point in their history and spend considerable funds to prevent, as well as to prepare and respond quickly to, future occurrences. The FSIS defines a *recall* as a firm's voluntary removal of distributed meat or poultry products from commerce when there is reason to believe that such products are adulterated or misbranded under the provisions of the Federal Meat Inspection Act or the Poultry Products Inspection Act (USDA–FSIS, 2007). Initiation of a recall is commonly due to the detection of microbial agents; undeclared allergens; chemical contamination; foreign materials such as glass, metal, and plastic; undercooking of product; and misinformation on the product label (Table 1).

Recent incidents of foodborne disease caused by pathogenic bacteria have increased consumer concerns and interest in meat safety. As a result, the FSIS has implemented a new inspection regulation that requires meat and poultry plants to establish sanitation standard operating procedures, operate under the hazard analysis and critical control point (HACCP) system, and meet microbial performance criteria and standards for *Salmonella* and *E. coli* (FSIS, 1996).

TABLE 1 USDA Recalls by Cause, 2000 to 2007

Cause	2000	2001	2002	2003	2004	2005	2006	2007
L. monocytogenes	35	26	42	16	14	28	6	9
E. coli O157:H7	20	25	35	11	2	0	0	1
Salmonella	4	2	4	2	6	5	8	19
Allergens	5	11	20	14	1	4	9	8
Other [a]	11	28	27	28	25	13	11	15

Source: Compiled from the USDA–FSIS recall archive.
[a]This category contains recalls due to foreign materials, mislabeling, undercooking, and chemical contamination.

Although these requirements have resulted in improved product safety in many cases, meat recalls have continued to be necessary.

Meat and poultry recalls have a direct economic and public perception effect on the industry. Research has shown that when meat recalls are announced, there is a direct negative effect on demand for the products and a move toward nonmeat products (Marsh et al., 2004). The growing number of recalls has changed the public perception of the meat and poultry supply in the United States. Media coverage of a few large outbreaks in recent years, due to pathogen contamination, has generated public concern regarding the industries' ability to provide safe and wholesome food products. Survey data of the U.S. Food Marketing Institute (FMI) indicate that the top food safety concerns of grocery shoppers in 2006 were, in descending order: bacterial contamination, pesticide residues, product tampering, and bioterrorism (Tucker et al., 2006).

MICROORGANISMS ASSOCIATED WITH POULTRY PRODUCTS

Listeria monocytogenes

Listeria monocytogenes is present in soil, water, vegetables, and intestinal contents of a variety of birds, fish, insects, and other animals (Mahmood et al., 2003). Due to its ubiquitous character, it easily enters the human food chain and may multiply rapidly (Farber and Peterkin, 1991). Evidence that *L. monocytogenes* could be foodborne was first reported in 1981 in an investigation of an outbreak in Nova Scotia, Canada, which implicated cabbage (Varma et al., 2007). However, it was not until an outbreak of listeriosis in California linked to inadequately pasteurized soft cheese that *L. monocytogenes* became a major concern of the food industry (MMWR, 1985; Voetsch et al., 2007). The U.S. Food and Drug Administration (FDA) began monitoring dairy products in 1986. This monitoring was later expanded to include RTE foods such as cold meat and poultry products, seafood, and salads (Tappero et al., 1995).

The CDC estimates that about 2500 cases of listeriosis occur annually in the United States, resulting in 500 deaths (Voetsch et al., 2007). The frequent occurrence of *L. monocytogenes* in RTE poultry may pose a potential risk for consumers. Those particularly at risk include pregnant women, neonates, adults with underlying disease (cancer, AIDS, diabetes, chronic hepatic disorder, transplant recipients), the elderly, and other immunocompromized persons (Mahmood et al., 2003). In humans, the illness can range from a mild flulike sickness to severe manifestations. The severe forms of human listeriosis present as meningoencephalitis followed by septic infections and occasionally isolated organ involvement (Farber and Peterkin, 1991). The ubiquitous nature of the bacterium, together with a varied incubation period (1 to >90 days), means that identifying specific food vehicles can be problematic (Voetsch et al., 2007). An important factor in foodborne listeriosis is that the pathogen can grow to significant numbers at refrigeration temperatures when given sufficient time. Although this foodborne illness is less common than other foodborne pathogens, *L. monocytogenes* accounts for 4% of all hospitalizations and 28% of all deaths from foodborne disease in the United States (Varma et al., 2007). Death is rare in healthy adults but can occur in as many as 30% of those who are at highest risk (Demetrios and Antoninos, 1996).

In 1989, after a case of listeriosis was linked to turkey frankfurters, the USDA adopted a zero-tolerance policy for *L. monocytogenes* in RTE meats and poultry (Varma et al., 2007). This resulted in about a 40% drop in listeriosis cases between 1989 and 1994 as companies adopted stricter sanitation procedures (Griffiths, 2001). Listeriosis incidence did not, however, decrease as markedly during 1996–2003, and large multistate outbreaks of *L. monocytogenes* infection continued to occur (Varma et al., 2007). In response to these concerns, the CDC launched a multicenter case–control study of sporadic listeriosis, and the FDA and the USDA conducted a *Listeria* Risk Assessment and revised the National *Listeria* Action Plan. The revised *Listeria* Action Plan focuses on high-risk foods such as RTE poultry products and includes strategies for guidance, training, research, education, surveillance, and enforcement (Voetsch et al., 2007).

After implementation of several initiatives by the FSIS to reduce *L. monocytogenes*, contamination of RTE foods decreased from 7.9 cases per million population in 1989 to 4.2 cases per million in 1993 (Tappero et al., 1995). The Healthy People 2010 national health objective for listeriosis was to achieve a 50% reduction in listeriosis incidence, from 5 cases per million population in 1997 to 2.5 cases per million population in 2010 (USDHHS, 2000). After a highly publicized listeriosis outbreak linked to turkey delicatessen meat in 2000, the government pledged to achieve this goal by 2005 (Olsen et al., 2005).

Regardless of national goals, listeriosis linked to RTE poultry products continues to be reported. Deli turkey and chicken meat have been a particular problem. A multistate outbreak of *L. monocytogenes* in 2000 (MMWR, 2000; Olsen et al., 2005) linked to consumption of deli turkey meat resulted in 29 illnesses, 4 deaths, and 3 miscarriages. A subsequent *L. monocytogenes* outbreak due to consumption of deli turkey or chicken meat was reported in 2002 (MMWR, 2002; Gottlieb et al., 2006) and resulted in the largest U.S. meat recall in history with more than 30 million pounds of product being recalled by the processors. In response to such large-scale outbreaks, the FSIS issued more stringent guidelines for sampling and testing RTE products, including verification testing to be conducted by the FSIS and HACCP and sanitation standard operating procedures verification steps to be taken by the agency in production establishments (MMWR, 2002).

L. monocytogenes is the only one of six species in the genus *Listeria* that is of concern to human health. Although infrequently identified in food microbiological surveys, serotype 4b is the most common among patients in the United States (Varma et al., 2007). In a study, five strains of *L. monocytogenes* were added to eight processed meats that were stored at 4.4°C for up to 12 weeks. The organisms survived on all products and increased in numbers by 3 to 4 logs in most samples (Glass and Doyle, 1990). The best growth occurred in chicken and turkey products. Because of its ability to survive and proliferate at refrigeration temperatures, *L. monocytogenes* may cause disease through frozen foods (Schillinger and Lucke, 1991). To minimize human listeriosis, foods should be cooked to an internal temperature of 70°C for more than 20 min to ensure its destruction. Cooked food should be reheated thoroughly to 70°C, and proper handling and storage of leftovers is necessary to prevent environmental contamination. Proper cold storage of meat and meat products (freezing −18°C) and proper personal hygiene of food handlers is advisable.

Mahmood et al. (2003) collected 320 samples from fresh and fresh boneless poultry meat, frozen poultry meat, frozen chicken nuggets, frozen chicken burgers, chopping boards, mincing machines, and cleaning cloths. Results revealed that *L. monocytogenes* could be isolated from 23.75% of the samples of poultry meat and poultry products. This research confirmed previous research that species of *Listeria* were higher in frozen meat than in fresh meat. This study also found higher incidence of *Listeria* from dead stock (i.e., chopping boards, mincing machines, and cleaning cloths). Higher incidence of *Listeria* in chicken nuggets and chicken burgers may also be attributed to contamination caused by dead stock. The serotype isolated from samples were *L. monocytogenes* serotypes 1 and 4, which are reported to be pathogenic to humans and animals (Mahmood et al., 2003).

Salmonella

Salmonella is the most commonly diagnosed bacterial agent causing foodborne illness in the United States, with 1.4 million illnesses, 16,000 hospitalizations, and 600 deaths estimated to be caused by *Salmonella* spp. each year (Kimura et al., 2004). Symptoms of salmonellosis include diarrhea, abdominal pain with cramps, and fever. Symptoms usually begin within 12 to 72 h but can begin up to a week after exposure. Chicken nuggets and strips have been identified as a significant risk factor for developing salmonellosis (Kimura et al., 2004). Bucher et al. (2007) assessed *Salmonella*-contaminated chicken nuggets and strips and pelleted feeds in an attempt to demonstrate whether the same *Salmonella* strains present in broiler feed could be isolated from chicken nuggets and strips. The data showed that *Salmonella* strains isolated from broiler feed were indistinguishable from strains isolated from chicken nuggets and strips. The results did not rule out that the *Salmonella* contamination may have occurred during processing or from the breeding stock (Bucher et al., 2007).

Salmonella is sometimes present in raw chicken, which is why it is important for consumers to follow safe food-handling practices. This includes cooking all raw poultry products to an internal temperature of at least 74°C. Participants' survey responses demonstrate that consumers do not perceive, handle, or prepare chicken nuggets and strips as they would raw, unprocessed chicken (Currie et al., 2005). This may be due to the appearance of frozen, breaded, and par-fried products not being similar to raw whole meat in terms of color, juice, or texture. The problem arises when consumers do not realize that they are preparing a raw product. As a result, consumers do not handle or prepare these products as they would raw, unprocessed chicken (MacDougall et al., 2004). Manufacturers of chicken nuggets or strips are required to include oven-cooking instructions on the product packaging, but labeling seldom indicates whether the product is raw or fully cooked and does not necessarily provide advice for safe handling or the use of avoidance of microwave cooking. The USDA is in the process of requiring all manufacturers to change the labels of these products to better inform consumers and requiring companies to validate cooking instructions (USDA–FSIS, 2007). Product labels will soon be required to indicate clearly when a product is raw or uncooked.

In 2002, researchers in Canada surveyed 106 chicken nugget samples originating from 14 different manufacturers and found that 30% were positive for *Salmonella* (Currie et al., 2005). This same study found a significant association between consumption of frozen chicken nuggets and/or strips and *S. heidelberg* infection. Currie et al. (2005) concluded that chicken nuggets and strips prepared at home were more likely to contain *S. heidelberg* than those prepared in a commercial establishment. A study by MacDougall et al. (2004) of chicken nuggets found *S. heidelberg* in both opened and unopened products. A study conducted in the United States examining risk factors for sporadic *S. heidelberg* infection did not identify chicken nuggets and/or strips as an important source of infection (CDC, 1999). However, the questionnaire employed did not ask specifically about chicken nugget or strip consumption—only chicken consumption in general.

Aerobic plate counts for *Salmonella* and *E. coli* associated with frozen chicken nuggets were evaluated over a 4-year period by Eglezos et al. (2008). The mean plate count was 5.4 $\log_{10}$ CFU/cm^2. The maximum number of bacteria detected was 6.6 $\log_{10}$ CFU/cm^2 with an *E. coli* prevalence found in 47% of the 300 samples. A correlation was also found between the season in which samples were collected and the prevalence of *E. coli*.

Minnesota reported 26 outbreaks of salmonellosis during 2006. These cases were linked to frozen, prebrowned, single-serving, microwavable stuffed chicken entrées (USDA–FSIS, 2007). Since *Salmonella* is not considered an adulterant in raw poultry, no recall was required, according to the USDA–FSIS. Even though these products are labeled as microwavable, microwaves vary in strength and tend to cook products unevenly. Almost 30% of those surveyed reported using the microwave sometimes or always when cooking these products. Microwave cooking is not recommended because it may not cook the product thoroughly, due to uneven heating (MacDougall et al., 2004). The cooking instructions for many of these products may not be sufficient for killing *Salmonella*; therefore, consumers should ensure that they have fully cooked the products before eating them. Since cases involving individually wrapped entrées continue to be a problem, the USDA is considering requiring more prominent label information, stating clearly that the entrées are raw products (USDA–FSIS, 2007).

Evaluation of foodborne pathogens found in household refrigerators by Jackson et al. (2007) noted that food products may be at high risk for cross-contamination. General hygienic status was estimated using total viable counts and total coliform counts on the interior surfaces of household refrigerators. Although *Campylobacter* spp., *Salmonella* spp., and *E. coli* O157:H7 were not recovered, *Staphylococcus aureus* was found in 6.4%, *L. monocytogenes* and *E. coli* in 1.2%, and *Yersinia enterocolitica* in 0.6% of the refrigerators examined. These findings are of particular concern for RTE foods since additional cooking before consumption may not occur.

CONCLUSIONS

Microbial hazards and associated issues will continue to be major challenges to RTE poultry safety into the future. It is important to realize that management of these products should be based on an integrated effort and approach that applies to all sectors, from the producer through the processor, distributor, packer, retailer, food service worker, and consumer. Most foodborne illness is due to mishandling of foods, including improper handling and inadequate heating. Therefore, the goal of consumer education must be addressed to improve food safety.

REFERENCES

Allos BM, Moore MR, Griffin PM, Tauxe RV. 2004. Surveillance for sporadic foodborne disease in the 21st century: the FoodNet perspective. Clin Infect Dis 38:115–120.

Bucher O, Holley RA, Ahmed R, Tabor H, Nadon C, Ng LK, D'Aoust JY. 2007. Occurrence and characterization of *Salmonella* from chicken nuggets, strips, and pelleted broiler feed. J Food Prot 70(10):2251–2258.

CDC (Centers for Disease Control). 1999. *Foodborne Diseases Active Surveillance Network (FoodNet): Population Survey Atlas of Exposures: 1993–1999*. Atlanta, GA: CDC, p. 72.

Currie A, MacDougall L, Aramini J, Gaulin C, Ahmed R, Isaacs S. 2005. Frozen chicken nuggets and strips and eggs are leading risk factors for *Salmonella heidelberg* infections in Canada. Epidemiol Infect 10:1–8.

Demetrios KM, Antoninos M. 1996. Growth of *Listeria monocytogenes* in the whey cheeses, Myzitheria, Anthotyros, and Manouri during storage at 5, 12, and 22°C. J Food Prot 59:1193–1199.

EC (European Commission). 2003. Opinion of the scientific committee on veterinary measures relating to public health on *Listeria monocytogenes*, Sept. 23, 1999. European Commission, Health and Consumer Protection Directorate-General (SANCO).

Eglezos S, Dykes GA, Huang B, Fegan N, Stuttard ED. 2008. Research note: bacteriological profile of raw, frozen chicken nuggets. J Food Prot 71:613–615.

Farber JM, Peterkin PI. 1991. *Listeria monocytogenes*, a food borne pathogen. Microbiol Rev 55:476–511.

Frazoo E. 1999. *America's Eating Habits: Changes and Consequences*. Agriculture Bulletin 750. Washington, DC: USDA Economic Research Service.

FSIS (Food Safety and Inspection Service). 1996. Pathogen reduction: hazard analysis and critical control point (HACCP) systems. Fed Reg 61:38805–38989.

Glass KA, Doyle MP. 1990. Fate of *Listeria monocytogenes* in processed meat products during refrigerated storage. Appl Environ Microbiol 55:1565–1569.

Gottlieb SL, Newborn EC, Griffin PM, et al. 2006. Multistate outbreak of listerosis linked to turkey deli meat and subsequent changes in US regulatory policy. Clin Infect Dis 42:29–36.

Griffiths MW. 2001. Current issues in HACCP application to poultry processing. Unpublished.

Harris JM. 2002. *Food Manufacturing: The US Food Marketing System*. Report AER-811. Washington, DC: U.S. Department of Agriculture Economic Research Service.

Jackson V, Blair IS, McDowell DA, Kennedy J, Bolton DJ. 2007. The incidence of significant foodborne pathogens in domestic refrigerators. Food Control 18(4):346–351.

Kimura AC, Reddy V, Marcus R, Cieslak PR, Mohle-Boetani JC, Kassenborg HD, Segler SD, Hardnett FP, Barrett T, Swerdlow DL. 2004. Chicken consumption is a newly identified risk factor for sporadic *Salmonella enterica* serotype *enteritidis* infections in the United States: a case–control study in FoodNet sites. Clin Infect Dis 38:244–252.

Kramer J, Scott WG. 2004. Food safety knowledge and practices in ready-to-eat food establishments. Int J Environ Health Res 14(5):343–350.

MacDougall L, Fyfe M, McIntyre L. 2004. Frozen chicken nuggets and strips: a newly identified risk factor for *Salmonella heidelberg* infection in British Columbia. J Food Prot 67:1111–1115.

Mahmood MS, Ahmed AN, Hussain I. 2003. Prevalence of *Listeria monocytogenes* in poultry meat, poultry meat products and other related inanimates at Faisalabad. Pak J Nutr 2(6):346–349.

Marsh LT, Mintert CT, Mintert J. 2004. Impact of meat product recalls on consumer demand in the USA. Appl Econ 36:897–909.

MMWR. 1985. Listeriosis outbreak associated with Mexican-style cheese—California. MMWR 34:357–359.

MMWR. 2000. Multistate outbreak of listeriosis—United States, 2000. MMWR 49:1129–1130.

MMWR. 2002. Public health dispatch: Outbreak of listeriosis—northeastern United States, 2002. MMWR 51:950–951.

Olsen SJ, Patrick M, Hunter SB, et al. 2005. Multistate outbreak of *Listeria monocytogenes* infection linked to delicatessen turkey meat. Clin Infect Dis 40:1569–1572.

Schillinger UM, Lücke FK. 1991. Behaviour of *Listeria monocytogenes* in meat and its control by a bacteriocin-producing strain of *Lactobacillus*. J Appl Bacteriol 70:473–478.

Tappero JW, Schuchat A, Deaver KA, Mascola L, Wenger JD. 1995. Reduction in the incidence of human listeriosis in the United States: effectiveness of prevention efforts. JAMA 273:1118–1122.

Tucker M, Whaley SR, Monto AS. 2006. Consumer perceptions of food-related risks. Int J Infect Dis 194:65–69.

USDA–FSIS (U.S. Department of Agriculture–Food Safety and Inspection Service). 2007. *Product Recall Guidelines for Firms*. FSIS Directive 8080.1 revision 7. http://www.fsis.usda.gov/oppde/rdad/fsisdirectives/8080_1/8080.1rev7_attach1.pdf. Accessed Feb. 2008.

USDHHS (U.S. Department of Health and Human Services). 2000. *Healthy People 2010: Understanding and Improving Health*. Washington, DC: USDHHS.

Varma JK, Samuel MC, Marcus R, Hoekstra RM, Medus C, Segler S, Anderson BJ, Jones TF, Shiferaw B, Haubert N, et al. 2007. *Listeria monocytogenes* infection from foods prepared in a commercial establishment: a case–control study of potential sources of sporadic illness in the United States. Clin Infect Dis. 44:521–528.

Voetsch AC, Angulo FJ, Jones TF, Moore MR, Nadon C, McCarthy P, Shiferaw B, Megginson MB, Hurd S, Anderson BJ, et al. 2007. Reduction in the incidence of invasive listeriosis in foodborne disease active surveillance network sites, 1996–2003. Clin Infect Dis 44:513–520.

Woteki E, Kineman BD. 2003. Challenges and approaches to reducing foodborne illness. Annu Rev Nutr 23:315–344.

35

CHEMICAL ANALYSIS OF POULTRY MEAT

MARÍA DE LOURDES PÉREZ-CHABELA

Departamento de Biotecnología, Universidad Autónoma Metropolitana–Unidad Iztapalapa, México D.F., Mexico

INTRODUCTION

Meat safety has been a main social concern in recent years, indicating that the challenge to improve it will continue in the future. *Residues* or *chemical*

Handbook of Poultry Science and Technology, Volume 2: Secondary Processing, Edited by Isabel Guerrero-Legarreta and Y.H. Hui

TABLE 1 Main Contaminants in Meat and Meat Products

Environmental	Animal Production
Pesticides	Coccidiostats Anthelmintics Antibiotics Anabolics

contaminants are any compound or substance present or formed in edible animal tissues resulting from the use of a xenobiotic or its metabolites. A *xenobiotic* is a chemical compound, such as a drug, a pesticide, or a carcinogen, not naturally part of a living organism. Residues in poultry are generated by compounds used in production to reduce or eliminate diseases, microorganisms, or parasites. Coccidiostats, anthelmintics, antimicrobials, pesticides, and β-agonists used for mites and/or lice control are potential residue producers. Commercial chicken and turkey flocks are tested for certain substances, especially drug residues. Table 1 shows the main groups of residues in meat and meat products. Although residues in animal tissues are rarely present at levels considered to be toxic for humans, high concentrations probably kill the animal. From this point of view, it is necessary to develop sensitive and easy-to-use analytical methods for rapid detection of chemicals and veterinary drug residues in poultry meat.

PESTICIDES

Pesticides are usually selected for their persistence. This means that a pesticide should not be sensitive to inactivation or degradation, and have low volatility, so it can exert its action for a considerable time. According to their persistency and degradability, organochlorides are divided into four groups: highly accumulative (HCB, β-HCH), moderately accumulative (Endrin, Heptachlor), low accumulative (α-HCH, γ-HCH), and very low accumulative (Methoxychlor) (Ruiter, 1981). Organochlorides are lipophilic substances that accumulate primarily in the fat. Residues are extracted from rendered animal fat or by direct extraction from low-fat meat products using organic solvents. This isolation procedure is generally carried out by solvent extraction and analyzed by gel permeation chromatography. Marvel et al. (1978) developed a simple apparatus for quantitative studies of extraction rate and degradation of C-labeled pesticides in soil under aerobic conditions. Other methods use gas chromatography for the quantitative determination of organochloride residues in poultry fat. The results indicated accuracy and precision similar to those of other official techniques. The method has been officially adopted as a first action (Ault and Spurgeon, 1984). In the same manner, Kim and Smith (2001), using gas chromatography, analyzed soil samples from rice fields in South Korea for the presence of organochloride. The main pesticides found were γ- and δ-Hexachlorocyclohexane, Heptachlor, Epoxide, and Dieldrin. However, the results showed inherent variability in this analytical method. The highest values were found in rice-cropping soil. The study concluded that although organochlorides were banned in 1980, substantial residue concentrations, particularly of the oxidized form (heptachlor), remained in the soil, with the possibility of being transferred to poultry feed. Aulakh et al. (2006) studied the presence of organochlorides in poultry feeds, chickens, and eggs at select poultry farms. Samples were prepared by silica gel column chromatography, and the analysis was carried out by gas chromatography attached to an electron capture detector. Higher concentrations of this contaminant were found in eggs than in poultry muscle, although none of the muscles analyzed exceeded maximum residue limits (MRLs) for organochlorides. The results indicated that poultry feed could be one of the major sources of contamination for chickens and eggs.

COCCIDIOSTATS

Poultry are highly susceptible to parasitic disease, such as coccidiosis; the disease is carried by unicellular organisms belonging to the genus *Eimeria*, class Sporozoa. For this reason, veterinary drugs called coccidiostats are used routinely in intensive poultry production. According to Regulation 1831/2003/EC issued by the European Parliament, coccidiostats are at the moment licensed as feed additives, although toxicological information on these drugs is incomplete. While the probability of consumers being exposed to toxic levels is very low, good manufacturing practices indicate that poultry food products should not contain residues of these drugs.

In the past, coccidiostats analysis was carried out primarily by gas chromatography; more recent methods use high-performance liquid chromatography (HPLC) attached to ultraviolet detectors (Ellis, 1999). Dubois et al. (2004) developed and validated a selective and sensitive multiresidue method for analysis of nine coccidiostats based on electrospray liquid chromatography–mass spectrometry. The method allows the extraction and analysis of up to 24 samples per day; it can be applied to liver sample analysis. Mortier and others (2005) developed a sensitive and specific method for the quantitative detection of five chemical coccidiostats (i.e., Halofuginone, Robenidine, Diclazuril, Nicarbaxin, and Dimetridazole) in eggs and feed by tandem liquid chromatographic–mass spectrometry. The method was validated by European Commission decision 2002/657/EC (EC, 2007).

Huet et al. (2005) used enzyme-linked immunosorbent assay (ELISA) for the screening of the coccidiostats Halofuginone and Nicarbaxin in egg and chicken muscle. Rokka and Peltonen (2006) used a confirmatory method for the quantitative determination of four ionophoric coccidiostats in eggs and broiler meat; purified samples were analyzed by liquid chromatography–mass spectrometry. This method was proposed to replace previous methods using HPLC.

ANTHELMINTICS

Benzimidazoles are anthelmintics, agents used widely in the parasitic infection treatment of a wide range of species and as fungicides in crops during storage and transport. Nematodes are the most important group of helminthes infesting poultry. Thiabendazole was the first benzimidazole to be marketed, over 40 years ago. After its introduction, a number of alternative benzimidazoles offering similar action were commercialized, such as Parbendazole, Cambendazole, Mebendazole, and Oxibendazole (Danaher et al., 2007). Levamisole is the levo isomer of Ditetramisole, a racemic mixture; the parent compound, Tetramisole, was first marketed as an anthelmintic in 1965, but it was soon noticed that its anthelmintic activity resided almost entirely in the L-isomer, Levamisole.

El-Kholy and Kemppainen (2005) studied tissues of 32-week-old chickens treated with Levamisole using HPLC–ultraviolet detector. The authors concluded that the MRL (0.1 μg/g) in eggs was reached after 9 days of withdrawing the anthelmintic but that 18 withdrawal days were necessary before slaughtering to produce meat safe for human consumption.

ANTIMICROBIALS (ANTIBIOTICS)

Antimicrobials, including antibiotics, are used in poultry for curative, preventive, and nutritive purposes. In many cases they are administered as a feed additive, through medication or via the drinking water (Ruiter, 1981). Regulatory monitoring for most antibiotic residues in edible poultry tissues is often carried out by accurate, although expensive and technically demanding, chemical analytical techniques.

Chloramphenicol, Furazolidone, and Enrofloxacin are broad-spectrum veterinary drugs. Furazolidone, a nitrofuran, has mutagenic and carcinogenic properties, and its residues should be avoided in edible animal tissue. Low Chloramphenicol levels can produce irreversible bone marrow depression. Enrofloxacin can cause allergies and lead to the emergence of drug-resistant bacteria; this is the only fluoroquinolone approved for the use in broiler chickens in the United States, although other fluoroquinolones are in use worldwide. The MRLs in poultry established by the European Union for Furazolidone, Chloramphenicol, and Enrofloxacin are 5, 0.0, and 100 μg/kg, respectively. The FDA-established tolerance for Enrofloxacin is 300 ppb.

Schneider and Donoghue (2004) compared bioassay analysis and liquid chromatography–fluorescence–mass spectrometry for the detection of Enrofloxacin residues in chicken muscle; their results indicated that bioassay is the most suitable method for examining a large number of samples for regulatory monitoring (120 samples in 2 days). Positive samples must then be examined further by a more sensitive method, such as liquid chromatography–fluorescence–mass spectrometry, for confirmation or rejection.

Immunochemical screening using surface plasma resonance have been developed for Chloramphenicol and Chloramphenicol glucuronide in poultry muscle

(Ferguson et al., 2005). Jafaro et al. (2007) studied the capability of CD-IMS (positive corona discharge–ion mobility spectrometry) for the analysis of residual veterinary drugs, including Furazolidone, Chloranphenicol, and Enrofloxacin in poultry. Ion mobility spectrometry has also been used as an instrumental analytical technique for detecting and identifying volatile organic compounds based on the mobility of gas-phase ions in a weak electric field. This method can analyze Chloranphenicol and Enrofloxacin but not Furazolidone, which is unstable under the conditions of analysis and is converted to its metabolites. Marchesini et al. (2007) studied the feasibility of coupling the simultaneous screening of several antimicrobials using a dual-surface plasma resonance biosensor immunoassay in parallel with an analytical chemical methodology for their identification. Measurement of concentrated muscle samples with the dual biosensor immunoassay resulted in three unknown immunoactive peaks, showing the potential applicability of the system for finding unknown structurally related compounds.

β-AGONISTS

The use of hormones increasing the weight gain rate in meat animals is a controversial global issue. β-Agonists are a class of drugs where the health concern in not antibiotic resistance but acute poisoning from the drug residues themselves. Clenbuterol, the most commonly β-agonist supplied to animals, increases their muscle mass but has been reported to cause human illness (Turnipseed, 2001). It has been observed that the residual levels of β-agonists detected in urine and liver have decreased over the years, due to optimization of administration schemes to avoid detection. Also, application of co-medication may lower the residue levels of β-agonists. Many laboratories now use a combination of screening and confirmatory methods to increase the reliability of the final result (Schilt et al., 1994).

Malucelli et al. (1994) studied distribution in the tissues and residues after withdrawal of various β-agonists (i.e., Clenbuterol, Salbutamol, and Terbutaline) in 160 chickens. The extraction was carried out by a method using heterobifunctional solid-phase extraction. The amount of β-agonists in the extracts was measured by enzyme-linked immunoassay. Clenbuterol showed the highest accumulation in the tissues analyzed. Withdrawal periods of more than 2 weeks were necessary to lower the residue concentration below detectable levels. The authors concluded that the purification method was effective in detecting the presence of Clenbuterol, Salbutamol, and Terbutaline in broiler chicken tissue.

A number of commercial enzyme-linked immunoassays are now available. Most kits are designed for Clenbuterol or Salbutamol but can also detect other cross-reacting β-agonists. Results of screening analysis should be confirmed by gas chromatography–mass spectrometry analysis (Kuiper et al., 1998). Jones et al. (1999) studied the analysis of β-agonists by packed-column supercritical fluid chromatography with ultraviolet and atmospheric-pressure chemical ionization–mass spectrometric detection; they concluded that this method is very efficient, fast, selective, and sensitive for β-agonists. Van

TABLE 2 Methods Most Reported for Drugs in Poultry

Residue	Method	References
Pesticides	Gas chromatography with electron capture detector	Ault and Spurgeon (1984), Kim and Smith (2001), Aulakh et al. (2006)
Coccidiostats	Electrospray liquid chromatography	Dubois (2004)
	Liquid chromatography tandem mass spectrometry	Mortier et al. (2005), Rokka and Peltonen (2006)
	ELISA	Huet et al. (2005)
Anthelmintics	Liquid chromatography	El-Kholy and Kemppainen (2005)
Antibiotics	Liquid chromatography fluorescence	Schneider and Donoghue (2004)
	Immunochemical screening (surface plasma resonance)	Ferguson et al. (2005), Marchesini et al. (2007)
	Positive corona discharge ion mobility spectrometry	Jafaro et al. (2007)
β-Agonists	Enzyme immunoassay	Malucelli et al. (1994)
	Gas chromatography–mass spectrometry	Kuiper et al. (1998), Jones et al. (1999), Van Poucke et al. (2007)

Poucke et al. (2007) analyzed 19 different dietary supplements by means of liquid chromatography–tandem mass spectrometry for the presence of anabolic steroids. After methanol extraction, the samples were screened for the presence of 40 compounds. Of the 19 dietary supplements, 15 contained between one and five prohormones (i.e., a precursor of an anabolic steroid with minimal hormonal effect). Table 2 shows the methods most reported for drug analysis in poultry.

ADVANTAGES AND DISADVANTAGES OF ANALYTICAL METHODS

Most analytical methods currently available for drug residues are screening, immunology, and chromatographic techniques (Toldrá and Reig, 2006), as described below.

Screening Methods

The main requirements for a screening method are: ease in use, time saving and low running cost, sensitivity, specificity, and repeatability. Various screening techniques are available: ELISA test kits, radioimmunoassay, multiarray biosensors, high-performance thin-layer chromatography, and HPLC (Toldrá and Reig, 2006).

Immunological Techniques

ELISA is the most popular technique; the detection system is usually based on enzyme-labeled agents. Two examples of immunoassays used in meat include

TABLE 3 Types of Chromatography Separations

Name	Type	
	Stationary Phase	Mobile Phase
Gas–liquid	Liquid	Gas
Gel	Liquid	Liquid
Ion exchange	Solid	Liquid
Paper	Liquid	Liquid
Gas–solid	Solid	Gas
Partition	Liquid	Liquid

Source: Modified from Skoog and West (1975).

analysis of levamisole in meat (Silverlight and Jackman, 1994) and tetracyclines in pork and chicken meat (De Wasch et al., 1998). Immunoassay may offer a cost-effective and rapid alternative to conventional methods for drug residue screening. The main advantages of the ELISA technique are ease in to use, a large number of samples per kit can be analyzed for a single compound, and high sensitivity. Although this method has several disadvantages, its primary problems are limited storage time under refrigeration, and false-positive results, and expense (Toldrá and Reig, 2006).

Chromatographic Techniques

Chromatography encompasses a diverse group of separation methods (Table 3) that are of great importance to the analytical chemist, for they often make it possible to separate, isolate, and identify components of mixtures that might otherwise be resolved with great difficulty. Chromatography involves processes based on differences in migration of individual components at different rates through a stationary phase under the influence of a moving phase (Skoog and West, 1975). Some advantages of chromatographic methods are sensitivity, specificity, short time needed for analysis, and the possibility of automation. The main disadvantage is the expertise required for sample preparation and column costs (Toldrá and Reig, 2006).

Other Techniques

Recent advances in residue analysis offer several promising techniques as possible solutions to complex analytical methods. Solid-phase extraction, matrix solid-phase dispersion, and immunoaffinity are receiving particular attention, as they have the potential to greatly reduce costs of analysis and the generation of wastes and pollution (Turnipseed, 2001).

CONCLUSIONS

Chemical residues are undesirable at any concentration, whereas other residues have a certain level of legal acceptability. Strict monitoring procedures must be followed. Drugs are used in animal production to control infectious and parasitic diseases as well as to enhance growth efficiency. Although there are economic advantages in continuing with these practices, the effect on human health is related particularly to antibiotic resistance. Better analytical monitoring techniques and more global regulatory cooperation will be needed to effectively manage the use of drugs in poultry meat production.

REFERENCES

Aulakh RS, Gill JPS, Bedi JS, Sharma JK, Joia BS, Ockerman HW. 2006. Organochloride pesticide residues in poultry feed, chicken muscle and eggs at a poultry farm in Punjab, India. J Sci Food Agric 86:741–744.

Ault JA, Spurgeon TE. 1984. Multiresidue gas chromatographic method for determining organochloride pesticides in poultry fat: collaborative study. J Assoc Off Anal Chem 67(2):284–289.

Danaher M, De Ruyck H, Crooks SRH, Dowling G, O'Keeffe M. 2007. Review of methodology for the determination of benzimidazole residues in biological matrices. J Chromatogr B 845(1):1–37.

De Wasch K, Okerman L, Croubels S, de Brabander H, Van Hoof JD. 1998. Detection of residues of tetracycline antibiotics in pork and chicken: correlation between results of screening and confirmatory test. Analyst 123:2737–2741.

Dubois M, Pierret G, Delahault Ph. 2004. Efficient and sensitive detection of residues of nine coccidiostats in egg and muscle by liquid chromatography–electrospray tandem mass spectrometry. J Chromatogr B 813:181–189.

EC (European Commission). 2007. Health and Consumer Protection Directorate-General. Directorate E, Food Safety: production and distribution chain. E3, Chemical contaminants and pesticides. Guidelines for the Implementation of Decision 2002/657/EC, regarding some contaminants. http://ec.europa.eu/food/food/chemicalsafety/residues/sanco00895_2007_en.pdf. Accessed Aug. 22, 2008.

El-Kholy H, Kemppainen BW. 2005. Levamisole residues in chicken tissues and eggs. Poult Sci 84:9–13.

Ellis RL. 1999. Food analysis and chemical residues in muscle food. In: Pearson AM, Dutson TR, eds., *Quality Attributes and Their Measurement in Meat, Poultry and Fish Products*. Gaithersburg, MD: Aspen Publishing, pp. 441–474.

Ferguson J, Baxter A, Young P, Kennedy G, Elliot C, Weigel S, Gatermann R, Ashwin H, Stead S, Sharman M. 2005. Detection of chloranphenicol and chloranphenycol glucuronide residues in poultry muscle, honey, prawn and milk using a surface plasmon resonance biosensor and Qflex kit chloranphenicol. Anal Chim Acta 529:109–113.

Huet AC, Mortier L, Daeseleire E, Fodey T, Elliot C, Delahaut P. 2005. Screening for the coccidiostats halofuginone and nicarbazin in egg and chicken muscle: development of an ELISA. Food Addit Contam 22(2):128–134.

Jafaro MT, Khayamian T, Shaer V, Zarai N. 2007. Determination of veterinary drug residues in chicken meat using corona discharge ion mobility spectrometry. Anal Chim Acta 581:147–153.

Jones DC, Dost K, Davidson G, George MW. 1999. The analysis of β-agonists by packed-column supercritical fluid chromatography with ultra-violet and atmospheric pressure chemical ionization mass spectrometric detection. Analyst 124:827–831.

Kim J-H, Smith A. 2001. Distribution of organochloride pesticides in soils from South Korea. Chemosphere 43:137–140.

Kuiper HA, Noordam MY, van Dooren-Flipsen MMH, Schilt R, Roos AH. 1998. Illegal use of β-adrenergic agonists: European Community. J Anim Sci 76:195–207.

Malucelli A, Ellendorff F, Meyer HHD. 1994. Tissue distribution and residues of clenbuterol, salbutamol and terbutaline in tissues of treated broiler chickens. J Anim Sci 72:1555–1560.

Marchesini GR, Haasnoot W, Delahaut P, Gercek H, Nielen MWF. 2007. Dual biosensor immunoassay-directed identification of fluoroquinolones in chicken muscle by liquid chromatography electrospray time-of-flight mass spectrometry. Anal Chim Acta 586:259–268.

Marvel JT, Brightwell BB, Malik JM, Sutherland ML, Rueppel ML. 1978. A simple apparatus and quantitative method for determining the persistence of pesticides in soil. J Agric Food Chem 26(5):1116–1120.

Mortier L, Daeseleire E, Van Peteghem C. 2005. Liquid chromatographic tandem mass spectrometric determination of five coccidiostats in poultry eggs and feed. J Cromatogr B 820:261–270.

Rokka M, Peltonen K. 2006. Simultaneous determination of four coccidiostats in eggs and broiler meat: validation of an LC-MS/MS method. Food Addit Contam 23(5):470–478.

Ruiter A. 1981. Contaminants in meat and meat products. In: Lawrie R, ed., *Developments in Meat Science*, vol. II. London: Applied Science Publishers, pp. 185–220.

Schilt R, Hooijerink H, Huf FA, Zuiderveld O, Bast A. 1994. Screening of cattle urine samples for the presence of β-agonists with a functional test: some preliminary results. Analyst 119:2617–2622.

Schneider MJ, Donoghue DJ. 2004. Comparison of a bioassay and a liquid chromatography–fluorescence–mass spectrometry method for the detection of incurred enrofloxacin residues in chicken tissues. Poult Sci 83:830–834.

Silverlight J, Jackman R. 1994. Enzyme immunoassay for the detection of levamisole in meat. Analyst 119:2705–2706.

Skoog DA, West DM. 1975. Fundamentails of analytical chemistry. In: *Analytical Separations*, 3rd ed. New York: Holt, Rinehart and Winston, pp. 624–671.

Toldrá F, Reig M. 2006. Methods for rapid detection of chemical and veterinary drugs residues in animal foods. Trends Food Sci Technol 17(9):482–489.

Turnipseed SB. 2001. Drug residues in meat: emerging issues. In: Hui YH, Nip W-K, Rogers RW, Young OA, eds., *Meat Science and Applications*. New York: Marcel Dekker, pp. 207–220.

Van Poucke C, Detaverniera C, Van Cauwenbergheb R, Van Peteghema C. 2007. Determination of anabolic steroids in dietary supplements by liquid chromatography–tandem mass spectrometry. Anal Chim Acta 586(1):35–42.

36

MICROBIAL ANALYTICAL METHODOLOGY FOR PROCESSED POULTRY PRODUCTS

Omar A. Oyarzabal
Department of Biological Sciences, Alabama State University, Montgomery, Alabama

Syeda K. Hussain
Department of Poultry Science, Auburn University, Auburn, Alabama

Handbook of Poultry Science and Technology, Volume 2: Secondary Processing, Edited by Isabel Guerrero-Legarreta and Y.H. Hui

INTRODUCTION

Different microbial methods have been described for the isolation and identification of pathogenic and spoilage bacteria present in poultry products. In the case of bacterial pathogens, most methods target the detection of a low number of cells. Therefore, enrichment procedures are used to allow bacterial cells to recover, multiply to large numbers, and be detectable. When using enrichment procedures, the results are *qualitative*: positive or negative. In the case of spoilage bacteria, it is very important to obtain the count of the different bacterial groups at different times during the storage of the product. Consequently, *quantitative* methods that yield numbers or counts per milliliter or gram of product are most commonly used.

The majority of the microbiological methods and most of the accumulated experiences are related to the analysis of commercial broiler meat. Accordingly, in this chapter we review the most successful methods used for the analysis of these products. However, these methods are also used in the microbial analysis of poultry products from other avian species, such as turkeys. The differences stem from the size of the poultry carcasses, and thus the collection of samples may vary from one poultry species to another. For example, turkey carcasses are very large to handle and require additional equipment to perform the initial sample collection (Dickens et al., 1986).

We first review some terms used to evaluate different methods. This topic is relevant to microbiologists trying to incorporate a new method in the laboratory and includes a brief summary from two organizations that are involved in the validation of new methods to assure performance. We then review sample collection methods, which have been discussed extensively in the literature (ICMSF, 1984), and the most commonly used isolation methods, including the most probable number technique and some of the newest developments in enumeration procedures. In the next sections of the chapter we describe techniques used for the detection—and enumeration when necessary—of bacterial pathogens (*Campylobacter* spp., *Clostridium perfringens, Listeria monocytognes*, and *Salmonella* spp.) and spoilage organisms (mesotrophic and psychrotrophic bacteria, and yeast and mold).

At the end of the chapter there is a brief description of the applications of molecular techniques for rapid identification purposes, and a few sentences on chromogenic agars and future trends in the isolation of bacterial pathogens. We believe that food microbiologists have to be aware of the major concepts driving the research in new detection methods, which happen to incorporate newly developed molecular techniques. These new approaches are expanding our understanding of the bacteria that are present in foods.

METHOD EVALUATION

An important decision to make when incorporating a method in a food microbiology laboratory is to determine the most appropriate method for each circumstance.

Although there is a plethora of scientific literature describing specific applications of the various methods, the search can be overwhelming and sometimes frustrating. Good resources are the organizations that evaluate different methods and compare the new methods with established ones. A couple of examples of those organizations are the Association of Official Analytical Chemists (AOAC) International (www.aoac.org) and the Association Française de Normalisation (AFNOR) (www.afnor.org).

Each organization has a series of validation procedures, and the names of the "validated" or "certified" methods are available on their Web sites. For example, the methods validated by AFNOR can be found at www.afnor-validation.com/afnor-validation-food-industry/food-industry.html. AOAC International has a Method Validation Program, which includes the Official Methods of Analysis (OMA) Program, the Peer-Verified Methods, and the Performance Tested (PT) Methods. These methods for comparison provide a range of accuracy that varies from "the highest degree of confidence in performance to generate credible and reproducible results" (OMA Program) to a "validation of performance claims where rapid validation and some degree of confidence is needed" (PT Methods).

For evaluation purposes, several terms are used to compare the efficacy among methods. The most important terms are:

- *Sensitivity:* the percentage of total positive samples (confirmed positive by one or more methods included in the study) that test positive (confirmed) by the test method
- *Specificity:* the percentage of total negative samples (confirmed negative by all methods included in the study) that test negative by the test method
- *Inclusivity:* the ability of the method to detect the target bacterium from a wide range of strains
- *Exclusivity:* the lack of interference on the method from a relevant range of nontarget strains which are potentially cross-reactive
- *Ruggedness:* the ability of the method to withstand perturbations to basic procedural specifications

When performing a comparison between two or more methods, a term that is usually included for the analysis is the *agreement between methods*, defined as the percentage of samples that test the same (confirmed positive or negative) by the two methods. Two other terms that are used to characterize the efficacy of a method are *false negative*, described as the probability that a test sample is a known positive but was classified as negative by the method, and *false positive*, defined as the probability that a test sample is a known negative but was classified as positive by the method.

Microbiologists should know these terms and should be able to compare different values obtained with different methods when performing small-scale evaluations in their laboratories to decide which method to incorporate. These are important terms because they may not correlate directly. For example, when attempts are made to reduce to a minimum the probability of false negatives in a new method, chances are that the probability of a false positive will increase.

SAMPLE COLLECTION

The collection of samples for microbiological analysis has the goal of obtaining a representative sample of a food lot. The sample should be kept under appropriate conditions—usually, refrigeration—to avoid changes from the time of collection to the start of the analysis. Before the actual microbial analysis takes place, it is important to determine the *sampling plan* to follow, which includes the number of samples and/or the size of the samples to collect, to ensure that the collected samples represent the food lot, and the appropriate method for *sample handling*: transportation and storage. The scope of this chapter does not include a review of the different statistical methods or transportation procedures available for the microbial analysis of foods, and therefore the reader is encouraged to review some relevant literature to incorporate the most appropriate sampling plan for each specific case (ICMSF, 1984; Messer et al., 1992; USDA–FSIS, 1998; FDA, 2000).

The sampling of poultry carcasses offers some difficulties because the whole carcass remains intact until the end of first processing (water cooling in the chiller). Because of the size, broiler carcasses in the United States have traditionally been sampled as a whole in plastic bags using the *carcass rinse method* (Cox and Blankenship, 1975). The amount of rinse used in this method has changed over the years, and with the introduction of hazard analysis and critical control points, the rinse volume has been established at 400 mL (USDA–FSIS, 1996). The rinse most commonly used is Butterfield's phosphate solution (0.00031 M KH_2PO_4, ca. pH 7.2), although buffered peptone water is preferred when collecting samples for *Salmonella* detection because this is the preferred medium for the preenrichment of the samples.

With other large carcasses the use of surface swabbing or tissue excision and maceration are common sampling procedures. Other sampling procedures include the collection of drip (weep), the spraying or scraping of skin areas, mainly on the breast, and agar contact plates (Cox et al., 1976). Because poultry carcasses are hung by the legs, the breast area has low bacterial counts, while the neck skin is the most contaminated area of the carcass (Barnes et al., 1973).

ISOLATION METHODS

Preenrichment

Preenrichment is the first step in the isolation of the target microorganism after the collection of the samples. There are several reasons to add this important step: (1) the organism of interest is present in low numbers or is distributed irregularly across the sample; (2) the food-processing methods and/or intrinsic factors of the food may have injured the organism; and (3) competing organisms are precent. Hence, the preenrichment step favors multiplication of the organisms present in the samples, without selecting for particular microorganisms. Therefore, preenrichment media are very nutritious and nonselective for the targeted microorganism.

It is important to highlight that all preenrichment and enrichment procedures used in food microbiology are in the liquid form (broths), and they are all intended to help bacterial cells recover from injuries, or sublethal stress, and reproduce to higher, detectable numbers. Traditionally, a ratio of 1 to 9 (where 1 is the amount of sample and 9 is the amount of enrichment) has been used to enrich food samples. In the United States, the suggested amount of food product to sample is 25 g or mL, and therefore the amount of enrichment to add to food samples is 225 mL. But recent studies using simulation and actual sampling programs have shown that the incidence and distribution of pathogens in broiler meat increases in a nonlinear manner as a function of sample size. For example, the incidence of *Salmonella* in 25-g samples has been estimated at 16%, while in 100-g samples the incidence estimated is 51%. We currently do not know the most appropriate sample size for pathogen identification based on risk analysis models. Besides, the ideal sample size may be different according to the targeted pathogen, and thus a linear extrapolation of enumeration results, a common practice in microbial risk assessment, may not appear to be the most appropriate approach (Oscar, 2004).

Enrichment

Enrichment is the step after preenrichment intended to facilitate the growth of the organism of interest to detectable numbers while inhibiting the growth of unwanted competing organisms. Sometimes, the preenrichment is bypassed and samples are only enriched before transferring to agar plates (e.g., for *Campylobacter* isolation). Therefore, enrichment broths contain antimicrobials to selectively isolate the target microorganism. Different types of enrichment broths with different incubation temperatures are employed for the detection of different organisms, or even for the detection of the same organism. For example, tetrathionate (TT) and Rappaport–Vassiliadis (RV) broths are common enrichment media for *Salmonella* isolation, and sometimes both broths are used to maximize *Salmonella* isolation.

Agar Plates

The third step of the isolation procedure includes the transfer of enriched samples to agar plates. Plating media provide a solid surface for growth of the organism and allows for the detection of individual colony-forming units (CFU), which are in most cases representative of the growth of a given bacterial clone. Plate media can also be used for direct isolation and enumeration of the target organism. In this case, the bacteria are in numbers high enough to grow directly on the agar plates and result in a count per milliliter or gram of food.

Methods for Bacterial Counts

As stated in the introduction, the counting of some bacterial groups present in food samples is an important tool to predict the shelf life of a product. Several attempts have been made in the last two decades to replace the *most probable number* (MPN) *technique* for bacterial enumeration. But the reality is that for samples contaminated with less than 1 CFU of the target organism per gram or milliliter of the product, a direct enumeration using plating media is not feasible. It is important to remember that the term *colony-forming unit* refers to the growth seen on agar plates and represents viable bacteria, which may be one or more bacteria growing to produce a single colony.

Some methods, however, concentrate the bacterial cells from a sample and increase the opportunity of detection. An example of these concentration methods is *hydrophobic grid-membrane filtration* (HGMF), which has had a major application for the enumeration of coliforms, *Escherichia coli, Salmonella*, and yeasts and molds in food products (Brodsky et al., 1982; Entis et al., 1982; Entis, 1984). Although filter membranes for the concentration of bacteria from water samples have been used for many decades (Goertz and Tsuneshi, 1951), the use of HGMF for counting bacteria in food samples was introduced in 1974 (Sharpe and Michaud, 1974). A specially constructed filter consisting of 1600 wax grids on a single membrane confines the growth of the organism to the square grid cell, restricts the colony size, and reduces the need for extensive dilutions of the sample (Sharpe and Michaud, 1974). Linear counting can result in up to 30,000 CFU per filter and as few as 10 cells/g can be enumerated within 24 h (Sharpe and Michaud, 1975; Sharpe et al., 1983).

When the numbers of bacterial cells in samples are higher than 100 CFU/mL or g, other methods can be used that are not as labor intensive as the MPN method but provide similar results. One of these methods is the *spiral plate method*, which allows for the enumeration of samples with large numbers of bacteria using very few plates and without extensive dilution schemes. A known amount of the sample is deposited on a rotating agar plate and the results are read by a laser eye that calculates the number of CFU based on the number of colonies found on the plate. The sensitivity of the method is higher with samples that have more than 1000 CFU/mL. It is important to obtain discrete colonies on the agar plates to avoid errors during the reading step. This method has cost advantages over the conventional method, which uses a serial dilution with the plating from each dilution for bacterial counts, and it can be done in a shorter time.

Another method is the *slim agar plate*. These agar plates have been developed to count different organisms, and the most common of these systems is Petrifilm (3M, Saint Paul, Mcrrie-sota). There is plenty of information validating the use of Petrifilm as equivalent to the traditional plate count agar method. Petrifilm consists of small, thin paper plates made of a water-soluble gelling agent, nutrients and indicators to facilitate enumeration. These films are cost-efficient and easy to use, and the inoculation and incubation steps are similar to those of agar plates in Petri dishes. An advantage of these plates is that the counting is simplified by a grid on the film background that helps divide the total inoculated areas in quadrants. They can also be read using a colony counter. During incubation, Petrifilm plates can be stacked, similar to other agar plates, but use much less space. Currently, Petrifilm has been developed to enumerate coliforms, Enterobacteriaceae, *Listeria, Staphylococus*, and total aerobic plate counts.

PATHOGENIC BACTERIA

Although the last few years have unwrapped an entire generation of advanced methods and technologies for pathogen detection and enumeration, conventional enumeration methods are still indispensable. In the case of presumptive positive samples, the pathogen has to be isolated from the food. In addition, these conventional or "traditional" procedures are still cost-efficient for small laboratories.

In this section we review the methods recommended in the *Microbiology Laboratory Guidebook* (MLG) from the U.S. Department of Agriculture's Food Safety and Inspection Service (USDA–FSIS), which is the federal agency regulating food products that contain 2% of red meat or poultry in their composition (USDA–FSIS, 1998). We also mention the methods recommended by the *Bacteriological Analytical Manual* (BAM) of the U.S. Food and Drug Administration (FDA) (FDA, 2000) and the International Organization for Standardization (ISO) directives in Europe. Within the various protocols we emphasize the methods that have been most accepted in food microbiology laboratories, an acceptability that may be related to the reliability, simplicity, and cost of the method. These more accepted methods may sometimes depart from the methods suggested by government agencies, especially in the isolation and detection of nonregulated bacterial pathogens.

Campylobacter

Campylobacter spp. are the most prevalent bacterial foodborne pathogens isolated from broiler meat in the United States. *C. jejuni* is isolated two to three times more frequently than *C. coli*. The prevalence of *Campylobacter* spp. in commercially processed broiler carcasses is around 80% (Figure 1), with a count averaging 10 or fewer CFU/mL of the rinse (Oyarzabal, 2005), and in retail products is about 60 to 82% (Willis and Murray, 1997; Zhao et al., 2001; Dickins et al., 2002; Oyarzabal et al., 2007), with a count averaging 0.7 CFU/g of product (Oyarzabal et al., 2007).

Campylobacter does not grow below 30°C. For the isolation of *Campylobacter* from poultry meat, 42°C (±1) has been used as the temperature of choice. Yet this temperature allows only for the isolation of *C. jejuni, C. coli*, and *C. clari*, which together with *C. upsaliensis* make up the *thermotolerant group*, and which have been the only species isolated from broiler meat until now. With the development of DNA-based methods for the identification of isolates, *C. lari* has not been reported from broiler samples for more than 10 years in the United States, which

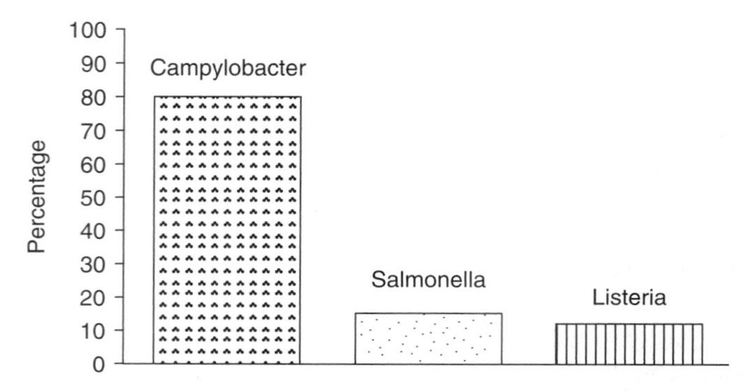

FIGURE 1 Prevalence of *Campylobacter, Listeria*, and *Salmonella* in processed broiler carcasses. [Modified from USDA–FSIS (1996) and USDA–FSIS data (www.fsis.usda.gov/Science/Progress_Report_Salmonella_Testing_Tables/index.asp).]

suggests that previous reports may have been misidentifications from biochemical tests (Oyarzabal et al., 1997).

There have been some discussions in the last 10 years on the best methodology for isolation and identification of *Campylobacter* from broiler meat, and although numerous enrichment broths, isolation media, and methods to generate a microaerobic atmosphere have been developed in the past 30 years, few methods have been validated in large-scale studies to guarantee reproducibility of results.

The best enrichment broth for *Campylobacter* spp. isolation from broiler meat is Bolton broth (Bolton and Robertson, 1982), although other broths, such as Preston broth, Park and Sanders (De Boer and Humphrey, 1991), and buffered peptone water (Oyarzabal et al., 2007) may also be used. The addition to the enrichment of 5 to 10% lysed blood appears to increase the likelihood of isolation. The use of cefoperazone (32 mg/L), trimethoprim (10 mg/L), and vancomycin (10 mg/L) appears to be the best combination to inhibit the growth of contaminants without affecting substantially the recovery and growth of *Campylobacter* cells. Bolton broth is also suggested as the enrichment of choice for *Campylobacter* isolation by the USDA–FSIS (1998) and by the protocol for the international standard, ISO 10272 (ISO, 2006), with an incubation of up to 48 h. The ISO protocol suggests an initial incubation at 37°C for 4 to 6 h to allow stressed or injured cells to recover.

It has also been suggested that an initial incubation without antimicrobials, similar to the protocol for the initial isolation of *Listeria monocytogenes* recommended by the FDA, may increase the likelihood of isolation. Yet a larger food sample and a larger incubation time do play a more important role and correlate directly with an increased probability of isolating *Campylobacter*. In addition, recent findings suggest that an enrichment ratio of 1:5 compares similarly to a ratio of 1:10. This lower ratio may allow for an increase in sample size without increasing, substantially, the overall volume during the enrichment step (Oyarzabal et al., 2007).

All the plate media used for isolation of *Campylobacter* spp. from food samples are adaptations, or direct compositions, of media originally designed to isolate *Campylobacter* spp. from clinical, mainly fecal, samples. The plate agar media used for the isolation of *Campylobacter* spp. can be divided into two groups. In one group we have plate media that have blood as a supplement, and in the other group we have plates that have charcoal as a supplement. Both blood and charcoal are added to help reduce the oxygen tension and create the microaerobic environment that allows campylobacters to grow. Although some agar plates have been developed that do not contain blood or charcoal in their composition, the effectiveness of these plates appear to be lower than that of plates with either blood or charcoal.

The most common plate agar used worldwide is the modified charcoal cefoperazone deoxycholate agar (mCCDA) (Bolton and Robertson, 1982; Hutchinson and Bolton, 1984). This plate is available commercially and has been validated in many different studies. The only limitation with this plate is that sometimes *Campylobacter* colonies stick to the surface of the plate and are very difficult to harvest. A plate that has been used extensively in the United States is Campy-Cefex (Stern et al., 1992), a blood-based plate, and its modification (Oyarzabal et al., 2005). Both CCDA and m-Campy-Cefex compare similarly for the direct enumeration of *Campylobacter* spp. in carcass rinses collected from postchilled, processed broiler carcasses. For a review of the most common enrichment broths and plate agars used for isolation of *Campylobacter* from foods, refer to the article by Corry et al. (1995).

The best isolation procedures still rely on the replacement of the air in jars or plastic bags with a microaerobic atmosphere (10% CO_2, 5% O_2, and 85% N_2) to guarantee the survival and reproduction of *Campylobacter* cells. This microaerobic atmosphere can be generated with pouches that are added to the jars or by extracting the air with a vacuum pump and replacing it with a commercial microaerobic mix.

Although *C. jejuni* has usually been found at high, countable numbers in broiler carcasses after processing, the counts have been decreasing in recent years and the current numbers in commercial broiler carcasses are low enough to justify the use of enrichment to detect positive carcasses (Oyarzabal, 2005; Oyarzabal et al., 2005). In retail poultry, the numbers are very low and enrichment is indispensable for isolation (Oyarzabal et al., 2007).

The fact that campylobacters are inert, which means that they do not use or ferment sugars to produce energy, has been a major drawback for the use of biochemical tests for identification. The best identification to the genus level is achieved with latex, enzyme-linked immunosorbent assay (ELISA), or DNA-based tests. But identification to the species level requires several physiological tests that are time consuming and have a low degree of reproducibility. Because few research laboratories perform routine physiological testing of *Campylobacter* isolates, the use of DNA-based methods have become indispensable for species identification in research laboratories working with *Campylobacter* spp.

Clostridium perfringens

Small numbers of *Clostridium perfringens* are present in the alimentary tract skin and in the feathers of a chicken, and therefore fecal contamination during processing is one of the reasons for the appearance of this organism in broiler chicken meat. There is also a high chance of occurrence of *C. perfringens* in meats and poultry that are cooked and not maintained at proper temperatures prior to serving. *C. perfringens* is a spore-forming anaerobic bacterium that produces an enter toxin that results in acute abdominal pain and diarrhea. The typical foodborne associated strains are type A.

When analyzing a food for *C. perfringens*, it is important to analyze the samples promptly to avoid the loss of viability of the *Clostridium* cells present when samples are frozen or refrigerated for lengthy periods. Hence, food samples should be treated with buffered glycerol salt solution (10% glycerol) prior to freezing and shipping.

The USDA–FSIS method for analysis of *C. perfringens* recommends the use of Butterfield's phosphate diluent, and plating of the dilutions on tryptose sulfite cycloserine (TSC) supplemented with egg yolk and overlaid with egg yolk–free TSC agar. The plates are incubated anaerobically at 35°C for 24 h. Each of the presumptive colonies (black with a halo) is inoculated into thioglycollate broth and incubated at 35°C overnight (USDA–FSIS, 1998).

The FDA method requires that the sample be mixed with peptone dilution fluid and the mix be blended with molten TSC agar without egg yolk and poured on a petri plate (FDA, 2000). Colonies that are black with opaque white zones are inoculated into chopped liver broth and incubated at 35°C for 24 to 48 h. *C. perfringens* is a gram-positive bacterium that will appear as short rods under microscopy. For further confirmation, biochemical tests such as the motility and nitrate reduction tests can be done. *C. perfringens* is nonmotile and can reduce nitrate to nitrite.

Listeria

Listeria monocytogenes is a psychrotrophic foodborne pathogen that can live in microaerobic or even anaerobic environments. *L. monocytogenes* can produce severe diseases in immunocompromized persons and pregnant women. The most recent outbreak of listeriosis in the United States, reported by the Massachusetts Department of Public Health on December 28, 2008, resulted in the death of two infected people and it was traced to pasteurized milk produced locally.

The USDA–FSIS has maintained a zero-tolerance policy on the detection of *L. monocytogenes* in ready-to-eat products (i.e., products that may be consumed without any further cooking or reheating). This policy means that if the product is contaminated with *L. monocytogenes*, the product is considered "adulterated" under the provisions of the Federal Meat Inspection Act and the Poultry Inspection Act, 21 U.S.C. 601(m) or 453 (g), respectively (U.S. Code, 1994). Many media have been formulated for the selective isolation of *Listeria* species, although the success in isolation is highly dependent on identification of the low numbers of *Listeria* cells present in the food. Experiments have shown that direct plating is not a good method for isolation of *Listeria*; therefore, preenrichment and enrichment steps are employed. The preenrichment broth has fewer amounts of selective agents, permitting the revival of injured cells; the enrichment medium contains acriflavin and nalidixic acid to select for *Listeria*. Modified Oxford (MOX) and PALCAM are selective plating media in widespread use. The agar turns black around the areas where the colonies grow, but this change in color may also occur around non-*Listeria* colonies growing on the plate.

The USDA–FSIS method uses University of Vermont (UVM) broth as the primary enrichment medium and Fraser broth as the secondary enrichment medium. Isolation is carried out on MOX agar plates. The FDA's BAM method requires use of buffered *Listeria* enrichment broth for nonselective enrichment at 30°C for 4 h and then addition of selective agents with incubation at 30°C for 48 h. The method for detecting *L. monocytogenes* in foods of international standard ISO 11290 (ISO, 1996) includes streaking on a chromogenic agar plate (ALOA) and streaking on a second agar medium from primary enrichment, and transfer followed by to half-strength Fraser broth and secondary enrichment in Fraser broth. Confirmation of *Listeria* spp. is done with biochemical tests, Gram staining, and microscopy to determine motility.

Listeria spp. can be identified using ELISA (polyclonal antibodies) and DNA-based methods. For the identification of species within the genus *Listeria*, several test strips have been developed that differentiate species based on a biochemical test. These test strips have been validated and are reliable for the identification of the species found most commonly in foods.

Salmonella

Even though the numbers of *Salmonella* have been decreasing over the last 10 years, keeping the organism numbers low is still a challenge in broiler processing plants. When small numbers of *Salmonella* are expected, the sampling method has a major influence on the identification of *Salmonella*-positive carcasses (Simmons et al., 2003). Four steps are involved in the isolation and identification of *Salmonella*. They include (1) preenrichment in a nonselective, nutritious medium; (2) selective enrichment; (3) plating on selective agars; and (4) confirmation by biochemical and serological tests. The FDA's BAM method (FDA, 2000) uses lactose broth as a preenrichment medium; RV broth and TT as selective enrichment; and Hektoen Enteric agar, xylose–lysine–deoxycholate (XLD) agar, and bismuth–sulfite agar as the selective plate media for isolation. The method suggested by the USDA–FSIS (1998) uses BPW for preenrichment, modified RV for selective enrichment, and xylose–lysine tergitol 4 (XLT4) agar or double-modified lysine–iron agar plates for selective isolation.

The international standard protocol ISO 6579 uses BPW as the preenrichment; RV supplemented with soya and TT supplemented with novobiocin for enrichment; XLD agar for selective isolation (ISO, 2002); and confirmation of presumptive *Salmonella* colonies with suitable biochemical tests. The identification of *Salmonella* to the genus level can be done using test strips that have a battery of biochemical tests. These tests are very reliable and have been used for several decades. The identification of specific O antigens for most common serotypes can be done with commercial agglutination tests. However, a thorough serotyping scheme requires testing with a large panel of antisera, and few laboratories in the world can do that.

SPOILAGE MICROORGANISMS

Mesotrophic Bacteria

The enumeration of viable bacterial or fungal cells in a food matrix is important and necessary to monitor the microbiological quality and safety of food. The most common plate for enumeration of mesotrophic bacteria is the *aerobic plate count* (APC) *agar*, a term that refers to agar plates originally composed of trypticase soy agar (containing glucose) (TSA) or tryptic soy agar (without glucose). TSA is highly nutritious because the combination of the soy and casein provides organic nitrogen from amino acids and longer chain peptides, and the sodium chloride provides osmotic balance. This medium can also be use for bacterial enumeration and as a base agar for other media (Forbes et al., 1998). TSA is used in the analysis of water, wastewater, and foods specified in FDA's BAM (Clesceri et al., 1998; FDA, 2000; Downes and Ito, 2001).

Currently, most APCs contain glucose, which is added to avoid undercounts when analyzing meats and using an incubation period of fewer than 4 to 5 days. Other agar plates that have been used for APC analysis in poultry meats include Standard 1 nutrient agar (Vorster et al., 1994), plate count agar (Warburton et al., 1988), tryptone glucose and yeast extract (van der Marel et al., 1988), and Petrifilm (Chain and Fung, 1991).

A plate that bas been used for the bacterial count in milk and dairy products, but that can also be used for analysis of poultry products, is the *standard plate count* (SPC) *agar*. The formulation has been developed by the American Public Health Association and contains the enzymatic digest of casein, which provides amino acid and other complex nitrogenous substances for bacterial growth. The yeast extract supplies vitamins, and the dextrose is the source of carbon to provide energy. The triphenyltetrazolium chloride present in the medium is reduced to the insoluble formazan inside the bacterial cell, and colonies appear in red.

Although APCs have long been used in food microbiology, there is a lack of agreement about the temperature and length of incubation. Most of the current literature on APCs includes psychrotropic and mesotrophic bacteria, which makes a comparison of the different results quite difficult. Since 1985, different temperature–time combinations have been used for APC, ranging from 20°C for 120 h to 37°C for 24 h. Under these different conditions, the isolates grown at one temperature do not grow at other temperatures and reflect different bacterial populations (Jay, 2002).

For direct enumeration of mesotrophic bacteria, various dilutions of the sample are plated and the plates are incubated at 30°C for 48 h to count the numbers of aerobes. An elaborate procedure for determining the APC of various foods has been developed by the Association of Official Analytical Chemists (AOAC) and the American Public Health Association (APHA, 1984).

Psychrotrophic Bacteria

Psychrotrophic bacteria include a wide variety of bacteria (gram-positive, gram-negative, aerobic, anaerobic, facultative, sporeformers, and non-sporeformers) which grow at temperatures between -5 and 30°C, with an optimum temperature of 21°C. Psychrotrophic bacteria and yeasts and molds make up for the majority of food-spoilage organisms. The enumeration of psychrotrophic bacteria can be done by plate count using trypticase soy agar or Petrifilm. The plates are usually incubated at 7°C for 10 days. A more rapid method called *modified psychrotrophic bacteria count* has been formulated to enumerate bacteria after incubation at 21°C for 25 h (Oliveria and Parmelee, 1976).

Pseudomonas species are some of the common spoilage bacteria of aerobically stored poultry meat (Arnaut-Rollier et al., 1999). Several studies have shown a direct correlation between the starting numbers of *Pseudomonas* and the shelf life of the product at refrigeration temperatures (Barnes et al., 1979). At refrigerated temperatures, spoilage occurs when their numbers reach 10^7 to 10^8 per cm^2. Other gram-negative bacteria, such as *Acinetobacter* and *Psychrobacter* spp., are also important spoilage organisms, and poultry meat in modified-atmosphere packages develops large numbers of lactic acid bacteria and *Carnobacterium* spp. These bacteria are usually accompanied by pseudomonads, Enterobacteriaceae, and *Brochothrix thermosphacta* (Jones et al., 1982; Mead et al., 1986).

Rapid methods have been developed to overcome the long incubation periods necessary to obtain psychrotrophic counts. Some of these methods are based on quantitative analysis of enzymes, such as catalase and cytochrome *c* oxidase. Psychrotrophic bacteria may become a concern in minimally processed foods if proper processing and handling conditions are not observed. An example is deli-type foods, where low heat and vacuum, instead of sterilization, are used to process the food. This processing results in the survival of cells or spores and leads to food spoilage.

Yeasts and Molds

Yeasts and molds have a wide range of temperature (5 to 35°C) and pH (4 to 6.5) requirements for their growth. The numbers of yeast cells on raw poultry meat can reach up to 10^4 per milliliter of carcass rinse at the end of the shelf life. High numbers of *Candida* spp. (*C. zeylanoides*) and *Yarrowia lipolytica* can be found in the products (Gallo et al., 1988; Ismail et al., 2000; Hinton et al., 2002). There are numerous media, depending on the type of food and the type of fungus being isolated. Acidified media were used traditionally, but better antibiotic media developed recently are prevalent these days. These media prevent unwanted bacterial growth, enhance revival of injured fungi, and minimize food particle precipitation.

The general-purpose media for yeast and mold enumeration are dichloran rose bengal, which restricts excessive mycelial growth, and antibiotic-supplemented plate count agar. This is done by dilution and surface-spread plating methods to expose the cells to atmospheric oxygen and to avoid the heat stress of molten agar when using the pour plate method. Recovery of yeasts and molds from intermediate-moisture foods can be done by soaking the food for some time.

Another widely used medium for cultivation and enumeration of yeasts and molds is potato dextrose agar. Potato starch (potato infusion) and dextrose (corn sugar) provide nutrients for the elaborate growth of fungi. It is essential that the media be adjusted to a pH of 3.5 by the addition of sterile tartaric acid to inhibit the growth of unwanted bacteria. After inoculation, plates have to be incubated for 5 days at 22 to 25°C. The fungal growth can be confirmed by making wet mounts or by Gram staining (yeasts are gram-positive and mold mycelia are gram-negative).

APPLICATION OF MOLECULAR TECHNIQUES FOR IDENTIFICATION PURPOSES

Molecular techniques based on antibody detection of whole bacterial cells and DNA detection of specific DNA markers have been used for several years in food microbiology. The main advantage of these molecular-based methods is shortening of the time necessary to determine if a sample is presumptively positive. In the case of antibodies, monoclonal antibodies have not been very successful for food applications because their high specificity may not detect some bacterial strains and therefore result in false-negative samples. In other words, some strains of the same bacterial group that is targeted by the antibodies may not be detected. Polyclonal antibodies, on the other hand, are more inclusive in their specificity but lack sensitivity to detect bacterial numbers of less than 10^3 CFU/mL. Yet polyclonal-based assays such as ELISA and latex agglutination tests, can be used in some food matrices for a rapid screening of presumptive positives and for the analysis of a larger number of samples (high throughput).

Within the DNA-based methods, the tests more frequently used in food microbiology are based on isothermal hybridization or amplification with the polymerase chain reaction (PCR) technique. The few commercial tests based on PCR assays are very reliable and specific. Yet PCR assays still have a limitation in their sensitivity and cannot detect the target organism if it is present at less than 10^3 CFU/g or mL. In addition, in meat products and other complex food matrices, the presence of organic compounds inhibitory to PCR reactions has been described, which limits the use of PCR in those food samples.

FUTURE TRENDS IN THE IDENTIFICATION OF BACTERIA

In the last 20 years, food microbiologistS have seen the appearance of chromogenic agars to facilitate the differentiation of presumptive positives on agar plates. These media detect the presence of a specific enzyme using suitable substrates, such as fluorogenic or chromogenic enzyme substrates (Manafi, 2000). The enzymes targeted depend on the medium and the manufacturer's selection. Some examples are β-galactosidase, β-glucosidase, β-glucuronidase, and tryptophan deaminase (Edberg and Kontnique, 1986; Manafi, 2000). There is a change in color around the suspect colonies, and sometimes the colonies have a unique color. These media have brought a more objective way of identifying bacterial colonies on agar plates. For example, one of the first chromogenic media developed to detect *Salmonella* spp. uses a chromogenic substrate for β-galactosidase that in conjunction with propylene glycol generates acids and changes, the color of the media around *Salmonella* colonies (Rambach, 1990). Currently, chromogenic agars have been developed for *Bacillus cereus, Clostridium perfringens*, Enterobacteriaceae, enterococci, *Escherichia coli, E. coli* O157:H7, *Listeria monocytogenes, Salmonella* spp., and *Staphylococcus aureus* (Manafi 2000).

With the advancements in molecular techniques and in our understanding of the genes involved in pathogenicity, there are now more opportunities to develop identification techniques that target specific virulence genes, unique to bacterial pathogens. Although these genetic markers are not present in all clones of the same species, at least the ones that carry them are candidates of concern from a human health standpoint.

Finally, the history of the development of microbiology tests for a given pathogen depends on many variables, such as difficulties of isolation, prevalence in food products, and pathogenicity for humans. An important factor to remember is the effect of established regulations on the development of methodologies. There are pathogens for which regulations have been established. These regulations may be in the form of complete absence after testing (no tolerance), as in the case of *L. monocytogenes* in RTE products or in the allowance of an incidence (prevalence or counts), as for *Salmonella* in process poultry meat. The pathogens that are under regulations have promoted more research, and more methods are available in the market for their testing.

REFERENCES

APHA (American Public Health Association). 1984. *Compendium of Methods for the Microbiological Examination of Foods*, 2nd ed. Washington, DC: APHA.

Arnaut-Rollier I, De Zutter L, Van Hoof J. 1999. Identities of the *Pseudomonas* species in flora from chilled chickens. Int J Food Microbiol 48:87–89.

Barnes EM, Impey CS, Parry RT. 1973. The sampling of chicken, turkeys, ducks, and game birds. In: Board RG, Lovelock DW, eds., *Sampling: Microbiological Monitoring of Environments*. Society for Applied Bacteriology Technical Series 7. London: Academic Press, pp. 63–75.

Barnes EM, Mead GC, Impey CS, Adam BW. 1979. Spoilage organisms of refrigerated poultry meat. In: Russell AD, Fuller R, eds., *Cold Tolerant Microbes in Spoilage and the Environment*. Society for Applied Bacteriology Technical Series 13. London: Academic Press, pp. 101–116.

Bolton FJ, Robertson L. 1982. A selective medium for isolation *Campylobacter jejuni/coli*. J Clin Pathol 35:462–467.

Brodsky MH, Entis P, Entis MP, Sharpe AN, Jarvis GA. 1982. Determination of aerobic plate and yeast and mold counts in foods using an automated hydrophobic grid membrane filter technique. J Food Prot 45:301–304.

Chain VS, Fung DYC. 1991. Comparison of Redigel, Petrifilm™, Spiral Plate System, Isogrid, and aerobic plate count for determining the numbers of aerobic bacteria in selected foods. J Food Prot 54:208–211.

Clesceri LS, Greenberg AE, Eaton AD. 1998. *Standard Methods for the Examination of Water and Wastewater*, 20th ed. Washington, DC: American Public Health Association.

Corry JEL, Post DE, Colin P, Laisney MJ. 1995. Culture media for isolation of campylobacters. Int J Food Microbiol 26:43–76.

Cox NA, Blankenship LC. 1975. Comparison of rinse sampling methods for detection of salmonellae on eviscerated broiler carcasses. J Food Sci 40(6):1333–1334.

Cox NA, Mercuri AJ, Thompson JE, Chew V. 1976. Swab and excise tissue sampling for total and Enterobacteriaceae counts of fresh and surface-frozen broiler skin. Poult Sci 55:2405–2408.

De Boer E, Humphrey TJ. 1991. Comparison of methods for the isolation of thermophilic campylobacters from chicken products. Microb Ecol Health Dis 4:S43.

Dickens JA, Cox NA, Bailey JS. 1986. Evaluation of a mechanical shaker for microbiological rinse sampling of turkey carcasses. Poult Sci 65:1100–1102.

Dickins MA, Franklin S, Stefanova R, Shutze GE, Eisenach KD, Wesley IV, Cave D. 2002. Diversity of *Campylobacter* isolates from retail poultry carcasses and from humans as demonstrated by pulsed-filed gel electrophoresis. J Food Prot 65:957–962.

Downes and Ito, eds. 2001. *Compendium of Methods for the Microbiological Examination of Foods*, 4th ed. Washington, DC: American Public Health Association.

Edberg SC, Kontnique CM. 1986. Comparison of β-glucuronidase based substrate systems for identification of *E. coli*. J Clin Microbiol 24:368–371.

Entis P. 1984. Enumeration of total colifoms and *Escherichia coli* in foods by hydrophobic grid membrane filter: collaborative study. J Assoc Anal Chem 67:812–823.

Entis P, Brodsky MH, Sharpe AN, Jarvis GA. 1982. Rapid detection of *Salmonella* spp. in food by use of the iso-grid hydrophobic grid membrane filter. Appl Environ Microbiol 43:261–268.

FDA (Food and Drug Administration). 2000. *Bacteriological Analytical Manual*. Online edition. http://www.cfsan.fda.gov/~ebam/bam-mm.html. Accessed Feb. 25 2008.

Forbes BA, Sahm DF, Weissfeld AS. 1998. *Bailey & Scott's Diagnostic Microbiology*, 10th ed. St. Louis, MO: Mosby.

Gallo L, Schmitt RE, Schmidt-Lorenz W. 1988. Microbial spoilage of refrigerated fresh broilers: I. Bacterial flora and growth during storage. Lebensml-Wiss Technol 21:216–223.

Goertz A, Tsuneshi N. 1951. Applications of molecular filter membranes to the bacteriological analysis of water. J Am Water Works Assoc. 43:943–945.

Hinton A, Cason JA, Ingram KD. 2002. Enumeration and identification of yeasts associated with commercial poultry processing and spoilage of refrigerated broiler carcasses. J Food Prot 65:993–998.

Hutchinson DN, Bolton FJ. 1984. An improved blood-free selective medium for isolation of *Campylobacter jejuni* from faecal specimens. J Clin Pathol 37:956–995.

ICMSF (International Commission on Microbiological Specifications for Foods). 1984. Sampling for microbiological analysis: principles and specific applications. In: *Microorganisms in Foods*, 2nd ed. Toronto, Ontario, Canada: University of Toronto Press.

Ismail SAS, Deak T, Abd. el-Rahman HA, Yassien MAM, Beuchat LR. 2000. Presence and changes in populations of yeast on raw and processed poultry products stored at refrigeration temperature. Int J Food Microbiol 62:113–121.

ISO (International Organization for Standardization). 1996. Microbiology of food and animal feeding stuffs—horizontal method for the detection and enumeration of *Listeria monocytogenes*: 1. Detection method. Added in 2004: Amendment 1: Modification of the isolation media and the haemolysis test, and inclusion of precision data. ISO 11290–1. Geneva, Switzerland: ISO.

ISO. 2002. Microbiology of food and animal feeding stuffs—horizontal method for the detection of *Salmonella* spp. ISO 6579. Geneva, Switzerland: ISO.

ISO. 2006. Microbiology of food and animal feeding stuffs—horizontal method for the detection and enumeration of *Campylobacter* spp.: 1. Detection method. ISO 10272. Geneva, Switzerland: ISO.

Jay JM. 2002. A review of aerobic and psychrotrophic plate count procedures for fresh meat and poultry products. J Food Prot 65:1200–1206.

Jones JM, Mead GC, Griffiths NM, Adams BW. 1982. Influence of packaging on microbiological, chemical and sensory changes in chill-stored turkey portions. Br Poult Sci 23:25–40.

Manafi M. 2000. New developments in chromogenic and fluorogenic media. Int J Food Microbiol 60:205–218.

Mead GC, Griffiths NM, Grey TC, Adam BW. 1986. The keeping quality of chilled duck portions in modified atmosphere packs. Lebensml-Wiss Technol 16:142–146.

Messer JW, Midura TF, Peeler JT. 1992. Sampling plans, sample collection, shipment, and preparation for analysis. In: Vanderzant C, Splitstoesser DF, eds., *Compendium of Methods for the Microbiological Examination of Foods*, 3rd ed. Washington, DC: American Public Health Association, pp. 25–49.

Oliveria JS, Parmelee CE. 1976. Rapid enumeration of psychrotrophic bacteria in raw and pasteurized milk. J Milk Food Technol 39:269.

Oscar TP. 2004. Predictive simulation model for enumeration of *Salmonella* on chicken as a function of polymerase chain reaction detection time score and sample size. J Food Prot 67:1201–1208.

Oyarzabal OA. 2005. Review: Reduction of *Campylobacter* spp. by commercial antimicrobials applied during the processing of broiler chickens: a review from the United States perspective. J Food Prot 68:1752–1760.

Oyarzabal OA, Wesley IV, Barbaree JM, Lauerman LH, Conner DE. 1997. Specific detection of *Campylobacter lari* by PCR. J Microbiol Methods 29:97–102.

Oyarzabal OA, Macklin KS, Barbaree JM. 2005. Evaluation of agar plates for direct enumeration of *Campylobacter* spp. from poultry carcass rinses. Appl Environ Microbiol 71:3351–3354.

Oyarzabal OA, Backert S, Nagaraj M, Miller RS, Hussain SK, Oyarzabal EA. 2007. Efficacy of supplemented buffered peptone water for the isolation of *Campylobacter jejuni* and *C. coli* from broiler retail products. J Microbiol Methods 69:129–136.

Rambach A. 1990. New plate medium for facilitated differentiation of *Salmonella* spp. from *Proteus* spp. and other enteric bacteria. Appl Environ Microbiol 56:301–303.

Sharpe AN, Michaud GL. 1974. Hydrophobic grid-membrane filters: new approach to microbiological enumeration. Appl Microbiol 28:223–225.

Sharpe AN, Michaud GL. 1975. Enumeration of high numbers of bacteria using hydrophobic grid-membrane filters. Appl Microbiol 30:519–524.

Sharpe AN, Diotte MP, Dudas I, Malcolm S, Peterkin PI. 1983. Colony counting on hydrophobic grid-membrane filters. Can J Microbiol 29:797–802.

Simmons M, Fletcher DL, Berrang ME, Cason JA. 2003. Comparison of sampling methods for the detection of *Salmonella* on whole broiler carcasses purchased from retail outlets. J Food Prot 66(10):1768–1770.

Stern NJ, Wojton B, Kwiatek K. 1992. A differential-selective medium and dry ice–generated atmosphere for recovery of *Campylobacter jejuni*. J Food Prot 55:515–517.

U.S. Code. 1994. *Meat Inspection. Inspection Requirements; Adulteration and Misbranding*. 21 CFR 601. Washington, DC: U.S. Government Printing Office.

USDA–FSIS (U.S. Department of Agriculture–Food Safety and Inspection Service). 1996. Pathogen reduction; hazard analysis and critical control point (HACCP) systems; final rule. 9 CFR 304, et al. Fed Reg 61(144):38806–38943.

USDA–FSIS. 1998. *Microbiology Laboratory Guidebook*, 3rd ed. Washington, DC: USDA–FSIS. http://www.fsis.usda.gov/Science/Microbiological_Lab_Guidebook/. Accessed Feb. 25 2008.

van der Marel GM, Van Logtestijn JG, Mossel DAA. 1988. Bacteriological quality of broiler carcasses as affected by in-plant lactic acid decontamination. Int J Food Microbiol 6:31–42.

Vorster SM, Greebe RP, Nortj GL. 1994. Incidence of *Staphylococcus aureus* and *Escherichia coli* in ground beef, broilers and processed meats in Pretoria, South Africa. J Food Prot 57:305–310.

Warburton DW, Weiss KF, Lachapelle G, Dragon D. 1988. The microbiological quality of further processed deboned poultry products sold in Canada. Can Inst Food Sci Technol 21:84–89.

Willis WL, Murray C. 1997. *Campylobacter jejuni* seasonal recovery observations of retail market broilers. Poult Sci 76:314–317.

Zhao C, Ge B, De Villena J, Sudler R, Yeh E, Zhao S, White DG, Wagner D, Meng J. 2001. Prevalence of *Campylobacter* spp., *Escherichia coli*, and *Salmonella* serovars in retail chicken, turkey, pork, and beef from the Greater Washington, D.C., area. Appl Environ Microbiol 67:5431–5436.

PART VII

SAFETY SYSTEMS IN THE UNITED STATES

37

SANITATION REQUIREMENTS

Y.H. Hui
Science Technology System, West Sacramento, California

Isabel Guerrero-Legarreta
Departamento de Biotecnología, Universidad Autónoma Metropolitana, Mexico D.F., Mexico

INTRODUCTION

Before 1986, the U.S. Food and Drug Administration (FDA) used umbrella requirements to help the food industries to produce wholesome food. In 1986

Handbook of Poultry Science and Technology, Volume 2: Secondary Processing, Edited by Isabel Guerrero-Legarreta and Y.H. Hui

the first umbrella requirements were established, under the title *good manufacturing practices* (GMPs). Since then, many aspects of the requirements have been revised.

For many years, the U.S. Department of Agriculture (USDA) has regulated meat and poultry for human foods in the United States, both in state and interstate. The specific agency within the USDA with this responsibility is the Food Safety and Inspection Service (FSIS). The FSIS issues sanitation requirements for establishments that process meat and poultry. Although many of them are similar to the GMPs, those of the FSIS are more diverse and complex because of the clauses in the statutes that it is enforcing. The underlying principles of sanitation are the same.

Currently, both agencies are implementing preventive measures to assure the highest safety for food, including meat and poultry. As a result, the food safety program *hazard analysis and critical control points* (HACCP) was born. Information on this program is available in many formats for many categories of food industries and has been discussed at length in other chapters.

In this chapter we discuss basic sanitation requirements in the processing of poultry and poultry products. The information given is based on the following premises:

1. It has been modified from the Web sites of the FSIS (www.fsis.usda.gov) and the FDA (www.fda.gov).
2. When the word *requirement(s)* is used, it means those of the FSIS and/or the FDA.
3. All legal citations have been removed to facilitate discussion. For legal information, a user should visit the Web sites.

It is important to realize that the foundation of any HACCP plan is a comprehensive sanitation program. The information in this chapter supplements Chapters 32 and 33 in Volume 1 and Chapter 38.

GENERAL GOOD MANUFACTURING PRACTICES

Personnel

Plant management should take all reasonable measures and precautions to ensure compliance with the following requirements.

Disease Control Any person who, by medical examination or supervisory observation, is shown to have an illness or open lesion, including boils, sores, or infected wounds, by which there is a reasonable possibility of food, food-contact surfaces, or food-packaging materials becoming contaminated, should be excluded from any operations that may be expected to result in such contamination until the condition is corrected. Personnel should be instructed to report such health conditions to their supervisors.

Cleanliness All persons working in direct contact with food, food-contact surfaces, and food-packaging materials should conform to hygienic practices while on duty. The methods for maintaining cleanliness include, but are not limited to, the following:

1. Wearing outer garments suitable to the operation to protect against the contamination of food, food-contact surfaces, or food-packaging materials.
2. Maintaining adequate personal cleanliness.
3. Washing hands thoroughly (and sanitizing if necessary to protect against contamination with undesirable microorganisms) in an adequate hand-washing facility before starting work, after each absence from the workstation, and at any other time when the hands may have become soiled or contaminated.
4. Removing all unsecured jewelry and other objects that might fall into food, equipment, or containers, and removing hand jewelry that cannot be adequately sanitized during periods in which food is manipulated by hand. If such hand jewelry cannot be removed, it may be covered by material that can be maintained in an intact, clean, and sanitary condition and which effectively protects against contamination of the food, food-contact surfaces, or food-packaging materials.
5. Maintaining gloves, if they are used in food handling, in an intact, clean, and sanitary condition. The gloves should be of an impermeable material.
6. Wearing, where appropriate, hairnets, headbands, caps, beard covers, or other effective hair restraints.
7. Storing clothing or other personal belongings in areas other than where food is exposed or where equipment or utensils are washed.
8. Confining the following personal practices to areas other than where food may be exposed or where equipment or utensils are washed: eating food, chewing gum, drinking beverages, or using tobacco.
9. Taking any other necessary precautions to protect against contamination of food, food-contact surfaces, or food-packaging materials with microorganisms or foreign substances, including, but not limited to, perspiration, hair, cosmetics, tobacco, chemicals, and medicines applied to the skin.

Education and Training Personnel responsible for identifying sanitation failures or food contamination should have a background of education or experience, to provide a level of competency necessary for the production of clean and safe food. Food handlers and supervisors should receive appropriate training in proper food-handling techniques and food-protection principles and should be informed of the danger of poor personal hygiene and insanitary practices.

Supervision Responsibility for assuring that all personnel comply with all legal requirements should be assigned clearly to competent supervisory personnel.

Plant and Grounds

Grounds The grounds surrounding a food plant that are under the control of the plant manager should be kept in a condition that will protect against the contamination of food. The methods for adequate maintenance of grounds include, but are not limited to, the following.

1. Storing equipment properly, removing litter and waste, and cutting weeds or grass within the immediate vicinity of plant buildings or structures that may constitute an attractant, breeding place, or harborage for pests.

2. Maintaining roads, yards, and parking lots so that they do not constitute a source of contamination in areas where food is exposed.

3. Draining areas that may contribute contamination to food by seepage or foot-borne filth, or by providing a breeding place for pests.

4. Operating systems for waste treatment and disposal so that they do not constitute a source of contamination in areas where food is exposed. If the plant grounds are bordered by grounds not under the operator's control and not maintained in an acceptable manner, steps must be taken to exclude pests, dirt, and filth that may be a source of food contamination. Implement inspection, extermination, or other countermeasures.

Plant Construction and Design Plant buildings and structures should be suitable in size, construction, and design to facilitate maintenance and sanitary operations for food-manufacturing purposes. The plant and facilities should:

1. Provide sufficient space for such placement of equipment and storage of materials as is necessary for the maintenance of sanitary operations and the production of safe food.

2. Take proper precautions to reduce the potential for contamination of food, food-contact surfaces, or food-packaging materials with microorganisms, chemicals, filth, or other extraneous material. The potential for contamination may be reduced by adequate food safety controls and operating practices or effective design, including the separation of operations in which contamination is likely to occur, by one or more of the following means: location, time, partition, airflow, enclosed systems, or other effective means.

3. Taking proper precautions to protect food in outdoor bulk fermentation vessels by any effective means, including:

   a. Using protective coverings
   b. Controlling areas over and around vessels to eliminate harborages for pests
   c. Checking on a regular basis for pests and pest infestation
   d. Skimming the fermentation vessels, as necessary

4. Be constructed in such a manner that floors, walls, and ceilings may be cleaned and kept clean and kept in good repair; that drip or condensate from

fixtures, ducts, and pipes does not contaminate food, food-contact surfaces, or food-packaging materials; and that aisles or working spaces are provided between equipment and walls and are adequately unobstructed and of sufficient width to permit employees to perform their duties and to protect against contaminating food or food-contact surfaces with clothing or personal contact.

5. Provide adequate lighting in hand-washing areas, dressing and locker rooms, and toilet rooms and in all areas where food is examined, processed, or stored and where equipment or utensils are cleaned; and provide safety-type light bulbs, fixtures, skylights, or other glass suspended over exposed food in any step of preparation or otherwise protect against food contamination in case of glass breakage.

6. Provide adequate ventilation or control equipment to minimize odors and vapors (including steam and noxious fumes) in areas where they may contaminate food; and locate and operate fans and other air-blowing equipment in a manner that minimizes the potential for contaminating food, food-packaging materials, and food-contact surfaces.

7. Provide adequate screening or other protection against pests.

Sanitary Operations

General Maintenance Buildings, fixtures, and other physical facilities of the plant should be maintained in a sanitary condition and should be kept in sufficient repair to prevent food from becoming adulterated within the meaning of the regulations. Cleaning and sanitizing of utensils and equipment should be conducted in a manner that protects against contamination of food, food-contact surfaces, or food-packaging materials.

Substances Used in Cleaning and Sanitizing; Storage of Toxic Materials

1. Cleaning compounds and sanitizing agents used in cleaning and sanitizing procedures should be free from undesirable microorganisms and should be safe and adequate under the conditions of use. Compliance with this requirement may be verified by any effective means, including purchase of these substances under a supplier's guarantee or certification, or examination of these substances for contamination. Only the following toxic materials may be used or stored in a plant where food is processed or exposed:

a. Those required to maintain clean and sanitary conditions

b. Those necessary for use in laboratory testing procedures

c. Those necessary for plant and equipment maintenance and operation

d. Those necessary for use in the plant's operations

2. Toxic cleaning compounds, sanitizing agents, and pesticide chemicals should be identified, held, and stored in a manner that protects against contamination of food, food-contact surfaces, or food-packaging materials.

Pest Control No pests should be allowed in any area of a food plant. Guard or guide dogs may be allowed in some areas of a plant if the presence of the dogs is unlikely to result in contamination of food, food-contact surfaces, or food-packaging materials. Effective measures should be taken to exclude pests from processing areas and to protect against the contamination of food on the premises by pests. The use of insecticides or rodenticides is permitted only under precautions and restrictions that will protect against the contamination of food, food-contact surfaces, and food-packaging materials.

Sanitation of Food-Contact Surfaces All food-contact surfaces, including utensils and food-contact surfaces of equipment, should be cleaned as frequently as necessary to protect against contamination of food.

1. Food-contact surfaces used for manufacturing or holding low-moisture food should be in a dry, sanitary condition at the time of use. When the surfaces are wet-cleaned, they should be sanitized and dried thoroughly before subsequent use.

2. In wet processing, when cleaning is necessary to protect against the introduction of microorganisms into food, all food-contact surfaces should be cleaned and sanitized before use and after any interruption during which the food-contact surfaces may have become contaminated. Where equipment and utensils are used in a continuous production operation, the utensils and food-contact surfaces of the equipment should be cleaned and sanitized as necessary.

3. Non-food-contact surfaces of equipment used in the operation of food plants should be cleaned as frequently as necessary to protect against contamination of food.

4. Single-service articles (such as utensils intended for one-time use, paper cups, and paper towels) should be stored in appropriate containers and should be handled, dispensed, used, and disposed of in a manner that protects against contamination of food or food-contact surfaces.

5. Sanitizing agents should be adequate and safe under conditions of use. Any facility, procedure, or machine is acceptable for cleaning and sanitizing equipment and utensils if it is established that the facility, procedure, or machine will routinely render equipment and utensils clean and provide adequate cleaning and sanitizing treatment.

Storage and Handling of Cleaned Portable Equipment and Utensils Cleaned and sanitized portable equipment with food-contact surfaces and utensils should be stored in a location and manner that protects food-contact surfaces from contamination.

Sanitary Facilities and Controls

Each plant should be equipped with adequate sanitary facilities and accommodations including, but not limited to, those stated here.

Water Supply The water supply should be sufficient for the operations intended and should be derived from an adequate source. Any water that contacts food or food-contact surfaces should be safe and of adequate sanitary quality. Running water at a suitable temperature, and under pressure as needed, should be provided in all areas where required for the processing of food, for the cleaning of equipment, utensils, and food-packaging materials, or for employee sanitary facilities.

Plumbing Plumbing should be of adequate size and design and adequately installed and maintained to:

1. Carry sufficient quantities of water to required locations throughout the plant.
2. Properly convey sewage and liquid disposable waste from the plant.
3. Avoid constituting a source of contamination to food, water supplies, equipment, or utensils or creating an unsanitary condition.
4. Provide adequate floor drainage in all areas where floors are subject to flooding-type cleaning or where normal operations release or discharge water or other liquid waste on the floor.
5. Assure that there is no backflow from, or cross-connection between, piping systems that discharge wastewater or sewage and piping systems that carry water for food or food manufacturing.

Sewage Disposal Sewage disposal should be made into an adequate sewerage system or disposed of through other adequate means.

Toilet Facilities Each plant should provide its employees with adequate, readily accessible toilet facilities. Compliance with this requirement may be accomplished by:

1. Maintaining the facilities in a sanitary condition.
2. Keeping the facilities in good repair at all times.
3. Providing self-closing doors.
4. Providing doors that do not open into areas where food is exposed to airborne contamination, except where alternative means have been taken to protect against such contamination (such as double doors or positive-airflow systems).

Hand-Washing Facilities Hand-washing facilities should be adequate and convenient and be furnished with running water at a suitable temperature. Compliance with this requirement may be accomplished by providing:

1. Hand-washing and, where appropriate, hand-sanitizing facilities at each location in the plant where good sanitary practices require employees to wash and/or sanitize their hands.
2. Effective hand-cleaning and sanitizing preparations.

3. Sanitary towel service or suitable drying devices.

4. Devices or fixtures, such as water control valves, designed and constructed to protect against recontamination of clean, sanitized hands.

5. Readily understandable signs directing employees handling unprotected food, unprotected food-packaging materials, of food-contact surfaces to wash and, where appropriate, sanitize their hands before they start work, after each absence from their duty post, and when their hands may have become soiled or contaminated. These signs may be posted in the processing room(s) and in all other areas where employees may handle such food, materials, or surfaces.

6. Refuse receptacles that are constructed and maintained in a manner that protects against contamination of food.

Rubbish and Offal Disposal Rubbish and any offal should be so conveyed, stored, and disposed of as to minimize the development of odor, minimize the potential for the waste becoming an attractant and harborage or breeding place for pests, and protect against contamination of food, food-contact surfaces, water supplies, and ground surfaces.

Equipment and Utensils

1. All plant equipment and utensils should be so designed and of such material and workmanship as to be adequately cleanable, and should be maintained properly. The design, construction, and use of equipment and utensils should preclude the adulteration of food with lubricants, fuel, metal fragments, contaminated water, or any other contaminants. All equipment should be so installed and maintained as to facilitate the cleaning of the equipment and of all adjacent spaces. Food-contact surfaces should be corrosion resistant when in contact with food. They should be made of nontoxic materials and designed to withstand the environment of their intended use and the action of food, and, if applicable, cleaning compounds and sanitizing agents. Food-contact surfaces should be maintained to protect food from being contaminated by any source, including unlawful indirect food additives.

2. Seams on food-contact surfaces should be smoothly bonded or maintained so as to minimize the accumulation of food particles, dirt, and organic matter and thus minimize the opportunity for growth of microorganisms.

3. Equipment that is in the manufacturing or food-handling area and that does not come into contact with food should be so constructed that it can be kept in a clean condition.

4. Holding, conveying, and manufacturing systems, including gravimetric, pneumatic, closed, and automated systems, should be of a design and construction that enables them to be maintained in an appropriate sanitary condition.

5. Each freezer and cold storage compartment used to store and hold food capable of supporting the growth of microorganisms should be fitted with an

indicating thermometer, temperature-measuring device, or temperature-recording device so installed as to show the temperature accurately within the compartment, and should be fitted with an automatic control for regulating temperature or with an automatic alarm system to indicate a significant temperature change in a manual operation.

6. Instruments and controls used for measuring, regulating, or recording temperatures, pH, acidity, water activity, or other conditions that control or prevent the growth of undesirable microorganisms in food should be accurate and adequately maintained, and adequate in number for their designated uses.

7. Compressed air or other gases mechanically introduced into food or used to clean food-contact surfaces or equipment should be treated in such a way that food is not contaminated with unlawful indirect food additives.

Equipment Specific to the Processing of Poultry

All food processors using a variety of equipment, especially custom-made equipment, should focus on correcting problems during the initial development of equipment instead of resolving problems that may result when improperly designed or constructed equipment is put into widespread use. This preventive mode of action benefits equipment manufacturers, food processors, state and federal regulators, and consumers.

Standardized and Basic Equipment

Simple hand tools
Equipment used to prepare packaging materials
Equipment used on fully packaged products
Equipment used on operations involving inedible products that will not be mixed with edible products
Central cleaning system
Utensil and equipment cleaning machinery
Pails, buckets, etc. (contact area material must be approved)
Pallets for packaged products (contact area material must be approved)
Picking fingers (contact area material must be approved)
Tanks for fully finished oils
Simple can openers
Chutes, flumes, hangback racks, supporting stands, and brackets
Equipment used for storing and transporting vegetable oils
Vegetable cleaning equipment (not applicable to spin-type washers and dryers)
Insect control units
Shipping containers
Pressure storage vessels for refrigerants (not applicable to CO_2 snow-making equipment)

Water softeners, water heaters, water meters, and chemical dispensers

Can and jar washers and cleaners

Dry spice mixing equipment

Hot-air shrink tunnels

Air and water filters

Devices for measuring physical characteristics (temperature, pressure, etc.)

Casing preparation equipment

Rubber floor mats

Nonexempt Conditions

1. All food-processing equipment.

2. *New establishments*. Equipment intended for use in newly constructed establishments should take into consideration all aspects of good manufacturing practices before their construction or purchase.

3. *Custom-made equiment*. Plant personnel may build their own equipment or have an outside contractor fabricate equipment for them. Even though it is custom made and not intended for resale, such equipment should be built to comply with good manufacturing practices. The same standards are applicable to custom-made equipment as are applicable to commercially available equipment.

4. Some equipment manufacturers or brokers are sometimes not interested in complying with FDA and USDA GMPs. In such an event, equipment is considered the same as custom made, and food establishment operators should be aware of this responsibility before they purchase equipment.

Acceptable Materials Equipment should be constructed of materials that will not deteriorate from normal use under the environment anticipated. For example, equipment must be constructed of materials that will withstand one category of environment (e.g., generally, a humid operating environment and high-pressure, hot-water cleaning with strong chemical cleaning agents). Of course, there are other categories of food-processing environments. In addition, all equipment surfaces should be smooth, corrosion and abrasion resistant, shatterproof, nontoxic, nonabsorbent, and not capable of migrating into food product (staining).

Aluminum may pit and corrode when exposed to certain chemicals. When friction occurs between aluminum and fats, a black oxide is produced which discolors the product. Anodizing the aluminum does not eliminate this problem. Therefore, the use of aluminum is limited to applications where the metal does not contact the product or in which the product is suspended in water.

Surface coatings and platings may be used if the base material is nontoxic and rendered noncorrosive and the plating material is USDA and FDA acceptable. Chrome, nickel, tin, and zinc (galvanization) platings will generally be acceptable for most appropriate applications. USDA and FDA clearance of other plating materials and processes can be obtained by receiving a favorable opinion for the intended use from the FDA's Office of Premarket Approval. Surface coatings and platings must remain intact. If a surface coating or plating begins to peel or crack, the FDA or USDA inspector will request correction from management and may even disallow use of the equipment.

Hardwood may be used for dry curing. In addition, solid (unlaminated) pieces of hardwood are acceptable as removable cutting boards provided that the wood is maintained in a smooth, sound condition and is free of cracks. Hardwood cutting boards must be the shortest dimension that is practical, preferably not exceeding 3 or 4 ft (0.91 or 1.22 m).

Unacceptable Materials

1. Cadmium, antimony, and lead are toxic materials that cannot be used as materials of construction either as a plating or as plated base material. Lead may be used, however, in acceptable alloys in an amount not exceeding 5%.

2. Enamelware and porcelain are not acceptable for handling and processing food products unless management provides reasons why they are needed.

3. Copper, bronze, and brass are not acceptable for direct product contact. These materials may be used in air and water lines or for gears and bushings outside the product zone. Brass is acceptable for potable water systems and direct contact with brine, but not for brine, or any solution, that is recirculated.

4. Leather and fabric are not acceptable materials unless management provides reasons why they are needed.

Design and Construction

1. Equipment should be designed so that all product-contact surfaces can be cleaned readily and thoroughly with high-temperature, high-pressure water and a caustic soap solution. Components such as electric motors and electric components which cannot be cleaned in this manner should be completely enclosed and sealed.

2. All product-contact surfaces should be visible (or easily made visible) for inspecIion.

3. All product-contact surfaces should be smooth and maintained free of pits, crevices, and scale.

4. The product zone should be free of recesses, open seams, gaps, protruding ledges, inside threads, inside shoulders, bolts, rivets, and dead ends.

5. Bearings (including greaseless bearings) should not be located in or above the product zone. In addition, bearings should be constructed so that lubricants will not leak or drip or be forced into the product zone.

6. Internal corners or angles in the product zone should have a smooth and continuous radius of $\frac{1}{4}$ in. (6.35 mm) or greater.

7. Equipment should be self-draining or designed to be evacuated of water.

8. Framework of equipment (if not completely enclosed and sealed) should be designed to use as few horizontal frame members as possible. Furthermore,

these components should be rounded or tubular in construction. Angles are not acceptable except in motor supports.

9. Equipment should be designed, constructed, and installed in a manner to protect personnel from safety hazards such as sharp edges, moving parts, electric shocks, excessive noise, and any other hazards. Safety guards should be removable for cleaning and inspection purposes.

10. All welds, in both product- and non-product-contact areas, should be smooth, continuous, even, and relatively flush with adjacent surfaces.

11. Equipment should not be painted on areas that are in or above the product zone.

12. External surfaces should not have open seams, gaps, crevices, and inaccessible recesses.

13. Where parts must be retained by nuts or bolts, fixed studs with wing nuts should be used instead of screws to a tapped hole.

14. Gasketing, packing materials, O-rings, and the like, must be nontoxic, nonporous, nonabsorbent, and unaffected by food products and cleaning compounds.

Installation Stationary equipment or equipment not easily movable (i.e., no casters) should be installed far enough from walls and support columns to allow thorough cleaning and inspection. In addition, there must be ample clearance between the floor and the ceiling. If these clearances are not possible, equipment should be sealed watertight to the surfaces. All wall-mounted cabinets; electrical connections; and electronic components should be at least 1 in. from the wall or sealed watertight to the wall.

Equipment and Water Use

1. *Water wasting equipment*. Water wasting equipment should be installed so that wastewater is delivered into the drainage system through an interrupted connection without flowing over the floor, or is discharged into a properly drained curbed area. Wastewater from cooking tanks, soaking tanks, chilling tanks, and other large vessels may be discharged for short distances across the floor to a drain after operations have ceased and all products have been removed from the area.

2. *Protection of water supply*. An airgap should be provided between the highest possible level of liquids in equipment and a directly connected water supply line(s). The airgap must be at least twice the diameter of the supply-side orifice. If submerged lines are unavoidable due to design considerations, the equipment must include a functional vacuum breaker which will, without fail, break the connection in the event of water pressure loss.

3. *Recirculation of water*. Equipment that recirculates water as part of its intended function should be equipped with sanitary recirculating components if the water contacts food- or product-contact surfaces directly or indirectly. For example, recirculating pumps should be accepted for direct product contact, and piping must be easily demountable, with quick-disconnect mechanisms at

each change of direction. In addition, establishment operators using equipment or systems that reuse water may be required to have written approval within a water reuse procedure. However, the requirement is mandatory for meat and poultry processors controlled by the USDA. Although the FDA does not require written approval at this stage, its GMP regulations make it clear that there must be built-in safeguards in the reuse of water in a food plant.

4. Valves on drainage outlets should be easily demountable to the extent necessary for thorough cleaning. Overflow pipes should be constructed so that all internal and external surfaces can be cleaned thoroughly.

Clean-in-Place (CIP) Systems *CIP* is defined as follows:

- CIP refers to cleaning in place by circulation or flowing by mechanical means through a piping system of a detergent solution, water rinse, and sanitizing solution onto or over equipment surfaces that require cleaning.
- CIP does not include the cleaning of equipment such as bandsaws, slicers, or mixers that are subjected to in-place manual cleaning without the use of a CIP system.

Sanitation procedures for CIP systems must be as effective as those for cleaning and sanitizing disassembled equipment. Only equipment that meets the following criteria may be cleaned in place. Any equipment or portions of equipment not meeting these requirements should be disassembled for daily cleaning and inspection.

1. Cleaning solutions, sanitizing solutions, and rinse water should contact all interior surfaces of the system.

2. All internal surfaces should be either designed for self-draining or physically disassembled for draining after rinsing.

3. Pipe interiors should be highly polished (120 to 180-grit abrasive) stainless steel or some other acceptable, smooth-surfaced material that is easy to inspect.

4. Easily removable elbows with quick-disconnect mechanisms should be located at each change of direction.

5. All sections of the system should be capable of being disassembled completely for periodic inspection of all internal surfaces.

6. All sections should be available for inspection without posing any safety hazard to the inspector.

Piping Systems Piping systems used to convey edible product (including pickle solutions) should be readily disassembled for cleaning and inspection. Pumps, valves, and other such components should comply with the sanitary requirements of good manufacturing practices promulgated by USDA and FDA. Piping systems must be designed so that product flow will be smooth and continuous (i.e., no traps or dead ends). Pipes must be either stainless steel or a

USDA/FDA-acceptable plastic. Clear demountable rigid plastic piping may be used for two-way flow provided that it is chemically and functionally acceptable. Opaque plastic piping may be used for one-way purposes only.

The foregoing requirements apply to systems for conveying raw fat and to recirculate cooking and frying oils. Black iron pipes with threaded or welded joints are acceptable for conveying completely finished, rendered fats. Continuous rendering is not considered complete until after the final centrifuge. Pipeline conveying systems for aseptic processing and packaging should comply with the requirements of the FDA and USDA.

Magnetic Traps and Metal Detectors The extensive exposure of some products to metal equipment such as grinders, choppers, mixers, shovels, and so on, causes the possibility of metal contamination. Magnetic traps have been found effective in removing iron particles from chopped or semiliquid products. However, these magnetic traps are not useful for removing nonmagnetic metals such as stainless steel or aluminum. Therefore, the use of electronic metal detectors is highly recommended for sausage emulsions, can filling lines (especially baby foods, etc.). Metal detectors are usually installed so that an alarm (a bell, a light, or both) is activated when a metal fragment is in the detection zone. The production line should stop automatically when the detector is activated. Alternatively, some systems are arranged so that the portion of the product containing the metal contaminant(s) is automatically removed from the production line.

Conveyor Belts Conveyor belts used in direct contact with food products must be moisture resistant and nonabsorbent. Conveyor belts should have the edges sealed with the same material as is used for the food-contact surface. In addition, belting material must be chemically acceptable and approved by the FDA or USDA. The belt-tensioning mechanism of conveyors with trough-like sides and bed should have a quick-release device to allow cleaning under the belt.

Jet-Vacuum Equipment Equipment used for cleaning jars or cans should have safety devices to indicate malfunction of either jet or vacuum elements. If necessary, vents to the outside should be provided to control exhaust currents and to prevent dust and/or paper particles from being blown back into cleaned containers.

Hoses Hoses used for product contact should comply with recommendations of trade associations or be accepted by both the FDA and the USDA. The hose material must be installed in a manner that allows for inspection of the interior surface. Sanitary connectors can be installed at appropriate intervals to allow breakdown for visual inspection or use of inspection devices. Hoses without sanitary connectors are acceptable for steam and water lines where breakdown for cleaning and inspection is not necessary. However, hoses used for recirculating water into and out of product-contact areas must satisfy the requirement for product-contact hoses.

Pickle Line Pickle lines should be either stainless steel or some other USDA-acceptable material. If recirculatod, pickle should be filtered and recirculated through a system that can be disassembled to the extent necessary for thorough cleaning and inspection.

Smokehouses and Ovens Smokehouses or ovens must be designed for easy cleaning and inspection of all inner and outer surfaces. Ducts should be designed to be disassembled easily to the extent necessary for thorough cleaning and inspection. Spray heads for dispensing liquid smoke must be mounted below the level of the rails and trolleys. If liquid smoke is to be recirculated, the pump and pipelines must be of sanitary-type construction. Liquid smoke cannot be recirculated if products are on rack trucks.

Screens and Filters Screens and straining devices should be readily removable for cleaning and inspection and should be designed to prevent incorrect installation. Permanent screens should be constructed of noncorrosive metals. Synthetic filter materials should have clearance from trade associations. The same applies to filters intended for direct product contact. Filter paper should be single service. Filter cloths should be washable. Asbestos is not acceptable for use as filtering material or for any other purpose.

Vent Stacks from Hoods Vent stacks from covered cooking vats or hoods over cook tanks and CO_2 equipment should be arranged or constructed so as to prevent drainage of condensate back into the product zone.

Ultraviolet (UV) Lamps UV lamps that generate ozone have specific restrictions for their use. UV lamps that do not produce ozone may be used in any area provided that shields are used to prevent exposure of workers to direct or reflected UV rays. Otherwise, rooms in which unshielded UV lights are used should be equipped with switches at all entry points so that the units may be turned off before workers enter. These switches should be identified with suitable placards such as "Ultraviolet Lights." Employees should not enter areas where unshielded UV lights are burning because of possible damage to skin and eyes.

Heat Exchangers Heat exchangers may be used to heat or cool product. Heat exchangers may also be used to heat or cool gases or liquids that contact product directly. However., extreme caution should be exercised to prevent contamination. Inspectors and plant personnel should be alert to the following conditions and requirements:

1. Only heat exchanger media authorized by trade associations, the FDA, the USDA, and other standardization bodies in the United States can be used for applications involving food products. Common materials such as brine or ammonia need not be submitted for review. Under no circumstances can toxic materials be used.

2. Heat exchangers should be pressure-tested routinely to ensure that pinholes, hairline cracks, loose fittings, and similar defects are not present. The presence

of off-color, off-odor, and/or off-flavor may indicate leakage. Frequent depletion of heat exchanger media may also indicate leakage.

3. Pressure on the product side should be higher than pressure on the media side.

In-Plant Trucks Trucks used to transport products within a plant should be constructed of stainless steel. However, galvanized metal is acceptable provided that it is maintained in a good state of repair and is regalvanizcd when necessary. Trucks should be free of cracks and rough seams. Metal wheels should be avoided, as they cause deterioration of the floor surfaces. All trucks should have some means of affixing a tag. This can be accomplished by drilling two holes approximately 1 in. (25.4 mm) apart in the lip of the truck to accommodate string or wire.

Air Compressors Compressed air may be used to contact products and/or product-contact surfaces directly provided that the air is filtered before entering a compressor and it is clean and free of moisture, oil, or other foreign material when contacting products or product-contact surfaces. Lubricants and coolants contacting air directly should be authorized by trade associations, the FDA, the USDA, and other standardization bodies. Compressed air storage tanks should have a drain. Water and oil traps must be located between storage tanks and the point of use. Spent air must be exhausted in a manner to prevent product contamination. Air contacting products or product-contact surfaces directly should be filtered as near the air outlet as feasible. Filters should be readily removable for cleaning or replacement and should be capable of filtering out 50-μm particles (measured in the longest dimension). Air intake on votators should also be filtered.

Product Reconditioning Equipment Products that are soiled accidentally may be cleaned on a separate, conveniently located wash table or sink. The wash station should be properly equipped with sprays and a removable, perforated plate to hold products off the bottom. The station should be identified as a "product wash station" and cannot be used for hand or implement washing.

Electric Cords Accepting the use of electric cords should be based on both sanitary and safety considerations. Drop cords suspended from the ceiling may be retractable and used to connect portable equipment on an as-needed basis if the cords are properly wired to a power source. Electric cords should not be strung across the floor, even on a temporary basis.

Electric Insect Traps Electric insect traps may be used in edible product handling and storage areas provided that the following conditions are met:

1. The equipment should be made of acceptable noncorrosive materials.

2. The traps must not be placed above uncovered products or above uncovered product trafficways.

3. The electrified components are either apparent or properly identified; insulated from nonelectrified components; and covered with a protective grille to prevent electric shock hazard.

4. The equipment should have a removable shelf or drawer that collects all trapped insects.

5. The equipment is designed and constructed so that all dead insects are trapped in the removable shelf or drawer. (Insects must not collect on the protective grille.)

Removable drawers or shelves should be emptied as often as necessary. If the drawer or shelf becomes full of dead insects, the fourth requirement above cannot be met, so the equipment should be rejected for use. Dead insects must be removed from the unit before they create an odor problem. They cannot be left in the unit as "bait."

Inedible Product Equipment Containers for handling and transporting inedible products should be watertight, maintained in a good state of repair (no rust or corrosion), and clearly marked with an appropriate identification. All inedible product containers in the plant should be uniformly identified. Inedible product containers should be cleaned before being moved into an edible products department.

Metal barrels, tanks, or trucks may be used for holding inedible poultry products in specially designated inedible product rooms. Alternatively, the containers may be stored outside the building provided that the storage area is paved, drained, and conveniently located. These storage areas should also be equipped with nearby hose connections for cleanup.

Automatic Poultry Eviscerating Equipment All surfaces of automatic poultry eviscerating equipment that come in contact with raw products must be sanitized between each use. Sanitization may be accomplished by flushing the contact surfaces with either 180°F (82.2°C) water or with water containing 20 ppm residual chlorine at the point of use.

The hock blow-off system should not be used as a substitute for good dressing procedures. Dressing contamination more than 2 cm in length in its longest dimension must be handled as usual and trimmed. Foreign material (e.g., din, specks, and hairs) should be removed and not spread to the round area. The compressed-air system should not be used on carcasses with fecal contamination. Such carcasses are to be handled and trimmed in a sanitary manner.

Processes and Controls

All operations in the receiving, inspecting, transporting, segregating, preparing, manufacturing, packaging, and storing of food should be conducted in accordance with adequate sanitation principles. Appropriate quality control operations should be employed to ensure that food is suitable for human consumption and that food-packaging materials are safe and suitable. Overall sanitation of the plant should be under the supervision of one or more competent persons assigned the responsibility for this function. All reasonable precautions should be taken to ensure that production procedures do not contribute contamination from any source. Chemical, microbial, or extraneous material testing procedures should be used where necessary to identify sanitation failures or possible food contamination. All food that has become contaminated to the extent that it is adulterated within the meaning of the regulations should be rejected or, if permissible, treated or processed to eliminate the contamination.

Raw Materials and Other Ingredients

1. Raw materials and other ingredients should be inspected and segregated or otherwise handled as necessary to ascertain that they are clean and suitable for processing into food and should be stored under conditions that will protect against contamination and minimize deterioration. Raw materials should be washed or cleaned as necessary to remove soil or other contamination. Water used for washing, rinsing, or conveying food should be safe and of adequate sanitary quality. Water may be reused for washing, rinsing, or conveying food if it does not increase the level of contamination of the food. Containers and carriers of raw materials should be inspected on receipt to ensure that their condition has not contributed to the contamination or deterioration of food.

2. Raw materials and other ingredients should either not contain levels of microorganisms that may produce food poisoning or other disease in humans, or they should be pasteurized or otherwise treated during manufacturing operations so that they no longer contain levels that would cause the product to be adulterated. Compliance with this requirement may be verified by any effective means, including purchasing raw materials and other ingredients under a supplier's guarantee or certification.

3. Raw materials and other ingredients susceptible to contamination with aflatoxin or other natural toxins should comply with current FDA and USDA requirements, guidelines, and action levels for poisonous or deleterious substances before these materials or ingredients are incorporated into finished food. Compliance with this requirement may be accomplished by purchasing raw materials and other ingredients under a supplier's guarantee or certification, or may be verified by analyzing these materials and ingredients for aflatoxins and other natural toxins.

4. Raw materials, other ingredients, and rework susceptible to contamination with pests, undesirable microorganisms, or extraneous material should comply with applicable requirements and guidelines. Compliance with this requirement may be verified by any effective means, including purchasing the materials under a supplier's guarantee or certification, or examination of these materials for contamination.

5. Raw materials, other ingredients, and rework should be held in bulk, or in containers designed and constructed so as to protect against contamination and should be held at such temperature and relative humidity as to prevent the food from becoming adulterated. Material scheduled for rework should be identified as such.

6. Frozen raw materials and other ingredients should be kept frozen. If thawing is required prior to use, it should be done in a manner that prevents the raw materials and other ingredients from becoming adulterated.

7. Liquid or dry raw materials and other ingredients received and stored in bulk form should be held in a manner that protects against contamination.

Manufacturing Operations

1. Equipment and utensils and finished food containers should be maintained in an acceptable condition through appropriate cleaning and sanitizing, as necessary. Insofar as necessary, equipment should be taken apart for thorough cleaning.

2. All food manufacturing, including packaging and storage, should be conducted under such conditions and controls as are necessary to minimize the potential for the growth of microorganisms or for the contamination of food. One way to comply with this requirement is careful monitoring of physical factors such as time, temperature, humidity, a_w (water activity), pH, pressure, flow rate, and manufacturing operations such as freezing, dehydration, heat processing, acidification, and refrigeration to ensure that mechanical breakdowns, time delays, temperature fluctuations, and other factors do not contribute to the decomposition or contamination of food.

3. Food that can support the rapid growth of undesirable microorganisms, particularly those of public health significance, should be held in a manner that prevents the food from becoming spoiled. Compliance with this requirement may be accomplished by any effective means, including:

a. Maintaining refrigerated foods at 45°F (7.2°C) or below as appropriate for the particular food involved
b. Maintaining frozen foods in a frozen state
c. Maintaining hot foods at 140°F (60°C) or above
d. Heat-treating acid or acidified foods to destroy mesophilic microorganisms when those foods are to be held in hermetically sealed containers at ambient temperatures

4. Measures such as sterilizing, irradiating, pasteurizing, freezing, refrigerating, controlling pH or controlling a_w, which are taken to destroy or prevent the growth of undesirable microorganisms, particularly those of public health significance, should be adequate under the conditions of manufacture, handling, and distribution to prevent food from being adulterated.

5. Work-in-process should be handled in a manner that protects against contamination.

6. Effective measures should be taken to protect finished food from contamination by raw materials, other ingredients, or refuse. When raw materials, other ingredients, or refuse are unprotected, they should not be handled simultaneously in a receiving, loading, or shipping area if that handling could result in contaminated food. Food transported by conveyor should be protected against contamination as necessary.

7. Equipment, containers, and utensils used to convey, hold, or store raw materials, work-in-process, rework, or food should be constructed, handled, and maintained during manufacturing or storage in a manner that protects against contamination.

8. Effective measures should be taken to protect against the inclusion of metal or other extraneous material in food. Compliance with this requirement may be accomplished by using sieves, traps, magnets, electronic metal detectors, or other suitable effective means.

9. Food, raw materials, and other ingredients that are adulterated should be disposed of in a manner that protects against the contamination of other food. If the adulterated food is capable of being reconditioned, it should be reconditioned using a method that has been proven to be effective or it should be reexamined and found not to be adulterated before being incorporated into other food.

10. Mechanical manufacturing steps such as washing, peeling, trimming, cutting, sorting and inspecting, mashing, dewatering, cooling, shredding, extruding, drying, whipping, defatting, and forming should be performed so as to protect food against contamination. Compliance with this requirement may be accomplished by providing adequate physical protection of food from contaminants that may drip, drain, or be drawn into the food. Protection may be provided by adequate cleaning and sanitizing of all food-contact surfaces, and by using time and temperature controls at and between each manufacturing step.

11. Heat blanching, when required in the preparation of food, should be effected by heating the food to the required temperature, holding it at this temperature for the required time, and then either rapidly cooling the food or passing it to subsequent manufacturing without delay. Thermophilic growth and contamination in blanchers should be minimized by the use of adequate operating temperatures and by periodic cleaning. Where the blanched food is washed prior to filling, water used should be safe and of adequate sanitary quality.

12. Batters, breading, sauces, gravies, dressings, and other similar preparations should be treated or maintained in such a manner that they are protected against contamination. Compliance with this requirement may be accomplished by any effective means, including one or more of the following:

a. Using ingredients free of contamination
b. Employing adequate heat processes where applicable
c. Using adequate time and temperature controls
d. Providing adequate physical protection of components from contaminants that may drip, drain, or be drawn into them
e. Cooling to an adequate temperature during manufacturing
f. Disposing of batters at appropriate intervals to protect against the growth of microorganisms

13. Filling, assembling, packaging, and other operations should be performed in such a way that the food is protected against contamination.

Compliance with this requirement may be accomplished by any effective means, including:

a. Use of a quality control operation in which the critical control points are identified and controlled during manufacturing
b. Adequate cleaning and sanitizing of all food-contact surfaces and food containers
c. Using materials for food containers and food-packaging materials that are safe and suitable
d. Providing physical protection from contamination, particularly airborne contamination
e. Using sanitary handling procedures

14. Food such as, but not limited to, dry mixes, nuts, intermediate-moisture food, and dehydrated food, which relies on the control of a_w for preventing the growth of undesirable microorganisms, should be processed to and maintained at a safe moisture level. Compliance with this requirement may be accomplished by any effective means, including employment of one or more of the following practices:

a. Monitoring the a_w value of food
b. Controlling the soluble solids/water ratio in finished food
c. Protecting finished food from moisture pickup, by use of a moisture barrier or by other means, so that the a_w value of the food does not increase to an unsafe level

15. Food, such as, but not limited to, acid and acidified food that relies principally on the control of pH for preventing the growth of undesirable microorganisms should be monitored and maintained at a pH of 4.6 or below. Compliance with this requirement may be accomplished by any effective means, including employment of one or more of the following practices:

a. Monitoring the pH of raw materials, food-in-process, and finished food.
b. Controlling the amount of acid or acidified food added to low-acid food.

16. When ice is used in contact with food, it should be made from water that is safe and of adequate sanitary quality, and should be used only if it has been manufactured in accordance with current good manufacturing practice.

17. Food-manufacturing areas and equipment used for manufacturing human food should not be used to manufacture nonhuman-food-grade animal feed or inedible products, unless there is no reasonable possibility of the contamination of the human food.

Receiving, Warehousing, and Distribution

Incoming Product Shipments The integrity of the food warehouse sanitation program requires that the materials, including foods and their packaging, that are received into the warehouse do not expose the warehouse to contamination by reason of infestation by insects, birds, rodents, or other vermin, or by introduction of filth or other contaminants. It is often useful, when practical, to work with suppliers and shippers in advance to establish guidelines for acceptance, rejection, and where appropriate, reconditioning of a particular product, taking into consideration factors such as the nature, method of shipment, and ownership of the product, in order to facilitate the effective implementation of these programs.

1. Place foods received at the food warehouse for handling or storage in a manner that will facilitate cleaning and the implementation of insect, rodent, and other sanitary controls and will maintain product wholesomeness.

2. Adopt and implement effective procedures to provide stock rotation appropriate to the particular food.

3. Unless repaired or corrected promptly and adequately at or near the point of detection, promptly separate from other foods all foods that are identified as being damaged or are otherwise suspect, for further inspection, sorting, and disposition. Destroy or remove from the food warehouse promptly products determined to present a hazard of contamination to foods in the warehouse.

4. Handle and store nonfood products that present hazards of contamination to foods stored in the same food warehouse by reason of undesirable odors, toxicity of contents, or otherwise, in a manner that will keep them from contaminating the foods. Take special measures to safeguard from damage and infestation those foods that are particularly susceptible to such risks.

5. Exercise care in moving, handling, and storing product to avoid damage to packaging that would affect the contents of food packaging, would cause spillage, or would otherwise contribute to the creation of insanitary conditions.

Shipping

1. Prior to loading with foods, inspect railcar and truck and trailer interiors for general cleanliness and for freedom from moisture; from foreign materials that would cause product contamination (such as broken glass, oil, toxic chemicals) or damage to packaging and contents (e.g., nails, boards, harmful protrusions); and from wall, floor, or ceiling defects that could contribute to insanitary conditions.

2. Clean, repair, or reject transport interiors as necessary to protect foods before loading. Exercise care in loading foods to avoid spillage or damage to packaging and contents. Maintain docks, rail sidings, truck bays, and driveways free from accumulations of debris and spillage.

Warehouse Conditions and Management

1. Maintain warehouse temperatures (particularly for refrigerated and frozen food storage areas) that are in compliance with applicable governmental temperature requirements, if any, for maintaining the wholesomeness of the particular foods received and held in such areas.

2. Assign responsibility for the overall food warehouse sanitation program and authority commensurate with this responsibility to persons who, by education,

training, and/or experience are able to identify sanitation risks and failures and food contamination hazards.

3. Instruct employees in the sanitation and hygienic practices appropriate to their duties and the locations of their work assignments within the food warehouse. Instruct employees to report observations of infestations (e.g., evidence of rodents, insects, or harborages) or construction defects permitting entry or harborage of pests, or other development of insanitary conditions.

4. Exercise programs of follow-up and control to ensure that employees, consultants, and outside services are doing their jobs effectively.

5. To ensure product wholesomeness and proper sanitation, the food warehouse sanitation program must have the commitment of top management, must be implemented by operating supervisors, and must be supported by the entire food warehouse staff. Preventive sanitation—the performance of inspection, sanitation, building maintenance, and pest control functions designed to prevent insanitation in preference to correcting it—should be an important goal of food warehouse management and operations.

Water, Plumbing, and Waste

Drinking water should be obtained from an approved source: a public water system, or a private water system that is constructed, maintained, and operated according to law. Unless it comes from a safe supply, water may serve as a source of contamination for processing operations, raw ingredients, food products, equipment, utensils, and hands. The major concern is that water may become a vehicle for the transmission of disease organisms. Water can also become contaminated with natural or human-made chemicals. Therefore, for the protection of consumers and employees, water must be obtained from a source regulated by law and must be handled, transported, and dispensed in a sanitary manner. A drinking water system should be flushed and disinfected before being placed in service after construction, repair, or modification and after an emergency situation, such as a flood, that may introduce contaminants to the system.

During construction, repair, or modification, water systems may become contaminated with microbes from soil because pipes are installed underground or by chemicals resulting from soldering and welding. Floods and other incidents may cause water to become contaminated. Chemical contaminants such as oils may also be present on or in the components of a system. To render the water safe, the system must be properly flushed and disinfected before being placed into service. Bacteriological and chemical standards have been developed for public drinking water supplies to protect public health. All drinking water supplies must meet standards required by law.

If the use of a nondrinking water supply is approved by the regulatory authority, the supply should be used only for purposes such as air conditioning, nonfood equipment cooling, fire protection, and irrigation, and may not be used so that the nondrinking water is allowed to contact, directly or indirectly, food-processing operations, equipment, or utensils.

Food plants may use nondrinking water for purposes such as air-conditioning or fire protection. Unlike drinking water, nondrinking water is not monitored for bacteriological or chemical quality or safety. Consequently, certain safety precautions must be observed to prevent the contamination of processing operations, raw ingredients, food products, drinking water, or food-contact surfaces. Identifying the piping designated as nondrinking waterlines and inspection for cross connections are examples of safety precautions.

Except when used as nondrinking water, water from a private water system should be sampled at least annually and sampled and tested as required by state water quality regulations. Wells and other types of individual water supplies may become contaminated through faulty equipment or environmental contamination of groundwater. Periodic sampling is required by law to monitor the safety of the water and to detect any change in quality. The controlling agency must be able to ascertain that this sampling program is active and that the safety of the water is in conformance with the appropriate standards. Laboratory results are only as accurate as the sample submitted. Care must be taken not to contaminate samples. Proper sample collection and timely transportation to the laboratory are necessary to assure the safety of drinking water used in the establishment.

The water source and system should be of sufficient capacity to meet the water demands of the operations in the food plant. Availability of sufficient water is a basic requirement for proper sanitation within a food establishment. An insufficient supply of safe water will prevent the proper cleaning of items such as equipment and utensils and of food employees' hands.

Water under pressure should be provided to all fixtures, equipment, and nonfood equipment that are required to use water except that water supplied as specified to a temporary facility or in response to a temporary interruption of a water supply need not be under pressure. Inadequate water pressure could lead to situations that place the public health at risk. For example, inadequate pressure could result in improper handwashing or equipment operation. Sufficient water pressure assures that equipment such as mechanical warewashers operate according to manufacturer's specifications.

Hot-water generation and distribution systems should be sufficient to meet the daily demands throughout the food plant in terms of operations and needs of employees. Hot water required for washing items such as equipment and utensils and employees' hands must be available in sufficient quantities to meet demand during peak water usage periods. Booster heaters for warewashers that use hot water for sanitizing are designed to raise the temperature of hot water to a level that assures sanitization. If the volume of water reaching the booster heater is not sufficient or hot enough, the required temperature for sanitization cannot be reached. Manual washing of certain processing equipment and utensils is most effective when hot water is used. Unless utensils are clean to sight and touch, they cannot be sanitized effectively.

Inadequate water systems may serve as vehicles for contamination of raw ingredients, food products, or contact surfaces. This requirement is intended to assure that sufficient volumes of water are provided from supplies shown to be safe, through a distribution system that is protected. Water from an approved source can be contaminated if conveyed inappropriately. Improperly constructed and maintained water mains, pumps, hoses, connections, and other appurtenances, as well as transport vehicles and containers, may result in contamination of safe water and render it hazardous to human health.

Plumbing systems and hoses conveying water must be made of approved materials and be smooth, durable, nonabsorbent, and corrosion resistant. If not, the system may constitute a health hazard because unsuitable surfaces may harbor disease organisms or it may be constructed of materials that may, themselves, contaminate the water supply. Water within a system will leach minute quantities of materials out of the components of the system. To ensure that none of the leached matter is toxic or in a form that may produce detrimental effects, even through long-term use, all materials and components used in water systems must be of an approved type. New or replacement items must be tested and approved based on current standards.

Improperly designed, installed, or repaired water systems can have inherent deficiencies, such as improper access openings, dead spaces, and areas difficult or impossible to clean and disinfect. Dead spaces allow water quality to degrade since they are out of the constant circulation of the system. Fixtures such as equipment and warewashing sinks that are not easily cleanable may lead to the contamination of raw ingredients and food products. An airgap between the water supply inlet and the flood-level rim of the plumbing fixture, equipment, or nonfood equipment should be in compliance with construction codes.

During periods of extraordinary demand, drinking water systems may develop negative pressure in portions of the system. If a connection exists between the system and a source of contaminated water during times of negative pressure, contaminated water may be drawn into and foul the entire system. Standing water in sinks, dipper wells, steam kettles, and other equipment may become contaminated with cleaning chemicals or food residue. Various means may be used to prevent the introduction of this liquid into the water supply through back siphonage.

The water outlet of a drinking water system must not be installed so that it contacts water in sinks, equipment, or other fixtures that use water. Providing an airgap between the water supply outlet and the flood-level rim of a plumbing fixture or equipment prevents contamination that may be caused by backflow. A person may not create a cross-connection by connecting a pipe or conduit between the drinking water system and a nondrinking water system or a water system of unknown quality. The piping of a nondrinking water system should be durably identified so that it is readily distinguishable from piping that carries drinking water.

Nondrinking water may be of unknown or questionable origin. Wastewater is either known or suspected to be contaminated. Neither of these sources can be allowed to contact and contaminate the drinking water system. A person should operate a water tank, pump, and hoses so that backflow and other contamination of the water supply is prevented. When a water system includes a pump, or a pump is used in filling a water tank, care must be taken during hookup to prevent negative pressure on the supplying water system. Backflow prevention to protect the water supply is especially necessary during cleaning and sanitizing operations on a mobile system.

Improper plumbing installation or maintenance may result in potential health hazards such as cross-connections, back siphonage, or backflow. These conditions may result in the contamination of ingredients, food products, utensils, equipment, or other contact surfaces. It may also adversely affect the operation of equipment.

SUGGESTIONS

Any sanitation program in poultry processing must be comprehensive and must be tested and evaluated continuously. Constant internal inspection is recommended, and occasional external inspection by consulting firms is a good practice. State and federal inspections by regulators serve as a reminder of satisfactory implementation of sanitation programs.

38

HACCP FOR THE POULTRY INDUSTRY

Lisa H. McKee

Department of Family and Consumer Sciences, New Mexico State University, Las Cruces, New Mexico

INTRODUCTION

Poultry products have been implicated in numerous foodborne illness outbreaks. These outbreaks have been traced to everything from barbecued chicken (Allerberger et al., 2003) and chicken nuggets (Kenny et al., 1999; MacDougall et al., 2004) to chicken salad (Dewaal et al., 2006; Mazick et al., 2006) and deli turkey meat (Olsen et al., 2005; Gottlieb et al., 2006). Consumers preparing food at home, food handlers in restaurants and other food service operations, and food processors have all been implicated in poultry-borne outbreaks.

The need to control the contamination of meat and poultry has been recognized in the United States since the passage of the Federal Meat Inspection Act in 1906 and the Poultry Products Inspection Act of 1957. Inspection mandated in these

Handbook of Poultry Science and Technology, Volume 2: Secondary Processing, Edited by Isabel Guerrero-Legarreta and Y.H. Hui

acts involved primarily sensory and visual analysis of carcasses to determine problems. Such analysis, however, cannot detect microbial loads or the presence of pathogens on the meat. The National Academy of Sciences (NAS, 1985, 1987), the General Accounting Office (GAO, 1993), and other experts, recognizing the need for a more comprehensive food safety system, encouraged the Food Safety and Inspection Service (FSIS) of the U.S. Department of Agriculture (USDA) to move toward a more scientific, risk-based inspection system for meat and poultry. The method chosen by the FSIS was the hazard analysis and critical control points (HACCP) system.

HISTORY OF HACCP

As the National Aeronautics and Space Administration (NASA) worked toward the goal of manned space flight in the 1950s, two primary problems arose related to feeding astronauts. First, food particles or crumbs floating in the zero gravity of space could pose a potential threat to the operation of delicate equipment and had to be eliminated. Second, the risk of foodborne illness from microbes or toxins in the food posed a potential catastrophic scenario that had to be prevented. These hazards, one physical and one biological, led NASA to the Pillsbury Company, where a system to address these and other food safety issues was developed.

The first hazard, crumbs, was controlled relatively simply through the use of bite-sized foods, the development of specialized coatings to contain foods, and the use of specialized packaging to minimize food exposure during all phases of the mission (Stevenson, 1995). The need for absolute microbiological safety of the food, however, posed a much more complex problem. Quality control practices in the food industry at that time involved primarily testing raw ingredients, followed by attribute sampling and destructive testing of the final product. These methods were, however, found to be economically impractical for the production of food for space flight and brought into question the possibility that many food safety issues were being missed between the raw ingredients and the final product. The need for a more efficient and comprehensive means of ensuring the safety of the food at the level required by NASA became apparent.

Team members recognized the need for a program that would control all aspects of the food process, from raw ingredients to processing to consumption. Initial exploration of NASA's program to ensure zero defects in hardware was found to be impractical when applied to food since it involved destructive testing. Developers eventually turned to a concept called *modes of failure*, constructed by the U.S. Army Natick Laboratories. In this concept, information was gathered about all areas of the production of a food. Using this information combined with previous experience on food manufacturing, a processor could predict potential problems as well as where and when those problems might occur. Measures could then be devised to control these critical points before problems could occur. The HACCP system was ready to apply to food production.

The initial HACCP system concept involved three principles (Stevenson, 1995). The first involved identifying all potential food safety problems that might occur, from growing and harvesting through consumption. The second principle involved determining points in the production process that needed to be controlled to prevent the occurrence of the problems identified. The final principle was to establish procedures to monitor the critical control points to determine if control was being maintained. Presented to the public for the first time in 1971, the new HACCP system gained favor with the U.S. Food and Drug Administration and the food industry for a short period, but the complexity of developing and maintaining a comprehensive HACCP plan prevented widespread use of the system.

Interest in HACCP revived in 1985 when the Subcommittee of the Food Protection Committee of the National Academy of Sciences issued a report on microbiological criteria that contained an endorsement of the system. Based on recommendations in the report, an expert scientific advisory committee was established. This committee, named the National Advisory Committee on Microbiological Criteria for Foods (NACMCF), was charged with creating the guidelines for the application of HACCP (Stevenson, 1995). The resulting document, entitled *HACCP Principles for Food Production*, was presented in November 1989. A review of the document in 1991 resulted in several revisions and a hazard analysis and critical control point system was adopted by NACMCF in March 1992 (NACMCF, 1995).

THE SEVEN HACCP PRINCIPLES

The HACCP method is a systematic approach to ensuring the safety of food at every point from growing and harvesting through processing, preparation, and consumption (NACMCF, 1992). Although each company must develop its own unique HACCP plan tailored to the requirements of the particular establishment, the seven underlying principles of HACCP allow standardization of the process throughout the food industry. The overarching goal of all HACCP plans is the prevention of problems before they occur.

Preliminary Steps The development of a comprehensive HACCP plan is a massive undertaking, particularly since a separate plan is required for each food being produced in an establishment. Although one person may have overall responsibility for a HACCP program, development and maintenance of each plan is generally the responsibility of a multidisciplinary team of experts. This team may include both employees from within the company as well as outside people with expertise in an area of interest. Knowledgeable persons external to the development team who can verify the completeness and accuracy of the plan are also needed.

The HACCP team should complete several activities prior to applying the seven specific principles to the plan development. A description of the food, including formulation and preparation steps, provides a basis for establishment of the plan, and a description of the distribution method(s) allows for consideration of potential abuses that might occur after the food has left the facility. Brief descriptions of the intended use and consumers of the product can provide information that may affect the steps outlined in the plan. A flow diagram detailing all of the steps in the process, verified with a walk-through of the actual production, provides both the HACCP team and outside inspectors or others verifying the plan's effectiveness with a guide to the subsequent HACCP plan.

Principle 1: Conduct a hazard analysis, identifying steps in the process where potential hazards might occur. Assessment of hazards is one of the most challenging parts of developing a HACCP plan (Kvenberg and Schwalm, 2000). A *hazard* is defined as any biological, chemical or physical property of a food that may result in harm when consumed (NACMCF, 1992). Information from a variety of sources may be used to establish the existence of a potential hazard. The flow diagram prepared in the preliminary steps as well as internal company records related to the processing can help identify a variety of potential hazards.

Chemical hazards are incidental additions to food products and may come from raw materials, production processes, or plant maintenance and sanitation operations. When used properly, chemicals in food production facilities should not pose a hazard. However, the proximity to the food of pesticides, cleaning solutions, lubricants, paints, and other chemicals used in the plant can be a safety concern and must be considered during construction of the HACCP plan. Although often less of a concern than biological hazards, chemicals have resulted in poultry-borne illnesses. Dworkin et al. (2004), for example, reported on a foodborne illness outbreak in Illinois schoolchildren which was eventually traced to exposure of the chicken tenders served at lunch to a liquid ammonia spill at the warehouse where the tenders had been stored.

Foreign objects such as glass shards, metal shavings, rock pieces, wood splinters, nails, hair, jewelry, and bones are considered physical hazards. Proper plant design and maintenance, strict control of processing, training employees, and regular inspection can all help control physical hazards. In-line equipment such as metal detectors, screens, and bone separators is also used to help prevent the occurrence of physical hazards in a food.

Information on biological hazards may come from in-plant microbiological data if available or from published literature, the Internet, or the government. A wide variety of microorganisms, including *Campylobacter, Salmonella, Listeria*, and *Escherichia coli*, at levels ranging from less than $\log_{10}1.0$ colony-forming units (CFU)/cm^2 to more than $\log_{10}7.0$ CFU/cm^2, have been reported in poultry products. The pathogen reduction and hazard analysis and critical control point systems (PR-HACCP) final rule (USDA–FSIS, 1996), published in 1996 by the USDA's Food Safety and Inspection Service (FSIS), sets *Salmonella* reduction standards as well as requirements for generic *E. coli* testing for producers of whole broilers and turkeys as well as processors of ground chicken and turkey meat.

Schlosser et al. (2000) reported on *Salmonella* levels and serotypes in raw ground poultry products sampled prior to implementation of the PR-HACCP rule. Of the 80 isolates from raw ground chicken serotyped, 30.0% (24 samples) were *Salmonella heidelberg* and 13.8% (11 samples) were *S. kentucky*. These were also the most common serotype isolates detected in chicken carcasses. In raw ground turkey, 76 of 319 isolates (23.8%) were *S. hadar*, 28 of 319 (8.8%) were *S. agona*, and *S. muenster* and *S. senftenberg* each accounted for 23 of 319 (7.2%) isolates.

Results of USDA–FSIS *Salmonella* testing in raw meat and poultry products from 1998 to 2003 was reported by Eblen et al. (2006) and Naugle et al. (2006). Results indicated that small establishments were more likely to experience a failed sample set than were larger or very small establishments. Failed sampled sets tended to occur early in the testing time frame for all non-broiler-producing companies, indicating that the implementation of HACCP requirements was effective. Failed sample sets for broiler processors occurred throughout the data collection time and even increased over time, suggesting the need for more investigation as to the cause of the continuing failures and development of procedures to alleviate the problem.

FSIS data on *Salmonella* contamination in broiler establishments collected between 1998 and 2006 indicated increasing levels of contamination, with 11.5% positive sample sets in 2002, 12.8% positive sets in 2003, 13.5% positive sets in 2004, and 16.3% positive sets in 2005 (USDA–FSIS, 2007). As a result, in 2006 the FSIS instituted new initiatives to help processors achieve and maintain consistent process control and food safety. These new initiatives included increased testing frequency for establishments with highly variable process control, risk-based rather than random testing schedules, and the start of turkey carcass sampling for *Salmonella*. The FSIS also began grouping establishments within product classes into those with consistent process control (two most recent *Salmonella* sample sets equal to or less than 50% of the performance standard), those with variable process control (at least one of the two most recent *Salmonella* sample sets greater than 50% of the performance standard), and those with highly variable process control (two most recent *Salmonella* sample sets greater than the performance standard). In 2006, 11.4% positive sample sets were recorded for broiler establishments and 7.1% were recorded for turkey establishments. A total of 45% positive *Salmonella* tests were recorded for ground chicken, while a 20.3% positive rate was recorded for ground turkey in 2006 (USDA–FSIS, 2007).

Principle 2: Identify the critical control points in the process. The NACMCF (1992) defines a *critical control point* (CCP) as "any point, step or procedure at which control can be applied and a food safety hazard can be prevented, eliminated or reduced to acceptable levels." Designation of CCPs can be a difficult choice but should be facilitated by the information gathered during the hazard analysis phase. The use of a CCP decision tree is also recommended to help determine the CCPs for a particular process (NACMCF, 1992).

González-Miret et al. (2006) conducted a study to determine if the washing and air-chilling steps in poultry meat production should be designated as CCPs. Total microbial loads, *Pseudomonas* counts, and Enterobacteriaceae loads were used to verify the need for control at the two processing steps. Multivariate statistical analysis indicated that the washing stage produced a significant decrease in microbial loads and thus must be considered a CCP. Air chilling was found to maintain the decrease in microbial loads as long as the temperature was controlled. It was therefore recommended that air temperature of the chiller be considered a CCP.

Vadhanasin et al. (2004) identified four critical control points (washing, chilling, deboning, and packing) from the generic HACCP plan for broiler slaughter and processing (NACMCF, 1997) that could be monitored to determine reduction in *Salmonella* during commercial frozen broiler processing in Thailand. A target maximum *Salmonella*-positive value of 20% was used to determine effectiveness of the CCPs. The *Salmonella* prevalence of 20.0% prior to washing was reduced to 12.5% after that CCP. All other CCPs failed to meet the 20.0% critical limit, with a *Salmonella* prevalence of 22.7% after chilling, 33.3% after deboning, and 23.3% after packing. Replacement of chlorine in the chiller water with hydrogen peroxide, peracetic acid, or ozone resulted in fewer *Salmonella*, with 0.5% peracetic acid producing a significant reduction in loads compared to chlorine. While the institution of HACCP was found to reduce the pathogen loads during frozen broiler processing, the need for proper disinfection at the chiller CCP was stressed.

Simonsen et al. (1987) identified a number of CCPs in cooked turkey, including cooking, hot holding, cooling, reheating, handling cooked product, and cleaning after handling raw product. Although not identified as a CCP, thawing was considered an important point in the process that could influence the microbial loads prior to preparation and result in cross-contamination during subsequent processing steps.

Nganje et al. (2007) conducted an economic analysis aimed at determining the most cost-effective number of CCPs needed in turkey processing. A stochastic optimization framework that included consideration of food safety risks, intervention costs, and risk reduction was developed and used to analyze production in an established Midwestern turkey-processing plant. Eight potential CCPs (prescalding, postscalding, preevisceration, prewash, postwash, postexamination for visible fecal contamination, postchill, and postremoval of the back frame) were identified and 20 product samples were collected from 20 birds at each of the eight points over five plant visits. Probability of *Salmonella* contamination at each of the eight points was tested at the 29%, 15%, and 5% tolerance levels. At the 29% tolerance level utilized by many processors following USDA recommendations, five CCPs (postscalding, preevisceration, prewash, postchill, and postremoval of back frame) were found to be the most effective. An additional CCP (prescalding) was added when the 15% tolerance level was used. Testing at all eight CCPs was found to be required at a 5% tolerance level.

Principle 3: Establish critical limits that will be used in monitoring the critical control points Once a CCP is identified, critical limits are established to provide the safety boundaries that indicate whether that CCP is in or out of control.

Critical limits are generally dictated by the characteristic of the CCP and can range from pH, temperature, water activity, moisture, viscosity, color, or salt concentration measurements to evaluation of sensory properties. Previous production information, experts, governmental regulations, standards and guidelines, and information in published literature may all be used as sources to determine appropriate critical limits.

Simonsen et al. (1987) reported several critical limits for cooked turkey. An end-product temperature of at least 74°C was specified for initial cooking of all noncured poultry. The temperature could be monitored continuously, taken at the time the poultry is removed from the heat source, or taken during the rise in temperature that occurs immediately after removal from the heat source. Temperatures of at least 55°C for hot holding and at least 74°C for reheating were also specified. Issues such as the depth of turkey in the container, the size and shape of the container, and the maintenance of spacing and stacking of containers were critical limits associated with cooling turkey.

The NACMCF (1997) specified several critical limits for handling raw broilers and broiler parts in food service and retail food establishments. A temperature of 41°F or less was specified for both receiving and storage of the poultry products. Critical limits related to preparation included a minimum internal temperature of 165°F as taken in the thickest portion of the breast muscle for cooking whole broilers; a minimum internal temperature of 165°F for 15 s for ground/restructured broiler products; a minimum internal temperature of 140°F for cooked poultry products held for hot display; and a minimum internal temperature of 165°F within 2 h for reheating precooked broiler and poultry products. Refrigeration of leftover poultry products in containers less than 2 in. in height and cutting large poultry parts into pieces 4 lb or smaller were also listed as critical limits.

Principle 4: Establish monitoring requirements for each CCP as well as procedures for use of monitoring data to adjust production processes so as to maintain control. Monitoring requirements are scheduled observations and/or measurements used to assess the effectiveness of each CCP. The monitoring procedures and the frequency associated with each monitoring procedure must be listed in the HACCP plan (USDA–FSIS, 1998). Monitoring provides a means of tracking operation of the HACCP system, provides a method for rapidly determining deviation from a CCP critical limit, and results in a written record that can be used in auditing and verifying the HACCP plan (NACMCF, 1992). Monitoring may be by continuous or discontinuous means.

Continuous monitoring utilizing automated equipment, in-line sensors, recording charts, and other methods of analysis is the preferred method of tracking CCPs. Temperatures, processing times, pH, and water activity are examples of monitoring information that can be collected on a continuous basis. Equipment calibration and continued accuracy is of critical importance in continuous monitoring.

Discontinuous monitoring, also known as *attribute sampling*, may be used in situations where continuous monitoring is not feasible. Potential uses for discontinuous monitoring procedures include testing ingredients where one or more microbiological, chemical, and/or physical characteristic is unknown, testing ingredients for approval prior to use in processing, troubleshooting out-of-control CCPs, and lot sampling of products placed on hold for safety evaluation. Since every unit within a lot cannot be sampled in discontinuous monitoring, a sampling plan is needed to determine statistically which units will be sampled. This is a major disadvantage of discontinuous monitoring, since the chances of detecting a defect are related directly to the level of the defect in the lot. Since many defects are present at low levels, the probability of missing a defect can be quite high with discontinuous monitoring.

Monitoring procedures must generally be rapid to accommodate on-line processes. Although CCPs are typically aimed at controlling a microbial hazard, microbiological analysis is often unsuitable for use as a CCP monitoring procedure because such tests are generally time consuming and frequency of sampling to detect low levels and sporadically occurring pathogens is often costly (NACMCF, 1992; Kvenberg and Schwalm, 2000). The continuing development of rapid microbiological methods, however, could allow the use of microbiological monitoring as one method of gathering information on the effectiveness of a CCP (Northcutt and Russell, 2003).

Principle 5: Establish corrective actions that will be taken if monitoring procedures indicate that a CCP is out of control. A written specification of corrective actions to be taken if a CCP is found to be out of control is required in the HACCP plans for meat and poultry processors (USDA–FSIS, 1998). For each CCP, a specific corrective action plan must be developed that includes guidelines for disposition of the noncompliant product, actions to be taken to bring the CCP back into compliance, and maintenance of records documenting the corrective actions taken (NACMCF, 1992). Corrective actions might include immediate process adjustment, which keeps the product compliant within the critical limits, with no product placed on hold; stopping production, removing noncompliant product, correcting the problem, and continuing with production; and solving the problem with a quick-fix solution while a long-term solution is sought (Stevenson et al., 1995).

Principle 6: Establish comprehensive recordkeeping procedures that document the HACCP system. Records provide permanent, written documentation that the processes and procedures outlined in the HACCP plan are being followed and help establish the continuing safety of the foods being produced. Records also allow tracking product history as well as traceback in the event of a problem.

Several types of records are required by the PR-HACCP final rule (USDA–FSIS, 1996). The written hazard analysis and any supporting documentation must be kept on file along with a HACCP manual documenting all parts of the formal plan. Documents related to selection and development of CCPs and critical limits as well as monitoring and verification procedures should also be kept on file.

Records documenting the monitoring of the CCPs and their critical limits are crucial to the continuing evaluation of the effectiveness of the HACCP plan. Employee training is an important aspect of CCP monitoring since line workers are often responsible for monitoring activities and must be able to determine the difference between normal, acceptable fluctuations in a process and a loss of control. Printed charts, checklists, record sheets, lab analysis sheets, and other forms developed for each CCP can assist those responsible for CCP monitoring in making such determinations and result in a written record for future verification and validation studies and audits. Information on the monitoring forms can include the title of the CCP; the date; lot number, code date, or other means of product identification; critical limits associated with the CCP; corrective action(s) to be taken if the CCP is deemed out of control; and spaces for data collection, operator initials, and reviewer's initials.

Principle 7: Establish verification procedures to ensure that the HACCP system is operating correctly. FSIS requires that all food production facilities validate their HACCP plans to ensure that the plans are successful in maintaining the safety of the food (USDA–FSIS, 1998). Such procedures provide a means for both food manufacturers and regulatory agencies to evaluate the day-to-day viability of HACCP plans. Validation procedures include initial validation and reassessment studies.

An initial validation is necessary to ensure that the HACCP plan is scientifically and technically sound (Kvenberg and Schwalm, 2000). Activities involved in this initial validation are primarily reviews of the information and data used to construct the HACCP plan. Validation studies may also need to be conducted, particularly if unusual or unique control measures and/or critical limits are being used.

Reassessment studies should be conducted as needed but no less than annually. Examples of things that should trigger a revalidation include regularly occurring critical limit deviations, any significant change in ingredients, processing, or packaging, and any HACCP system failure (Kvenberg and Schwalm, 2000). Prerequisite programs should also be reassessed on an annual or more frequent basis.

Verification of HACCP plans is a continuous process and includes both internal and external verification activities. In some cases, in-house verification should be conducted on a daily basis. CCP records, for example, should be reviewed frequently to ensure that any problems are investigated and corrected immediately. Any product holds should be followed up regularly. Other records, such as those related to sanitation procedures, may be reviewed less frequently but on a preplanned schedule. Spot checks and product sampling may be useful verification procedures in some cases. If in-house capabilities are available, microbiological testing may also play a role in verification of HACCP plans. Although not required as a part of HACCP verification procedures, results of the FDA HACCP pilot program confirmed that many processors were using microbiological testing to (1) verify the effectiveness of sanitation procedures, (2) ensure that incoming ingredients were of good quality, (3) determine whether or not a product put on hold was safe, (4) verify the safety of a product to buyers, and (5) verify that standard operating procedures were being followed (Kvenberg and Schwalm, 2000).

Regulatory agencies also play an important role in verifying the effectiveness of HACCP plans. Verification activities of regulatory agencies may include review of all records as well as the HACCP plan itself, on-site reviews of procedures, visual inspections, and random sampling and analysis (NACMCF, 1994). Inspections by regulatory agencies might be triggered by (1) implication of a food in a foodborne disease outbreak, (2) the need for greater coverage of a food due to additional food safety information, (3) a request for consultation, or (4) to verify that appropriate procedures have been implemented when HACCP plans are modified (NACMCF, 1994).

RESOURCES FOR DEVELOPMENT OF HACCP PLANS

Although the seven HACCP principles have been standardized throughout the food industry, the implementation of those principles is unique within each food production facility. Development of such plans is a massive undertaking requiring the participation of everyone from management to line workers. For small establishments, the need for a HACCP plan can be a daunting task that is often put off. There are resources, however, to help poultry producers develop and implement their own specific HACCP plans.

The USDA–FSIS (www.usda.gov) has developed a number of generic HACCP plans to assist food manufacturers, particularly small and very small establishments who have the least HACCP experience, in the development of individual HACCP plans. These generic plans provide guidance to manufacturers on HACCP concepts and the steps involved in constructing an appropriate plan. Each generic plan contains definitions, explanations for regulatory requirements found in the PR-HACCP final rule, and step-by-step discussion of each of the seven HACCP principles required in a plan. Two appendixes are also included in each generic plan. Appendix A contains reference lists for HACCP systems and related regulatory requirements, microbiological principles, and processing procedures for the specific poultry product outlined in the generic plan. Appendix B contains examples of process flow diagrams, product descriptions, a complete HACCP plan, including a hazard analysis, and monitoring logs. Generic models are available for several poultry products, including raw ground poultry (model 3); raw, not ground poultry (model 4); mechanically separated/mechanically deboned poultry (model 6); thermally processed, commercially sterile poultry products (model 7); and heat-treated, shelf-stable poultry products (model 10).

Stevenson and Katsuyama (1995) provided an example HACCP plan for battered and breaded chicken pieces. A process flow description and diagram provide the relevant information necessary to complete the HACCP plan. Critical control points and their associated critical limits, as well as monitoring information, corrective actions to be taken, records to be completed, and verification methods, are listed in a series of master worksheets. Similar examples and other resources for food manufacturers constructing HACCP plans can be found at many university cooperative extension service Web sites as well as Web sites for organizations such as the American Meat Science Association (www.meatscience.org).

REFERENCES

Allerberger F, Al-Jazrawi N, Kreidl P, Dierich MP, Feierl G, Hein I, Wagner M. 2003. Barbecued chicken causing a multi-state outbreak of *Campylobacter jejuni enteritis*. Infection 31(1):19–23.

Dewaal CS, Hicks G, Barlow K, Alderton L, Vegosen L. 2006. Foods associated with foodborne illness outbreaks from 1990 through 2003. Food Prot Trends 26(7):466–473.

Dworkin MS, Patel A, Fennell M, Vollmer M, Bailey S, Bloom J, Mudahar K, Lucht R. 2004. An outbreak of ammonia poisoning from chicken tenders served in a school lunch. J Food Prot 67(6):1299–1302.

Eblen DR, Barlow KE, Naugle AL. 2006. U.S. food safety and inspection service testing for *Salmonella* in selected raw meat and poultry products in the United States, 1998 through 2003: an establishment-level analysis. J Food Prot 69(11):2600–2606.

GAO (General Accounting Office). 1993. Building a scientific, risk-based meat and poultry inspection system: testimony before the Subcommittees on Livestock and Department Operations and Nutrition, Committee on Agriculture, House of Representatives; statement of John W. Harmon, Director, Food and Agriculture Issues; Resources, Community, and Economic Development Division. GAO/T-RECE-93-22. Washington, DC: U.S. GAO.

González-Miret ML, Escudero-Gilete ML, Heredia FJ. 2006. The establishment of critical control points at the washing and air chilling stages in poultry meat production using multivariate statistics. Food Control 17:935–941.

Gottlieb SL, Newborn EC, Griffin PM, Graves LM, Hoekstra RM, Baker NL, Hunter SB, Holt KG, Ramsey F, Head M, et al., the Listeriosis Outbreak Working Group. 2006. Multistate outbreak of listeriosis linked to turkey deli meat and subsequent changes in US regulatory policy. Clin Infect Dis 42:29–36.

Kenny B, Hall R, Cameron S. 1999. Consumer attitudes and behaviours: key risk factors in an outbreak of *Salmonella typhimurium* phage type 12 infection sourced to chicken nuggets. Aust NZ J Public Health 23:164–167.

Kvenberg JE, Schwalm DJ. 2000. Use of microbial data for hazard analysis and critical control point verification: Food and Drug Administration perspective. J Food Prot 63(6):810–814.

MacDougall L, Fyfe M, McIntyre L, Paccagnella A, Cordner K, Kerr A, Aramini J. 2004. Frozen chicken nuggets and strips: a newly identified risk factor for *Salmonella heidelberg* infection in British Columbia, Canada. J Food Prot 67(6):1111–1115.

Mazick A, Ethelberg S, Møller Nielsen E, Mølbak K, Lisby M. 2006. An outbreak of *Campylobacter jejuni* associated with consumption of chicken, Copenhagen, 2005. Euro Surveill 11(5):137–139.

NACMCF (National Advisory Committee on Microbiological Criteria for Foods). 1992. Hazard analysis and critical control point system. Int J Food Microbiol 16:1–23.

NACMCF. 1994. The role of regulatory agencies and industry in HACCP. Int J Food Microbiol 21:187–195.

NACMCF. 1995. Hazard analysis and critical control point system. In: *HACCP: Establishing Hazard Analysis Critical Control Point Programs*. Washington, DC: National Food Processors Association, pp. 2-1 to 2–26.

NACMCF. 1997. Generic HACCP application in broiler slaughter and processing. J Food Prot 60(5):579–604.

NAS (National Academy of Sciences). 1985. *Meat and Poultry Inspection: The Scientific Basis of the Nation's Program*. Washington, DC: National Academies Press.

NAS. 1987. *Poultry Inspection: The Basis for a Risk-Assessment Approach*. Washington, DC: National Academies Press.

Naugle AL, Barlow KE, Eblen DR, Teter V, Umholtz R. 2006. U.S. Food Safety and Inspection Service testing for *Salmonella* in selected raw meat and poultry products in the United States, 1998 through 2003: analysis of set results. J Food Prot 69(11): 2607–2614.

Nganje WE, Kaitibie S, Sorin A. 2007. HACCP implementation and economic optimality in turkey processing. Agribusiness 23(2):211–228.

Northcutt JK, Russell SM. 2003. *General Guidelines for Implementation of HACCP in a Poultry Processing Plant*. Bulletin 1155. Athens, GA: University of Georgia Cooperative Extension Service. http://www.caes.uga.edu/publications. Accessed Jan. 2008.

Olsen SJ, Patrick M, Hunter SB, Reddy V, Kornstein L, MacKenzie WR, Lane K, Bidol S, Stoltman GA, Frye DM, et al. 2005. Multistate outbreak of *Listeria monocytogenes* infection linked to delicatessen turkey meat. Clin Infect Dis 40:962–967.

Schlosser W, Hogue A, Ebel E, Rose B, Umholtz R, Ferris K, James W. 2000. Analysis of *Salmonella* serotypes from selected carcasses and raw ground products sampled prior to implementation of the pathogen reduction; hazard analysis and critical control point final rule in the U.S. Int J Food Microbiol 58:107–111.

Simonsen B, Bryan FL, Christian JHB, Roberts TA, Tompkin RB, Silliker JH. 1987. Prevention and control of food-borne salmonellosis through application of hazard analysis and critical control point (HACCP). Int J Food Microbiol 4:227–247.

Stevenson KE. 1995. Introduction to hazard analysis critical control point systems. In: *HACCP: Establishing Hazard Analysis Critical Control Point Programs*. Washington, DC: National Food Processors Association, pp. 1-1 to 1–5.

Stevenson KE, Katsuyama AM. 1995. Workshop flow diagrams and forms. In: *HACCP: Establishing Hazard Analysis Critical Control Point Programs*. Washington, DC: National Food Processors Association, pp. 11-1 to 11–29.

Stevenson KE, Humm BJ, Bernard DT. 1995. Critical limits, monitoring and corrective actions. In: *HACCP: Establishing Hazard Analysis Critical Control Point Programs*. Washington, DC: National Food Processors Association, pp. 9-1 to 9–10.

USDA–FSIS (U.S. Department of Agriculture–Food Safety and Inspection Service). 1996. Pathogen reduction; hazard analysis and critical control point (HACCP) systems, final rule. Fed Reg 61(144):38806–38989.

USDA–FSIS. 1998. Key facts: The seven HACCP principles. http://www.fsis.usda.gov/oa/background/keyhaccp.htm. Accessed Jan. 2008.

USDA–FSIS. 2007. Progress report on *Salmonella* testing of raw meat and poultry products, 1998–2006. http://www.fsis.usda.gov/science/progress_report_salmonella_testing/index.asp. Accessed Sept. 2007.

Vadhanasin S, Bangtrakulnonth A, Chidkrau T. 2004. Critical control points for monitoring salmonellae reduction in Thai commercial frozen broiler processing. J Food Prot 67(7):1480–1483.

39

FSIS ENFORCEMENT TOOLS AND PROCESSES

Y.H. Hui
Science Technology System, Sacramento, California

Isabel Guerrero-Legarreta
Departamento de Biotecnología, Universidad Autónoma Metropolitana, México D.F., México

Handbook of Poultry Science and Technology, Volume 2: Secondary Processing, Edited by Isabel Guerrero-Legarreta and Y.H. Hui

INTRODUCTION

The U.S. Department of Agriculture's (USDA's) Food Safety and Inspection Service (FSIS) is charged with ensuring that meat, poultry, and egg products are safe, wholesome, and properly labeled. The FSIS, in cooperation with state counterparts, inspects, monitors, and verifies the proper processing, handling, and labeling of meat and poultry products, from the delivery of animals to the slaughterhouse to when the products reach consumers. In cooperation with the U.S. Food and Drug Administration (FDA) and the states, the FSIS provides similar coverage for egg products, the processed whole egg ingredients used in manufacturing other foods. This regulatory oversight generally reflects compliance by the large majority of businesses. However, if the FSIS detects problems at any step along the way, it can use a number of product control and enforcement measures to protect consumers.

The USDA has traditionally focused much of its effort on the plants that slaughter food animals and process products. The USDA ensures that products at these establishments are produced in a sanitary environment in which inspectors or plant employees identify and eliminate potential food safety hazards. These establishments must apply for a grant of inspection from the FSIS and demonstrate the ability to meet certain requirements for producing safe, wholesome, and accurately labeled food products. Requirements include meeting sanitation, facility, and operational standards and, through new requirements now being implemented, having preventive systems in place to ensure the production of safe and unadulterated food. Products from official establishments are labeled with the mark of inspection, indicating that they have been inspected and passed by the USDA and can be sold in interstate commerce.

The FSIS uses compliance officers throughout the chain of distribution to detect and detain potentially hazardous foods in commerce to prevent their consumption and to investigate violations of law. Even if products are produced under conditions that are safe and sanitary, abuse on the way to the consumer, for example, if transported in trucks that are too warm or if exposed to contamination, can result in products that can cause illness or injury. The FSIS has recognized a need to spend increasing amounts of its energy on activities to promote safe transport, warehousing, and retailing of meat, poultry, and egg products, and is moving forward with these efforts.

The FSIS also works closely with the USDA's Office of Inspector General (OIG), which assists the FSIS in pursuing complex criminal cases. In addition, many state and local jurisdictions have enforcement authorities that apply to USDA-regulated products. The FSIS cooperates with these other jurisdictions in investigations and case presentations and participates with the OIG and the U.S. Department of Justice in monitoring conditions of probation orders and pretrial diversion agreements developed to resolve cases.

In January 1997, the FSIS began implementing new requirements in plants that produce meat and poultry. New regulations, entitled "Pathogen Reduction; Hazard Analysis and Critical Control Point (HACCP) Systems," require that federally inspected meat and poultry plants (1) develop and implement a preventive

HACCP plan; (2) develop and implement sanitation standard operating procedures (SSOPs); (3) collect and analyze samples for the presence of generic *Escherichia coli*, and record results; and (4) meet *Salmonella* performance standard requirements. These new requirements are designed to help target and reduce foodborne pathogens. This report provides a discussion of the regulatory and enforcement actions, including actions that address the pathogen reduction/HACCP regulatory requirements. The FSIS has undertaken to ensure that products that reach consumers are safe, wholesome, and properly labeled.

NONCOMPLIANCE RECORDS AND APPEALS

FSIS inspection program personnel perform thousands of inspection procedures each day in federally inspected establishments to determine whether or not inspected plants are in compliance with regulatory requirements. Each time that inspection program personnel make a noncompliance determination, they complete a noncompliance record (NR). An NR is a written report that documents noncompliance with FSIS regulations and notifies the establishment of the noncompliance that it should take action to remedy the situation and prevent its recurrence. Noncompliance reported on NRs varies from nonfood safety issues to serious breakdowns in food safety controls. When noncompliance occurs repeatedly, or when a plant fails to prevent adulterated product from being produced or shipped, the FSIS takes action to control products and may take enforcement action, such as to suspend inspection.

PORT-OF-ENTRY REINSPECTION

The FSIS conducts port-of-entry reinspections of imported meat, poultry, and egg products. This activity is a reinspection of products that have already been inspected and passed by an equivalent foreign inspection system. Thus, imported product reinspection is a monitoring activity for verifying on an ongoing basis the equivalence of a foreign country's inspection system. Port-of-entry reinspection is directed by the Automated Import Information System (AIIS), a centralized computer database that generates reinspection assignments and stores results. After clearing the Department of Homeland Security, Customs and Border Protection, and the Animal and Plant Health Inspection Service, every imported meat, poultry, or egg product shipment must be presented to the FSIS. When a meat or poultry shipment is presented for reinspection, the AIIS verifies that the product is from an eligible country and certified establishment. Shipments are refused entry if the foreign country, or the foreign establishment that produced the product, is not eligible to export to the United States.

All imported product shipments are reinspected for general condition, labeling, proper certification, and accurate count. In addition, other types of inspection may be generated by the AIIS. These could include a physical examination of the product for visible defects, collection of samples for microbiological analysis, samples for food chemistry analyses, and samples to be analyzed for drug and chemical residues. Shipments are randomly selected for reinspection using a statistical sampling plan that allocates samples by HACCP process categories. The level of sampling is based on the volume imported from a country within each category. Products that fail reinspection are rejected and must be reexported, converted to nonhuman food, or destroyed. Product rejections cause the AIIS automatically to generate an increased rate of reinspection for future shipments of like product from the same establishment.

PRODUCT CONTROL ACTIONS

FSIS takes product control actions to gain physical control over products when there is reason to believe that they are adulterated, misbranded, or otherwise in violation of the FMIA (Federal Meat Inspection Act), PPIA (Poultry Products Inspection Act), or EPIA (Egg Products Inspection Act). These actions are designed to ensure that those products do not enter commerce or, if they are already in commerce, that they do not reach consumers.

Retentions and Condemnations

In official establishments, FSIS inspection program personnel may retain products that are adulterated or mislabeled when there are insanitary conditions, for inhumane slaughter or handling, when conditions preclude the FSIS from determining that the product is not adulterated or misbranded, or for other reasons authorized by the statutes. FSIS inspection program personnel condemn animals for disease, contamination, or other reasons, to prevent their use as human food.

Detentions

FSIS investigators, enforcement, investigation, and analysis officers, import officers, and other designated FSIS inspection program personnel will detain products that may be adulterated, misbranded, or otherwise in violation of the law when found in commerce. Most detentions result in voluntary action by the product owner or custodian, such as voluntary disposal of the product. If detained product cannot be disposed of within 20 days, the FSIS may request, through the OGC and the U.S. Attorney's office, that a U.S. district court enter an order to seize the product as provided for in the FMIA, PPIA, and EPIA.

FOOD RECALLS

A food recall is a voluntary action by a manufacturer or distributor to protect the public from products that may cause health problems or possible death. A recall is intended to remove food products from commerce when there is reason to believe the products may be adulterated, misbranded, or otherwise in violation of the FMIA, PPIA, or EPIA. Recalls are initiated by the manufacturer or distributor of the meat, poultry, or egg products, sometimes at the request of FSIS. All recalls are voluntary. However, if a company refuses to recall its products, the FSIS has the legal authority to detain and seize those products in commerce.

The FSIS classifies food recalls as follows:

- A class I recall involves a health hazard situation in which there is a *reasonable* probability that eating the food will cause health problems or death.
- A class II recall involves a potential health hazard situation in which there is a *remote* probability of adverse health consequences from eating the food.
- A class III recall involves a situation in which eating the food will not cause adverse health consequences.

Next, we describe some important recalls and alerts for poultry and poultry products. All examples are "class I recall, health risk: high.

Public Health Alert for Frozen, Stuffed Raw Chicken Products, June 9, 2008 The USDA–FSIS issued the following alert:

> Gourmet Foods, Inc., a Rancho Dominquez, Calif., firm, is recalling approximately 130 pounds of various ready-to-eat chicken products that may be contaminated with *Listeria monocytogenes*,.
>
> The following products are subject to recall:
>
> 17-ounce packages of "Famima!! CHICKEN TERIYAKI BOWL WITH SAUCE ON RICE, KEEP REFRIGERATED/PERISHABLE." Each package bears the establishment number "EST. P-7738" inside the USDA mark of inspection, as well as a "SELL BY" date of "06/04/08," "06/05/08," "06/06/08" or "06/07/08."
>
> 12.85-ounce packages of "Famima!! THAI STYLE CHICKEN WITH RICE, PEANUT SAUCE AND VEGETABLES, KEEP REFRIGERATED/PERISHABLE." Each package bears the establishment number "EST. P-7738" inside the USDA mark of inspection, as well as a "SELL BY" date of "06/04/08," "06/05/08," "06/06/08" or "06/07/08."
>
> These ready-to-eat chicken products were distributed to retail establishments in the Los Angeles, Calif., area. Consumers that may have purchased these ready-to-eat chicken products at retail establishments between June 2 and June 7 are urged to check their refrigerators and freezers and discard them if found.
>
> The problem was discovered through microbiological sampling by FSIS. FSIS has received no reports of illnesses associated with consumption of these products. Anyone with signs of symptoms of foodborne illness should consult a medical professional.

Public Health Alert for Frozen, Stuffed Raw Chicken Products, March 29, 2008 The announcement was as follows:

> The FSIS/USDA issued a public health alert due to illnesses from *Salmonella* associated with frozen, stuffed raw chicken products that may be contaminated with *Salmonella*.

Frozen raw chicken breast products covered by this alert, and similar products, may be stuffed or filled, breaded or browned such that they appear to be cooked. These items may be labeled "chicken cordon bleu," "chicken kiev" or "chicken breast stuffed with" cheese, vegetables or other items.

This public health alert was initiated after an investigation and testing conducted by the Minnesota Department of Health and Minnesota Department of Agriculture determined that there is an association between the products listed below and 2 illnesses. The illnesses were linked through the epidemiological investigation by their PFGE pattern (DNA fingerprint).

Products linked to the illnesses were produced by Serenade Foods, a Milford, Ind., establishment. Products include "Chicken Breast with Rib Meat Chicken Cordon Bleu" and "Chicken Breast with Rib Meat Buffalo Style" sold under the brand names "Milford Valley Farms," "Dutch Farms" and "Kirkwood." The individually wrapped, 6-ounce products were produced on January 21, 2008 (date code C8021 is printed on the side of the package).

Each of these packages bears the establishment number "Est. P-2375" inside the USDA mark of inspection. These specific products were distributed to retail establishments in Illinois, Indiana, Minnesota, North Dakota, Vermont and Wisconsin.

Ohio Firm Recalls Frozen Chicken Products due to Mislabeling, March 29, 2008
The USDA–FSIS issued the following alert:

Koch Foods, a Fairfield, Ohio, establishment, is recalling approximately 1,420 pounds of frozen chicken breast products because they were packaged with the incorrect label. The frozen, pre-browned, raw products were labeled as "precooked" and therefore do not provide proper preparation instructions. These raw products may appear fully cooked.

The following product is subject to recall:

10-pound cases of "Koch Foods Fully Cooked Breaded Chicken Breast Fillet with Rib Meat" containing two 5-pound bags. Each case bears the establishment number "P-20795" inside the USDA mark of inspection, a production code of "24837-2", a date code of "B03982" and as well as a product code of "86861" printed on the label.

The frozen chicken products were produced on Feb. 8, 2008, and were shipped to distribution centers in Connecticut, Maine, Massachusetts, Michigan, Pennsylvania and Tennessee, intended for use by food service institutions.

The problem was discovered by the company. FSIS has received no reports of illness due to consumption of these products. Anyone concerned about an illness should contact a physician.

Alabama Firm Recalls Poultry Giblets That May Be Adulterated, March 14, 2008
The USDA–FSIS issued the following alert:

Cagle's Inc., a Collinsville, Ala., establishment, is voluntarily recalling approximately 943,000 pounds of various fresh and frozen poultry giblets and fresh carcasses with giblets inserted that may be adulterated due to improper disposition of the giblets.

Although carcasses were condemned, FSIS could not verify that the associated viscera, including the giblets, were condemned and diverted for inedible purposes, and they are therefore adulterated.

Examples of products subject to recall include:

50-lb. bulk packages of "*Cagle's* FRYING CHICKEN LIVERS." Each label bears a product code of "62150."

13-lb. bulk packages of "OUR PREMIUM DELI PRE-BREADED CHICKEN GIZZARDS." Each label bears a product code of "21210."

33-lb. bulk packages of "*Cagle's* FRYING CHICKEN SKINLESS NECKS." Each label bears a product code of "63191."

Each shipping package bears the establishment number "P-548" inside the USDA mark of inspection, however these products were repackaged for consumer sale and will therefore not include the establishment's number.

The products were produced on various dates between Dec. 3, 2007 and March 12, 2008, and were distributed to institutions and restaurants nationwide.

The problem was discovered through FSIS inspection. In November 2007, the plant installed new evisceration sorting equipment which changed the previous practice of condemning all viscera. FSIS has been unable to confirm that the plant had properly sorted or disposed of viscera from condemned carcasses and therefore some of the inspected and passed products may have been commingled with viscera from condemned carcasses. FSIS has received no reports of illness at this time.

The frozen chicken entrées were produced on Oct. 18, 2007, and were exported to the United States and then sent to distributors and retail establishments nationwide.

The problem was discovered through FSIS microbiological sampling. FSIS has received no reports of illnesses associated with consumption of this product.

Connecticut Firm Recalls Chicken and Pasta Product for Possible Listeria Contamination, October 9, 2007 The USDA–FSIS issued the following alert:

Aliki Foods, Inc., an Old Lyme, Connecticut firm, is voluntarily recalling approximately 70,400 pounds of a chicken and pasta product that may be contaminated with *Listeria monocytogenes*.

The following product is subject to recall:

5-pound boxes containing two 2.5-pound trays of "Aliki Chicken Broccoli Fettuccine Alfredo Made with White Chicken." Each box bears the establishment number "Est. 219" inside the Canadian Food Inspection Agency mark of inspection as well as a best if used by date of "SEPT 08" printed on the top of the box

The chicken and pasta product was produced on Sept. 28, 2007 and was distributed to retail establishments in CT, DE, ME, MD, MA, NH, NJ, NY, Oh, PA, RI and VA.

The problem was discovered through routine FSIS microbiological sampling at the import establishment. FSIS has received no reports of illnesses associated with consumption of this product.

NOTICES OF PROHIBITED ACTIVITY

Firms that recall products are expected to provide notification to their consignees concerning the recalled product and request that customers review inventory records and segregate, hold, or destroy product. Firms that have already shipped or sold recalled product are to retrieve and control the product, prevent further distribution, and contact their consignees and have them retrieve and control product that is part of a recall. FSIS program personnel conduct *effectiveness checks* to ensure that the recalling firm and firms that received or distributed product subject to a recall take effective action to notify all consignees of the recalled product that there is a need to control and remove the recalled product from commerce.

In situations where the FSIS determines that a federally inspected establishment or a firm operating in commerce has not taken responsibility to remove or control adulterated, misbranded, or other unsafe product in commerce or to advise its consignees of product that is subject to recall, it may issue prohibited activity warning notices to the firm. FSIS issues prohibited activity notices for the following:

- Failure of a recalling firm to notify its consignees of recalled product
- Failure of a consignee to notify its customers of recalled product
- Recalling firm or consignee found offering for sale recalled product

NOTICES OF WARNING

A *notice of warning* (NOW) provides notice of violations to firms and responsible individuals. The FSIS issues a NOW to firms and individuals to notify them of prohibited acts or other conduct that violates FSIS statutes or regulations. Generally, the FSIS issues a NOW for minor violations of law that are not referred to a U.S. Attorney for prosecution or other action. FSIS may also issue a NOW when a U.S. Attorney declines to prosecute a case or bring action against a specific business or person. A NOW identifies the violative conduct, condition, practice, or product, and the statutory or regulatory provisions violated. It advises the firm or individual that the agency will not pursue further action for the violation and warns that FSIS may seek criminal prosecution or other action for continued or future violations. Notices of warning may be issued to any individual or business, including federal plants, wholesalers, distributors, restaurants, retail stores, and other entities that process, store, or distribute meat, poultry, and egg products.

ADMINISTRATIVE ACTIONS

The FSIS *rules of practice*, which are set out in 9 CFR Part 500, define the type of administrative enforcement actions that FSIS takes, the conditions under which these actions are appropriate, and the procedures FSIS will follow in taking these actions. These regulations provide notice to establishments of FSIS enforcement actions, criteria, and processes, and ensure that all establishments are afforded due process. Administrative actions in the rules of practice include regulatory control action, withholding action, and suspension. These actions are defined in 9 CFR 500.1. FSIS takes these actions to prevent preparation and shipment of adulterated products, when products are produced under insanitary conditions, and for other reasons described in the rules of practice. When there is an imminent threat to public health or safety, such as the shipment of adulterated product, FSIS takes immediate enforcement action. In other situations, FSIS provides the establishment prior notification of intended enforcement action and the opportunity to demonstrate or achieve compliance. This is called a *notice of intended enforcement action* (NOIE). In appropriate situations, the FSIS may defer an enforcement decision based on corrections submitted by the establishment. The FSIS also may place a suspension action, if taken, in abeyance if an establishment presents and puts into effect corrective and preventive actions. FSIS Office of Field Operations district offices monitor and verify an establishment's implementation of corrective and preventive actions, and take follow-up action if needed to protect the public health.

Formal Adjudicatory Actions for Food Safety

In some situations it is necessary to withdraw inspection from an establishment based on the failure of a recipient of inspection service to meet critical sanitation and food safety regulatory requirements [e.g., sanitation standard operating procedures (SSOPs) or hazard analysis and critical control point (HACCP) system regulations] necessary to protect the public health. In these cases, the FSIS files an administrative complaint with the USDA hearing clerk. The plant may request a hearing before a USDA administrative law judge. If the action is based on insanitation or other imminent threats to public health or safety, the plant may remain closed while proceedings go forward. In cases that do not involve a threat to public health, operations may continue. These actions may be resolved by the FSIS and the plant entering into a consent decision, which allows the plant to operate under certain specified conditions. If inspection service is withdrawn, a closed plant must reapply to receive federal inspection. The FSIS may also take enforcement action, by filing an administrative complaint, to deny federal inspection service to an applicant. These actions are taken in accordance with the FSIS rules of practice (9 CFR 500) and department regulations governing formal adjudicatory proceedings (7 CFR 1).

Withdrawal for Unfitness

The FSIS can move to withdraw or deny inspection, after an opportunity for a hearing, based on the unfitness of recipient or applicant for inspection because of a felony conviction, more than one violation involving food, or certain other violations set out in the statutes. Actions pending or taken (other than outstanding consent decisions) are reported.

Removal of Exempt Privilege

The meat and poultry laws exempt certain custom, retail, or other operations from inspection, such as facilities that slaughter animals or poultry, or process meat or poultry, for owners of the animals. When insanitary conditions create health hazards at these businesses, the FSIS may remove custom or other exempt privileges through issuance of a *notice of ineligibility*, and require the business to cease operations until sanitary conditions are restored. The FSIS can also take action when custom or other exempt facilities fail to properly label product as "not for sale." These exempt businesses have the opportunity to correct violations before or after such actions, contest the basis for such actions at hearings, and to enter settlement agreements to resolve actions.

CRIMINAL ACTIONS

If evidence is found that a person or business has engaged in violations of the Federal Meat Inspection Act, Poultry Products Inspection Act, or Egg Products Inspection Act, the USDA may refer the case, through the Office of the General Counsel (OGC) or the Inspector General (OIG), to the appropriate U.S. Attorney to pursue criminal prosecution. Conviction for a criminal offense can result in a fine, imprisonment, or both. In certain situations, U.S. Attorneys may enter into pretrial diversion (PTD) agreements with alleged violators in lieu of actual prosecution. Under these agreements, the government agrees not to proceed with criminal prosecution if the alleged violator meets certain terms and conditions. The terms and conditions of a PTD may be tailored to each individual case. If the divertee completes the program successfully, no criminal charges are filed. If, on the other hand, the divertee does not complete the program successfully, criminal charges may be reinstated. The FSIS frequently monitors these agreements so that it can assist U.S. Attorneys in determining whether the terms have been met or that prosecution should be reinstituted. As an illustration, a court case in 2002 is provided here by the FSIS and the U.S. Food and Drug Administration (FDA).

On April 24, 2002, Hop Kee, Inc., a Chicago food distribution firm, and Thomas M. Lam, its chief executive officer, were sentenced in federal court for violating federal food safety inspection laws administered by the FSIS and the FDA. Thomas M. Lam, chief executive officer, was sentenced on two misdemeanor counts for storing adulterated poultry and food products. He was placed on probation for one year, fined $2928, and ordered to perform 100 hours of community service.

Hop Kee, Inc. was sentenced on the same two misdemeanors, plus a felony count for selling adulterated poultry. Hop Kee pled guilty Sept. 26, 2001, in federal court and was ordered to serve four years' probation, pay special assessment fees of $650, and pay a $300,000 fine. Special terms of Hop Kee's four-year probation include:

1. Repairing and maintaining its facilities in compliance with federal and state laws applicable to sanitation and pest control

2. Implementing a sanitation program and regularly checking and disposing of adulterated or spoiled meat and poultry products
3. Using an outside pest control service and consenting to periodic reviews by the USDA and the FDA of the records and recommendations of the service
4. Notifying the court or probation department of any material change in its financial condition, such as bankruptcy, criminal prosecution, or civil litigation

The defendants stored approximately 98,000 lb of meat and poultry products under insanitary conditions and sold adulterated poultry products in commerce. An additional 87,000 lb of food products under the jurisdiction of the FDA were also held under insanitary conditions, causing them to become adulterated. The conditions included live and dead rats, cockroaches, birds, as well as rodent nests, excreta, and rodent-gnawed food products.

CIVIL ACTIONS

FSIS also has authority to seek a variety of civil actions and case dispositions in federal court.

Seizures

When the FSIS has reason to believe that distributed products are adulterated, misbranded, or otherwise in violation of law, the agency will, through the USDA Office of the General Counsel and the U.S. Attorney's office, institute a seizure action against the product. The product is held pending an adjudication of its status. If the court finds that the product is adulterated, misbranded, or otherwise in violation of FSIS laws, it will condemn the product. Condemned product cannot be further processed to be used for human food.

Injunctions

The FSIS, through the U.S. Attorney's office, may request a U.S. district court to enjoin firms or persons who engage in repetitive violations of the FMIA, PPIA, or EPIA, or whose actions pose a threat to public health and safety. An injunction requires a person or firm to take certain action or to refrain from doing acts that violate the law. Injunctions may be resolved by a consent decree.

False Claims Act and Other Actions

The FSIS also works with the OGC, the OIG, and U.S. Attorneys to obtain other civil case outcomes. The Department of Justice Affirmative Civil Enforcement program is used by U.S. Attorneys to recover damages when a violation of law involves fraud against the federal government. Case examples where civil action may be appropriate include cases involving products not in compliance, sold to the military, to public schools engaged in the school lunch program, or to other federal institutions.

OUTBREAKS

Restaurants provide opportunities for outbreaks of foodborne disease because large quantities of different foods are handled in the same kitchen. Failure to wash hands, utensils, or countertops can lead to contamination of foods that will not be cooked. The regulatory environment for food preparation in this country has changed considerably in recent decades. At present, the FDA and health authorities of most states have developed guidelines for food handlers to prevent cross-contamination of foods. Also, most states have required safety training programs for food preparation handlers.

States can reduce the risk for foodborne illness in restaurants by ensuring that restaurant employees receive training in food safety. For example, food handlers should be aware that pathogens can be present on raw poultry and meat and that foodborne disease can be prevented by adhering to the following measures:

1. Raw poultry and meat should be prepared on a separate countertop or cutting board from other food items.
2. All utensils, cutting boards, and countertops should be cleaned with hot water and soap after preparing raw poultry or meat and before preparing other foods.
3. Hands should be washed thoroughly with soap and running water after handling raw poultry or meat.
4. Poultry should be cooked thoroughly to an internal temperature of 180°F (82°C) or until the meat is no longer pink and juices run clear.

Campylobacter jejuni

On August 29, 1996, the Jackson County Health Department in southwestern Oklahoma notified the Oklahoma State Department of Health of a cluster of *Campylobacter jejuni* infections that occurred August 16 to 20 among persons who had eaten lunch at a local restaurant on August 15. This report summarizes the investigation of these cases and indicates that *C. jejuni* infection was most likely acquired from eating lettuce cross-contaminated with raw chicken. This report also emphasizes the need to keep certain foods and cooking utensils separate during food handling.

All patients reported diarrhea, fever, abdominal cramps, nausea, vomiting, and visible blood in their stools. Health department staff visited the restaurant to obtain information about menu items, to observe food preparation, and to inspect the kitchen. Inspection of the restaurant indicated that the countertop surface area was too small to separate raw poultry and other foods adequately during preparation. The cook reported cutting up raw chicken for the dinner meals before preparing salads, lasagna, and sandwiches as luncheon menu items. Lettuce for salads was shredded with a knife, and the cook wore a towel around her waist that she used frequently to dry her hands. Bleach solution at the appropriate temperature [>75°F (>24°C)] and concentration (>50 ppm) was present to sanitize tables surfaces, but it was uncertain whether the cook had cleaned the countertop after cutting up the chicken. The lettuce or lasagna was probably contaminated with *C. jejuni* from raw chicken through unwashed or inadequately washed hands, cooking utensils, or the countertop. The health department recommended that the restaurant enlarge its food-preparation table and install a disposable hand towel dispenser and that food handlers wash hands and cooking utensils between uses while preparing different foods.

Campylobacter is one of the most common causes of foodborne disease in the United States, causing approximately 2 million cases of gastroenteritis each year. Illness associated with *Campylobacter* infection is usually mild, but can be severe and even fatal. *Campylobacter* has been found in up to 88% of broiler chicken carcasses in the United States. The infectious dose of *Campylobacter* is low; ingestion of only 500 organisms, easily present in one drop of raw chicken juice, can result in human illness. Therefore, contamination of foods by raw chicken is an efficient mechanism for transmission of this organism.

Cryptosporidium parvum

On September 29, 1995, the Minnesota Department of Health (MDH) received a report of acute gastroenteritis among an estimated 50 attendees of a social event in Blue Earth County on September 16. This report summarizes the epidemiologic and laboratory investigations of the outbreak, which indicate that the probable cause for this foodborne outbreak was *Cryptosporidium parvum*. Symptoms included watery diarrhea, abdominal cramps, and chills. Based on the case–control study, only consumption of chicken salad was associated with increased risk for illness. Water consumption at the event was not associated with illness. The chicken salad was prepared by the hostess on September 15 and was refrigerated until served. The ingredients were cooked chopped chicken, pasta, peeled and chopped hard-boiled eggs, chopped celery, and chopped grapes in a seasoned mayonnaise dressing. The hostess operated a licensed day-care home (DCH) and prepared the salad while attendees were in her home. She denied having recent diarrheal illness and refused to submit a stool specimen. In addition, she denied knowledge of diarrheal illnesses among children in her DCH during the week before preparation of the salad. She reported changing diapers on September 15 before preparing the salad and reported routinely following hand-washing practices.

Stool specimens from two of the persons whose illnesses met the case definition were obtained by the MDH 7 days after resolution of their symptoms; one sample was positive for oocysts and *Cryptosporidium* sporozoites on acid-fast staining, but the DFA test was negative. The presence of oocysts containing sporozoites was confirmed by acid-fast tests at two other reference laboratories. Stool specimens obtained from a third person, the spouse of a case patient, who did not attend the event but had onset of diarrhea 8 days after onset of diarrhea in his spouse was positive for *C. parvum* by acid-fast staining and DFA. All stools obtained by MDH were negative for bacteria and for parasites. No chicken salad was available for testing.

Listeriosis

On November 15, 2002, the CDC made the following announcement.

> CDC, state and local health departments, and the USDA–FSIS have been investigating an outbreak of listeriosis primarily affecting persons in the northeastern United States. Thus far, 52 ill persons infected with the outbreak strain of *Listeria* have been identified since mid-July; most were hospitalized, seven have died, and three pregnant women have had miscarriages or stillbirths. Epidemiologic data indicate that precooked, sliceable turkey deli meat is the cause of this outbreak.
>
> As part of the ongoing outbreak investigation, USDA–FSIS has been investigating turkey processing plants. *Listeria* bacteria have been found in turkey products from two plants. USDA–FSIS laboratories performed DNA fingerprinting on these bacteria.
>
> From Pilgrim's Pride Corporation, located in Franconia, Pennsylvania, one ready-to-eat turkey product and 25 environmental samples tested positive for *Listeria*. The turkey product had a strain of *Listeria* different from the outbreak strain. Of the 25 environmental *Listeria* strains fingerprinted, 2 matched that of the turkey product and 2 matched that of patients in the current outbreak. On October 12, the plant voluntarily shut down operations and issued a recall of approximately 27 million pounds of fresh and frozen ready-to-eat turkey and chicken products produced since May 1, 2002.
>
> From Jack Lambersky Poultry Company, located in Camden, New Jersey, some ready-to-eat poultry products were contaminated with a strain of *Listeria* that is indistinguishable from that of the outbreak patients. In addition, one environmental sample from the plant tested positive for a strain of *Listeria* different from the outbreak strain. On November 2, the plant voluntarily suspended operations and recalled approximately 200,000 pounds of fresh and frozen ready-to-eat poultry products
>
> Listeriosis is a serious foodborne disease that can be life-threatening to certain individuals, including the elderly or those with weakened immune systems. It can also cause miscarriages and stillbirths in pregnant women. The affected patients live in 9 states: Pennsylvania (14 cases), New York (12 cases in New York City, 9 in other locations), New Jersey (5 cases), Delaware (4 cases), Maryland (2 cases), Connecticut (1 case), Michigan (1 case), Massachusetts (3 cases), and Illinois (1 case).
>
> Thirty-two patients were male and 20 were female. Sixteen patients were age 65 or above, 16 patients were age 1 to 64 years and had an immunocompromising medical condition, eight others were pregnant, and four were neonates; seven patients were age 1 to 64 years and were not pregnant or known to have an immunocompromising

condition. No medical information was available for one patient. Of the seven patients who died, six had immunocompromising conditions (three of these patients were also age 65 or older), and one was a neonate. The most recent patient became ill on October 9. In addition to the patients whose illnesses have been confirmed as part of the outbreak, CDC and state and local health departments have learned about other cases of *Listeria* infection in the same region during the outbreak time period. DNA fingerprinting has shown that strains from 94 patients in these same states are different from the outbreak strain and 23 of these patients have died; these illnesses are part of the "background" of sporadic *Listeria* infections and are likely due to a variety of different foods. In addition, testing of strains from several additional persons is ongoing; some of these may be identified as the outbreak strain. Because pregnant women, older adults, and people with weakened immune systems are at higher risk for listeriosis, we recommend the following measures for those persons:

- Do not eat hot dogs and luncheon meats, unless they are reheated until steaming hot.
- Avoid cross-contaminating other foods, utensils, and food preparation surfaces with fluid from hot dog packages, and wash hands after handling hot dogs.
- Do not eat soft cheeses such as Feta, Brie and Camembert cheeses, blue-veined cheeses, and Mexican-style cheeses such as "queso blanco fresco." Cheeses that may be eaten include hard cheeses; semi-soft cheeses such as mozzarella; pasteurized processed cheeses such as slices and spreads; cream cheese; and cottage cheese.
- Do not eat refrigerated pâtés or meat spreads. Canned or shelf-stable pâtés and meat spreads may be eaten.
- Do not eat refrigerated smoked seafood, unless it is contained in a cooked dish, such as a casserole. Refrigerated smoked seafood, such as salmon, trout, whitefish, cod, tuna or mackerel, is most often labeled as "nova-style," "lox," "kippered," "smoked," or "jerky." The fish is found in the refrigerator section or sold at deli counters of grocery stores and delicatessens. Canned or shelf-stable smoked seafood may be eaten.
- Do not drink raw (unpasteurized) milk or eat foods that contain unpasteurized milk.

About 2500 cases of listeriosis occur each year in the United States. The initial symptoms are often fever, muscle aches, and sometimes gastrointestinal symptoms such as nausea or diarrhea. The illness may be mild and ill persons sometimes describe their illness as flulike. If infection spreads to the nervous system, symptoms such as headache, stiff neck, confusion, loss of balance, or convulsions can occur. Most cases of listeriosis and most deaths occur in adults with weakened immune systems, the elderly, pregnant women, and newborns. However, infections can occur occasionally in otherwise healthy persons. Infections during pregnancy can lead to miscarriages, stillbirths, and infection of newborn infants. Previous outbreaks of listeriosis have been linked to a variety of foods, especially processed meats (such as hot dogs, deli meats, and paté) and dairy products made from unpasteurized milk.

INDEX

Handbook of Poultry Science and Technology, Volume 2: Secondary Processing, Edited by Isabel Guerrero-Legarreta and Y.H. Hui

CONTENTS OF VOLUME 1: Primary Processing

Handbook of Poultry Science and Technology, Volume 2: Secondary Processing, Edited by Isabel Guerrero-Legarreta and Y.H. Hui